BLACK HOLES, COSMOLOGY AND EXTRA DIMENSIONS

Second Edition

Other Related Titles from World Scientific

The WSPC Handbook of Astronomical Instrumentation
(In 5 Volumes)
Volume 1: Radio Astronomical Instrumentation
Volume 2: UV, Optical & IR Instrumentation: Part 1
Volume 3: UV, Optical & IR Instrumentation: Part 2
Volume 4: X-Ray Astronomical Instrumentation
Volume 5: Gamma-Ray and Multimessenger Astronomical Instrumentation
Editor-in-chief: David N Burrows
edited by Alex Wolszczan and Anna M Moore
ISBN: 978-981-4644-31-0 (Set)
ISBN: 978-981-4644-35-8 (Vol. 1)
ISBN: 978-981-4644-36-5 (Vol. 2)
ISBN: 978-981-4644-37-2 (Vol. 3)
ISBN: 978-981-4644-38-9 (Vol. 4)
ISBN: 978-981-4644-39-6 (Vol. 5)

The Invisible Universe: Dark Matter, Dark Energy, and the Origin and End of the Universe
by Antonino Del Popolo
ISBN: 978-981-122-943-5

A Cabinet of Curiosities: The Myth, Magic and Measure of Meteorites
by Martin Beech
ISBN: 978-981-122-491-1

BLACK HOLES, COSMOLOGY AND EXTRA DIMENSIONS

Second Edition

Kirill A Bronnikov
Russian Research Institute of Metrological Service, Russia

Sergey G Rubin
National Research Nuclear University "MEPhI", Russia

World Scientific

NEW JERSEY • LONDON • SINGAPORE • BEIJING • SHANGHAI • HONG KONG • TAIPEI • CHENNAI • TOKYO

Published by

World Scientific Publishing Co. Pte. Ltd.

5 Toh Tuck Link, Singapore 596224

USA office: 27 Warren Street, Suite 401-402, Hackensack, NJ 07601

UK office: 57 Shelton Street, Covent Garden, London WC2H 9HE

Library of Congress Cataloging-in-Publication Data

Names: Bronnikov, Kirill A., author. | Rubin, Sergei G., author.

Title: Black holes, cosmology and extra dimensions /
Kirill A. Bronnikov, Russian Research Institute of Metrological Service, Russia,
Sergey G. Rubin, National Research Nuclear University "MEPhI," Russia.

Description: Second edition. | Hackensack, NJ : World Scientific Publishing Co. Pte. Ltd., [2022] |
Includes bibliographical references and index.

Identifiers: LCCN 2021017847 | ISBN 9789811233449 (hardcover)

Subjects: LCSH: Black holes (Astronomy) | Gravitation. | Cosmology.

Classification: LCC QB843.B55 B76 2022 | DDC 523.1--dc23

LC record available at https://lccn.loc.gov/2021017847

British Library Cataloguing-in-Publication Data

A catalogue record for this book is available from the British Library.

For any available supplementary material, please visit
https://www.worldscientific.com/worldscibooks/10.1142/12186#t=suppl

Desk Editor: Ng Kah Fee

Typeset by Stallion Press
Email: enquiries@stallionpress.com

To our wives — Lena and Natasha

Acknowledgments

We are thankful to our coauthors, friends and colleagues Konstantin Belotsky, Sergei Bolokhov, Heinz Dehnen, Irina Dymnikova, Júlio Cesar Fábris, Vladimir Ivashchuk, Stepan Grinyok, Sung-Won Kim, Maxim Khlopov, Alexander Kirillov, Rostislav Konoplich, Roman Konoplya, Vladimir Krechet, Jose P. S. Lemos, Boris Meierovich, Vitaly Melnikov, Valery Nikulin, Sergei Odintsov, Polina Petriakova, Arkady Popov, Alexander Sakharov, Georgy Shikin, Milena Skvortsova, Sergey Sushkov, Artyom Yurov, Oleg Zaslavskii, Alexander Zhidenko and many others for helpful discussions and support without which this book could never be written.

Notations

In the whole book, $\hbar = c = 1$ (except for Sect 2.1);
Signature of the metric: $(1, -1, -1, -1, ...)$.

The cosmological metric with synchronous time:

$$ds^2 = dt^2 - \gamma_{\alpha\beta} dx^\alpha dx^\beta.$$

Units

1 light year $\sim 10^{18}$ cm.
1 parsec (pc) ~ 3 light years.
The Sun's mass: $M_\odot \sim 2 \cdot 10^{33}$ g.
The Sun's luminosity: $L_\odot \sim 4 \cdot 10^{33}$ erg/s.

The Planck units:

$M_{\rm Pl}^2 = 1/G$, or $M_{\rm Pl}^2 = 1/(8\pi G)$.
$M_{\rm Pl} \sim 10^{-5}$ g $\sim 10^{19}$ GeV.
$l_{\rm Pl} = 1/M_{\rm Pl} \sim 10^{-33}$ cm.
$t_{\rm Pl} = 1/M_{\rm Pl} \sim 10^{-44}$ s.

The main parameters of our Universe

Size: $\sim 10^{28}$ cm, or ~ 6000 Mpc.
Age: ~ 14 billion years.
The Hubble parameter: $H_0 = 100h$ km s^{-1} Mpc^{-1}; $\quad h \simeq 0.72$.
Composition: $\Omega_\Lambda \simeq 0.72$; $\Omega_{\rm DM} \simeq 0.24$; $\Omega_b \simeq 0.04$.
$\sim 10^{11}$ of galaxies concentrated in clusters, with voids between them.

Our Galaxy, the Milky Way

Size: ~50 kpc.
$\sim 10^{11}$ of stars; neutron stars, white dwarfs, dark stars (brown dwarfs), black holes, hidden mass (dark matter).

History of the Universe, main landmarks

Time since the Big Bang	Temperature, K	Redshift	Comments
$10^{-43} - 10^{-37}$ s	$>10^{26}$		Inflation
10^{-6} s	$>10^{12}$		Quark-gluon plasma, electrons, neutrinos and their antiparticles
$3 \cdot 10^{-5}$ s	10^{12}		Baryosynthesis, nucleon formation
10^{-4} s – 3 min.	10^{12}–10^{9}		D, He, Li formation
15,000 yrs	10^{4}	3000	Equal matter and radiation densities
300,000 yrs	4000	1300	Recombination, transparency of the Universe
$15 \cdot 10^{6}$ yrs	300	100	Room temperature in the Universe
$1 - 3 \cdot 10^{9}$ yrs	20	6	Birth of the first stars
$3 \cdot 10^{9}$ yrs	10	3	Formation of heavy nuclei in Supernovae
$3 - 14 \cdot 10^{9}$ yrs	3	1	Emergence of life and intelligence
10^{14} yrs	—	—	Stellar formation stops
10^{37} yrs	—	—	The last stars die
10^{100} yrs	—	—	All black holes have evaporated

Abbreviations

BD	=	Brans-Dicke
BH	=	black hole
CMB	=	cosmic microwave background
DE	=	dark energy
DEC	=	dominant energy condition
DM	=	dark matter
FPC	=	fundamental physical constant
FRW	=	Friedmann-Robertson-Walker
GR	=	general relativity
HMPG	=	hybrid metric-Palatini gravity
IRF	=	inertial reference frame
NEC	=	null energy condition
PBH	=	primordial black holes
RF	=	reference frame
RS	=	Randall-Sundrum
SEC	=	strong energy condition
SMBH	=	super massive black hole
SR	=	special relativity
STT	=	scalar-tensor theory
WEC	=	weak energy condition

Contents

Chapter 1

Modern ideas of gravitation and cosmology — a brief essay

At the beginning of the 20th century, only two physical fields were known, the electromagnetic and gravitational ones. Einstein's theory of special relativity (SR), created in 1902–1905, described quite well the mechanical and electromagnetic phenomena at velocities up to the velocity of light, which had been impossible in the framework of Galileo and Newton's classical approach. However, Newton's theory of gravity, which was almost a perfect basis for celestial mechanics and all terrestrial physics, was formulated using the old notions of absolute space and time and could not be reconciled with SR where space and time were unified in Minkowski's four-dimensional geometry.

After the advent of SR there were numerous attempts to describe the gravitational field in Minkowski space in the hope to include rapidly moving gravitating objects into the theory. Newtonian gravity in such cases ought to be recovered in the limit of small velocities.

Henri Poincaré, great French mathematician who had actually "discovered" SR simultaneously with Einstein, was the first to try to extend it to gravity, assuming a finite propagation velocity of the gravitational field. The idea that gravity is transferred with the speed of light had been expressed before, but without SR there was no proper geometric language for such ideas to underlie a consistent theory. But the fate of special-relativistic theories of gravity was also not too happy. Despite being intrinsically free of contradiction, they still faced serious problems. One of them was their inability to explain the anomalous secular perihelion advance of planet Mercury's orbit, about $43''$ per century, which could not be explained by Newton's theory as well. Such a small perihelion shift had been still quite confidently measured in astronomical observations.

By the way, Mercury is the fastest planet of the Solar system, and it was natural to expect a manifestation of the new "gravity of high velocities" in the peculiarities of its motion. Even more than that: as we know today, relativistic gravity is better visible in strong gravitational field, and Mercury experiences the strongest field among the planets.

By Kepler's laws that follow from Newton's universal law of gravity, the planets of the Solar system move along closed elliptic orbits. This ideal picture would be true if each planet interacted with the Sun only. In reality, the mutual influence of the planets slightly changes their orbits, so that their ellipses slowly rotate from one revolution to another, thus becoming unclosed spiral-like curves. These effects are calculated in Newtonian celestial mechanics with high accuracy. The anomalous perihelion shift is obtained by subtracting all "normal" shifts (explained in Newtonian physics by coordinate effects and gravitational perturbations due to other planets, whose joint effect amounts to about 5558 angular seconds per century) from the observed value of $5601''$ per century (see, e.g., [554] for details).

One more circumstance of theoretical nature made any attempts to describe gravity in the framework of SR, so to say, less attractive. Since Galilean times it has been well known that, if one excludes air resistance, quite different bodies — a bit of fluff, a wooden bar, a brick, a lead ingot, a can of water, etc. — fall to Earth with precisely the same acceleration. The universality of free-fall acceleration was confirmed with high accuracy (up to 10^{-9}) in the Eötvös experiment with a torsion balance at the end of the 19th century: in fact, it established an equivalence between the Earth's attraction force and the inertial centrifugal force due to the Earth's diurnal rotation. In Newton's equations it is expressed as an equality between the inertial and gravitational masses, the so-called equivalence principle (EP).

Newton's theory itself cannot explain this equality, and the same is true for all its generalizations in the framework of SR.

The inertial mass that appears in the second law of Newtonian mechanics $\vec{a} = \vec{F}/m$ (acceleration is equal to force divided by mass) and the gravitational mass that appears in the law of gravity are, in essence, quantities of absolutely different physical nature. It was clear to Einstein that an equality between them cannot be a mere coincidence and should have deep reasons. The universality of the action of gravity on all bodies led him to the idea that became the basis of general relativity (GR): *the gravitational field is a property of space itself,* and this property should change from point to point since the gravitational field is, in general, inhomogeneous. Therefore, Minkowski space, which is flat, homogeneous (the same at all its points) and isotropic (the same in all directions) is not suitable; gravity should warp and curve it. That is how emerges the idea of curvature of physical space–time. And since the gravitational field is created by heavy bodies, the same must be true for the curvature.

Certainly the main idea of GR, as any fundamental idea, had its forerunners and proclaimers. Even N.I. Lobachevsky, discoverer of non-Euclidean geometry, spoke in 1826 of a possible experimental determination of the world's geometry. Riemann (1854) and Clifford (1876) assumed that the curvature of space must depend on the properties of matter that fills it, and Clifford even expressed the idea that curvature may propagate by waves. So the ideas as though wandered in air. But it was Einstein (in contact with Hilbert, Poincaré and other outstanding mathematicians and physicists of that time) who converted them into an elegant and logically consistent theory.

The fruit was ripe by 1915. GR became one more step aside from the simple and clear views of classical physics: the four-dimensional space–time (or simply space, as is often said for brevity) became curved. Riemannian geometry, the geometry of curved spaces, already invented by that time, became a mathematical tool and a language of the new physical theory.

In Riemannian geometry, hence also in GR, the basic quantity used to describe the space–time is the symmetric metric tensor $g_{\mu\nu}(x^\alpha)$ (the metric), which depends on the four coordinates x^μ and consequently changes from point to point. The metric carries information on intervals between close space–time points, or events, and in its terms one expresses the quantities that directly characterize the space–time curvature, the Riemann and Ricci tensors. It is the metric tensor components that are the unknowns in the dynamic equations of GR, the Einstein equations (or Hilbert–Einstein equations, as they are often called to emphasize Hilbert's substantial contribution in GR creation):

$$R_{\mu\nu} - \frac{1}{2} g_{\mu\nu} R = -\frac{8\pi G}{c^4} T_{\mu\nu}. \tag{1.1}$$

In the general case, it is a set of 10 nonlinear partial differential equations with respect to 10 unknown functions of four space–time coordinates. Their basic meaning is a direct relationship between the space–time curvature (the left-hand side of the equations, called the Einstein tensor) and the distribution and motion of matter (the right-hand side of the equations, called the stress-energy tensor). Thus, "matter tells space how to curve". Any solution to the Einstein equations describes some possible configuration of matter and the gravitational field.

Just as SR did not cancel Newton's mechanics (remaining quite suitable at small velocities), GR did not cancel SR, which is still valid in any small region of a curved but smooth space–time. The smaller the size of such a region, the more precisely this region coincides with a certain region of Minkowski flat tangent space, and with greater accuracy hold SR and its numerous consequences. The Newtonian theory of gravity also follows from GR under proper conditions: Newton's equations are obtained from those of GR in the limit of small curvature (that is, in weak gravitational fields) and small relative velocities of the gravitating bodies. Most of the observable phenomena belong to this "weak regime". But GR interprets the gravitational forces in quite a different manner: these are now not forces but certain geometric characteristics of the world lines along which the bodies move in four-dimensional space–time. From the viewpoint of GR, a body falling freely in a gravitational field moves without any external forces, and its world line is a geodesic, a direct analogue of a straight line in flat space.

A phenomenon of utmost importance, absent in Newton's theory but predicted by GR, is the gravitational waves. Their existence directly follows from the wave nature of GR equations. It was first confirmed theoretically by the numerous wave solutions of GR and many theories that generalize and further develop it. Next, there emerged indirect observational data confirming their existence on the basis on an analysis of pulsar dynamics: a binary pulsar loses its energy just as predicted by the general-relativistic radiation formulas. And quite recently (in 2015) gravitational waves were directly detected [1], which became one of the greatest discoveries of the present century.

But let us return to our historical reminissence. As soon as GR was finally formulated (in 1915), it readily responded to the observational challenge and surprisingly precisely explained the above-mentioned anomaly in Mercury's motion. Another classical effect of GR available to verification is the effect of gravity on light rays, leading to light bending in the field of a celestial body. By Einstein's calculation, light passing near the Sun should be bent by an angle of $1.75''$. A similar effect is obtained in Newtonian theory if one represents light as a flow of particles flying, naturally, with the speed of light. But then the calculated bending will be half as much as in GR: approximately $0.87''$ for a ray passing near the edge of the solar disk.

The total eclipse of the Sun on 29 May 1919 made it possible to measure this effect by comparing the photos of stars near the solar disk closed by the Moon with usual night photos of the same part of the sky. As was expected, for pictures taken during the eclipse, the stars turned out to be slightly shifted from the disk edge as compared to their night positions. The deflection angle varied in different observers' data from $1.61''$ to $1.8''$ near the disk edge, gradually diminishing in the outward direction, at an error within $30''$. The sky itself confirmed the rightness of Einstein's prediction.

It was a true triumph: a theory created on the tip of a pen, swiftly won a place under the Sun. And what is even more amazing is that it generally preserves its leading position even now, in more than a century, successfully withstanding all challenges of time and lots of experiments.

1.1 Einstein after Einstein

But let us again avoid "running ahead of the engine" and return to the 20s and 30s of last century. It was the time of physics' active penetration into the micro-world and formation of a language adequate to its properties — quantum mechanics, later quantum electrodynamics and, wider, quantum field theory. The quantum theory was first built in the framework of the old, Newtonian concepts of absolute time and absolute space (nonrelativistic quantum mechanics), and it required substantial effort to extend it to the world of large velocities and high energies and to formulate its content in Minkowski's four-dimensional space–time.

The understanding of gravity as space–time curvature endowed GR with an exclusive character as compared to all the rest of physics, and this was in conflict with the feeling of unity of the material world, important for both philosophers and physicists. On the other hand, in GR itself there emerge quite a number of important problems, one of which is known as the problem of energy. The notions of energy and other conserved quantities play a very significant part in the structure of quantum theory. In flat space, one easily formulates the energy, momentum and angular moment conservation laws due to the symmetry of Minkowski space with respect to temporal and spatial translations and rotations, forming the 10-parameter Poincaré group. In curved space–time there are no such symmetries at all, and it is therefore quite difficult to define the energy and momentum of the gravitational field in GR in a consistent way.

For this and some other reasons, not all physicists have agreed with GR, and even now there are repeated attempts to build a theory of gravity in Minkowski space. Unlike the first attempts of this kind, the new authors have learned to explain the classical observed effects of GR, and gravitation in such theories is represented by a field with normal conservation laws and hopes to be quantized on equal grounds with other physical fields. According to Will's book [563], as early as in 1960 the number of such theories exceeded 25. But neither then nor afterwards did such theories give rise to really substantial interest (though certainly their followers would not agree with such a conclusion).

Going the other way, the contrary trend "to reduce all physics to geometry" created a number of new ideas which even now remain topical in theoretical physics. In this connection, GR was considered (and is now considered) as a basis for extension which can be achieved by introducing more complicated geometries than the Riemannian one (Weyl, Eddington, Cartan), space–time dimension larger than four with additional invisible coordinates (Kaluza, Klein), and new requirements to the symmetry of the initial formulation of the theory (Weyl's gauge symmetry principle). An ambitious task was then formulated, going far beyond a simple unification of the gravitational and electromagnetic fields: to obtain, from a unified field, all characteristics of the small number of elementary particles known at that time. And Einstein was not aside from all this effort, he was a leader of the program aimed at building a unified field theory on the basis of GR, and he remained that leader to the end of his life.

A description of these attempts would lead us too much away from our main subject, gravitation. Therefore we would like only to refer to Heisenberg's words said in the early 60s: "It has been in essence a splendid attempt. But at the time when Einstein developed his unified field theory, more and more new elementary particles were discovered, with their corresponding new fields. So there did not exist a steady empirical basis for carrying out Einstein's program, and his attempt did not lead to any convincing results." [Back translation from Russian.]

Even today the problem of creating a "theory of everything" remains a central problem of theoretical physics.

1.2 The technological breakthrough

By the end of the 1950s, physics already knew four rather than two basic interactions: the gravitational, electromagnetic, strong nuclear (due to which protons and neutrons join to form atomic nuclei) and weak nuclear (which is responsible for many particle transmutations and nuclear reactions of which the most well-known is beta decay). Among them, the gravitational interaction appeared to be something of minor importance: for particles, being much weaker than even the weak interaction, it seemed absolutely insignificant in micro-world physics. The accelerators supplied more and more new experimental data about the other three interactions, and quantum field theory in flat Minkowski space experienced rapid progress, formulating and solving various problems of particle physics. Against such a background, gravitational studies seemed to be something of extravagance. Yes, GR was recognized as a fundamental, almost philosophical theory, important for the world outlook, but its experimental basis was too poor: one effect, Mercury's perihelion advance, was checked up to 1%, and another, light bending near the Sun, up to roughly 30%. The cosmological observations could only testify to a nontrivial geometry of the Universe but could tell us nothing about the validity of particular gravitational equations... Kip Thorne, at that time a student and now one of the most famous gravitational physicists, was advised by his professors not to deal with GR, which, in their opinion, was a theory very weakly connected with the rest of physics and astronomy. He did not obey such an advice and, as far as we can guess, hardly regrets now.

The situation began to change only in the late 50s and early 60s. The development of experimental technology made it possible to plan and carry out a number of new tests of gravity while astronomical observations gave more and more evidence on the existence of real sources of strong gravitational fields in space. As the number of alternative theories of gravity grew, tens of new effects were predicted along with suggestions to test them.

Amazingly it is GR that is being confirmed now with greater and greater accuracy. Very briefly, that is how the present experimental status of GR looks.

One of the fundamentals of GR, the equivalence principle, has been confirmed, using torsion balances and various test bodies, to an impressing accuracy of 10^{-12} by now [70]. It seems that the limit of experiments on Earth's surface has been reached by this result, due to inevitable atmospheric, seismic, and technological noise. The planned satellite experiment STEP (Satellite Test of the Equivalence Principle) [569] will raise the accuracy to 10^{-17}–10^{-18}. The equivalence principle is predicted by all metric theories of gravity, including GR and its numerous extensions where gravity is identified with space–time curvature.

Another effect, which is also universal and common to a large class of theories is the so-called gravitational redshift. Its essence is quite simple: a photon, moving away from a gravitating centre, loses energy, hence its wavelength grows (the photon "reddens"), while if it approaches the center, it gains energy and becomes "more blue". In the same way a stone thrown up loses its speed but gains it while falling down. In GR this effect is also related to the clock slowing-down: the more we approach a source of gravity, the slower are clocks of the same physical nature. This effect has been checked both for photons (the experiments of Pound, Rebka and Snider with resonance photon absorption by atomic nuclei) [452, 453], and directly for clocks (deflections in precision atomic clock readings in air travels around the world) [278].

By the way, it is this effect that makes GR not only an abstract theory but really a working tool. And, it should be stressed, very well working. The global positioning systems using satellites, of which the one named GPS is the most well-known and widespread, are more and more actively used in diverse military and civil applications and by lots of individuals. The system includes high-precision clocks, whose rate depends on both the satellite velocity (an effect of special relativity) and the Earth's gravitational field (an effect of GR), therefore corrections taking into account these effects are included in the signal calculation software: the on-board clocks are periodically "slowed down" adjusting them to clocks located on Earth. One satellite revolution around the Earth yields a clock difference that, if neglected, leads to an error of 50–100 m in determining the coordinates of a receiver on Earth; meanwhile, the positioning accuracy is on the order of a meter or better.

The light (and radio signal) bending effect has been checked many times and with high accuracy, approaching 10^{-4} arcseconds. It became a basis for the theory of gravitational lensing, the basic method used for discovering otherwise invisible heavy bodies in the Universe.

One more confirmation of GR is a measurement of radar echo delay in the solar gravitational field (the Shapiro effect, sometimes called the fourth classical effect of GR, in addition to the perihelion advance, light bending and redshift). The delay is not caused by lower velocity of the signals (the speed of light in vacuum is the same everywhere since special relativity holds in a small neighborhood of every point) but by their longer distance as compared to the calculated trajectory in flat space. The experiments consisted of active radar tracking of spacecrafts, and the most precise results have been obtained by using the Viking orbital and landing components that reached the planet Mars in 1977 [498]. The effect itself amounted to about 250 microseconds while the signals travelled for about an hour in the interplanetary space.

A measurement of a qualitatively different effect of GR at last produced its final result in 2011: the mission is called Gravity Probe B [235, 564]. It was an amazing project that took 47 years and 750 million dollars [564]. It was designed to test two fundamental predictions of GR, connected with precession of a gyroscope in the gravitational field of a rotating body (this time, the Earth) that had never been measured before, using four high precision gyroscopes on board a satellite orbiting the Earth at about 640 km above its surface. One of the effects is called geodetic precession: the space-time curvature exerts a torque on the gyroscope so that its axis slowly precesses, in the relevant case, by about 6.6 arcseconds per year, in the plane of the satellite's orbit. The other phenomenon, called frame dragging, or the Lense–Thirring effect, causes a precession of the gyro axis perpendicular to the orbital plane and amounts to 40 milliarcseconds per year. The results once again well confirm the GR predictions: $6.602 \pm 0.018''$ vs. predicted $6.606''$ for the first effect (a relative uncertainty of about 0.27%), and $0.0372 \pm 0.0072''$ vs. predicted $0.0392''$ for the second effect (a relative uncertainty near 18%) [235].

All effects of GR in the Solar system are small corrections to the predictions of classical physics, which have been confidently checked up to 10^{-3} by now, and a number of modern projects suggest an improvement by a few orders of magnitude, reaching the second post-Newtonian approximation.

Much more interesting phenomena might be expected in strong gravitational fields. Observations bring us more and more new data on the manifestations of strong gravitational fields in the Universe.

Thus, neutron stars, whose existence had been predicted by Oppenheimer and Volkoff as early as in the 1930s (by the way, on the basis of GR) [437], were discovered in 1967 as radio pulsars [289]. A rapid development of pulsar astronomy and physics has led, among other things, to new tests of GR. The pulsars are superdense objects with masses of the order of the solar mass and size of a few kilometers; quite often they are components of binary systems, sometimes rather close. In such systems there "operate" gravitational fields stronger than those of the Solar system by factors of hundreds and even thousands. The outstandingly high stability of pulsar "clocks" allows for accurate tracking of the celestial mechanics of such a binary. This, in particular, has brought about new confirmations of the GR prediction of secular pericentre (analogue of perihelion) shifts in binary systems.

An even more substantial contribution to gravitation theory from pulsar observations is related to gravitational waves. Despite enormous effort, they were not directly observed until recently [1], but binary pulsar observations unambiguously pointed at their trace: a slow decrease in the pulsar orbital period indicates an energy loss, and, as obtained from a calculation by general-relativistic formulas, such a loss can be explained with reasonable accuracy by gravitational wave emission. Nowadays there are hardly any people who would doubt the reality of gravitational waves; on the contrary, there has emerged a new area of science, gravitational-wave astronomy, as an important source of data on the most violent events that occur in the Universe [541]. On the other hand, the observational results immediately rule out many alternative theories which predict the same effects as GR in the Solar system but a more rapid energy loss by binary stars due to gravitational-wave emission.

The most popular and exotic prediction of GR is certainly the existence of black holes. The image of huge insatiable mouths that devour everything and return nothing has become, no exaggeration, a part of human culture, from science fiction to folklore. Modern astrophysics considers black holes as quite real objects in space, emerging as a result of gravitational collapse of massive stars, while many phenomena in the cores of galaxies and quasars are well explained by the existence of supermassive black holes of billions of solar masses.

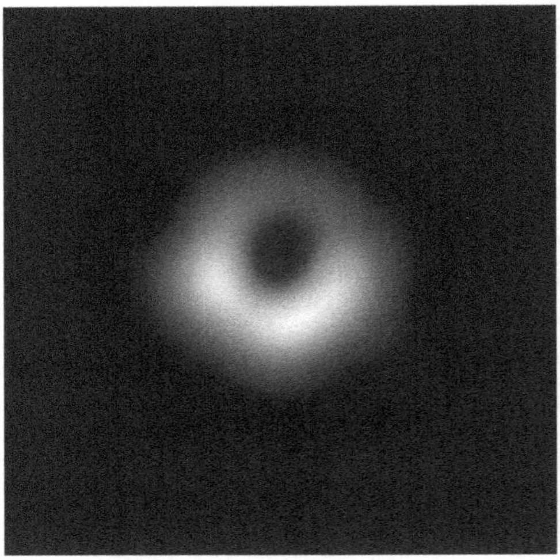

Figure 1.1 The image of M87 black hole obtained by the EHT.

A great success of modern radio astronomy has been the observation of the supermassive black hole in the giant elliptic galaxy Messier 87 (M87*) by the Event Horizon Telescope (EHT) collaboration [233]. The EHT is an array consisting of a large number of radio telescopes that form a global network and employ very-long-baseline interferometry (VLBI). The image obtained (Fig. 1.1) and the data anaysis as a whole have also led to new estimates of the mass ($(6.5 \pm 0.7) \cdot 10^9$ solar masses) and size of the M87 black hole: its event horizon radius was estimated as about 270 astronomical units (close to the orbital radius of Uranus). The shadow (the black spot at the image center) is larger by a factor of $\sim$2.5.

Both the gravitational wave detection and the EHT results added much confidence in the black hole existence. Thus, black hole mergers are the most plausible explanation of most of the observed gravitational wave bursts. In addition, one of the bursts, interpreted as a result a merger of two neutron stars [2] made it possible to detect electromagnetic signals from the same event almost simultaneously with the gravitational waves (1.74 ± 0.05 seconds later), which proved that these waves propagate with the velocity of light with an incredible accuracy, up to 10^{-15}. Such a result rules out many theories of gravity alternative to GR, predicting a different gravitational wave velocity. On the other hand, this observation gave rise to multi-messenger astronomy, which promises plenty of important and interesting new knowledge.

Curiously, the first exact solution of the Einstein equations, the Schwarzschild solution (1916, [493]), characterizing the static field of a gravitating centre, describes the simplest black hole. Though, the properties of the Schwarzschild solution were entirely understood only in the 50s [364, 524], and its peculiarities are sometimes a subject of controversy even now.

Since the early 60s, black hole physics has been developing as an independent line of investigations, and it has led to many interesting and unexpected results. Thus, it has been discovered that it is possible to mine energy from rotating black holes by launching small bodies in their neighborhood (thereby slightly slowing down their rotation) [445]; that black holes can be at full rights considered as thermodynamic objects with certain temperature and entropy [38]; that black holes "evaporate", emitting energy to the ambient space precisely as usual bodies heated to the corresponding temperature [284]. The Hawking evaporation process is connected with quantum particle production in the hole's classical gravitational field. The black holes of stellar and larger masses evaporate extremely slowly, and this does not affect any observable processes. Under real conditions, on the contrary, a black hole increases its mass at the expense of matter falling. But this process is of utmost importance for small black holes which could have survived since the first stages of the Universe expansion. Their evaporation gradually becomes stronger and stronger (since the black hole temperature grows as the mass decreases) and ends with an explosion.

In the third and fifth chapters of this book we introduce the basic notions of black hole physics and briefly discuss their main properties. A comprehensive presentation of this area of physics can be found, e.g., in the books [250, 549].

There is one more, no less intriguing consequence of the description of gravity in terms of space–time curvature. Indeed, if it is curved in principle, it is quite natural to assume that under certain conditions the curvature will be very strong — for example, it may lead to something like narrow bridges between different weakly curved universes, or "handles" connecting remote regions of the same Universe. Such geometric structures have been named wormholes. If they really exist, then, at least in principle, both time machines and interstellar travel become possible. It is not surprising that all this is readily employed in science fiction — but nowadays the properties of wormholes and the conditions for their existence and construction are broadly discussed in special physical literature, see, e.g., Visser's book [539] and the review [385]. We will touch upon some of these questions in Chapter 6.

1.3 To quantize or not?

As already pointed out, the experiment entirely supports GR. However, the picture is not so unclouded in theory. We have previously mentioned the gravitational energy problem. Another well-known difficulty of GR is the existence of singularities which emerge in the majority of exact solutions to the Einstein equations and, in particular, they are hidden beyond black hole horizons and occur at the beginning and, in some models, at the end of the cosmological evolution as well as in modeling isolated bodies. These are, to put it simply, the points, lines or surfaces where the space–time loses smoothness and the quantities characterizing the curvature become either indefinite or infinite. Singularities may be connected with infinite matter densities or pressures, but there also exist purely geometric singularities, such as those in solutions to the Einstein equations in vacuum, in the absence of any matter. The inevitability of singularities in GR solutions has been proved under very general and reasonable conditions in a number of theorems, and this clearly indicates that GR is apparently not very precise in its description of extremely strong gravitational fields.

Unlike event horizons, i.e., black hole boundaries (distinguished but quite regular surfaces that work according to the remarkable principle "everybody can come in, nobody can go out"), singularities represent a real problem for the theory since they indicate, on the basis of the theory itself, the places where it does not work any more. Thus GR itself prompts to the necessity of going out of its own framework. How to do that? It is a question of great importance, the subject of many studies and discussion, maybe a question outside the framework of not only gravitation theory but physics as a whole.

It seems natural, for example, to try to account for quantum phenomena.

The relationship between gravity and quantum theory is a separate, long and very intricate story. On the one hand, as any wave field, gravitation should possess quantum properties at small scales. On the other hand, the mathematical procedures of quantization, working well in particle theory, cope very poorly with space-time curvature. There are a few ways of obtaining quantum versions of GR, but they lead to fundamentally different results. For this reason it is often claimed that quantum gravity must not be built on the basis of GR, but one should instead begin with formulating a deeper and more general theory unifying gravity with other interactions. It is well known that Einstein was for many years Bohr's opponent in discussions on the interpretation of quantum mechanics ("God does not play dice"). Quite possible, the nonacceptance of a probabilistic picture of the world by Einstein was related to its incompatibility with GR.

Some qualitative arguments prompt what might be expected from quantum gravity. Thus, from the three fundamental constants: the Planck constant $\hbar$ (quantum of action, the attribute inherent to quantum mechanics), the speed of light c (the equally fundamental constant of relativity, the maximum velocity of matter and interaction transfer) and the gravitational constant G (it is the same in Newton's theory and GR) one can construct a quantity with the dimension of length. It is called the Planck length ($l_{\rm Pl}$) and is equal to approximately 10^{-33} cm — it is probably this characteristic scale at which the space–time begins to display quantum properties. It is a very small quantity: it is smaller that an atomic nucleus (about 10^{-13} cm) by roughly the same factor by which such a nucleus is smaller than the globe ($\sim 10^7$ cm).

At such lengths there must be significant fluctuations of the metric and even topology. Accordingly, at such lengths one can expect a foam-like structure of space, which permanently changes, "breathes" at the Planck time rate $l_{\rm Pl}/c$. At still smaller lengths, the very notion of smooth space-time becomes inapplicable, and instead of it, very likely, there is some discrete (point) structure. All this is certainly nothing more than hypotheses, but they have already given rise to numerous attempts of mathematical calculations and observational predictions.

1.4 The zoo of theories

To have a general idea of what concerns gravitational physicists nowadays, it makes sense to look through the program of some recent large gravitational conference. About a third of all talks presented will probably belong to classical GR, its astrophysical and cosmological applications. The mathematical tools are being refined, including the methods of solving the Einstein equations, new solutions are found and old ones are analyzed, questions of principle are discussed and new observational effects are suggested or predicted. In the experimental section, there are numerous works on gravitational wave detection and suggestion of measurements in space. There is inevitably a section or a few sections devoted to alternative theories of gravity, where the grand trend belongs to multidimensional theories and unification of interactions including gravity. (Let us note that the very word "alternative" — naturally, with respect to GR — stresses the outstanding role of GR among gravitational theories.) A quantum section is also certainly there.

Those who work on alternative or generalized theories pursue quite diverse objectives. There are attempts to overcome the difficulties of GR (such as, for instance, the gravitational energy problem) while preserving or strengthening its advantages; there are attempts to take into consideration principles and phenomena absent in GR. But probably the main point in all new theories is an approach to gravity as a constituent of a future "theory of everything" (or much more if not everything). The unified theories that include gravity, as a rule, use more complex geometric structures than 4D Riemannian geometry and new physical fields apart from the metric. Many of them employ ideas put forward as long ago as in the 1920s. Each of them reduces to GR under certain conditions or restrictions. And, as in GR, one seeks there solutions of physical interest (such as black holes, cosmologies, etc.) and observational predictions.

Let us mention some examples, by no means exhausting the whole diversity of approaches.

Scalar-tensor theories (STT). As is clear from the name, in these theories gravity is characterized, apart from the metric that determines the space-time curvature, by one or a few scalar fields. Let us present the Lagrangian of the Bergmann–Wagoner–Nordtvedt general STT with a single scalar field [48, 428, 548]:

$$L_{\rm STT} = f(\phi)R + h(\phi)g^{\mu\nu}\phi_{,\mu}\phi_{,\nu} - 2U(\phi) + L_m, \tag{1.2}$$

where R is the scalar curvature, ϕ is the scalar field, f, h, U are arbitrary functions of ϕ, and L_m is the Lagrangian of all other matter. The STT make the mathematically simplest extension of GR, which predicts, in general, the dependence of the gravitational constant on the scalar field and consequently on the position in space and time; it also predicts different values of classical gravitational effects from GR and more types of gravitational waves. The Lagrangian (1.2) reduces to that of GR, $L_{\rm GR} = (R - 2\Lambda)/(16\pi G)$, if the field ϕ is constant in the whole space–time. (Here, G is the gravitational constant and Λ is the cosmological constant.)

The observational data strongly constrain the choice of admissible STT. Without touching upon the motivations that led Jordan, Brans and Dicke in the 50s and 60s to formulation of the first STT [74, 310], let us note that numerous scalar fields naturally emerge in various multidimensional theories, in particular, those following from unification ideas, when reduced to four dimensions.

Curvature-nonlinear theories. Another important class of extended theories are those whose Lagrangian includes, apart from the scalar curvature R, some other functions of R (the so-called $f(R)$ theories) as well as other curvature invariants. Their distinguishing feature is that their gravitational field equations include third-order and higher derivatives of the metric tensor whereas this order is no more than second in GR. As a result, the set of solutions of these equations becomes much wider (though, together with the difficulty of finding them).

It should be noted that the advent of curvature-nonlinear terms in the gravitational Lagrangian is a direct consequence of quantum field theory in curved space since such terms inevitably emerge in the regularization and renormalization procedures necessary to give sense to the results of calculations [51, 263]. In other words, any theories which do not include curvature-nonlinear contributions, should be regarded as approximate ones, able to describe reality at small curvature only (certainly, to the extent that we trust the methods of quantum field theory in curved space-time).

Gauge theories. The basic idea of gauge theories traces back to the works of Hermann Weyl of 1918–1922, where he suggested using the equations of gravity and electromagnetism with a richer set of symmetries than the one contained in the Einstein equations. The additional symmetry includes some transformations of the fields themselves. Beginning with the 50s (the works of Yang and Mills), such symmetries, called local gauge symmetries, are widely used for the description of particles' interactions. It is this basic idea that has led to a unification of the weak nuclear interaction, responsible, e.g., for beta decay, with the familiar electromagnetic interaction (Weinberg, Salam, Glashow). There are promising versions that unify the resulting "electroweak" interaction with the strong nuclear one: such theories are called Grand Unification Theories (GUTs). But all this does not concern gravity. What is important is that gauge symmetries can be described in terms of geometries of some special spaces (fiber bundles), thus continuing the trend of geometrization of physics, see, e.g., the books [54, 198, 346].

Attempts to include gravity in the general scheme of gauge theories have led to different variants of gravitational theories with torsion which, in addition to the curvature, is one of the forms of "warping" the flat geometry (the Einstein–Cartan theory, the Poincaré gauge theory and others [248, 305, 450]). It has turned out that in such theories one can get rid of many kinds of singularities which exist in GR solutions as well as give new formulations of the problems of energy and quantization.

Multidimensional theories, models of superstring origin. One more idea related to geometrization of physics rests on the possible existence of extra space–time dimensions. It traces back to the pioneering works of T. Kaluza and O. Klein (1921) related to attempts to unify gravity and electromagnetism. In recent decades this idea has become a necessary ingredient of almost all attempts to unify all the four physical interactions.

Among the candidate "theories of everything", probably the most popular are the so-called superstring theories [261, 322]. Strings are one-dimensional micro-objects which, like, say, guitar strings can oscillate with a certain spectrum of frequencies. These frequencies correspond to energies of different particles. The prefix "super" designates the presence of the so-called supersymmetry, a symmetry between bosons (particles with integer spin) and fermions (particles with half-integer spins). Supersymmetry requires that each boson have a fermion counterpart. Superstrings "live" in curved spaces of dimensions 10 or 11 (depending on the particular version of the theory) and, under certain conditions, lead to a 10- or 11-dimensional version of GR. Classical GR is then obtained after distinguishing four "usual" coordinates in such spaces and inherits a large set of diverse fields from its multidimensional prototype.

Brane worlds, wells and trenches. Let us say a few words on one of the "fashionable" trends in the theory of gravity, the brane-world concept, related to unification ideas and superstrings but being of interest of their own as such.

All theories formulated in space-times of dimension higher than four have to answer the question of why the extra dimensions are still not observed. In the majority of models, beginning with those due to Kaluza and Klein, the answer is like that: the extra dimensions are closed, or compact, and have an extremely small size. A tube (a two-dimensional surface) of a micron in diameter cannot be distinguished from a geometric line, a one-dimensional object, without a microscope. The same happens to the fifth or tenth dimension: since our instruments have not reached a sufficiently small size (or, from the viewpoint of particle physics, a sufficiently high energy), the world for us is four-dimensional.

But another answer is also possible: the fifth dimension is not small, it can even be infinite, but ourselves and all observable particles and fields are "confined" in a single four-dimensional surface, or a thin layer, and a huge energy is required in order to leave it. It is like sitting on the bottom of a deep well having no power to climb out. The well is extended in three spatial dimensions, so it is more like a trench than a well: you can move along it where you wish, but not up or aside.

Such a distinguished surface or layer is called a brane (the cut-off word "membrane"), and the whole concept is called the "brane world". The first models of this sort were suggested in the 1980s [13, 17, 441, 475] and became very popular in a few recent years. It became clear that they lead to new approaches in quite a number of problems of particle physics and cosmology. From the viewpoint of the theory of gravity, it is of interest that the gravitational field equations on the brane (the only field that we can observe) are more complicated than the Einstein equations and lead to somewhat different predictions. Thus, Newton's law of gravity should be violated at distances smaller than a fraction of a millimeter: instead of a squared distance, there should appear its higher power in the denominator (see, e.g., the review [117]). A discovery of such a modification of Newton's law would be a weighty argument in favor of extra dimensions in reality. Brane worlds can contain nonsingular black holes [98], wormholes without exotic matter [139] and other nonstandard phenomena.

It seems to be a good time to stop and look at the situation as a whole.

We have seen that, despite the splendid observational status of GR, most of the specialists do not believe that it is the last word in this area of physics, but only a low-energy limit of a so far unknown fundamental theory, most probably, a multidimensional one, unifying all interactions and free of such difficulties as singularities, the problem of energy and ambiguity of quantization.

Variable "constants". Among experimental predictions of the alternative theories of gravity there are quite a number of effects, such as the already-mentioned difference in the properties of gravitational waves. But probably the most striking new prediction as compared to GR is common for most of the new theories, is the variability of quantities which had been regarded before as Fundamental Physical Constants (FPCs). In the first place, it is the gravitational constant G that appears both in Newton's law of gravity and in the Einstein equations. The electron charge, the elementary particle masses and many other constants can also vary, though the characteristic times of their variation should be comparable with the age of the Universe or even exceed it, otherwise such a variability would have been discovered long ago.

To date, the experiments and observations give only upper limits of such FPC variations. But there are two exceptions. First, beginning with the 70s, now and then there appear data on variations of G on the level $\sim 10^{-12}$ per year, but there is no confident result confirmed by different research groups. Second, observations of optical spectra of remote quasars [552] testify to slow variations of the fine structure constant $\alpha = e^2/(\hbar c)$, where e is the electron charge [116]. Some observations [553] point at spatial variations of α, more precisely, that quasar observations in the Northern hemisphere (mostly by the Keck telescope, Hawaiian islands) give slightly smaller values of α in the remote past while the opposite part of the celestial sphere shows values of α slightly larger than the modern one, α_0, measured on Earth (by approximately 10^{-5}). The anisotropy has a dipole nature and has received the name "Australian dipole" [45]. The existing data are summarized by the following expression for deflections of α from α_0 at any point of space:

$$\delta\alpha/\alpha_0 = (1.10 \pm 0.25) \times 10^{-5}\, r \cos\psi,$$

where ψ is the angle between the measurement direction and the axis of the dipole (declination -61 ± 9 degrees, right ascension 17.3 ± 0.6 hours), and the distance r is measured in billions of light years. There have already appeared theoretical models trying to explain both spatial and temporal variations of α [175, 436]. These models include a scalar field interacting with electromagnetism (thus affecting the value of α) and forming a cosmic domain wall due to the properties of its potential. But these observations themselves are not confirmed by other groups and cannot be regarded completely confident.

More details on the theoretical grounds for FPC variability and the experimental situation can be found in books and review papers [176, 344, 345, 511, 534, 535, 563].

1.5 Gravitation and the Universe

For gravitational theory, the application area of utmost importance is cosmology, the science on the Universe as a whole or on its part available to observation. Modern cosmology is a rapidly developing field of knowledge: the wealth of observational data is impetuously growing, and a lot of most diverse models are being developed. Some of the models will be discussed in detail in this book.

For a long time the Universe was considered as a kind of "vessel" containing different object: particles, stars, planets etc. It seemed that there is no relationship between the properties of this "vessel" and its content. The situation began to change with the advent of GR whose equations just established a relationship between matter and geometry. In cosmology, this relationship became even more evident after A.A. Friedmann's discovery that a stationary state of the Universe was unstable, so that it must either expand or contract. It inevitably followed from an analysis of the Einstein equations in a cosmological context. The expansion or contraction rate depends on the density and other properties of matter. The properties of the "box" have turned out to depend on its content. Further studies have led to the conclusion that the presently observed part of the Universe some time ago (about 14 billion years) had a size of about 10^{-27} cm or even smaller. But it is smaller than the size of an atom by 19 orders of magnitude. Such a small region certainly could not contain the whole wealth of particles making the stars. It means that the Universe and the particles were born simultaneously or almost simultaneously and undoubtedly exerted influence on each other. By now it is quite clear that the Universe is not a "vessel" able to contain anything but a complex organism whose parts are all mutually intertwined and interrelated. Everything is of importance here: the particle properties, gravitational physics, statistical physics, electrodynamics... and frequently all that at the same time.

A fundamental feature of the present epoch is the establishment of cosmology as a science. We can now answer many questions with proper grounds, we can explain a lot of observational data and theoretically predict many effects. The general path of the Universe's evolution is becoming clear, beginning with its creation and further into the future. Certainly not all details are clear, but it seems that almost all possible variants of the Universe's evolution have been named.

Gravity plays a key role not only in the physics of local objects, such as stars, galaxies, black holes, etc., but also in the properties of the Universe as a whole. Thus, the size of our Universe (more precisely, of its visible part) unambiguously depends on the value of the gravitational constant. The development of cosmology on the basis of GR is gradually establishing the features of the Standard Cosmological Model (SCM) whose basic significance begins to approach that of the Standard Model of particle physics.

The SCM on the basis of GR explains, above all, the following phenomena:

1. The homogeneity and isotropy of the Universe;
2. Fluctuations of the Cosmic Microwave Background (CMB) temperature;
3. The primordial nucleosynthesis.

One can assert that the history of cosmology as a science studying the Universe as a whole began in 1929 with E. Hubble's discovery. He found that the numerous nebulae, which looked on the photographs as small spots of unclear origin, are galaxies located millions of light years from our Milky Way Galaxy. It also turned out that these galaxies fly away from ours, leading to the conclusion that the whole Universe is expanding and hence there has been some time in the past when the Universe was born. Further studies have confirmed this conjecture. The abundance of helium and the existence of the CMB, discovered in 1965, also indicated that the primordial Universe consisted of a very hot plasma. It is the essence of the idea of the Big Bang.

The Universe as a whole is to a large extent homogeneous and isotropic, which is by itself hard to explain, but the large-scale structure, consisting of galaxies and clusters of galaxies, could only appear due to primordial energy density fluctuations. A detailed analysis has shown that only baryonic fluctuations are insufficient for modern structure formation. There should be some "invisible" component of energy density ("dark matter") in order that the gravitational forces be able to compress the domains of enhanced energy density and to launch stellar formation. The existence of dark matter has been by now practically proven, but its precise nature and its composition are still unclear. It is only clear that its total mass is larger than that of baryons, the particles that form stars, by approximately a factor of five.

A discovery of fundamental significance was made in 1998. It was discovered that our Universe is expanding not with a deceleration, as had been believed before, but with an acceleration. In the framework of GR it can only be explained by that the main part of the energy density is not concentrated in baryons or even dark matter but in some strange form of matter possessing a large negative pressure. It cannot consist of particles, otherwise it would form inhomogeneous structures, which is not confirmed by observations. This strange kind of matter, which has been called "dark energy", is distributed in space in a surprisingly homogeneous way, but its presence affects the galaxies' recession rate. The origin of dark energy is still unclear. The simplest explanation is that the minimum of the potential energy of a certain hypothetic field is not strictly zero but is a small positive constant. This very fact is not too surprising: what is strange is that this minimum is extremely close to zero. There is so far no widely accepted theory explaining this smallness.

There are now many open questions, but there also exist substantial advances. The most impressive step forward has occurred with the advent of the idea of inflation, i.e., a very rapid expansion of space at a primordial stage of the Universe evolution. According to the inflationary scenario, immediately after the end of inflation there happened abrupt heating of matter, reproducing the already well-known picture of a hot Universe which had been considered in pre-inflationary models as the initial stage. There are an enormous number of specific realizations of the inflationary scenario, and it is the matter of future studies to choose the correct model. Their common feature is that all of them predict an extremely rapid expansion of space in the framework of that or other theory of gravity.

One should bear in mind that cosmological models crucially depend on the choice of a theory of gravity. Therefore one can hope that cosmology, as greater and greater amounts of knowledge are stored, will not only help us to make clear the fate of our Universe but also to come to a better understanding of the mechanisms of gravity.

Part I

Gravitation

Chapter 2

Fundamentals of general relativity

The basic content of general relativity (GR) is very widely presented in quite a number of well-known and well-written textbooks and monographs (e.g., [366, 409, 525, 549, 554] and many more). Assuming that the reader is aware of these fundamentals, in this chapter, mostly for reference purposes, we will mention some of its basic facts and relations. Many geometric notions (such as those of a vector and a tensor, co- and contravariant components of vectors and tensors, contraction of tensors and so on) are supposed to be known and are used without explanation.

The space–time in GR is a four-dimensional manifold with a pseudo-Riemannian metric (the prefix "pseudo" is often omitted to say simply "Riemannian space–time"). The gravitational field is described in GR in terms of space-time curvature, which is expressed using the metric tensor and its derivatives with respect to the coordinates. Thus GR belongs to the class of metric theories of gravity [563]; it is historically the first, the simplest, and the most well-elaborated theory of this class.

We begin this presentation by recalling the basic facts of special relativity (SR), bearing in mind that in the close neighborhood of any point a Riemannian space–time coincides with its tangent flat (Minkowski) space–time, in which SR is formulated. So, GR is not only a generalization of SR including gravity: SR holds locally in all cases as long as gravitational effects can be neglected.

2.1 Special relativity — Minkowski geometry

2.1.1 *Geometry*

The simplest example of a Riemannian geometry is Minkowski flat space-time. The basic geometric invariant in this space is the interval between two arbitrary points — events. There is a class of preferable coordinates (the Minkowski coordinates) x^μ, and their each possible choice corresponds to a certain inertial reference frame (IRF). The summed force applied to a body at rest in a certain IRF is equal to zero, or, in other words, such a body moves by inertia, uniformly and straightly with respect to any other IRF. In any IRF, the squared four-dimensional "distance" (interval) between the events 1 $[x_1^\mu = (ct_1, \vec{x}_1)]$ and 2 $[x_2^\mu = (ct_2, \vec{x}_2)]$ is written in the form

$$s^2(1,2) = c^2(t_2 - t_1)^2 - (\vec{x}_2 - \vec{x}_1)^2, \tag{2.1}$$

where c is a universal constant coinciding with the propagation velocity of electromagnetic waves (light, in particular) in vacuum and called the speed of light. For close events 1 and 2, the interval (2.1) can be written as

$$ds^2 = \eta_{\mu\nu} dx^\mu dx^\nu, \quad \mu = 0, 1, 2, 3, \tag{2.2}$$

where the tensor with the covariant components

$$\eta_{\mu\nu} = \mathrm{diag}(1, -1, -1, -1) \tag{2.3}$$

is called the Minkowski metric tensor (the Minkowski metric). As usual, summing is assumed over repeated indices if one of them is covariant and the other contravariant. The matrix (2.3), together with its inverse matrix

$$\eta^{\mu\nu} = \mathrm{diag}(1, -1, -1, -1) \tag{2.4}$$

of the contravariant components of the Minkowski tensor are used for raising and lowering vector and tensor indices, so that, e.g., for an arbitrary vector $a = (a^\mu)$ we have $a^\mu = \eta^{\mu\nu} a_\nu$, $a_\mu = \eta_{\mu\nu} a^\nu$. The Minkowski tensor defines a scalar product (ab) of two arbitrary 4-vectors a^μ and b^μ as follows:

$$(ab) = \eta_{\mu\nu} a^\mu b^\nu = \eta^{\mu\nu} a_\mu b_\nu = a_\mu b^\mu = a^\mu b_\mu. \tag{2.5}$$

In particular, the squared interval is a scalar square of the vector $x_1^\mu - x_2^\mu$.

Evidently, the scalar square $a^2 = a_\mu a^\mu = g_{\mu\nu} a^\mu a^\nu$ of any 4-vector a is an invariant which can take a positive, negative or zero value because the matrix $\eta_{\mu\nu}$ is not positive- or negative-definite. In the case $a^2 > 0$ the 4-vector a is called timelike, if $a^2 < 0$, it is called spacelike, and if $a^2 = 0$ it is called null, light, or lightlike. These notions also refer to the vectors $x_1 - x_2$ that connect events 1 and 2 as well as intervals between events.[1]

Let us fix an arbitrary point O in Minkowski space and place there the origin of the coordinate system of a certain IRF. Then the whole set of ends of all null vectors x^μ beginning at O form a conic surface called the *light cone* (or *null cone*) of the point O. Its equation follows from (2.1):

$$(x^0)^2 - \vec{x}^2 \equiv c^2 t^2 - x^2 - y^2 - z^2 = 0 \tag{2.6}$$

(where, as usual, $x^0 = ct, x^1 = x, x^2 = y, x^3 = z$, and $\vec{x}^2 = x^2 + y^2 + z^2$ is the three-dimensional scalar square). All timelike straight lines that contain the point O are located inside the light cone (2.6) while all spacelike ones are located outside it.

The trajectories (world lines) of particles of nonzero mass are timelike at their any points. Since the speed of light in SR is the maximum velocity of body motion and signal propagation, two events may be causally connected only if they are connected by a timelike or null interval, and one can use for them such notions as "earlier than" and "later than". In such cases, one of the events is located either inside the light cone of the other or on the cone itself. The light cone of any event splits into the future and past light cones.

---

[1]In some books, e.g., [366], the Minkowski metric and coordinate systems are strangely called "Galilean", despite the fact that, in Minkowski space, the (special) Einstein relativity principle is valid rather than the Galilean relativity principle, and the latter is only restored at small velocities. The usage of the word "Galilean" instead of "Minkowski" is still more confusing in connection with some new ideas in the theory of gravity, involving the Galilean transformations in some multidimensional space-times or special spaces.

2.1.2 *Coordinate transformations*

A transition from one IRF to another is described in the simplest way if the velocity $\vec{v}$ of the system S' with respect to the system S is directed along one of the coordinate axes of the latter, for instance, along the axis Ox, i.e., in the system S the origin of the system S' moves according to the law $x = vt$. In this case, the coordinate transformation (the special Lorentz transformation) that leaves the interval (2.1) invariant, has the form

$$x' = \frac{x - vt}{\sqrt{1 - v^2/c^2}}, \quad y' = y, \quad z' = z, \quad t' = \frac{t - vx/c^2}{\sqrt{1 - v^2/c^2}}, \tag{2.7}$$

where the primed coordinates belong to the IRF S'.

A general Lorentz transformation, connecting any two IRFs, includes a boost (a transition of the type (2.7) with a vector $\vec{v}$ of arbitrary direction) and an arbitrary rotation of the spatial coordinate axes. All Lorentz transformations form a six-parameter group called the Lorentz group. In addition to general Lorentz transformations, the interval (2.1) is invariant under space and time translations

$$x'^\mu = x^\mu + a^\mu, \quad a^\mu = \text{const}, \tag{2.8}$$

forming the four-parameter group of translations. Thus the complete group of *isometries* (coordinate transformations leaving invariant the metric tensor contains ten parameters. It is called the Poincaré group.

The matrix of an arbitrary Lorentz transformation $A = (A^\nu_\mu)$ has the definitive property of pseudo-orthogonality. It is this property that expresses the invariance of the Minkowski metric under such transformations. Namely, let the coordinates x^μ of the system S and the coordinates y^μ of the system S' be connected by the linear transformation

$$x^\mu = A^\mu_\alpha y^\alpha + a^\mu, \quad A^\mu_\alpha,\ a^\mu = \text{const}. \tag{2.9}$$

According to (2.9), $A^\mu_\alpha = \partial x^\mu/\partial y^\alpha$. Substituting (2.9) to the expression for the interval (2.2), we obtain

$$ds^2 = \eta_{\mu\nu} A^\mu_\alpha A^\nu_\beta dy^\alpha dy^\beta. \tag{2.10}$$

The invariance of the metric $\eta_{\mu\nu}$ means that in the right-hand side of this equality there stands the expression $\eta_{\alpha\beta}dy^{\alpha}dy^{\beta}$, where the matrix $\eta_{\alpha\beta}$ is the same as $\eta_{\mu\nu}$ and has the form $\mathrm{diag}(1,-1,-1,-1)$. The condition

$$\eta_{\mu\nu}A^{\mu}_{\alpha}A^{\nu}_{\beta} = \eta_{\alpha\beta} \tag{2.11}$$

is the condition that the matrix A is pseudo-orthogonal. The arbitrary constants a^{μ} in the transformation (2.9), forming the translation vector, are absent in the condition (2.11) and play the part of integration constants for the condition (2.11) treated as a set of differential equations for the functions $x^{\mu}(y^{\alpha})$.

2.1.3 *Kinematic effects*

An analysis of the Lorentz transformations leads to the most important kinematic effects of SR. Thus, any motion of a point particle at any fixed time instant can be considered to be approximately inertial, so one can introduce an IRF S' in which the particle is at rest at this time instant. Assuming, without loss of generality, that the motion occurs along the axis Ox, it is easy to find that the time increment dt' by a clock connected with the particle (and equal to ds/c) is related to the time increment dt by the clock of the "laboratory" IRF S according to

$$dt' = dt\sqrt{1 - v^2/c^2}. \tag{2.12}$$

Due to arbitrariness of choosing the axes, this formula is valid for a velocity of any direction, $\vec{v} = d\vec{x}/dt$. Consequently, for an arbitrary trajectory of motion $\vec{x}(t)$, the proper time interval τ of the particle (that is, the time elapsed according to a clock connected with the particle or any object whose size is insignificant) is determined by the relation

$$\tau(t_1, t_2) = \frac{1}{c}\int_{t_1}^{t_2} ds = \int_{t_1}^{t_2} dt\sqrt{1 - v^2(t)/c^2}, \tag{2.13}$$

if the time t_1 to t_2 has elapsed at the clock of an observer at rest.

From (2.12) and (2.13) it follows that the proper time interval of a moving object is always smaller than the time between the same events from the viewpoint of an observer at rest. It is the so-called *Lorentzian time slowing-down*; it leads, in particular, to the famous *twin paradox*. If one of the twins is at rest (or moves slowly) in a certain IRF while the other travels with relativistic velocities and, having completed his closed trajectory, meets his brother, then, at their meeting, their ages will be different: the traveler will be younger than the home-sitter. It could seem that if one considers the situation in the RF where the traveler is at rest, the result should be the opposite. However, a careful analysis, taking into account the fact that the traveler must have changed his IRF at least three times (at acceleration, at turning back and at final deceleration) shows that his calculated age will be smaller than that of his brother. The paradox is explained by the asymmetry of the situation: the integral (2.13) turns out to be smaller for an object (or subject in the present case) that has carried out noninertial motion.

Another effect, the *Lorentzian length contraction*, consists in the fact that the length of a ruler or a rod has different values while measured in different IRFs. The length is maximum in the RF where the rod is at rest and contracts in the ratio $\sqrt{1 - v^2/c^2}$ in an IRF moving with the velocity $\vec{v}$ parallel to the rod. This result is derived by finding the coordinates of the rod ends at the same time by the clock of a moving RF. The transverse size of a body does not change at its motion, and its volume changes in the same ratio as the length in the direction of motion. A body moving at relativistic velocities, comparable with the speed of light, becomes oblate, and in the ultra-relativistic limit $v \to c$ a spherical body acquires the shape of a disc or a pancake.

The third effect is the relativity of simultaneity: in different IRFs, the spacelike hypersurfaces $t = \text{const}$ are different. Even more than that: any two events separated by a spacelike interval can be made simultaneous by choosing a suitable IRF. This circumstance is illustrated by Fig. 2.1 showing a two-dimensional section of Minkowski space–time making the (t, x) plane. In the coordinates x, t belonging to the IRF S, depicted are the axes x', t' of another IRF, S'; the lines $t = \text{const}$ represent the spatial sections (surfaces of simultaneity) of

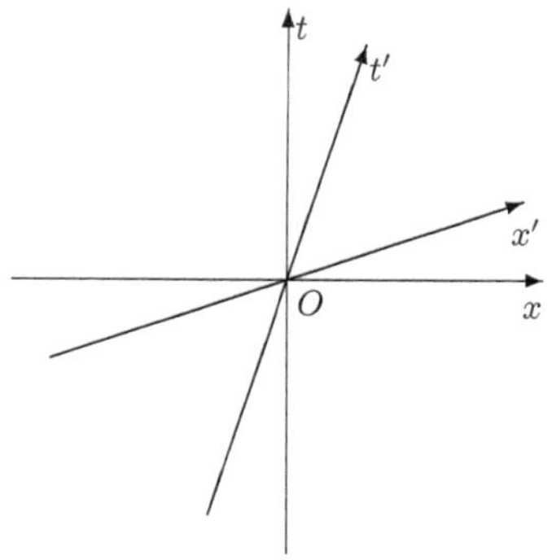

Figure 2.1 The x and t coordinate axes in different IRFs.

the IRF S, and the lines $t' = \text{const}$ (the x' axis and lines parallel to it) depict similar surfaces of the IRF S'.

There are other relativistic kinematic effects, e.g., light aberration (non-coincidence of directions to a distant source of light being observed from different IRFs), the relativistic formulas for velocity addition and the Doppler effect, different from their nonrelativistic counterparts etc. Their detailed analysis can be found in numerous textbooks on SR.

2.1.4 *Elements of relativistic point mechanics*

In the formulation of SR in terms of the Minkowski geometry, the three-dimensional quantities appearing in nonrelativistic mechanics are replaced by their four-dimensional counterparts, which leads to a more elegant, economic and transparent form of many equations. In particular, for the motion of a material point or an element of a medium, the 4-velocity is introduced

$$u^\mu = \frac{dx^\mu}{ds} = \left(\gamma, \frac{\gamma v^i}{c}\right), \quad \gamma \equiv \frac{1}{\sqrt{1 - v^2/c^2}}. \tag{2.14}$$

Here, $ds = c\,dt\sqrt{1 - v^2/c^2}$ is an element of the interval along the particle trajectory. It is easy to verify that the vector defined in this way is normalized

$$u_\mu u^\mu = 1. \tag{2.15}$$

The 4-acceleration of a material point or an element of a medium is

$$a^\mu = \frac{du^\mu}{ds} = \frac{\gamma}{c}\frac{du^\mu}{dt}. \tag{2.16}$$

Due to the normalization condition (2.15), the 4-velocity and 4-acceleration are mutually orthogonal: $a_\mu u^\mu = 0$.

The 4-momentum of a material point (particle) is, by definition,

$$p^\mu = m\, u^\mu = (E/c, \vec{p}), \qquad m = \text{const}, \tag{2.17}$$

where m is the particle rest mass, characterizing its inertial properties, $\vec{p} = (p^i)$ is the spatial momentum, and $E = \gamma mc^2$ is the particle energy in a given IRF. By definition, the squared momentum is $p^2 = p_\mu p^\mu = m^2c^2$.

At small velocities, $v \ll c$, expanding the expression for the energy in powers of v/c, we obtain

$$E \approx mc^2 + \frac{mv^2}{2} + O(v^2/c^2). \tag{2.18}$$

From this relation follows a conclusion of utmost importance: that the total energy of a particle includes, in addition to the classical kinetic energy, the rest energy mc^2.

The role of Newton's second law is played by an equality which can be considered as a relativistic definition of the force 4-vector f^μ:

$$f^\mu = \frac{dp^\mu}{ds} = mca^\mu = mc\frac{du^\mu}{ds}, \tag{2.19}$$

where the components f^μ are connected with the usual force 3-vector by the relations

$$f^0 = \frac{\gamma}{c^2}(\vec{F}\vec{v}), \quad f^i = \frac{\gamma}{c}F^i. \tag{2.20}$$

The action of a free particle is written in the form [366]

$$S = -mc\int ds = -mc^2\int\sqrt{1-\frac{v^2}{c^2}}dt, \tag{2.21}$$

and, since by definition $S = \int L\, dt$, the Lagrangian L is

$$L = -mc^2\sqrt{1-\frac{v^2}{c^2}}. \tag{2.22}$$

For a particle moving subject to an action of external forces, this action must be supplemented with other terms which include potentials of interaction between the particle and other bodies and actions of various physical fields.

For a free particle, variation of the action (2.21) with respect to the functions $x^i(t)$, describing the particle world line, is zero, from which we obtain the Lagrangian equation

$$\frac{d}{dt}\frac{v^i}{\sqrt{1-v^2/c^2}} = 0 \quad \Rightarrow \quad v^i = \text{const}, \tag{2.23}$$

so that, as should be the case, a uniform rectilinear motion is obtained.

Photons and other massless particles follow lightlike (null) rectilinear trajectories and can have an arbitrary energy E and spatial momentum $p^i = (E/c^2)v^i$, so that $|\vec{p}| = E/c$. The connection between the photon energy and frequency, $E = h\nu$ appears, as is well known, not in SR but in quantum mechanics.

The relativistic mechanics satisfies the correspondence principle: Newtonian mechanics for massive bodies is restored at velocities $v \ll c$.

2.2 Riemannian space–time — coordinate systems and reference frames

So far we have been using Minkowski coordinates in Minkowski space–time, where the metric has the form (2.2), and the inertial reference frames (IRFs) connected with these coordinates. However, nothing prevents us from describing physical processes in the framework of SR with the aid of other coordinate systems, even remaining in a fixed RF in Minkowski space. For example, one can introduce spherical or cylindrical coordinates.

In the most general cases, both in Minkowski space–time and in any Riemannian space–time, and wider, in any differentiable manifold, arbitrary coordinate transformations $x^\mu \mapsto y^\mu$ are possible, with arbitrary functions

$$x^\mu = x^\mu(y^0, y^1, y^2, y^3). \tag{2.24}$$

In the physical space–time, the coordinate transformations (2.24) lead in general to changes in the reference frame.

One should note that a relationship between the notions of coordinate systems and reference frames is rather a subtle question which sometimes leads to confusions and misconceptions. It is therefore reasonable to explain in which sense we shall use these notions.

In this discussion, we will mostly follow the books [542, 578, 579].

To begin with, omitting a number of mathematical details, we can say that a Riemannian space (or space–time) is a differentiable manifold of arbitrary dimension D equipped with a metric $g_{\mu\nu}$ of arbitrary signature. If it is positive-definite [the signature $(+\ +\cdots+)$], the space is called proper Riemannian, in other cases it is called pseudo-Riemannian, but the prefix "pseudo" is often omitted.

We will consider an arbitrary real four-dimensional (pseudo-) Riemannian space–time with the (infinitesimal) interval

$$ds^2 = g_{\mu\nu}dx^\mu dx^\nu, \quad \mu, \nu = 0, 1, 2, 3, \tag{2.25}$$

where the symmetric metric tensor $g_{\mu\nu}$ has the signature $(+\ -\ -\ -)$.[2]

Fixing an arbitrary point, we can bring the metric tensor at this point to a diagonal form by linear coordinate transformations, and then, by evident additional transformations, ordering the obtained diagonal elements and normalizing them to plus or minus unity, we will always obtain the Minkowski tensor (2.3) due to our choice of the signature. This clearly indicates that in a small neighborhood of each point the geometry of any Riemannian space–time coincides with the Minkowski geometry, and all phenomena for which the curvature is insignificant can be described in the framework of SR.

[2]Space–times with this signature are called Lorentzian. The same term is also applied to the signature $(-\ +\ +\ +)$; many authors use this signature. The physical content of the theory certainly does not depend on the choice of one plus and three minuses or vice versa, but the forms of some equations become slightly different which creates some inconvenience and requires certain care.

We will also use everywhere (unless otherwise indicated) the units in which the speed of light $c = 1$, so that $1\,\mathrm{s} \approx 3{\cdot}10^{10}\,\mathrm{cm}$.

2.2.1 *Covariance, maps and atlases*

Let us present the general definition of a coordinate system in a differentiable manifold. The definitions of all the corresponding notions and their detailed and rigorous discussions can be found in textbooks and monographs on global differential geometry, such as, e.g., [285, 375, 461].

A *coordinate system* in a certain region U of a differentiable manifold of class k and dimension D (or, which is the same, *a map* of the region U) is a one-to-one map of the region U to a certain region of the arithmetic space $\mathbb{R}^D$. The region U itself is then called the range of the map or the range of the coordinate system.

Thus each point x of the region U is put into correspondence with an ordered set of D real numbers, which numbers are called the coordinates of this point. Moreover, if U_1 and U_2 are ranges of two maps and $x \in U_1 \cap U_2$, then the coordinates of the point x in one of these maps are functions of class C^k of the coordinates of the point x in the other map, with a nonzero Jacobian of the transformation. Most frequently, manifolds of the class C^∞ are considered, with infinitely differentiable transformation functions.

The global properties of a manifold are described by sets of maps whose union of ranges covers the whole manifold (*atlases*). In other words, a union of ranges of maps belonging to a certain atlas is identical to the whole manifold.

Thus a coordinate system in a manifold and, in particular, in a four-dimensional Lorentzian space-time is simply a way to supply each point (event) with a certain "address" or label in the form of a set of numbers, and different ways of addressing must be related to each other by smooth transformations.

It is thus clear that the space–time coordinates in general do not bear any physical meaning if their values are not deliberately taken as functions of certain physical or geometric quantities (such as distances, time intervals, values of the electric field strength and so on). At the same time, the equations of a physical theory, describing objects in space–time, contain coordinates of points as variables or parameters. But neither the properties of any physical bodies or phenomena nor any measurable quantities can depend on which labels are ascribed to the points-events. Thus social life in a city cannot depend on whether the houses are numbered by order, by meters, miles, cubits or leagues and from which point the counting begins. But the numbering can be very convenient or quite clumsy. The same with coordinates in space–time.

It then follows that *physical laws should admit a universal formulation independent of a choice of the coordinate system* (the so-called generally covariant formulation). This statement is the content of the *General Covariance Principle.*

The General Covariance Principle may be called a mathematical expression of common sense: any physical theory pretending to be reasonable should admit a generally covariant formulation. For instance, Newton's theory of gravity, special and general relativity are all generally covariant.

2.2.2 *Reference frames and relativity*

Let us now pass on to the notion of a *reference frame* (RF). By definition, a reference frame is an imaginary, massless, in general, arbitrarily (but smoothly) deformable body (*the reference body*), existing in a certain region of space–time and equipped at each point with perfect rulers, allowing for length measurements, and perfect clocks, allowing for time measurements.

Thus, unlike coordinate systems, which notion is purely mathematical, a RF is a physical notion, quite necessary for connecting the theory with measurements. As is well known, real prototypes of perfect clocks are atomic clocks, while real prototypes of perfect rulers are the length standards attached to the wavelengths of certain spectral lines.

The *General Relativity Principle* asserts the equivalence of all RFs in formulations of the laws of nature. It is a consistent physical principle, from which follows the absence of preferred RFs in the nature. Let us note that even theories in which there are preferred RFs admit a generally covariant formulation of their equations. An evident example is SR, where inertial RFs (IRFs) play a distinguished role.

The absence of preferred RFs in the formulation of the laws of physics certainly does not mean that the results of physical measurements will be the same in all RFs: indeed, a measurement instrument is always situated in a certain RF, and this inevitably affects its readings. In particular, in SR, none of the IRFs is preferential, but measuring the length of the same rod by instruments situated in different RFs give, as we have seen, different results.

2.2.3 *Reference frames and chronometric invariants*

In principle, it is possible to use coordinate systems quite independently from RFs. For example, when describing the life of a city in its natural RF connected with its streets and houses, nothing and nobody forbids one from using coordinates connected with a certain system of regularly moving trams, or with the shadows of clouds floating in the sky.

In practice, however, it is much simpler and more convenient to attach the coordinates to a certain RF, assuming that the reference body is covered by a three-dimensional coordinate net x^i while the time coordinate x^0 changes along the world lines of fixed points of the reference body (they are naturally called time lines). It is said in this case that *the coordinate system belongs to a reference frame*. The world lines of particles at rest in a given RF are described by the equation $x^i = \text{const}$ (are x^0-independent). If one makes a coordinate transformation (2.24), then the condition that the new coordinates y^μ belong to the same RF as x^μ is that the new spatial coordinates y^i are x^0-independent.

Thus the transformations

$$y^i = y^i(x^1, x^2, x^3), \tag{2.26}$$

$$y^0 = y^0(x^0, x^1, x^2, x^3), \tag{2.27}$$

are the most general transformations between coordinate systems that belong to the same RF. Equation (2.26) describes three-dimensional coordinate transformations which change the spatial coordinate net. Equation (2.27) describes arbitrary chronometric transformations that change the course of arbitrary (coordinate) clocks as well as their synchronization from one spatial point to another.

In any coordinate system x^μ, one often considers *simulatneity surfaces, or spatial sections*, i.e., the sets of space-time points (events) defined by the condition $x^0 = \text{const}$.

It is clear from Eq. (2.27) that the condition $x^0 = \text{const}$ does not at all guarantee $y^0 = \text{const}$. In other words, even in coordinates belonging to the same RF, there can be different spatial sections, since there can be different sets of coordinate clocks (not to be confused with perfect clocks) which are synchronized in different ways and having different (and differently variable!) rates. A good example: different time scales on the Earth, i.e., manifestly in the same RF: Greenwich time, times in other time zones, the true solar time etc. Also, time according to clocks with pendulums of fixed length (whose rate depends on the gravity force, which in turn depends on latitude and altitude) and atomic time which is thought to be universal and is identified as the true physical time.

All measurable physical quantities in a given RF should be covariant under the transformations (2.26) and invariant under (2.27) (*chronometrically invariant*). This is the *chronometric invariance principle* introduced by Zel'manov [578]. It is dictated by the same reasoning as the general covariance principle but applied to a particular RF.

It is not hard to obtain expressions for chronometric invariants from the components of any four-dimensional tensors. Thus, any 4-vector A_μ has the following chronometrically invariant (ChI) components:

$$A^i, \quad i = 1, 2, 3; \qquad A_t = A_0/\sqrt{g_{00}}. \tag{2.28}$$

Dealing with higher-rank tensors, one must do the same with each index. More specifically, from the components of an arbitrary tensor $A_{\mu_1\mu_2\ldots\mu_r}$ of rank r one constructs the quantities

$$A_{0\ldots0}^{i_1 i_2 \ldots i_m}(g_{00})^{-(r-m)/2}, \tag{2.29}$$

where m ranges from zero to r while the number of lower indices (zeros) is equal to $r - m$. The expressions (2.28) are ChI and form contravariant tensors with respect to the spatial transformations (2.26). This result is verified in a straightforward manner. The metric coefficient g_{00} is always nonzero in coordinates belonging to a certain RF.

A very important example of a chronometric quantity is the 3D chronometric metric tensor h_{ik}: according to the above, its contravariant components coincide (up to the sign) with their 4D counterparts:

$$h^{ik} = -g^{ik}. \tag{2.30}$$

It can be shown [410, 578] that the covariant components of the chronometric spatial metric tensor and its determinant are expressed in terms of 4D quantities as follows:

$$h_{ik} = -g_{ik} + \frac{g_{i0}g_{k0}}{g_{00}}, \quad h = \det(h_{ik}) = -\frac{g}{g_{00}}. \tag{2.31}$$

The tensors h^{ik} and h_{ik} can be used to raise and lower indices in various chronometric vectors and tensors.

The ChI coordinate displacement vector dx^i and the ChI (physical) time interval $dt = \sqrt{g_{00}}dx^0$ can be used to obtain the expressions for the ChI spatial length element dl and the 4D interval:

$$dl^2 = h_{ik}dx^i dx^k, \quad ds^2 = dt^2 - dl^2, \tag{2.32}$$

the latter coinciding with the interval in SR. It is also clear that the ChI 3-velocity defined as dx^i/dt automatically turns to unity in the case of a null interval, $ds = 0$. This leads to the important result that *the locally measured velocity of light in vacuum in GR is the same in all RFs and is equal to the fundamental velocity*, as should be the case since a Riemannian space–time coincides at any point with its tangent Minkowski space (up to $O(dx^\mu)$, or, as is sometimes said, "in the first neighborhood").

A further development of the chronometric invariants theory includes a reformulation of all equations in terms of ChI quantities and ChI differential operators as well as a description of RFs employing their ChI characteristics: *acceleration, rotation and deformation* [410, 578, 579].

One should note that it is sometimes convenient to use coordinate systems that do not belong to any RF, e.g., null coordinates, and in such systems it is quite possible to have $g_{00} = 0$ in the whole space–time or its certain subsets.

An even more universal description of reference frames is realized using the so-called monad formalism [411, 542, 579], where a monad is the 4-velocity field of the reference body; this formalism considers chronometric invariants as one of its "gauges".

More well-known is the tetrad (vierbein, hedron, vielbein) formalism of RF description (see, e.g., [409, 410, 469]), where an RF is characterized with the aid of an orthonormalized hedron in Minkowski tangent space at each world point. Recall that, to specify an RF according to the above definition, it is sufficient to fix only one (timelike) vector at each point (i.e., the monad), while the other three vectors of a tetrad, characterizing the positions of spatial axes, are unnecessary. Therefore the tetrad formalism is partly redundant in many situations. However, this formalism is extremely useful for calculations of observable quantities, for the description of fermions in curved space-times and other purposes. For example, in Sec. 2.5.2 we will introduce a tetrad for a convenient description of anisotropic fluids. One should also mention that many authors consider tetrads, instead of the metric, as the dynamic variables describing the gravitational field in GR and its extensions.

2.2.4 *Covariance and relativity*

As we have seen, the general covariance principle is nothing else but an opportunity to use any coordinates for describing the space-time and events in it.

On the contrary, various forms of the relativity principle are important postulates of physical theories. Thus, the Galilean relativity principle that acts in Newtonian mechanics asserts that the laws of nature are independent of the choice of an IRF in Newtonian absolute space. The special relativity principle asserts quite the same but in Minkowski space, where IRFs are related by Lorentz transformations. In both cases, the theory contains a class of privileged RFs, and the transformations in question occur inside this class.

The Cartesian coordinates in Newtonian (Euclidean) space and the Minkowski coordinates in Minkowski space are privileged in two respects. They are distinguished mathematically since their coordinate lines coincide with the orbits of isometry groups of the corresponding spaces (though, this choice is not unique: spherical and cylindrical coordinates also possess this property). On the other hand, they are distinguished physically by the convenience of describing IRFs with their aid. It is well known, however, that, in many problems of classical mechanics, curvilinear coordinates are used, and physical phenomena in various noninertial RFs are studied. In SR, there is also a necessity to study accelerated RFs and to use coordinates belonging to them (e.g., uniformly accelerated frames and Rindler coordinates [467]). In such cases, quite helpful are the mathematical methods elaborated in GR and related to generally covariant formulations of all equations.

One should remark that the word "relativity" that is part of the names of both SR and GR may be said to have two faces. On the one hand, this word denotes principles which declare quite an opposite property of physical laws, their absolute nature, their RF-independence. However, one applies these laws in specific RFs, i.e., relative to a chosen class of observers. On the other hand, in both SR and GR, many things that have been absolute in classical physics, become relative, i.e., RF-dependent. In SR these are simultaneity, lengths, time intervals. In GR (as well as in other metric theories of gravity) one should add to this list the finiteness or infiniteness of the spatial volume and even the topology of 3-space. In what follows we will meet specific examples of such relativity.

2.3 Riemannian space–time — curvature

Let us present some definitions and relations, important for the further content of the book.

To begin with, the contravariant components of the metric $g^{\mu\nu}$ form a matrix reciprocal to the matrix $g_{\mu\nu}$:

$$g_{\mu\alpha}g^{\alpha\nu} = \delta^{\nu}_{\mu}, \tag{2.33}$$

where δ^ν_μ is the Kronecker symbol, a tensor whose components form the unity matrix: they are equal to unity for coinciding values of the indices and to zero for noncoinciding ones. The tensors $g_{\mu\nu}$ and $g^{\mu\nu}$ are used for raising and lowering arbitrary tensor indices.

Partial derivatives of any scalar function (that is, a function whose values are the same in all coordinate systems) $f(x^\mu)$ with respect to the coordinates, $\partial_\mu f$, form a covariant vector called the gradient of f. Partial derivatives of the components of a vector A_μ or A^μ, in general, do not form a tensor because coordinate transformations are, in general, nonlinear. To obtain a covariant form of physical equations and for many other purposes it is thus necessary to generalize the notion of a derivative, to make it a tensor. This goal is achieved by introducing the covariant derivatives

$$\nabla_\mu A_\nu = \partial_\mu A_\nu - \Gamma^\alpha_{\mu\nu} A_\alpha;$$
$$\nabla_\mu A^\nu = \partial_\mu A^\nu + \Gamma^\nu_{\mu\alpha} A^\alpha, \tag{2.34}$$

where the quantities $\Gamma^\alpha_{\mu\nu}$ (they do not form a tensor!) are called the Christoffel symbols, or affine connection coefficients agreeing with the metric (the metric connection, or the metric affinity). They are expressed in terms of the metric tensor and its first-order partial derivatives:

$$\Gamma^\sigma_{\mu\nu} = \frac{1}{2} g^{\sigma\alpha}(\partial_\nu g_{\alpha\mu} + \partial_\mu g_{\alpha\nu} - \partial_\alpha g_{\mu\nu}). \tag{2.35}$$

For the contraction $\Gamma^\alpha_{\mu\alpha}$, using (2.35), it is easy to obtain

$$\Gamma^\alpha_{\mu\alpha} = \partial_\mu(\ln\sqrt{-g}), \quad g := \det(g_{\mu\nu}). \tag{2.36}$$

The Christoffel symbols are symmetric in their lower indices, therefore, in general, there can be as many as 40 different components (ten variants of the lower pair and four values of the upper index).

One can directly verify the tensor nature for the transformations of the covariant derivatives under coordinate transformations.

For tensors of any rank, the covariant derivatives are calculated according to the methodology (2.34) applied to each upper or lower index separately. For instance, for the mixed tensor T^ν_μ we have

$$\nabla_\alpha T^\nu_\mu = \partial_\alpha T^\nu_\mu + \Gamma^\nu_{\alpha\beta} T^\beta_\mu - \Gamma^\beta_{\alpha\mu} T^\nu_\beta. \tag{2.37}$$

Due to (2.35) (in fact, since the connection agrees with the metric), the metric tensor is covariantly constant:

$$\nabla_\alpha g_{\mu\nu} = \nabla_\alpha g^{\mu\nu} = 0. \tag{2.38}$$

It then follows that the operation of covariant differentiation commutes with index raising and lowering, which is extremely convenient in transformations of complex tensor expressions.

The next point of importance concerns repeated application of covariant derivatives. Usual partial derivatives are known to commute with each other; the same is true for covariant derivatives of a scalar f. The first-order derivative, $\nabla_\mu f \equiv \partial_\mu f$, coincides with the partial one; the second nabla operator is already applied to a vector (gradient), but nevertheless

$$(\nabla_\alpha \nabla_\beta - \nabla_\beta \nabla_\alpha) f = 0.$$

Applied to a vector, commutation of covariant derivatives gives

$$(\nabla_\mu \nabla_\nu - \nabla_\nu \nabla_\mu) A_\rho = R^\alpha{}_{\rho\mu\nu} A_\alpha;$$
$$(\nabla_\mu \nabla_\nu - \nabla_\nu \nabla_\mu) A^\rho = -R^\rho{}_{\alpha\mu\nu} A^\alpha, \tag{2.39}$$

where the quantities $R^\rho{}_{\alpha\mu\nu}$ are components of a tensor called the curvature tensor, or *Riemann tensor* (sometimes also called the *Riemann–Christoffel tensor*):

$$R^\sigma_{\mu\rho\nu} = \partial_\nu \Gamma^\sigma_{\mu\rho} - \partial_\rho \Gamma^\sigma_{\mu\nu} + \Gamma^\sigma_{\alpha\nu} \Gamma^\alpha_{\rho\mu} - \Gamma^\sigma_{\alpha\rho} \Gamma^\alpha_{\mu\nu}. \tag{2.40}$$

The Riemann tensor plays a central role in Riemannian geometry since it is this tensor that characterizes the distinction of a given metric from that of flat space. For the latter, all components of the Riemann tensor are zero (and, due to its tensor nature, all its components are zero, being calculated in an arbitrary coordinate system, not only in Minkowski coordinates).

It is most convenient to describe the symmetry properties of the Riemann tensor if it is represented by its covariant components,

$R_{\mu\nu\rho\sigma} = g_{\mu\alpha}R^{\alpha}{}_{\nu\rho\sigma}$. By construction, it is symmetric under permutation of the first and second pair of indices and is antisymmetric within each pair

$$R_{\mu\nu\rho\sigma} = R_{\rho\sigma\mu\nu} = -R_{\nu\mu\rho\sigma} = -R_{\mu\nu\sigma\rho}. \tag{2.41}$$

Furthermore, the components of the Riemann tensor satisfy the Ricci identity

$$R_{\mu\alpha\beta\gamma} + R_{\mu\beta\gamma\alpha} + R_{\mu\gamma\alpha\beta} = 0, \tag{2.42}$$

where the first index is the same in all terms while the others are subject to cyclic permutation. Due to (2.41), one can rewrite this identity with any other fixed index.

Given the symmetry of the Riemann tensor, the number of its independent components in four-dimensional space is 20. In the general case of a D-dimensional space this number is equal to $D^2(D^2-1)/12$. In special examples of spaces possessing substantial symmetries, the number of independent components is much smaller: for example, in de Sitter space that has the maximum possible symmetry, all components of the Riemann tensor are expressed via only one constant.

Contractions of the Riemann tensor lead to the Ricci tensor $R_{\mu\nu}$ and the scalar curvature R, also called the Ricci scalar:

$$R_{\mu\nu} = R^{\alpha}_{\mu\alpha\nu}; \quad R = g^{\mu\nu}R_{\mu\nu} = R^{\alpha}_{\alpha}. \tag{2.43}$$

A direct inspection shows that the Riemann tensor, in addition to the algebraic identities (2.41) and (2.42), satisfies the differential identities

$$\nabla_{[\sigma}R_{\alpha\beta]\gamma\delta} = 0, \tag{2.44}$$

called the Bianchi identities.[3]

Their contraction by one pair of indices gives

$$\nabla_{\sigma}R^{\sigma}_{\alpha\beta\gamma} + \nabla_{\gamma}R_{\alpha\beta} - \nabla_{\beta}R_{\alpha\gamma} = 0, \tag{2.45}$$

[3] As usual, symmetrization is meant over indices taken in parentheses and alternation over those in square brackets.

while a further contraction leads to an equality of utmost importance in gravitation theory

$$\nabla_\alpha G^\alpha_\mu = 0, \quad G^\nu_\mu := R^\nu_\mu - \frac{1}{2}\delta^\nu_\mu R. \tag{2.46}$$

The tensor G^ν_μ is called the Einstein tensor.

In different parts of the book we will most often deal with space-time metrics which can be written in a diagonal form. Therefore, for reference purposes, we will here present compact expressions for the Riemann and Ricci tensor components assuming a general diagonal metric in a space-time of arbitrary dimension D and arbitrary signature,

$$ds^2 = g_{AB}dx^A dx^B = \sum_A \eta_A \, e^{2b_A(X)}(dx^A)^2, \tag{2.47}$$

where capital Latin indices take values from 0 to $D-1$, $b_A(X)$ are arbitrary functions of all coordinates x^A, and $\eta_A = \pm 1$.

Let us introduce the notations $b_{M,N} = \partial_N b_M$,

$$\Psi^{MN} = b_{M,N}\, e^{b_M - b_N} \quad \text{for} \quad M \neq N, \tag{2.48}$$

and let $\Psi_{MN} = 0$ for $M = N$. Then the components of the Riemann tensor can be written as

$$\begin{aligned}
&R^{\underline{AB}}{}_{AB} = e^{-b_A - b_B}\Big(\eta_B \Psi^{\underline{BA}}{}_{,A} + \eta_A \Psi^{\underline{AB}}{}_{,B} + \eta_M \Psi^{\underline{A}M}\Psi^{\underline{B}M}\Big),\\
&R^{\underline{A}N}{}_{AM} = \eta_N\, e^{-b_A - b_N}\Big(\Psi^{\underline{AN}}{}_{,M} - \Psi^{\underline{AM}}\Psi^{M\underline{N}}\Big) \quad \text{for} \quad M \neq N,\\
&R^{AB}{}_{MN} = 0 \quad \text{if all indices are different,}
\end{aligned} \tag{2.49}$$

where no summing is assumed over an underlined index. Furthermore, for the diagonal components of R^A_B we have

$$R^{\underline{M}}_{\underline{M}} = \sum_{A\neq M}\Big(\Box_A b_M + \Box_M b_A - \eta_M\, e^{-2b_M} b_{A,M} \sum_{B\neq A,M} b_{B,M}\Big), \tag{2.50}$$

where, for an arbitrary function $f(x^M)$,

$$\Box_M f(X) = \frac{1}{\sqrt{g}}\partial_M\Big(\sqrt{g}g^{\underline{MM}}\partial_M f(X)\Big), \quad g = |\det(g_{MN})|.$$

The off-diagonal components of the Ricci tensor are more conveniently written with lower indices, namely,

$$R_{MN} = \sum_{A\neq M,N} \left(b_{A,MN} + b_{A,M}b_{A,N} - b_{M,N}b_{A,M} - b_{N,M}b_{A,N}\right). \quad (2.51)$$

To conclude the section, we would like to present a formula for the invariant volume element in a Riemannian space of arbitrary dimension D: if we try to find the volume of a small parallelepiped specified by D vectors dx^{μ}_{a} (a is the number of a vector, and μ is, as usual, the number of its component), the invariant volume element is

$$dV = \sqrt{|g|}\,|\det(dx^{\mu}_{a})|, \quad g := \det(g_{\mu\nu}). \quad (2.52)$$

Consequently, an invariant integral of a scalar function $f(x)$ over a certain volume V is

$$\int_V \sqrt{|g|} f(x) d^D x, \quad d^D x := dx^1 dx^2 \cdots dx^D. \quad (2.53)$$

Evidently, similar formulae are valid for integration over surfaces of any dimension $d < D$ if x^{μ} denotes the coordinates specified on the surface while $g_{\mu\nu}$ is the internal metric of the surface, induced by the metric of ambient space.

2.4 Gravitational field action and dynamic equations

2.4.1 *The Einstein equations*

In GR, the dynamic variables characterizing the gravitational field are the metric tensor components $g_{\mu\nu}$. The dynamic equations of GR are derived from Hilbert's variation principle

$$\delta S = 0, \quad S = \int \frac{R - 2\Lambda}{2\varkappa}\sqrt{-g}\, d^4x + S_m, \quad (2.54)$$

where $S_m = \int L_m \sqrt{-g}\, d^4 x$ is the action of matter, i.e., substance and all fields except the gravitational field, and Λ is the cosmological constant, which is usually negligible when considering "local" configurations (up to the scale of a cluster of galaxies) but is manifestly important on the cosmological scale. The condition $\delta S = 0$ leads to the Hilbert–Einstein equations (more frequently they are simply called the Einstein equations)

$$R_{\mu\nu} - \frac{1}{2} g_{\mu\nu} R + g_{\mu\nu}\Lambda = -\varkappa T_{\mu\nu}. \tag{2.55}$$

Here, $\varkappa = 8\pi G/c^4$ is the Einstein gravitational constant (G is the Newtonian gravitational constant), and $T_{\mu\nu}$ is the (metric) stress-energy tensor (SET) of matter:

$$T_{\mu\nu} = \frac{2}{\sqrt{-g}} \frac{\delta S_m}{\delta g^{\mu\nu}}. \tag{2.56}$$

In (2.55), there are in general ten nonlinear partial differential equations. However, first, the freedom of choosing a coordinate system makes it possible to impose four arbitrary coordinate conditions which can be formulated as equalities involving the coefficients $g_{\mu\nu}$, and therefore remain only six independent equations. Second, among the remaining equations there are four differential dependences related to the identities (2.46), which results in only two real dynamic equations. The other four are constraint equations which do not contain second-order time derivatives. These circumstances are of importance for all dynamic processes in GR, and above all for gravitational waves, which can have only two independent polarizations.

Due to the contracted Bianchi identities (2.46) and the Einstein equations (2.55), the SET of matter obeys four differential equalities having the meaning of conservation laws[4]

$$\nabla_\alpha T^\alpha_\mu = 0, \tag{2.57}$$

from which one can obtain the equations of motion for matter. Thus the equations of motion follow from the field equations. This circumstance is a fundamental difference between GR and the majority of

[4]In the general case, these differential equalities do not lead to integral conservation laws because, to do so, they should have the form $\partial f^\nu_\mu / \partial x^\nu = 0$, and it would be more precise to say that they express the SET change related to the metric change. Nevertheless, the name "conservation laws" for the equalities (2.57) has become quite common, and even more, any tensor T^ν_μ, obeying the condition (2.57), is often called *conservative*.

classical field theories, including the Newtonian theory of gravity, where one has to postulate the equations of motion for field sources separately from the field equations.

Evidently, if the matter sources of gravity contain more than four degrees of freedom, their equations of motion can be only partly restored from the Einstein equations.

2.4.2 *Geodesic equations*

The equations of motion for free particles in Riemannian space-times can be obtained by varying the action (2.21) (now written for a Riemannian linear element). The variation equation, as in Minkowski space, has the meaning of a trajectory (geodesic) equation, describing the extremum of the world line length between two given points. It can be written as

$$\frac{du^\alpha}{ds} + \Gamma^\alpha_{\mu\nu} u^\mu u^\nu = 0. \tag{2.58}$$

Here, s is the interval which coincides with the proper time of an observer moving along the geodesic if it is timelike and the proper length along the geodesic if it is spacelike. In all cases s is a canonical parameter.[5]

Let us derive Eq. (2.58) from the conservation law (2.57). It will be a good illustration of the possibility to derive the equations of motion for matter from the Einstein equations.

Let us begin with the expression for the SET of a perfect fluid, which can be obtained as a natural extension of the corresponding expression from SR [366] to Riemannian spaces:

$$T_{\mu\nu} = (\varepsilon + p) u_\mu u_\nu - p\, g_{\mu\nu}, \tag{2.59}$$

---

[5] A canonical parameter is defined as the parameter along a curve with which the geodesic equation has the form (2.58) (see, e.g., [525]). Evidently, if s is a canonical parameter, then $\bar{s} = as + b$, where a and b are numbers, is also a canonical parameter; however, in gravitation theory it is reasonable to use just the parameter equal to the interval, and the freedom of its choice is restricted to constant shifts, $\bar{s} = s + b$. Usage of other, noncanonical parameters would complicate the form of Eq. (2.58).

where u_μ is the 4-velocity of particles of the fluid, $\varepsilon = \rho c^2$ is its energy density, and p its pressure. In particular, for dustlike matter, consisting of noninteracting particles,

$$p = 0, \quad T_{\mu\nu} = \varepsilon u_\mu u_\nu. \tag{2.60}$$

The dust grains move without influence of external forces other than gravity, therefore their equations of motion coincide with those of free particles. Let us obtain them by differentiating the tensor (2.60):

$$\nabla_\alpha T^\alpha_\mu = u_\mu \nabla_\alpha(\rho u^\alpha) + \rho\, u^\alpha \nabla_\alpha u_\mu = 0. \tag{2.61}$$

Multiplying this equality by u^μ, taking into account that $u^\mu u_\mu = 1 \Rightarrow \nabla_\alpha(u^\mu u_\mu) = 0$, we obtain from (2.61) the continuity equation $\nabla_\mu(\rho u^\mu) = 0$ (its physical meaning is mass conservation for the dust) and the equation of motion

$$u^\mu \nabla_\mu u_\alpha = 0, \tag{2.62}$$

which can be rewritten in the form

$$u^\mu \nabla_\mu u^\alpha = 0 \quad \Rightarrow \quad \frac{dx^\mu}{ds}\frac{\partial u^\alpha}{\partial x^\mu} + \Gamma^\alpha_{\mu\nu} u^\mu u^\nu = 0,$$

whence finally we obtain Eq. (2.58)

$$\frac{du^\alpha}{ds} + \Gamma^\alpha_{\mu\nu} u^\mu u^\nu = 0.$$

The geodesics can be spacelike, timelike and null, according to the sign of $u_\mu u^\mu$, and the nature of a particular geodesic cannot change along it because $\partial_\alpha(u_\mu u^\mu) = 0$.

2.4.3 *The correspondence principle*

Since a Riemannian space–time coincides with Minkowski space (its tangent space) in a small vicinity of each world point, the laws of SR are approximately valid at any point. It is possible to pass on to SR in the whole space in the limit of weak gravity; the weak gravity condition is formulated as the condition that the metric is only weakly deflecting from the flat metric (in any coordinates), or that the Riemann tensor is small in terms of a certain length scale.

It should be noted, however, that the formal transition $\varkappa \to 0$ in a solution to the Einstein equations does not generally lead to a flat metric: instead, it leads to a certain nonflat vacuum solution of the Einstein equations, that is, a solution with $T_{\mu\nu} \equiv 0$.

Newton's law of gravity follows from GR under the conditions of small velocities ($v \ll c$) and weak gravity (that is, the Newtonian gravitational potential, introduced in a proper way, should be small, $V \ll c^2$). In what follows, we shall verify the existence of such a transition using as an example the Schwarzschild metric, and also the validity of the relation $\varkappa = 8\pi G/c^4$.

A transition to Newtonian gravity can also be carried out under some additional conditions by the formal transition $c \to \infty$, and an expansion of the metric in powers of c^{-1} (more precisely, in powers of v/c and V/c^2). Such an expansion of the metric tensor is convenient for describing the observable effects of relativistic gravity in weak gravitational fields (for example, in the Solar system) by using a few first terms. This tool is applicable both in GR (the post-Newtonian approximation) and in other metric theories of gravity (the parametrized post-Newtonian (PPN) approximation) [563]. For example, in the first PPN approximation the component g_{00} should be calculated up to c^{-4} (since it is multiplied by c^2dt^2), the components g_{0i} up to c^{-3}, and the components g_{ij} up to c^{-2}. Higher-order PPN coefficients are also in demand owing to the increasing accuracy of measurements [565].

2.5 Macroscopic matter and nongravitational fields in GR

All kinds of matter other than the gravitational field admit a description in the framework of SR. For their description in GR (and, in general, in any theory formulated in a Riemannian space), most frequently (but not always), the so-called *minimal coupling principle* is used, according to which all equations known in SR are extended to curved space–time by replacing all partial derivatives with covariant derivatives. We note that this trick can even increase the freedom of calculations in the framework of SR without restriction to Minkowski coordinates, introducing curvilinear coordinates and invoking any accelerations, both translational and rotational ones.

We will present some relations valid for nongravitational matter in curved space–time according to the minimal coupling principle.

2.5.1 *Perfect fluids*

We have previously presented the expression (2.59) for the SET of a perfect fluid in Riemannian space–time in a derivation of the geodesic equations. Let us now consider a more general conservation law for a perfect fluid and derive the corresponding equations of motion, the general-relativistic analogues of the continuity equation and the Euler equation.

Rewrite the tensor (2.59) in mixed components,

$$T_\mu^\nu = (\varepsilon + p)u_\mu u^\nu - p\delta_\mu^\nu, \tag{2.63}$$

apply to it the operator ∇_ν and equate the result to zero

$$\nabla_\nu T_\mu^\nu = (\partial_\nu w)u_\mu u^\nu + w\nabla_\nu(u_\mu u^\nu) - \partial_\mu p = 0, \tag{2.64}$$

where $w := \varepsilon + p$ is the thermal function of the fluid. Contracting (2.64) with u^μ (i.e., finding its projection to the direction of u^μ) and recalling that $\partial_\mu(u_\nu u^\nu) = 0$, we obtain

$$\nabla_\nu(wu^\nu) - u^\mu\partial_\mu p = 0. \tag{2.65}$$

Let us now make a projection of (2.64) onto a direction perpendicular to u^μ. Such a projection has the form $\nabla_\nu T_\mu^\nu - u^\nu u_\mu \nabla_\lambda T_\nu^\lambda = 0$. As a result, we arrive at the perfect fluid equation of motion, the general-relativistic analogue of the Euler equation

$$wu^\nu\nabla_\nu u_\mu = \partial_\mu p - u_\mu u^\nu\partial_\nu p. \tag{2.66}$$

In nonrelativistic hydrodynamics, the continuity equation is known to represent the mass conservation law. In SR and, even more in GR, the mass is not conserved, and analogues of the continuity equation are only obtained for conserved quantities such as the number of particles if one can neglect their possible production and absorption. Then one can introduce the particle number current

$n^\mu = nu^\mu$, where n is the particle number density in the RF where the fluid formed by these particles is at rest, and u^μ is the 4-velocity of this fluid. The particle number conservation law (valid in the absence of their creation, annihilation and conversion) is expressed in the equality

$$\nabla_\mu(nu^\mu) = 0, \tag{2.67}$$

quite similar to the electric charge conservation law (2.79) (see below).

2.5.2 *Anisotropic fluids*

More general kinds of condensed matter than perfect fluids, such as anisotropic and viscous fluids, are also frequently used in gravitational problems related to astrophysics and cosmology. We will meet some of them in the subsequent chapters. Let us consider anisotropic fluids, which are important, among other reasons, because many field systems in particular problems with high symmetry behave as such fluids.

The general formalism of nondissipative anisotropic fluids in Riemannian space–time [194, 373] assumes matter with a certain energy density ρ and three principal pressures p_i in mutually orthogonal directions, so that the SET has the form (tetrad indices are written in parentheses)

$$T^{(\alpha\beta)} = \mathrm{diag}(\rho, p_1, p_2, p_3) \tag{2.68}$$

in an orthonormal tetrad corresponding to the fluid's comoving reference frame (its comoving nature is evident from the zero values of the energy flux components $T^{(0i)}$, $i = 1, 2, 3$)

$$\left(e^\mu_{(\alpha)}\right) = \left(u^\mu, \phi^\mu, \chi^\mu, \psi^\mu\right), \tag{2.69}$$

where u^μ is timelike and has the meaning of the fluid's 4-velocity while the other tetrad vectors $\phi^\mu, \chi^\mu, \psi^\mu$ are spacelike. The usual coordinate SET components are

$$T^{\mu\nu} = e^\mu_{(\alpha)} e^\nu_{(\beta)} T^{(\alpha\beta)} = \rho u^\mu u^\nu + p_1\phi^\mu\phi^\nu + p_2\chi^\mu\chi^\nu + p_3\psi^\mu\psi^\nu. \tag{2.70}$$

A more conventional form of $T^{\mu\nu}$ is obtained from (2.70) by adding and subtracting the quantity $p_1 g^{\mu\nu}$, where the choice of p_1 among the pressures is arbitrary, and taking into account that

$$g^{\mu\nu} = e^{\mu}_{(\alpha)} e^{\nu}_{(\beta)} \eta^{(\alpha\beta)} = u^{\mu}u^{\nu} - \phi^{\mu}\phi^{\nu} - \chi^{\mu}\chi^{\nu} - \psi^{\mu}\psi^{\nu}.$$

We thus obtain

$$T^{\mu\nu} = (\rho + p_1)u^{\mu}u^{\nu} - p_1 g^{\mu\nu} + (p_2 - p_1)\chi^{\mu}\chi^{\nu} + (p_3 - p_1)\psi^{\mu}\psi^{\nu}. \tag{2.71}$$

If $p_1 = p_2 = p_3 = p$, we obtain the usual (isotropic) perfect fluid SET, $T^{\mu\nu} = (\rho + p)u^{\mu}u^{\nu} - pg^{\mu\nu}$.

The mixed form of the SET T^{ν}_{μ} in a comoving reference frame has the form (2.68) with a minus at each pressure.

2.5.3 *Scalar fields*

For a scalar field ϕ with arbitrary self-interaction, described by a potential $V(\phi)$, if it is minimally coupled to gravity, the Lagrangian in curved space is written in precisely the same way as in Minkowski space

$$L_s = \frac{1}{2} g^{\mu\nu} \phi_{,\mu}\phi_{,\nu} - V(\phi), \tag{2.72}$$

and its variation with respect to ϕ leads to an equation that generalizes the Klein–Gordon relativistic equation

$$\Box\phi + dV/d\phi = 0, \tag{2.73}$$

with the general-relativistic d'Alembert operator applied to a scalar

$$\Box = \nabla^{\alpha}\nabla_{\alpha} = \frac{1}{\sqrt{-g}} \partial_{\alpha}\left(\sqrt{-g}g^{\alpha\beta}\partial_{\beta}\right) \tag{2.74}$$

(the same operator $\nabla^{\alpha}\nabla_{\alpha}$ applied to a vector or to a higher-rank tensor will have a more complicated expression due to more Christoffel symbols involved, see (2.34) and (2.37)).

The scalar field SET is obtained from the Lagrangian (2.72) by its variation according to (2.56):

$$T^{\nu}_{\mu\,s} = \phi_{\mu}\phi^{\nu} - \delta^{\nu}_{\mu} L_s. \tag{2.75}$$

More complex forms of scalar fields are also considered in the problems of gravitation and cosmology. For instance, the so-called k-essence with Lagrangians of the general form $L_s = L(\phi, X)$ (with $X = (\partial\phi)^2$) does not violate the minimal coupling principle.

Some of these Lagrangians as well as those with nonminimal coupling will be considered later.

2.5.4 *The electromagnetic field*

The electromagnetic (massless vector) field is characterized by the vector potential A_μ and by the strength tensor, also called the Maxwell tensor

$$F_{\mu\nu} = \partial_\mu A_\nu - \partial_\nu A_\mu. \tag{2.76}$$

The electromagnetic field Lagrangian directly generalizes the corresponding flat-space expression. For a field with sources, the Lagrangian reads

$$L_{\text{e-m}} = -\left(\frac{1}{4}F_{\mu\nu}F^{\mu\nu} - j^\mu A_\mu\right), \tag{2.77}$$

where j^μ is the electric charge current density. Its variation with respect to A_μ gives a dynamic equation that corresponds to the second pair of Maxwell equations in usual electrodynamics,

$$\nabla_\nu F^{\mu\nu} \equiv \frac{1}{\sqrt{-g}}\partial_\nu(\sqrt{-g}F^{\mu\nu}) = j^\mu, \tag{2.78}$$

and due to (2.78), the electric charge conservation law automatically holds:

$$\nabla_\mu j^\mu \equiv \frac{1}{\sqrt{-g}}\partial_\mu(\sqrt{-g}j^\mu) = 0. \tag{2.79}$$

The first pair of Maxwell equations finds its analogue in the identity that follows from (2.76),

$$\nabla_\mu F_{\nu\sigma} + \nabla_\sigma F_{\mu\nu} + \nabla_\nu F_{\sigma\mu} = 0, \tag{2.80}$$

or, equivalently,

$$\nabla^*_\mu F^{\mu\nu} = 0, \tag{2.81}$$

where ${}^*F^{\mu\nu}$ is the tensor dual to $F_{\mu\nu}$:

$$ {}^*F^{\mu\nu} = \frac{1}{2}\frac{\varepsilon^{\mu\nu\rho\sigma}}{\sqrt{-g}} F_{\rho\sigma}. \qquad (2.82) $$

The electromagnetic field SET is obtained by varying the action of the field $F_{\mu\nu}$ with respect to the metric $g^{\mu\nu}$:

$$ T^{\nu}_{\mu\,\mathrm{e-m}} = -F_{\mu\alpha}F^{\nu\alpha} + \frac{1}{4}\delta^{\nu}_{\mu}F_{\alpha\beta}F^{\alpha\beta}. \qquad (2.83) $$

The Maxwell field (2.76) does not change if one adds a gradient of any scalar function $f(x^\mu)$ to the vector potential A_μ. This property of *gauge invariance* is of utmost importance in electrodynamics. Gauge invariance is evidently preserved if one considers Lagrangians more general than (2.77), such as $L(F)$ and $L(F, G)$, where $F := F^{\mu\nu}F_{\mu\nu}$ and $G := {}^*F^{\mu\nu}F_{\mu\nu}$. Such forms of nonlinear electrodynamics, respecting the minimal coupling principle, are also used in various gravitational problems, such as finding nonsingular black holes and models of the early Universe.

In this book, we will not consider other fields, such as the spinor (spin 1/2), massive vector fields and fields with spin 3/2. The properties of such fields are discussed in detail, as well as their applications in the problems of gravity, particle physics and astrophysics, e.g., in the books [261, 324, 410].

The minimal coupling principle is not universal, and many researchers discuss nonminimal interactions between material fields and gravity. These interactions are introduced by adding such terms in the Lagrangian that contain both a material field and some invariants of the curvature tensor (most frequently the scalar curvature), or invariants involving both the material field and the curvature (like, for instance, $A^\mu\partial_\mu R$). Thus, a nonminimal interaction between a scalar field and gravity can be introduced by adding in (2.72) the term $\xi R\phi^2$, $\xi = \mathrm{const}$, while a nonminimal interaction between the electromagnetic field and gravity can be described by the term $\xi R\, F_{\mu\nu}F^{\mu\nu}$. In what follows, we will deal with nonminimal interactions of scalar fields and gravity in the framework of scalar-tensor and multidimensional theories of gravity.

2.6 Energy conditions

The energy conditions are certain invariant inequalities, mostly characterizing the known and conventional forms of matter, which are used in many studies and lead to many well-known results of outstanding significance, see, e.g., [230, 285, 549]. Let us enumerate and briefly discuss them.

Weak energy condition (WEC): Let ξ^μ be any timelike or null vector field ($\xi^\mu \xi_\mu \geq 0$), and $T_{\mu\nu}$ the SET of matter or a particular kind of matter, then

$$T_{\mu\nu}\xi^\mu\xi^\nu \geq 0. \tag{2.84}$$

If ξ^μ is timelike, we can choose the reference frame where this vector has the direction of 4-velocity of matter, and then (2.84) reduces to $T^0_0 \geq 0$, that is, the energy density of this kind of matter is nonnegative.

If ξ^μ is null, then the inequality (2.84) has its own separate name.

Null energy condition (NEC): It is the inequality whose violation is a necessary condition at wormhole throats in GR and other "exotic" objects and phantom dark energy in cosmology. If $T_{\mu\nu}$ is the SET of an anisotropic fluid (2.68) in its comoving frame, then (2.84) takes the form

$$\rho + p_i \geq 0, \quad i = 1, 2, 3. \tag{2.85}$$

Example: The cosmological constant $\Lambda > 0$, if treated as a kind of matter, satisfies the WEC, but respects the NEC only marginally since for it $\rho = \Lambda$ and $p = -\Lambda$.

Strong energy condition (SEC): This condition reads

$$\left(T_{\mu\nu} - \tfrac{1}{2} T g_{\mu\nu}\right)\xi^\mu\xi^\nu \geq 0, \quad \xi^\mu\xi_\mu > 0. \tag{2.86}$$

It is really more restrictive and often violated. For (comoving) anisotropic fluids it reduces to

$$\rho + p_1 + p_2 + p_3 \geq 0, \tag{2.87}$$

and if the fluid is isotropic, the SEC reads simply $\rho + 3p \geq 0$. It is violated by the cosmological constant and many field sources.

Dominant energy condition (DEC): It requires that if the vector ξ^μ is causal (timelike or null) and points to the future, the vector $T^\nu_\mu \xi^\mu$ is also causal and points to the future. For an anisotropic fluid it reduces to $\rho \geq |p_i|$.

2.7 The most symmetric spaces

2.7.1 *Isometry groups and Killing vectors*

Minkowski flat space–time possesses the greatest possible symmetry, expressed in the invariance of the interval with respect to the Poincaré group (G_{10}, i.e., a 10-parameter group). A general Riemannian space–time $\mathbb{V}_4$ does not possess any symmetry, while in special cases where such a symmetry does exist, it is described by an isometry group with a certain number of parameters, obviously no more than 10. This number is equal to the number of linearly independent solutions ξ_α of the Killing equation

$$\nabla_\alpha \xi_\beta + \nabla_\beta \xi_\alpha = 0. \tag{2.88}$$

The vectors ξ^μ satisfying Eq. (2.88) (the Killing vectors) possess the following property: if all points are shifted by $\xi^\alpha d\lambda$ (where $d\lambda$ is an infinitesimal quantity), all metric relations in $\mathbb{V}_4$ remain invariable. Consecutive shifts by $\xi^\alpha d\lambda$ lead to a motion of world points along an *orbit of the Killing vector*, whose points are all mutually equivalent. (Example: on an arbitrary surface of rotation in usual three-dimensional space, there is a Killing vector pointing along the "parallel" and describing shifts along it by small azimuthal angles, while the whole "parallel" is an orbit of this Killing vector.

From conventional geometry we know that there are three kinds of maximally symmetric 2-surfaces in Euclidean 3-space: in addition to planes, these are also spheres and pseudospheres with Lobachevsky geometry. The situation is quite similar in four-dimensional pseudo-Riemannian space–times: in addition to Minkowski space, there are maximally symmetric spaces with constant positive and negative curvature.

The maximally symmetric (G_{10}) metrics in $\mathbb{V}_4$ are known to be solutions to the Einstein equations in vacuum,

$$R_\mu^\nu - \delta_\mu^\nu \Lambda = 0, \tag{2.89}$$

with the cosmological constant Λ, which corresponds to the effective SET $T_\mu^\nu = \delta_\mu^\nu \Lambda/\varkappa$. These are spaces called the *constant curvature spaces*: the space of zero constant curvature is Minkowski space ($\Lambda = 0$), that of positive constant curvature is de Sitter (dS) space ($\Lambda > 0$), and that of negative constant curvature is the so-called anti-de Sitter (AdS) space ($\Lambda < 0$) (even though an author named Anti-de Sitter never existed). In constant curvature spaces, the Riemann tensor is determined by a single constant (Λ) and is expressed in terms of the metric tensor in an algebraic manner:

$$R_{\mu\nu\rho\sigma} = \frac{\Lambda}{3}(g_{\mu\rho}g_{\nu\sigma} - g_{\mu\sigma}g_{\nu\rho}), \tag{2.90}$$

whence for the Ricci tensor and scalar we obtain

$$R_{\mu\nu} = \Lambda g_{\mu\nu}, \qquad R = 4\Lambda. \tag{2.91}$$

Explicit forms of the de Sitter [504] and anti-de Sitter metrics will be presented a bit later.

2.7.2 *Isotropic cosmology — the dS and AdS spaces*

A smaller but still high symmetry (G_6) characterizes homogeneous isotropic cosmological models, to be discussed in detail later. In accordance with their name, these space–times are characterized by spatial sections satisfying the *cosmological principle,* that their metric properties are the same at all points (homogeneity) and in all directions (isotropy). This means that such space-times possess spatial sections of maximum possible symmetry, which, in 3 dimensions. is described by 6-parameter isometry groups, G_6.

Postponing a detailed discussion of cosmological models to other chapters, we will here restrict ourselves to the simplest cases, including those where the symmetries are higher than G_6 — these are, in particular, dS and AdS space–times, which admit a description as isotropic cosmologies.

The general form of the metric of a homogeneous and isotropic space–time, which is called the Friedmann–Robertson–Walker (FRW) metric, is

$$\begin{aligned} ds^2 &= dt^2 - a^2(t)dl^2, \\ dl^2 &= \frac{dr^2}{1-kr^2} + r^2(d\theta^2 + \sin^2\theta\, d\phi^2), \end{aligned} \tag{2.92}$$

where $k = 1, 0, -1$ correspond to closed (spherical), open flat and open hyperbolic models; t is the physical (proper) time measured by perfect clocks of comoving observers. It is the metric with maximally symmetric spatial sections: for $k = +1$, each spatial section $t = \text{const}$ is a three-dimensional sphere, for $k = 0$ it is a three-dimensional Euclidean space, and for $k = -1$ it is a three-dimensional Lobachevsky space. The isometry group G_6 of any of them, including translations in three directions and rotations around three axes, is at the same time the isometry group of the space–time as a whole.

The function $a(t)$ describes the common evolution of all spatial lengths in the course of time and is called the scale factor.

For the metric (2.92), the Einstein tensor has nonzero components G_0^0 and $G_1^1 = G_2^2 = G_3^3$ only

$$G_0^0 = -\frac{3}{a^2}(\dot{a}^2 + k); \quad G_{\underline{i}}^{\underline{i}} = -\frac{1}{a^2}(2a\ddot{a} + \dot{a}^2 + k), \tag{2.93}$$

where the dot denotes d/dt, and the underlined index assumes no summing.

Due to the Einstein equations, the SET of matter, irrespective of its particular nature, has the same structure as the Einstein tensor, therefore inevitably

$$T_\mu^\nu = \text{diag}(\rho, -p, -p, -p), \tag{2.94}$$

where ρ is the energy density, and p is the isotropic pressure. It is an example of how the space-time geometry prescribes the form of matter: in the FRW metric, any kind of matter has actually the perfect-fluid SET (2.94).

The $\binom{0}{0}$ and $\binom{1}{1}$ components of the Einstein equations are written in the form

$$\frac{3}{a^2}(\dot{a}^2 + k) = \varkappa\rho, \tag{2.95}$$

$$\frac{1}{a^2}(2a\ddot{a} + \dot{a}^2 + k) = -\varkappa p, \tag{2.96}$$

Equation (2.96) is a consequence of (2.95) if one takes into account the conservation law (2.57), which gives

$$\frac{d\rho}{\rho + p} = -\frac{3\,da}{a}. \tag{2.97}$$

Thus, we have a set of two independent equations for three unknown functions of time: $a(t)$, $p(t)$, and $\rho(t)$. To make the system determined, one should add one more equation, and it should bear information on the kind of matter, more precisely, it should be the equation of state of matter, i.e., a relation between p and ρ, and it will actually determine the form of the solution.

In the case of the simplest, so-called barotropic, equation of state $p = w\rho$ where $w =$ const, Eq. (2.97) gives the density evolution law

$$\rho = \text{const} \cdot a^{-3(1+w)}. \tag{2.98}$$

Some characteristic values of w are:

- $w = 0$ — dustlike matter, no pressure;
- $w = 1/3$ — disordered radiation, in particular, the cosmic microwave background;
- $w = 1$ — maximally stiff matter compatible with causality, for which the speed of sound is equal to the speed of light;
- $w = -1/3$ — disordered cosmic strings;
- $w = -1$ — the cosmological constant;
- $w < -1$ — phantom matter.

In the case of dustlike matter, $w = 0$, the law (2.98) corresponds to mass conservation in a given coordinate volume (proportional to a^3) in the course of expansion or contraction.

The hypothetic phantom matter is invoked by many authors to account for the modern accelerated expansion of the Universe. Looking at (2.98), it is easy to understand that it is indeed a very strange form of matter: as the volume $\sim a^3$ grows, its density also grows instead of decreasing. Such a behavior, being extrapolated to the future, really leads to a specific form of space-time singularity, called the Big Rip. Let us look how it happens due to Eqs. (2.95) and (2.98).

Suppose $k = 0$ (spatially flat cosmological models). Then, with (2.98), Eq. (2.95) is easily integrated to give (provided $w \neq -1$)

$$a(t) = \text{const} \cdot |t - t_0|^{2/(3+3w)}, \qquad t_0 = \text{const}. \tag{2.99}$$

For $w < -1$, we thus obtain an infinite value of a at some finite time $t = t_0$, and at this time the phantom matter density grows to infinity, while all forms of "normal" matter vanish.

The case $w = -1$, the cosmological constant, is also called the case of vacuum matter. One of the reasons is that the corresponding effective SET, $\Lambda\delta_\mu^\nu/\varkappa$, is invariant under all coordinate transformations, hence any RF is comoving to such matter — this property is incompatible with any kind of massive particles which are at rest only in some particular RF. Its another important feature is that it has a constant energy density $\rho = \text{const} = \Lambda/\varkappa$ irrespective of any evolution of the metric.

In the case $w = -1$, Eq. (2.95) leads to the following solutions at different k and Λ: for $\Lambda > 0$ we obtain

$$k = 0\colon\ a(t) = a_0\, \mathrm{e}^{\pm H_0 t}, \qquad a_0 = \text{const}; \tag{2.100}$$

$$k = 1\colon\ a(t) = H_0^{-1} \cosh[H_0(t - t_0)], \qquad t_0 = \text{const}; \tag{2.101}$$

$$k = -1\colon\ a(t) = H_0^{-1} \sinh[H_0(t - t_0)], \qquad t_0 = \text{const}, \tag{2.102}$$

where $H_0 := \sqrt{\Lambda/3}$. For $\Lambda < 0$, Eq. (2.95) has a solution for hyperbolic models only, $k = -1$:

$$a(t) = H_0^{-1} \sin[H_0(t - t_0)], \qquad t_0 = \text{const}, \quad l = \sqrt{-3/\Lambda}. \tag{2.103}$$

The metrics (2.100)–(2.102) describe the de Sitter space in different coordinates *belonging to different RFs*. It is evident that even the topology of spatial sections is different: it is $\mathbb{R}^3$ for the models (2.100) and (2.102) and $\mathbb{S}^3$ (a 3-sphere) for the model (2.101). An analysis shows (see, e.g., the book by Hawking and Ellis [285]) that dS space–time is described completely by the metric (2.101), whereas (2.100) and (2.102) only describe its certain parts.

The metric (2.103) describes AdS space–time.

Chapter 3

Spherically symmetric space–times: Black holes

In this chapter, we will begin considering some questions of black-hole physics. A black hole (BH) is, by definition, a space–time region where the gravitational field is so strong that no material bodies or light rays can leave this region and escape to infinity [250].

In the opinion of the great majority of astrophysicists, BHs are quite widespread in the Universe, a great number of BHs formed in the early Universe, they form as a result of massive star evolution, and supermassive BHs (10^6 to 10^{10} solar masses) are located in the central regions of many types of galaxies, including our own Milky Way Galaxy, quasars and active galactic nuclei. The recent detection of gravitational waves (GWs) has added even more confidence in the existence of BHs whose mergers are quite reasonably regarded as the most plausible sources of most of the detected GW bursts [1, 541], although alternative ideas on the nature of their sources are also actively discussed.

BHs possess many interesting and unusual properties, both theoretical and observational ones, described in detail in the books [250, 285, 409, 549] and others as well as a lot of review papers. We here only give an elementary introduction to this complicated and rapidly developing branch of physics, and we will be mostly restricted to the simplest case of spherical symmetry. Still many of the results to be presented here (e.g., the methodology of building

Carter–Penrose diagrams in the general case, exact solutions with scalar fields and others) have not yet been included in standard textbooks, and some of them, to our knowledge, have only been published in journal articles and appear in a book for the first time.

Before discussing BH configurations, we will consider some general facts and relations for spherically symmetric space–times that will be useful in further descriptions.

3.1 Spherically symmetric gravitational fields

Spherical symmetry is a natural assumption in describing the simplest isolated bodies and island-like configurations. Spherically symmetric space–times are invariant under spatial rotations forming the isometry group $G3$.

In the general case, a spherically symmetric metric can be written in the form (see, e.g., [366])

$$ds^2 = \mathrm{e}^{2\gamma} dt^2 - \mathrm{e}^{2\alpha} du^2 - \mathrm{e}^{2\beta} d\Omega^2, \quad d\Omega^2 = d\theta^2 + \sin^2\theta\, d\phi^2, \tag{3.1}$$

where α, β, γ are, in general, functions of the radial coordinate u and the time coordinate t. We will also use the notation $r \equiv \mathrm{e}^{\beta}$; thus r is the radius of a coordinate sphere u = const, t = const, and is also called the Schwarzschild radial coordinate, or the spherical radius, or, sometimes, the areal radius (since the area of such a coordinate sphere is equal to $4\pi r^2$). In the expression (3.1), there is a freedom of choosing a reference frame (RF): different RFs correspond to spherically symmetric reference bodies with different radial velocity distributions with respect to each other.

The following two remarks are here in order.

First, let us note from the very beginning that in curved space the spherical radius r has nothing to do with a distance to the center (as is the case in flat space), and in many spherically symmetric space–times there is no center at all.

Second, the "exponential" notations in (3.1) assume positive values of the corresponding metric coefficients. However, the quantities

g_{tt} and g_{uu} can change their sign, and we will then accordingly change the notations.

In the comparatively simple case of static space–times (the isometry group G_4, since rotations are supplemented with time translations), one can always choose the RF in such a way that α, β, γ depend on u only. But there still remains the opportunity of replacing the radial coordinate u with another one by transformations of the form $u = u(u_{\rm new})$. The choice of the radial coordinate can be fixed by postulating a relation between the functions α, β, γ or by choosing one of them (or a function of one of them) as the coordinate.

In solving different problems, different variants of such coordinate conditions can be convenient. Let us enumerate some of them.

1. $e^{\alpha} \equiv 1$, $du = dl$ — the Gaussian normal coordinates; the coordinate l is the true length along the radial direction, counted from a certain fixed sphere $l = 0$.
2. $u = r$, $\gamma = \gamma(r)$, $\alpha = \alpha(r)$ — the curvature (Schwarzschild) coordinates: r is the curvature radius of the sphere $r = \text{const}$.
3. $e^{2\beta(u)} = e^{2\alpha(u)} u^2$ — isotropic coordinates, in which the spatial part of the metric is written in a conformally flat form:

$$dl^2 = e^{2\alpha(u)}(du^2 + u^2 d\Omega^2) = e^{2\alpha(u)} d\vec{x}^{\,2}, \tag{3.2}$$

   where, in the three-dimensional linear element, Cartesian coordinates have been introduced: $\vec{x} = (x^1, x^2, x^3)$.
4. $\alpha(u) = 2\beta(u) + \gamma(u)$ — the harmonic coordinate u is particularly convenient for solving problems with massless scalar fields.
5. $\alpha = -\gamma$ — the quasiglobal coordinate $u = x$. This name is connected with the fact that (as we shall see later) it is more suitable than others for describing BH and other similar metrics on both sides of horizons.
6. $\alpha(u) = \gamma(u)$ — such $u = r_*$ is sometimes called the "tortoise coordinate" because in many important cases the metric functions change very slowly when expressed in terms of r_*. In this case,

the metric of the (t, u) subspace takes a conformally flat form, which simplifies a consideration of wave equations.

Other coordinate conditions can also be used.[1]

Let us present the expressions of some quantities for the metric (3.1) with arbitrary coordinates u and t, without specifying their choice (and even the choice of a RF) from the very beginning. To obtain them in specific coordinates like those enumerated above, one should simply substitute the corresponding coordinate condition.

For a static metric (3.1), the Ricci tensor is diagonal; if there is a t dependence, there appears the component $R_{01} \neq 0$. Let us present its nonzero components in the general case:

$$\begin{aligned}
R_0^0 &= e^{-2\gamma}[2\ddot{\beta} + \ddot{\alpha} + 2\dot{\beta}^2 + \dot{\alpha}^2 - \dot{\gamma}(2\dot{\beta} + \dot{\alpha})] \\
&\quad - e^{-2\alpha}[\gamma'' + \gamma'(2\beta' + \gamma' - \alpha')]; \\
R_1^1 &= e^{-2\gamma}[\ddot{\alpha} + \dot{\alpha}(2\dot{\beta} - \dot{\gamma} + \dot{\alpha})] \\
&\quad - e^{-2\alpha}[2\beta'' + \gamma'' + 2\beta'^2 + \gamma'^2 - \alpha'(2\beta' + \gamma')]; \\
R_2^2 &= e^{-2\beta} + e^{-2\gamma}[\ddot{\beta} + \dot{\beta}(2\dot{\beta} - \dot{\gamma} + \dot{\alpha})] \\
&\quad - e^{-2\alpha}[\beta'' + \beta'(2\beta' + \gamma' - \alpha')] = R_3^3; \\
R_{01} &= 2[\dot{\beta}' + \dot{\beta}\beta' - \dot{\alpha}\beta' - \dot{\beta}\gamma'].
\end{aligned} \tag{3.3}$$

Dots and primes stand for $\partial/\partial t$ and $\partial/\partial u$, respectively. The expressions of R_μ^ν for a static metric are obtained from here by putting all time derivatives equal to zero and are invariant (behave as scalars) under reparametrizations of the radial coordinate.

---

[1]In what follows, dealing with spherically symmetric geometries, we will use (unless specially indicated) the following notations for different radial coordinates:
u — the general notation, without any choice made, and also the harmonic coordinate, $\alpha = 2\beta + \gamma$;
l — the Gaussian normal coordinate, $\alpha \equiv 0$;
$r = e^\beta$ — the curvature (Schwarzschild) coordinate;
x — the quasiglobal coordinate, $\alpha = -\gamma$;
y — the isotropic coordinate, $\alpha = \beta - \ln y$;
r_* or z — the tortoise (wave) coordinate, $\alpha = \gamma$.

Among the components of the Einstein tensor, of greatest interest to us is the component G_1^1 since it does not contain second-order derivatives in u; for a static metric,

$$G_1^1 = -\,\mathrm{e}^{-2\beta} + \mathrm{e}^{-2\alpha}\beta'(\beta' + 2\gamma'). \tag{3.4}$$

The expressions (3.3) and (3.4) are necessary for substituting into the basic equations that determine the properties of self-gravitating systems in GR, the Einstein equations (2.50), which can be written in two equivalent forms:

$$G_\mu^\nu \equiv R_\mu^\nu - \frac{1}{2}\delta_\mu^\nu R = -\varkappa T_\mu^\nu \tag{3.5}$$

and

$$R_\mu^\nu \equiv R_\mu^\nu = -\varkappa\left(T_\mu^\nu - \frac{1}{2}\delta_\mu^\nu T\right), \tag{3.6}$$

where $T \equiv T_\alpha^\alpha$ is the trace of the stress-energy tensor (SET) of matter, and we have put $\varkappa = 8\pi G$ according to the Newtonian limit of GR.

When studying any space–time, it is above all important to know whether it is regular, which means that all curvature invariants are finite at all its points, or contains curvature singularities at which at least one such invariant is infinite. In many cases, it is most helpful to check the finiteness of the Kretschmann scalar (sometimes also called the Riemann tensor squared) $\mathcal{K} = R_{\alpha\beta\gamma\delta}R^{\alpha\beta\gamma\delta}$. For a static metric (3.1), the Kretschmann scalar is a sum of squares of all nonzero components $R_{\alpha\beta}{}^{\gamma\delta}$ of the Riemann tensor:

$$\begin{aligned}
\mathcal{K} &= 4K_1^2 + 8K_2^2 + 8K_3^2 + 4K_4^2,\\
K_1 &= \mathrm{e}^{-\alpha-\gamma}(\gamma'\,\mathrm{e}^{\gamma-\alpha})' = -R_{01}{}^{01},\\
K_2 &= \mathrm{e}^{-2\alpha}\beta'\gamma' = -R_{02}{}^{02} = -R_{03}{}^{03},\\
K_3 &= \mathrm{e}^{-\alpha-\beta}(\beta'\,\mathrm{e}^{\beta-\alpha})' = -R_{12}{}^{12} = -R_{13}{}^{13},\\
K_4 &= -\,\mathrm{e}^{-2\beta} + \mathrm{e}^{-2\alpha}\beta'^2 = -R_{23}{}^{23}.
\end{aligned} \tag{3.7}$$

This happens because in this case (and in many other important cases) the tensor $R_{\alpha\beta}{}^{\gamma\delta}$ is pairwise diagonal.

It is significant that all K_i are invariant under reparametrizations of the u coordinate, i.e., under the transformations $u \mapsto f(u)$, just as the mixed components of all second-rank tensors, including R^ν_μ and G^ν_μ — that is why, when writing the Einstein equations, we prefer to deal with these components rather than $R_{\mu\nu}$, $G_{\mu\nu}$ or $R^{\mu\nu}, G^{\mu\nu}$, which are sensitive to u reparametrizations. Let us note that a reparametrization $u \mapsto f(u)$ is a special case of a purely spatial transformation of coordinates, leaving the reference frame unchanged (see Chapter 2).

Since the scalar $\mathcal{K}$ is a sum of squares, for its finiteness it is necessary and sufficient that *all* its components K_i, or, in other words, *all* nonzero components of the Riemann tensor $R^{\mu\nu}{}_{\alpha\beta}$ are finite. Thus, if the scalar $\mathcal{K}$ is finite, then *all* invariants that can be constructed algebraically from the Riemann tensor and the metric tensor are finite, in particular, the scalar curvature R, the "Ricci tensor squared" $R_{\alpha\beta}R^{\alpha\beta}$, the invariant $R_{\alpha\beta\gamma\delta}R^{\alpha\gamma}R^{\beta\delta}$ and so on. So, the finiteness of $\mathcal{K}$ at some space–time point means that a curvature singularity is absent at this point. From the viewpoint of calculations, to verify regularity of some particular metric, it is easier to deal with simple expressions like K_i than with a bulky one for $\mathcal{K}$.

In the general case of a time-dependent metric (3.1), the situation with singularities is more involved.

It should be mentioned here that curvature singularities are not the only type of singularities that can appear in physically relevant space–times. In the most general form, a singularity is defined as a point or a set of points where geodesics terminate at a finite value of their canonical parameter. In other words, they are **places of geodesic incompleteness**. This can happen in any case where the metric loses its analyticity, for instance, where some of the metric functions behave like $(x - x_0)^a$ with fractional a in terms of a manifestly admissible coordinate x. In such cases one speaks of *weak, or mild singularities.*

3.1.1 *A regular center and asymptotic flatness*

A **center** in a static, spherically symmetric space–time is, by definition, a point, line or surface in its spatial section where $r \equiv e^\beta = 0$, that is, a place where coordinate spheres shrink to points. A center can be regular or singular; and regularity, as at any space–time point, is determined by finiteness of all K_i in the expression (3.7). It is necessary to note that there can be no center at all in a spherically symmetric space–time; this happens if the quantity r is nonzero in the whole space–time, or at least in its static region. We will encounter such behavior very soon, while discussing the properties of the Schwarzschild geometry.

With an arbitrary u coordinate, the necessary and sufficient conditions for regularity of the metric at the center ($r = 0$) are obtained in the form

$$\gamma = \gamma_0 + O(r^2), \qquad |\beta'|\,\mathrm{e}^{-\alpha+\beta} = 1 + O(r^2), \tag{3.8}$$

where γ_0 is a constant. The second condition is obtained from the finiteness requirement of the quantity K_4 in (3.7). Its meaning is that the circumference to radius ratio should take the correct value (2π) for small circles circumscribed around the center. This guarantees local flatness of space at the center and the existence of a tangent space. These are properties of any regular point — but for a center one has to introduce special regularity conditions because a center is a singular point of the class of spherical coordinate systems used.

Asymptotic flatness. As $r \to \infty$, far from gravitational field sources, in many cases (though not always) the space–time geometry should coincide with the Minkowski geometry. Such space–times are called asymptotically flat. It means, above all, that all components of the Riemann tensor turn to zero, i.e., all $K_i \to 0$. The latter condition is, however, not sufficient: for an asymptotically flat space–time it is also necessary to have a correct circumference to radius ratio (2π) for large coordinate circles, otherwise even with a zero curvature tensor the three-dimensional space will have a funnel-like asymptotic, with a deficit or excess solid angle as compared to its standard value of 4π. Such a geometry characterizes a specific type of configurations called *global monopoles.* (A two-dimensional analogue of such three-dimensional surfaces is a conical surface whose metric is flat at all points but the top, however, the length of a circle circumscribed around the top relates to its radius by a factor different from 2π.)

Just as a center can be lacking in a spherically symmetric space–time, a flat infinity can also be lacking; moreover, there can be no limit $r \to \infty$. We have already met such a geometry: Friedmann's closed world is spherically symmetric, and its every spatial section contains (in a given spherical coordinate system) two centers but no spatial infinity.

The necessary and sufficient conditions for the metric (3.1) to be asymptotically flat as $r \to \infty$ in terms of an arbitrary coordinate u are as follows:

$$\gamma = \gamma_\infty + O(1/r), \qquad |\beta'|\, \mathrm{e}^{-\alpha+\beta} = 1 + O(1/r), \tag{3.9}$$

where γ_∞ is a constant.

In the conditions (3.8) and (3.9), as everywhere, the notation $y = O(x)$ means that either x and y are quantities of the same order ($y \sim x$) or y is much smaller than x (the latter is denoted as $y \ll x$ or $y = o(x)$).

One more very useful relation for static, spherically symmetric space–times is the expression for the Schwarzschild mass, valid for any radial coordinate u, obtained by comparison of the asymptotic expansion of $\mathrm{e}^{2\gamma}$ at large radii ($u \to u_\infty$) with the Schwarzschild metric (see Sec. 3.2):

$$m = GM = \lim_{u \to u_\infty} \left(\mathrm{e}^{\beta} \frac{\gamma'}{\beta'} \right) = \lim_{u \to u_\infty} \left(r(u) \frac{\gamma'}{r'} \right). \tag{3.10}$$

3.2 The Reissner–Nordström–(anti-)de Sitter solution

3.2.1 *Solution of the Einstein equations*

Let us find an important class of exact static, spherically symmetric solutions to the Einstein equations, characterizing the gravitational fields in vacuum or in the presence of an electromagnetic field (without charges) and a cosmological constant. This class contains the metrics that have the greatest number of astrophysical applications among all spherically symmetric metrics; it will also provide us with explicit examples in our future discussion of general properties of spherically symmetric space–times, including those with BHs.

It proves to be convenient to solve the problem in the curvature coordinates, in which two independent Einstein equations can be written in the form (3.5):

$$G_0^0 + \Lambda = \mathrm{e}^{-2\alpha}\left(\frac{1}{r^2} - \frac{2\alpha'}{r}\right) - \frac{1}{r^2} + \Lambda = -\varkappa T_0^0, \tag{3.11}$$

$$G_1^1 + \Lambda = \mathrm{e}^{-2\alpha}\left(\frac{1}{r^2} + \frac{2\gamma'}{r}\right) - \frac{1}{r^2} + \Lambda = -\varkappa T_1^1, \tag{3.12}$$

where the prime denotes d/dr while the SET in the present case corresponds to the electromagnetic field.

The Lagrangian and SET of the electromagnetic field are

$$L_{\text{e-m}} = -\frac{1}{4}F_{\mu\nu}F^{\mu\nu}, \quad T_\mu^\nu = \frac{1}{4}[-4F_{\mu\alpha}F^{\nu\alpha} + \delta_\mu^\nu F_{\alpha\beta}F^{\alpha\beta}]. \tag{3.13}$$

The Maxwell equations $\nabla_\alpha F^{\alpha\beta} = 0$ must now be written for the spherically symmetric case, so that among the components of $F_{\mu\nu}$ only the ones describing a radial electric field ($F_{01} = -F_{10}$) and a radial magnetic field ($F_{23} = -F_{32}$) can be nonzero.[2] Let us restrict ourselves to an electric field. Then the only nontrivial Maxwell equation yields

$$(\sqrt{-g}F^{01})' = 0 \;\Rightarrow\; F^{01} = \frac{Q\,\mathrm{e}^{-\alpha-\gamma}}{\sqrt{4\pi}r^2}, \quad F_{10} = \frac{Q\,\mathrm{e}^{\alpha+\gamma}}{\sqrt{4\pi}r^2}, \tag{3.14}$$

where the constant Q is interpreted as an electric charge (we are using the Heaviside units in electrodynamics). The SET takes the form

$$T_\mu^\nu = \frac{Q^2}{8\pi r^4}\,\mathrm{diag}(+1, +1, -1, -1). \tag{3.15}$$

Let us now solve the Einstein equations. Equation (3.11) can be brought to the form

$$[r(A-1)]' = -(\Lambda + \varkappa T_0^0)r^2 \;\Rightarrow\; A = 1 - \frac{\Lambda r^2}{3} - \frac{G}{r}\int T_0^0\, r^2 dr, \tag{3.16}$$

[2]It is often claimed that a radial magnetic field is only possible if one assumes the existence of magnetic monopoles. However, in configurations like wormholes both electric and magnetic radial fields can exist without any charges, see Chapter 6.

where $A(r) = e^{-2\alpha}$, and we have recalled that $\varkappa = 8\pi G$. On the other hand, the difference of Eqs. (3.11) and (3.12), owing to the condition $T^0_0 = T^1_1$, gives

$$\alpha' + \gamma' = 0 \Rightarrow e^{\alpha+\gamma} = 1, \tag{3.17}$$

under a proper choice of scale along the time axis. As a result of integration in (3.16), the metric finally takes the form

$$\begin{aligned} ds^2 &= A(r)dt^2 - \frac{dr^2}{A(r)} - r^2 d\Omega^2, \\ A(r) &= 1 - \frac{\Lambda}{3}r^2 - \frac{2GM}{r} + \frac{GQ^2}{r^2}, \end{aligned} \tag{3.18}$$

where the integration constant M is interpreted as the active gravitational mass of the gravitational field source. The solution to the Einstein–Maxwell equations given by (3.14) and (3.18) is called the Reissner–Nordström-de Sitter (RNdS) solution.

3.2.2 *Special cases of the RNdS solution*

The (anti-)de Sitter metric

In the special case of zero mass and charge, the function $A(r)$ in Eq. (3.18) takes the simple form $A = 1 - (\Lambda/3)r^2$. At $\Lambda = 0$, the Minkowski metric is reproduced. At $\Lambda > 0$, it can be shown that the metric describes the de Sitter space–time, already discussed in the previous chapter in a cosmological context. But here we have its static representation, hence we obtain it here in a RF different from the previous ones. At $\Lambda < 0$ we accordingly obtain the AdS space–time. Explicit relations between the static and cosmological coordinates in de Sitter space–time will be presented later.

Consider the case $\Lambda = 3H^2 > 0$ in more detail. The metric is

$$ds^2 = (1 - H^2r^2)dt^2 - (1 - H^2r^2)^{-1}dr^2 - r^2 d\Omega^2. \tag{3.19}$$

At $r = 0$ it has a regular center. It must be so because de Sitter space-time, being a constant-curvature space, is homogeneous; all its points are equivalent and regular, and any of them can be "appointed" to be a center.

At $r = 1/H$ the metric coefficients g_{00} and g_{11} turn to zero (this sphere is called a cosmological horizon), while at $r > 1/H$ they take negative values, so that the coordinates r and t exchange their roles: the quantity r now becomes a temporal coordinate and t a spatial coordinate. Such a region is called a T-region; see a more general description of horizons, R- and T-regions in the next section. In this T-region, the metric can be rewritten as follows:

$$\begin{aligned} ds^2 &= (H^2r^2 - 1)^{-1}dr^2 - (H^2r^2 - 1)dx^2 - r^2 d\Omega^2 \\ &= d\tau^2 - \sinh^2(H\tau)dx^2 - H^{-2}\cosh^2(H\tau)d\Omega^2, \end{aligned} \qquad (3.20)$$

where the "former time" t is renamed x while the variable τ, introduced instead of r, has the meaning of proper time in this cosmological model, which is homogeneous but anisotropic. The homogeneity follows from the fact that, in addition to spherical symmetry, the metric coefficients are independent of the spatial coordinate x; the anisotropy manifests itself in the existence of two different scale factors, $|g_{xx}| = \sinh^2(H\tau)$ and $|g_{\theta\theta}| = r^2 = \cosh^2(H\tau)$, characterizing the expansion of the Universe in the x direction and in the two angular directions. Such spherically symmetric cosmological models, with the topology of the spatial section $\mathbb{R} \times \mathbb{S}^2$, whose special case is the model (3.20), are called Kantowski–Sachs models. Looking ahead, we note that Kantowski–Sachs models describe T-regions of any spherically symmetric BHs.

In the model (3.20), the cosmological expansion begins with the horizon $\tau = 0$, which is an extremely anisotropic state because there $g_{xx} = 0$, but becomes isotropic and exponential (with the Hubble constant H) at large τ. It is one more face of the de Sitter metric in addition to the three isotropic cosmological models and (3.19).

Let us stress that in this description the metrics (3.19) and (3.20) must be considered separately from each other because at $r = H^{-1}$, $A = 0$, so both metrics possess coordinate singularities. The same is true for all R- and T-regions up to the end of this section. At all horizons, corresponding to regular zeros of the function $A(r)$, as is easily verified using Eq. (3.7), the four-metric is regular, and it is natural to suppose that a transition to new coordinates can allow for avoidance of coordinate singularities and obtaining a complete space–time picture naturally including R- and T-regions. Such transformations will be described in the next sections.

As to the anti-de Sitter (AdS) metric, obtained from (3.19) by replacing H^2 with $-H^2 = \Lambda/3 < 0$, it possesses again a regular center at $r = 0$, has no horizons and a nonflat spatial infinity $r \to \infty$, which is repulsive for test particles due to $g_{tt} \to \infty$, because g_{tt} is acting similarly to the Newtonian gravitational potental.

The Schwarzschild metric and Newton's law

In the case $\Lambda = 0$, $Q = 0$, the metric (3.18) turns into the Schwarzschild metric

$$ds^2 = \left(1 - \frac{2GM}{r}\right) dt^2 - \left(1 - \frac{2GM}{r}\right)^{-1} dr^2 - r^2 d\Omega^2, \qquad (3.21)$$

where the integration constant M has the meaning of an active gravitational mass. To confirm this interpretation, let us use the geodesic equation (2.58) for the case of a test particle instantaneously at rest, at a large radius r (as compared to $2GM$):

$$\frac{d^2x^i}{ds^2} + \Gamma^i_{00}(u^0)^2 = 0,$$

where $ds^2 = g_{00}dt^2 \approx dt^2$ and $u^0 \approx 1$ due to the assumptions made. Among the Christoffel symbols Γ^i_{00} only one is nonzero, namely, $\Gamma^1_{00} = \gamma' \mathrm{e}^{2\gamma - 2\alpha}$ (a general expression in terms of the metric (3.1)), or, with (3.21), $\Gamma^1_{00} \approx GM/r^2$. Thus the particle experiences the acceleration $-GM/r^2$ in the direction of smaller radii, as was required and expected.

The Schwarzschild metric describes the gravitational field in vacuum around any isolated spherically symmetric body provided that the space-time far from the body can be regarded asymptotically flat — and this is the case for an overwhelming majority of phenomena of astrophysical interest. The properties of the Schwarzschild metric will be discussed in detail in Sec. 3.4. We will only mention here that, like the de Sitter metric (3.19), the Schwarzschild metric (3.21) contains a single horizon at $r = r_g = 2GM$ (the so-called gravitational radius), beyond which, at $r < r_g$, there is a T-region representing a certain Kantowski–Sachs model.

The region $r \leq r_g$ is the simplest example of a black hole.

The Reissner–Nordström metric

The Reissner–Nordström metric is obtained from (3.18) at $\Lambda = 0$ and describes the external gravitational field of a body of mass M having the electric charge Q. The geometric and hence physical properties of the Reissner–Nordström space–time depend on the charge to mass ratio. Let us introduce the "geometrized" mass and charge having the dimension of length

$$m = GM, \quad q = \sqrt{G}Q, \tag{3.22}$$

then under the condition $m^2 > q^2$ the quadratic trinomial (see (3.18))

$$r^2 A(r) = r^2 - 2mr + q^2, \tag{3.23}$$

has two positive roots

$$r_{\pm} = m \pm \sqrt{m^2 - q^2}, \tag{3.24}$$

so that $A(r) > 0$ (the metric is static) at $r_+ < r < \infty$ — it is the outer R-region, and at $0 < r < r_-$ — the inner R-region. The intermediate values $r_- < r < r_+$ correspond to a T-region, in which, as in the Schwarzschild and de Sitter space–times, the metric describes a certain homogeneous anisotropic Kantowski–Sachs model. The spheres $r = r_+$ and $r = r_-$, separating the R- and T-regions, are called the outer and inner horizons of a Reissner–Nordström BH, respectively. The outer horizon is also called the **event horizon**, the inner one the **Cauchy horizon**, both notions will be discussed later.

Under the condition $m^2 = q^2$, so that $r^2 A(r) = (r - m)^2$, the two horizons merge into a single one, $r = r_h = m = |q|$; it is called a **double, or extremal horizon**, and the region $r \leq m$ (or sometimes the whole configuration) is called an extremal Reissner–Nordström BH. At both sides of the sphere $r = r_h$, the function $A(r)$ is positive.

Lastly, for "large charges," $q^2 > m^2$, the whole space $r > 0$ is occupied by a single R-region.

At $r \to \infty$, the contribution of the charge to the function $A(r)$ can be neglected, and the metric approximately coincides with Schwarzschild's.

On the contrary, at small radii, $r \to 0$, the properties of the metric are completely determined by the charge. Since it is in all cases an R-region, the value $r = 0$ corresponds to a center, and, as shown by comparison with Eq. (3.7), this center is singular. Since $g_{00} = A \to \infty$ as $r \to 0$, it is easy to show that the center repels neutral test particles.

Metrics with a nonzero cosmological constant

At $\Lambda \neq 0$, the basic properties of the metric (3.18) are again determined by the behavior of the function $A(r)$ and, above all, by the number and disposition of its zeros, each of them corresponding to a horizon that separates R- and T-regions.

At $Q = 0$, $M > 0$ and $\Lambda > 0$ (the Schwarzschild–de Sitter metric, also called the Kottler metric [358]), the number of horizons can range from zero to two, but at small and large r, at $r \to 0$ and $r \to \infty$, there are T-regions. If there is, in addition, a nonzero charge, then there is an R-region at $r \to 0$, with a singular center just as in the Reissner–Nordström solution, while at $r \to \infty$ the metric behaves as in the de Sitter solution (a T-region), so that the number of horizons varies from one to three depending on the particular values of the three parameters m, q and Λ.

If $\Lambda < 0$, at large r the metric behaves as in AdS space–time ($A(r) \to \infty$), and, depending on particular values of the parameters, the number of horizons varies from zero to two if $Q \neq 0$, and there is a single horizon if $Q = 0$ and $M > 0$.

3.3 Horizons and geodesics in static, spherically symmetric space–times

After the first appearance of horizons in our discussion of the de Sitter, Schwarzschild and Reissner–Nordström metrics and their generalizations, it makes sense to introduce the notion of a horizon more exactly and to establish its close relationship with the notion of a black hole (BH). As was mentioned at the beginning of this chapter, a BH is by definition such a space–time region that no material body or light signal can leave it. Since the motion of massive and massless particles is described in metric theories of gravity by geodesic equations, the question of which space–time region is a BH is answered by studying the properties of geodesics.

It is for this reason that this section begins with a description of the general form of geodesics in an arbitrary static, spherically symmetric space–time; after that, we will discuss the behavior of geodesics in the vicinity of horizons and their relationship with the BH notion. We will conclude the section by briefly describing different kinds of horizons mentioned in papers on the theory of gravity.

3.3.1 *The general form of geodesic equations*

Let us return to the general static, spherically symmetric metric (3.1) with an arbitrary radial coordinate u and consider the geodesic equations (2.53) as equations with respect to the unknowns $x^0 = t(\lambda)$, $x^1 = u(\lambda)$, $x^2 = \theta(\lambda)$, $x^3 = \phi(\lambda)$, which are coordinates of a point on the trajectory, as functions of the canonical parameter λ.

For simplicity, and in connection with the symmetry of the problem, we will assume that the geodesic is located in the equatorial plane $\theta = \pi/2$. We will denote $d/d\lambda$ by a dot and d/du by a prime (as before). The equations, along with evident first integrals of two of them, have the form

$$\ddot{t} + 2\gamma'\dot{t}\dot{u} = 0 \ \Rightarrow \ \dot{t} = E\,\mathrm{e}^{-2\gamma}, \tag{3.25}$$

$$\ddot{u} + \gamma'\,\mathrm{e}^{2\gamma-2\alpha}\dot{t}^2 + \alpha'\dot{u}^2 - \beta'\,\mathrm{e}^{2\beta-2\alpha}\dot{\phi}^2 = 0, \tag{3.26}$$

$$\ddot{\phi} + 2\beta'\dot{\phi}\dot{u} = 0 \ \Rightarrow \ |\dot{\phi}| = L\,\mathrm{e}^{-2\beta}, \tag{3.27}$$

where E and L are integration constants. The equation that contains $\ddot{\theta}$ is trivial.

Equations (3.25)–(3.27) are not independent: there is the constraint $u_\alpha u^\alpha = k$, where $u^\mu = dx^\mu/d\lambda$; $k = +1$ for timelike geodesics (in this case u^μ is the 4-velocity and the parameter λ coincides with proper time), $k = 0$ for null geodesics, and $k = -1$ for spacelike ones. Let us write down the explicit form of this constraint, which represents an integral of Eq. (3.26):

$$e^{2\gamma}\dot{t}^2 - e^{2\alpha}\dot{u}^2 - e^{2\beta}\dot{\phi}^2 = k, \tag{3.28}$$

and substitute here the integrals (3.25) and (3.27). After multiplying the resulting relation by $e^{2\gamma}$ we obtain

$$e^{2\alpha+2\gamma}\dot{u}^2 + k\,e^{2\gamma} + L^2\,e^{2\gamma-2\beta} = E^2. \tag{3.29}$$

The latter relation has the form of an energy conservation law for a particle moving in a potential field along the u axis; the quantity $E^2 \geq 0$ plays the part of the total energy, the first term is an analogue of the particle's kinetic energy while a sum of the second and third ones makes an analogue of the potential energy; moreover, the effective potential

$$V(u) = e^{2\gamma}(k + L^2\,e^{-2\beta}) \tag{3.30}$$

plays the same role for geodesic motion as the potential in classical mechanics for one-dimensional motion of a point particle: the motion is only possible in a region where $E^2 \geq V(u)$, while the values of the coordinate u at which $E^2 = V$ correspond to turning points.

The constant L, related to changes in the azimuthal angle ϕ, can be interpreted as the particle's conserved angular momentum per unit mass.

3.3.2 *Horizons, geodesics and the quasiglobal coordinate*

In our further discussion, we will frequently use the quasiglobal coordinate x (see Sec. 3.1), which possesses some important advantages as compared to other choices of the radial coordinate.

Thus, we can say in advance that this variable x behaves at horizons in the same way as the coordinates used to carry out an analytic extension of the metric across these horizons, and therefore it can be used simultaneously on both sides of the horizon (hence the name "quasiglobal"). In addition, as we will see immediately, in some important cases the behavior of the x coordinate can be a criterion of geodesic completeness of the space–time and therefore the very necessity of its extension.

In terms of the quasiglobal coordinate $u = x$, the static metric (3.1) can be written as follows:

$$ds^2 = A(x)dt^2 - \frac{dx^2}{A(x)} - r^2(x)d\Omega^2, \tag{3.31}$$

where $A(x) \equiv e^{2\gamma}$.

We have previously introduced the notion of a horizon using as examples the de Sitter, Schwarzschild and other metrics in which the coordinate $r = x$ turned out to be simultaneously a curvature and quasiglobal coordinate. In agreement with what has been said, let us define a horizon in any space–time with the metric (3.31) as a regular sphere $x = x_h$, near which

$$A(x) \sim (x - x_h)^n, \tag{3.32}$$

where $n \in \mathbb{N}$ is the order of the horizon. Since $A(x)$ is equal to the norm $\xi^\alpha \xi_\alpha$ of the Killing vector $\xi^\alpha = (1, 0, 0, 0)$, regular surfaces where $A = 0$ are **Killing horizons** (surfaces where the timelike or spacelike Killing vector becomes null).

It is easy to notice that, in terms of the metric (3.31), an R-region is a region where $A > 0$, and a T-region is a region where $A < 0$. A horizon of odd order n separates an R-region from a T-region, while a horizon of even order separates two R-regions or two T-regions from one another.[3]

Consider the behavior of geodesics of the metric (3.31) near horizons. According to (3.29), in the notation of the metric (3.31) ($u = x$, $e^{2\gamma} = e^{-2\alpha} = A(x)$), and bearing in mind that $\dot{u} = du/d\lambda$, we obtain

$$\pm \frac{d\lambda}{dx} = \left[E^2 - A\left(\frac{L^2}{r^2} + k\right)\right]^{-1/2}. \tag{3.33}$$

[3]Noninteger n would violate the analyticity of the metric even if the curvature invariants are finite, thus making meaningless an analytic extension beyond the horizon. Thus, for instance, if $x > x_h$ in an R- or T-region, then at fractional n the metric loses sense at $x < x_h$. Although while considering only a static reference frame one could include fractional n into consideration, we here put $n \in \mathbb{N}$ for simplicity.

In particular, as $x \to x_h$ such that $A(x_h) = 0$ at $r \neq 0$ (a possible horizon), where the quantity x_h may be finite or infinite, we have

$$d\lambda \approx dx/E. \tag{3.34}$$

If E is a finite constant, then, evidently, near a possible horizon *the coordinate x behaves as a canonical parameter for any geodesic approaching it*, be it spacelike, timelike or null. For timelike geodesics one can state that *a horizon is achieved along a geodesic at finite proper time $\lambda \equiv \tau$ if and only if this horizon corresponds to a finite value of the quasiglobal coordinate x.*

If $A \to 0$ as $x \to \infty$, such a surface may be called a *remote horizon*: it is reached along a geodesic at infinite proper time, and apparently any nongeodesic trajectories cannot make this time finite.

The constant E can be zero only in the case (see (3.25)) of spacelike geodesics completely located in the spatial section $t =$ const. In this case the canonical parameter λ, coinciding with the proper length l, is determined by the integral

$$l = \int \frac{du}{\sqrt{A(1 - L^2/r^2)}}. \tag{3.35}$$

The integrand is meaningless at $r < L$. Therefore $r = L$ is the minimum value of r reached by purely spatial geodesics. This is the simple geometric meaning of the constant L.

According to (3.33), if $E = 0$, then in the limit $A \to 0$, if $x \to \infty$, then the canonical parameter λ along the geodesics tends to infinity, just as for nonzero values of E. Thus, we obtain the following important result for static, spherically symmetric space–time:

If $x \to \infty$ on a surface where $A \to 0$ (i.e., on a candidate horizon), then this surface (called a remote horizon) is the boundary of the space-time under consideration, and it cannot be reached by any geodesics at finite values of their canonical parameter.

So the space–time is geodesically complete at a remote horizon, and no extension is needed beyond it.

Let us now return to usual (not remote) horizons and look how the coordinate time t (it is also the time according to the clocks of a distant observer at rest in the case of an asymptotically flat space–time) behaves there on timelike and null geodesics approaching such a horizon. From (3.25) and (3.29) we obtain for the metric (3.31)

$$\frac{dx}{dt} = \pm A\sqrt{1 - V(x)/E^2}, \tag{3.36}$$

where $E > 0$, and the potential has the form $V = A(k + L^2/r^2)$. At the horizon, the potential turns to zero, and due to (3.32), we get from (3.36)

$$t \sim \pm \int \frac{dx}{(x - x_h)^n} \to \pm\infty. \tag{3.37}$$

Thus *the coordinate time t is infinite at the horizon for all geodesics that cross it, except for purely spatial ones*: it is equal to $+\infty$ for motion to the horizon and $-\infty$ for motion from the horizon. It means that a horizon is in absolute past or absolute future for any observer located in the static region, and it becomes clear that to cross a horizon and to gain a joint description of regions on its both sides it is necessary to pass on to coordinates which do not belong to the static RF. More than that: even from this consideration, completely restricted to the static RF, it follows that a horizon as a limiting surface in the set of spheres $r = \text{const}$ splits into two parts: *the past horizon* and *the future horizon.*

Equation (3.36) looks especially simple for radially moving photons: $dx/dt = \pm A$. It shows that *the coordinate velocity of photons (and the same for massive particles), from the viewpoint of any static observer, tends to zero as the photon approaches a horizon*, and tends so rapidly that the coordinate time t at which a photon reaches or leaves a horizon is infinite.

We see that if a given static region of space–time contains a future horizon, then this horizon and the region beyond it realize the notion of a BH from the viewpoint of this static region since, to put it simply, it is possible to get there but it is impossible in principle to return from there — from the absolute future. In a similar way, a past horizon and a region at and beyond it can be called a *white hole*, from which photons or any other particles can appear, but it is impossible in principle to get there from the static region because it is located in the absolute past.

3.3.3 *Null geodesics and gravitational lensing*

Let us now consider the motion of massless particles along null geodesics, so that $k = 0$ in Eqs. (3.29) and (3.30). While for timelike geodesics, as we remember, it is reasonable to identify the canonical parameter λ in the geodesic equations with proper time and in spacelike geodesics with proper length, for null geodesics we have no such evident identification, and λ is defined up to a linear substitution.

If we restrict ourselves to asymptotically flat space–times, so that in the metric (3.31) $A \to 1$ as $x \to \infty$, we can simplify some relationships, and in particular, fix the canonical parameter. Thus, Eq. (3.25) gives $dt/d\lambda = E/A(x)$, and identifying λ with the time coordinate of a distant observer at rest with respect to our RF, we simply obtain $E = 1$.

Next, if a photon moves from infinity and approaches our spherically symmetric gravitating body, we can, without loss of generality, assume that it moves in the equatorial plane $\theta = \pi/2$ and that the azimuthal angle ϕ is counted from the beam which is, at large x, parallel to the velocity of our photon, and we can write $\phi = b/r$, where the constant b is called the *impact parameter*, it is the initial distance between the photon path and the beam $\phi = 0$. We thus have $\dot{\phi} = -b\dot{r}/r^2$ and can compare it with the integral (3.27) of the geodesics equations, to obtain $-b\dot{r} = \pm L$. Lastly, since in the flat asymptotic region $x \to \infty$ for a photon moving with a finite impact parameter towards $r = 0$ it holds $\dot{r} \approx -c = -1$, and properly choosing the sign of L, we can put simply $L = b$.

Since b is a parameter inherent to a particular photon path, and L is its constant of motion, we can use the equality $L = b$ for the whole trajectory of any photon moving in any asymptotically flat static, spherically symmetric space–time.

According to Eq. (3.30), the photon motion is governed by the equation

$$\dot{r}^2 + V(x) = 1, \quad V(x) = A(x)b^2/r^2(x). \tag{3.38}$$

The photon motion is only possible in the range of x where $V(x) \leq E = 1$. In particular, circular photon orbits ($\dot{x} \equiv 0$) can take place on such spheres $x = x_c$ where $V' \equiv dV/dx = 0$, that is, according to (3.38),

$$A/r^2 = 1/b^2, \quad rA' - 2Ar' = 0. \tag{3.39}$$

Spheres (3.39) (the so-called *photon spheres*) play an important role in gravitational lensing since the deflection angle of a photon approaching such a sphere along its tangent tends to infinity.

To describe gravitational lensing, one can follow the approach of [69, 533] applicable to all asymptotically flat static, spherically symmetric space–times. Let us assume that both the source of the signals and the observer are located at very large distances from the deflecting body (the lens) as compared to the characteristic lengths of the lens itself. Therefore, in particular, all incident paths of photons are assumed to be initially parallel to each other and differ from each other only by their impact parameters.

From Eqs. (3.38) and (3.27) we find the derivative

$$\frac{d\phi}{dx} = \frac{b}{r^2(x)\sqrt{1 - V(x)}} = \frac{b}{r(x)\sqrt{r^2(x) - A(x)b^2}}, \tag{3.40}$$

and its integration over the relevant values of x gives the full angle ϕ covered by the photon. Taking into account that this angle would be equal to π if there were no deflection, we find the deflection angle α as

$$\alpha = \alpha(x_0) = I(x_0) - \pi,$$
$$I(x_0) = 2\int_{x_0}^{\infty} \frac{dx}{r(x)\sqrt{r^2(x)/b^2 - A(x)}}, \tag{3.41}$$

where x_0 is the coordinate value corresponding to the nearest approach of the photon to the strong field region, which can be found from the condition $dr/d\lambda = 0$ which leads to

$$A(x_0)b^2 = r^2(x_0), \tag{3.42}$$

and $I(x_0)$ in Eq. (3.41) can be rewritten in the form

$$I(x_0) = 2\int_{x_0}^{\infty} \frac{r_0 dx}{r(x)\sqrt{r^2(x)A(x_0) - r_0^2 A(x)}}, \tag{3.43}$$

where $r_0 = r(x_0)$. Equations (3.41) and (3.43) can be used for finding the lensing characteristics of different spherically symmetric objects.

What can be found immediately is the small deflection angle at large impact factors, such that the metric is approximately Schwarzschild and $b \gg 2m$. We thus take $r = x$, $A = 1 - 2m/r$, and use Eq. (3.43) which is more convenient than (3.41) because it contains only one parameter $x_0 = r_0 \gg 2m$ and does not contain b. It turns out, however, that the integral (3.43) leads to a very cumbersome expression in terms of elliptic integrals, therefore we will use the following trick. Since, as can be directly verified, the integral $I(r_0) = \pi$ for any r_0 if $A \equiv 1$ (which simply means that there is no light deflection in flat space–time), we can directly calculate the quantity $\alpha(r_0) = I(r_0) - \pi$ as

$$\alpha(r_0) = 2\int_{r_0}^{\infty} \frac{r_0 dr}{r}\left[\frac{1}{\sqrt{r^2 A(r_0) - r_0^2 A(r)}} - \frac{1}{\sqrt{r^2 - r_0^2}}\right]. \tag{3.44}$$

Under the assumption $r_0 \gg 2m$, and since integration is carried out over $r > r_0$, we can obtain the following approximate expression for the content of the square bracket in (3.44):

$$\frac{1}{\sqrt{r^2 - r_0^2}} \frac{m(r^2 + rr_0 + r_0^2)}{rr_0(r + r_0)},$$

and for $\alpha(r_0)$ it results in

$$\alpha(r_0) = 2m\int_{r_0}^{\infty} \frac{dr\ (r^2 + rr_0 + r_0^2)}{r^2(r + r_0)\sqrt{r^2 - r_0^2}} = \frac{4m}{r_0}. \tag{3.45}$$

The last expression is nothing else than Einstein's classical formula for light deflection in the gravitational field of a massive body. A more detailed derivation of this expression can be found, e.g., in Weinberg's book [554], along with the history of its verification by observations at solar eclipses.

It remains to find a relation between the radius of the closest approach r_0 and the impact parameter b. It is easily achieved using the condition $\dot{r} = 0$ for the radius r_0 in (3.38) with $A(r) = 1 - 2m/r$. Under the condition $b \gg 2m$ it gives approximately

$$r_0 = b - m. \tag{3.46}$$

and in the same approximation the expression $\alpha = 4m/r_0$ may be rewritten as $\alpha = 4m/b$.

3.3.4 *Transitions to Lemaître reference frames*

Let us now use the geodesic equations to make a transition to a nonstatic RF connected with a family of test particles moving in a static, spherically symmetric gravitational field. In other words, this "cloud" of test particles will then play the role of a reference body in such an RF.

Consider the general static metric (3.31) written in terms of the quasiglobal coordinate x. We shall see that a transition from the coordinates (x, t) of the static RF to the geodesic coordinates (R, τ), where τ is the proper time along a geodesic and the radial coordinate R is the congruence parameter, different for different geodesics, can be described in a general form. A radial timelike geodesic in the metric (3.31) satisfies the equations following from (3.25) and (3.29):

$$\left(\frac{dx}{d\tau}\right)^2 = E^2 - A(x), \qquad \frac{dt}{d\tau} = \frac{E}{A(x)}, \tag{3.47}$$

where the constant E is connected with the initial velocity of a particle moving along this particular geodesic at a given value of R. In general, $E = E(R)$, i.e., the particle energies are different for different geodesics.

Equation (3.47) gives two of the four components of the transition matrix

$$\|\partial(t, x)/\partial(\tau, R)\|, \tag{3.48}$$

namely, $\dot{x}$ and $\dot{t}$ (dots and primes here stand for $\partial/\partial\tau$ and $\partial/\partial R$, respectively) since this partial differentiation occurs along the geodesics

$$\dot{x} = \pm\sqrt{E^2(R) - A(x)}, \quad \dot{t} = E(R)/A(x). \tag{3.49}$$

A relation between the other two components, t' and x', can be found from the condition $g_{\tau R} = 0$ when we substitute $dt = \dot{t}d\tau + t'dR$ and $dx = \dot{x}d\tau + x'dR$ into the metric (3.31):

$$t' = \frac{\sqrt{E^2(R) - A(x)}}{E(R)A(x)} x'. \tag{3.50}$$

It remains to find $x'(R, \tau)$, which can be done by using the integrability condition $(\partial_\tau \partial_R - \partial_R \partial_\tau)x = 0$. The latter takes the form of a linear first-order differential equation with respect to x' considered as a function of R and x, $x' = y(R, x)$:

$$\partial_x y = -\frac{y\, \partial_x A}{2(E^2 - A)} + \frac{EE'}{E^2 - A}. \tag{3.51}$$

Solving it, we obtain

$$y = x'(R, \tau) = \sqrt{E^2 - A(x)} \left[f_0(R) + EE' \int \frac{dx}{\left(E^2 - A(x)\right)^{3/2}} \right], \tag{3.52}$$

where $f_0(R)$ is an R-dependent integration constant. The other integrability condition $(\partial_\tau \partial_R - \partial_R \partial_\tau)t = 0$ holds automatically due to (3.50). The functions $t(R, \tau)$ and $x(R, \tau)$ can now be found by further integration of Eqs. (3.49)–(3.52). The resulting metric can be written as follows:

$$ds^2 = d\tau^2 - \frac{x'(R, \tau)^2}{E^2(R)} dR^2 - r^2(x(R, \tau)) d\Omega^2. \tag{3.53}$$

One can see how this procedure works using the de Sitter space–time as an example. Its static form is (3.31) with $r \equiv x$ and $A(r) = 1 - H^2 r^2$, $H = \text{const} > 0$. Let us choose three different families of geodesics such that

$$E(R) = \sqrt{1 - kR^2}, \quad k = 0, \pm 1, \tag{3.54}$$

and show that their corresponding RFs represent the three well-known forms of the de Sitter metric as isotropic cosmologies with different signs of spatial curvature (see the end of the previous chapter). Indeed, integrating the first relation in (3.49) as an equation for $\tau = \tau(r, R)$ and properly choosing the arbitrary function of R that appears as an integration constant, we obtain the following expressions for r:

$$r(R, \tau) = (R/H) \times \{\cosh(H\tau),\ e^{H\tau}, \sinh(H\tau)\}, \tag{3.55}$$

where the expressions in the curly brackets are ordered according to $k = 1,\ 0, -1$. Substituting them into (3.53), we obtain the metric in the form

$$\begin{aligned} ds^2 &= d\tau^2 - a^2(t)\left(\frac{dR^2}{1 - kR^2} + R^2 d\Omega^2\right), \\ a(\tau) &= (1/H) \times \{\cosh(H\tau),\ e^{H\tau}, \sinh(H\tau)\}, \end{aligned} \tag{3.56}$$

as was intended. One can also verify that the expression (3.52) with $E(R)$ given by (3.54) (provided the function $f_0(R)$ is chosen properly) coincides with the expression for r' obtained directly from (3.55) in all three variants. So the transition has been completed.

We see that the de Sitter metric in the form of an isotropic cosmological model is written in geodesic RFs, comoving to some families of freely moving test particles in de Sitter space–time.

Thus for finding the metric in the Lemaître RF we needed only the transition matrix (3.48) consisting of derivatives whereas for writing explicitly the corresponding coordinate transformation it is necessary to perform further integration. In our example this integration has been performed for r as a function of R and τ but not for t. Let us find $t(R, \tau)$ for the geodesic family $k = 0$ that leads to a spatially flat cosmology. We have the expressions

$$\dot{t} = \frac{1}{1 - R^2 e^{2H\tau}}, \quad t' = \frac{R e^{2H\tau}}{1 - R^2 e^{2H\tau}}. \tag{3.57}$$

Their integration gives

$$t = t_0 + \tau - \frac{1}{2H} \ln(1 - R^2 e^{2H\tau}), \tag{3.58}$$

where t_0 is an integration constant corresponding to an arbitrary zero point of t. The corresponding expression for r from (3.55) is $r = (R/H)\,e^{H\tau}$.

The resulting metric (3.56) does not show any trace of the horizon at $r = H$ that restricted the static RF. This perfectly agrees with the previous result that a horizon is reached by a geodesically moving particle at its finite proper time. Such a particle (or observer) simply does not notice a horizon being crossed. On the other hand, according to (3.58), the coordinate time of the static RF makes sense only at $r < H$ and diverges where $r = H$. (This happens at different values of the proper time τ of different geodesics whose family parameter R characterizes their initial conditions.)

Meanwhile, the metric (3.56) makes sense at any r. Thus the transformation to a geodesic (Lemaître) RF is one of the ways of extending the metric beyond a horizon, and such an extension was really carried out not only for de Sitter but also for BH space–times. It is really a useful tool for many purposes, in particular, for comparison and matching vacuum metrics with those describing the gravitational field inside matter distributions (such as, e.g., the Tolman solution for a dust cloud, to be presented in Sec. 3.6.2). It is, however, not very convenient for extracting information on the global structure of space–time.

3.3.5 *Horizons, R- and T-regions*

The horizons described above in static (or homogeneous if a T-region is concerned) spherically symmetric space–times are special cases of more general kinds of horizons discussed in the literature on the theory of gravity. Let us mention some of them, only briefly explaining their meaning but without going into deep details and subtleties since it would require an introduction of many concepts from such advanced branches of mathematics as differential and algebraic topology and global differential geometry. An interested reader can find a detailed analysis of these concepts and their application to gravity in the books [250, 285, 409, 549] and in review articles.

An *event horizon* is a boundary of a black hole, and both these notions are rigorously defined in asymptotically flat space–times, while in other cases, such as spaces with a nonzero cosmological constant, natural extensions of these notions are used. It is known (Penrose's theorem) that an event horizon is formed by null geodesics.

An *apparent horizon* is defined as a two-dimensional surface, such that null geodesics which orthogonally leave it have zero expansion. In stationary BHs, the apparent horizon coincides with the event horizon; however, an advantage of the apparent horizon is that it is defined locally.

An *isolated horizon* is also a local notion, whose meaning is rather close to that of an apparent horizon but is introduced in a more complicated manner, not to be discussed here, see, e.g., [22, 23].

A *Cauchy horizon* is the boundary of a space–time region in which the evolution of physical fields and matter distributions can be found from their equations of motion with initial data specified on a certain Cauchy hypersurface (which can be null or spacelike). The emergence of Cauchy horizons is most frequently connected with singularities since they can be a source of impacts unpredictable in the framework of classical gravity.

A *Killing horizon* is, by definition, a surface whose normal is null and coincides with a Killing vector. In other words, it is a surface at which a certain Killing vector is null. Quite evidently, Killing horizons can occur only in space–times possessing isometries which are then characterized by Killing vectors.

For static space–time, of greatest importance are Killing horizons at which the timelike Killing vector $\xi^\mu = (1, 0, 0, 0)$ becomes null. This happens where $g_{00} = e^{2\gamma} = 0$. From the viewpoint of a static RF, there is a singularity at such a horizon, although the 4-curvature is certainly finite.[4]

Lastly, sometimes the notion of a horizon is introduced, being connected with a certain scalar. Thus, for instance, in spherically symmetric space–times with the metric (3.1), the quantity $r(u, t)$ is a scalar from the viewpoint of the two-dimensional subspace

[4]The definition of a Killing horizon implicitly assumes its regularity since otherwise such a definition would be meaningless: we would then deal with a singularity rather than a horizon.

parametrized by the coordinates u and t. On the basis of the behavior of the function $r(u,t)$, one can introduce the notions of R- and T-regions, generalizing these notions from static to arbitrary nonstatic space–times. By definition, an **R-region** is a space–time region where the gradient of the function $e^{\beta} = r(u,t)$, considered as a scalar in the (u,t)-subspace, is spacelike:

$$r^{,\alpha} r_{,\alpha} = e^{-2\gamma} \dot{r}^2 - e^{-2\alpha} r'^2 < 0. \tag{3.59}$$

In this case, the quantity r can be chosen as a spatial coordinate. A **T-region** is, by definition, a region where $r^{,\alpha} r_{,\alpha} > 0$, and therefore one can choose r as a time coordinate. Furthermore, an r**-horizon** is a surface on which $r^{,\alpha} r_{,\alpha} = 0$.

All the horizons considered previously are evidently r-horizons and Killing horizons. The sphere $r = 2m = 2GM$ in the Schwarzschild metric and the sphere $r = r_+$ in the Reissner–Nordström metric are Killing horizons and event horizons simultaneously, while $r = r_-$ in the Reissner–Nordström metric is a Cauchy horizon, like many horizons in metrics with $\Lambda \neq 0$.

3.4 Schwarzschild black holes — geodesics and a global description

3.4.1 *R- and T-regions*

As already noticed, in the Reissner–Nordström–de Sitter solution (3.18) the radial coordinate is simultaneously a curvature coordinate and a quasiglobal coordinate ($r \equiv x$). Therefore, we can directly use the results of the previous section in analyzing the properties of the solution.

Let us address the simplest special case, the Schwarzschild metric (3.21). At $r = 2GM = 2m = r_g$ (this value of r is called the gravitational radius corresponding to the mass M, or the Schwarzschild radius), there is a simple horizon, and in the region $r < 2GM$ (T-region) the coordinates r and t exchange their roles: r becomes a temporal coordinate and t a spatial one (just as was discussed for the de Sitter metric). The T-region is described by the Kantowski–Sachs metric

$$ds^2 = \frac{dr^2}{|A(r)|} - |A(r)|dy^2 - r^2 d\Omega^2, \tag{3.60}$$

where $|A(r)| = 2GM/r - 1$, and the spatial coordinate y replaces t. If we suppose that r decreases from the value $2GM$ to zero, then the homogeneous model (3.60) contracts (collapses) to zero in two angular directions along the coordinate spheres, whereas in the third direction (y) it stretches without limit. The final state $r = 0$ is a spacelike singularity, reached at a certain time instant.[5]

The inverse process of anisotropic evolution (expansion of coordinate spheres $\theta = \text{const}$, $\phi = \text{const}$ and contraction from infinity to zero along the y axis) is observed as r changes from $r = 0$ to $r = 2GM$; it is what happens in a white hole configuration.

The metric (3.60) describes both expanding (in the sense of growing r) and contracting cosmological models since the time direction is not unambiguously related to the growth or fall of r: the physical time τ is related to r by the integral

$$\tau = \pm \int dr/\sqrt{A(r)}, \tag{3.61}$$

whose convergence at both $r \to 2GM$ and $r \to 0$ indicates that the evolution from the horizon to the singularity (or conversely) occurs in a finite time interval according to a clock at rest in the metric (3.60) at a fixed spatial point (y, θ, ϕ).

Meanwhile, the expanding or contracting T-region is in no way connected with the initial R-region even though the curvature at

---

[5]It is sometimes claimed that the Schwarzschild space–time has a singularity at the center. As is clear from the above-said, this claim is completely wrong. Indeed, the singularity $r = 0$ is in a T-region, which represents a homogeneous cosmological model, whose all spatial points are equivalent, whereas the notion of a center assumes a certain distinguished spatial point. Moreover, in the coordinate system in which the metric (3.60) is written, all spatial points reach the singularity simultaneously.

the horizon $r = \tau = 2GM$ which separates them is finite: the Kretschmann scalar is

$$\mathcal{K} = 48(GM)^2/r^6. \tag{3.62}$$

3.4.2 *Geodesics in the R-region*

The behavior of geodesics near the horizon completely corresponds to the general description presented above. Thus, according to (3.29) and (3.25), in the Schwarzschild metric

$$\left(\frac{dr}{d\lambda}\right)^2 + \left(1 - \frac{2m}{r}\right)\left(\frac{L^2}{r^2} + k\right) = E^2,$$
$$\frac{dt}{d\lambda} = \frac{E}{1 - 2m/r}, \qquad m := GM. \tag{3.63}$$

(The quantity m, having the dimension of length, is sometimes called the geometrized mass.) The first of these equations gives near the horizon ($r \to 2m$), independently of the nature of the geodesic and of the angular momentum L, $dr/d\lambda \approx \pm E$, whence (assuming $E > 0$) $d\lambda \approx \pm dr/E$. Thus *all geodesics cross the horizon at a finite value of their canonical parameter* λ. In particular, for timelike geodesics it means that *the horizon is crossed at finite proper time* s.

In the exceptional case of purely spatial geodesics ($E = 0$), Eq. (3.35) implies:

$L > 2m$: $r = L$ is the minimum value of r, such geodesics do not reach the horizon.

$L = 2m$: the horizon is reached but the length $l \sim \int dr/(r - 2m)$ logarithmically diverges. In other words, the distance from any fixed point to the horizon along such a (tangent) geodesic is infinite.

$L < 2m$: the horizon is reached, and the integral (3.35) converges at $r = 2m$.

Let us dwell upon massive particle motion in the Schwarzschild field, i.e., on timelike geodesics ($k = 1$, $d\lambda = ds$), where ds is the increment of the particle's proper time. From (3.63) we obtain

$$\left(\frac{dr}{ds}\right)^2 + V(r) = E^2, \tag{3.64}$$

with the potential

$$V(r) \equiv \left(1 + \frac{L^2}{r^2}\right)\left(1 - \frac{2m}{r}\right). \tag{3.65}$$

Taking the derivative d/ds in (3.64), we obtain the equation of motion

$$\frac{d^2r}{ds^2} = -\frac{1}{2}V'(r). \tag{3.66}$$

Stable circular orbits with $r = \text{const}$ can be found from the conditions $V' = 0$, $V'' > 0$. Substituting the expression of the potential, we obtain

$$mr^2 - L^2 r + 3mL^2 = 0. \tag{3.67}$$

A solution of this quadratic equation gives extrema of the potential:

$$r_{\text{ext}}(L) = \frac{L^2}{2m}(1 \pm \sqrt{1 - 12m^2/L^2}). \tag{3.68}$$

The sign of the second-order derivative $V''(r)$ at $r = r_{\text{ext}}$

$$V''(r_{\text{ext}}) = \frac{2}{r^4}(2mr - L^2)\Big|_{r=r_{\text{ext}}}$$

distinguishes the orbital radii ($V''>0$ [the plus sign in (3.68)] and a maximum of the potential that occurs at smaller radii [the minus sign in (3.68)]:

$$r = r_{\text{max}}(L) = \frac{L^2}{2m}(1 - \sqrt{1-12\,m^2/L^2}) = \frac{6m}{1 + \sqrt{1-12\,m^2/L^2}}. \tag{3.69}$$

It is clear that stationary orbits are absent at small angular momenta L,

$$L \leq L_{\text{min}} = 2\sqrt{3}\,m.$$

The minimum orbital radius is

$$r_{\text{min}} = r_{\text{ext}}(L_{\text{min}}) = 6GM = 3r_g, \tag{3.70}$$

and this orbit is only marginally stable due to $V'' = 0$ (see Fig. 3.1).

Let us stress once again that in curved space–time the radius r of a coordinate sphere is almost never equal to its distance from the

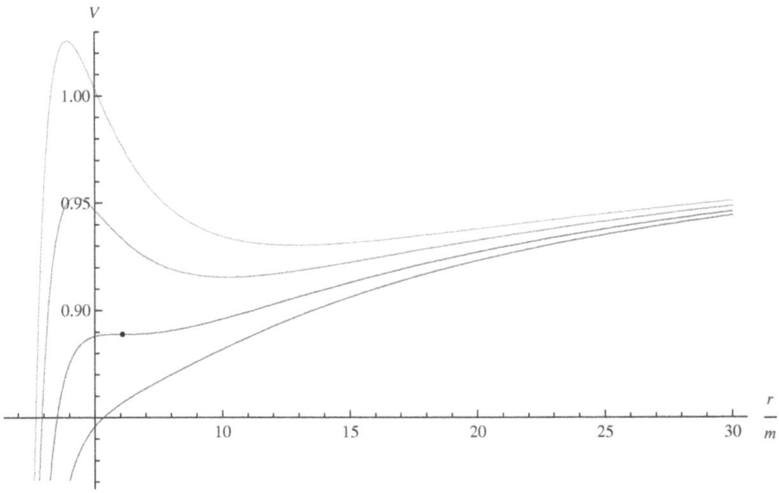

Figure 3.1 The potential (3.65) for $L/m = 3.2, 2\sqrt{3}, 3.8, 4.1$ (bottom-up). Stable circular orbits correspond to minima of the potential. The dot on the second curve at $r/m = 6$ corresponds to the smallest (marginally stable) circular orbit. The horizon is located at $r/m = 2$.

center even if there is a center. And in the Schwarzschild field, as we have seen, there is no center at all because the value $r = 0$ belongs to a T-region.

3.4.3 *Massive and massless particle capture by a black hole*

Let us return to Eq. (3.64) and consider the problem of particle capture by a Schwarzschild BH. Consider a massive particle moving from infinity with an "energy" $E \geq 1$ (where $E = 1$ corresponds to zero velocity at infinity). If it is not captured by the BH, it again escapes to infinity. Then there should be a maximum proximity position such that

$$\dot{r} = 0 \Rightarrow V(r) = E^2.$$

On the contrary, if it is captured, the equation $V(r) = E^2$, with $V(r)$ given by (3.65), must have no solution.

For a particle of unit mass moving from infinity with the initial velocity $v_\infty \ll 1$, we have the following expressions for the energy and angular momentum corresponding to nonrelativistic mechanics in flat space: $E_\infty = 1 + v_\infty^2/2 \simeq 1$ and $L = bv_\infty$, where b is the impact parameter. We arrive at the algebraic equation

$$\left(1 + \frac{L^2}{r^2}\right)\left(1 - \frac{r_g}{r}\right) \simeq 1,$$

whose solution is absent if

$$L \lesssim 4m \ \Rightarrow b \lesssim 4m/v_\infty.$$

Consequently, the capture cross-section for a nonrelativistic particle is

$$\sigma = \pi b^2 = 4\pi r_g^2/v_\infty^2 = 16\pi m^2/v_\infty^2. \tag{3.71}$$

In the general, relativistic case, the capture condition is the same: the absence of solutions to Eq. (3.64) with $dr/ds = 0$. By definition of the angular momentum (see (3.27)),

$$L \equiv r^2 \frac{d\phi}{ds}.$$

At large distances from the BH, $\phi \approx b/r$, hence $d\phi = -dr(b/r^2)$ (where again b is the impact parameter), and therefore $dr/ds = L/b$. Substituting it into Eq. (3.64), at large r, we obtain

$$E^2 = 1 + \frac{L^2}{b^2}.$$

Now, for arbitrary r, (3.64) takes the form

$$\left(1 + \frac{L^2}{r^2}\right)\left(1 - \frac{r_g}{r}\right) = 1 + \frac{L^2}{b^2}.$$

In the ultrarelativistic limit, $L/r \gg 1$, $L/b \gg 1$, and the equation is simplified:

$$\frac{b^2}{r^2}\left(1 - \frac{r_g}{r}\right) \simeq 1.$$

This equation has no solution if

$$b \le \frac{3\sqrt{3}}{2} r_g.$$

Thus the capture cross-section for ultrarelativistic particles moving near a Schwarzschild black hole is

$$\sigma = \pi b^2 = \frac{27}{4}\pi r_g^2 = 27\pi m^2. \tag{3.72}$$

The above derivations did not take into account that the interaction of a particle with a BH is accompanied by gravitational wave (GW) emission. Most frequently this emission is really negligible, but it cannot be neglected at interactions of two BHs and, in such cases, it substantially increases the cross-section of their mutual capture. More precisely, the cross-section of gravitational capture of two BHs with masses M_0 and M_1 to a close binary is expressed as follows [419]:

$$\sigma = 2\pi \left(\frac{85\pi}{6\sqrt{2}}\right)^{2/7} \frac{G^2(M_0 + M_1)^{10/7} M_0^{2/7} M_1^{2/7}}{c^{10/7} v_{\rm rel}^{18/7}}, \tag{3.73}$$

where $v_{\rm rel}$ is the relative velocity of the two BHs. After the capture, the BHs rapidly merge due to energy loss from gravitational wave emission, and a new mighty GW outburst is produced at the event of merging. The above cross-section is larger than the cross-section of direct BH collisions taking into account the gravitational focusing in a wide range of BH masses and velocities.

This result is directly related to the GW signals many of which have been detected in a few recent years [1, 541]. Another very important observable phenomenon, gravitational lensing, depends on the behavior of null geodesics that describe the motion of photons near BHs (or other celestial bodies) in the geometric optics approximation, i.e., assuming that the photon wavelengths are much smaller than the characteristic size of the body in question. We have discussed gravitational lensing at $r \gg 2m$ in Sec. 3.3.3. If the photon approaches closer to the BH, we can use Eqs. (3.41)–(3.43) with $r = x$ and $A = 1 - 2m/r$ without assuming $r \gg 2m$.

But what happens when the photon comes closer to the horizon? The potential for photon motion has the form given in (3.38), that is,

$$V(r) = \frac{b^2}{r^2}\Big(1 - \frac{2m}{r}\Big), \quad \frac{dV}{dr} = \frac{2b^2}{r^4}(3m - r). \tag{3.74}$$

This potential has a maximum at $r = 3m = 3r_g/2$, which is thus a photon sphere, along which the photon motion is unstable since $d^2V/dr^2 < 0$. Furthermore, according to (3.38), a turning point $r = r_0$ for a photon coming from outside (the radius of its closest approach to the BH) is characterized by $\dot{r} = 0$, $\ddot{r} \geq 0$, hence $dV/dr \leq 0$. Therefore, by (3.74), there are no turning points inside the sphere $r = 3m$, so any photon that gets there is captured by the BH.

Since by (3.38) at a turning point $r = r_0$ we have $V(r) = 1$, we have the following relation between r_0 and the impact parameter b:

$$r_0^3 = b^2(r_0 - 2m) = 0. \tag{3.75}$$

For the minimum value of r_0, that is, $r_0 = 3m$, we find $b = b_0 = 3\sqrt{3}m$, and then the photon capture cross-section turns out to be

$$\sigma = \pi b_0^2 = 27\pi m^2, \tag{3.76}$$

which coincides with (3.72), as should be the case since ultrarelativistic particles move like photons.

A detailed discussion of gravitational lensing in Schwarzschild space–time can be found in [52].

3.4.4 *A global description — the Kruskal metric*

A transition from the Schwarzschild static metric to coordinates providing a complete description of the Schwarzschild space–time by a single map was independently formulated in 1960 by Kruskal and Szekeres [364, 526].

Let us describe this transition. The spherical symmetry implies that the transformation must involve only the (r, t) subspace while the angular coordinates θ, ϕ will remain the same.

As the first step, we transform r passing on to the "tortoise" coordinate r_*, such that the two-dimensional metric of the (r, t) subspace becomes conformally flat:

$$\begin{aligned} r_* &= \int \left(1 - \frac{2m}{r}\right)^{-1} dr = r + 2m \ln \left|\frac{r}{2m} - 1\right|, \\ ds_2^2 &= \left(1 - \frac{2m}{r}\right) dt^2 - \left(1 - \frac{2m}{r}\right)^{-1} dr^2 \\ &= \left(1 - \frac{2m}{r}\right)(dt^2 - dr_*{}^2). \end{aligned} \tag{3.77}$$

The horizon $r = 2m$ corresponds to the limit $r_* \to -\infty$.

Next, in Eq. (3.77) it is convenient to pass on to the null coordinates V, W:

$$2t = V + W, \quad 2r_* = V - W. \tag{3.78}$$

At fixed V, the horizon corresponds to $W \to \infty$, hence $t \to \infty$. On the contrary, at fixed W the horizon corresponds to $V \to -\infty$, hence $t \to -\infty$. Thus (in agreement with the previous discussion of geodesics), the horizon is either in absolute past or in absolute future with respect to an observer at rest and splits into a past horizon and a future horizon. So, in terms of V and W, the Schwarzschild metric has the form

$$ds^2 = \left(1 - \frac{2m}{r}\right) dV\, dW - r^2 d\Omega^2. \tag{3.79}$$

Here, we have not yet removed the singularity at $r = 2m$. To this end, one more step is necessary: a transformation of the form $V = V(v)$, $W = W(w)$. Let us for certainty consider a transition through the future horizon: we fix V and focus on the limit $W \to \infty$. We wish to choose such a function $W(w)$ that the coefficient of $dV dw$ in the metric be finite. It is easy to verify that a suitable choice is $W = -4m \ln |w|$, because $1 - 2m/r = (r - 2m)/r \sim \mathrm{e}^{-r_*/(2m)} \sim \mathrm{e}^{W/(4m)}$ as $r \to 2m$.

In a similar way the metric is regularized at a transition through the past horizon. It is not hard to see that the transformation

$$V = 4m \ln |v|, \quad W = -4m \ln |w| \tag{3.80}$$

leads to the relation

$$1 - \frac{2m}{r} = -vw\,\mathrm{e}^{-r/(2m)}, \tag{3.81}$$

and the metric acquires the regular form

$$ds^2 = \frac{32m^3}{r}\,\mathrm{e}^{-r/(2m)}dv\,dw - r^2 d\Omega^2. \tag{3.82}$$

It is now easy to further transform the null coordinates into spatial (R) and temporal (T) ones by assuming

$$v = \frac{1}{2}(T+R), \qquad w = \frac{1}{2}(T-R), \qquad dv\,dw = dT^2 - dR^2, \tag{3.83}$$

whence

$$ds^2 = \frac{32m^3}{r}\,\mathrm{e}^{-r/(2m)}(dT^2 - dR^2) - r^2 d\Omega^2. \tag{3.84}$$

This is the Kruskal metric giving a complete picture of the Schwarzschild space-time. The initial region $r > 2m$ is mapped in it into the right quadrant $v > 0, w < 0$, or $R > 0$, $-R < T < R$ (region R_+, see Fig. 3.2). In addition to it, there are two T-regions T_- and T_+ (the first one is obtained by transition through the future horizon, the second one through the past horizon), and one more R-region, R_-.

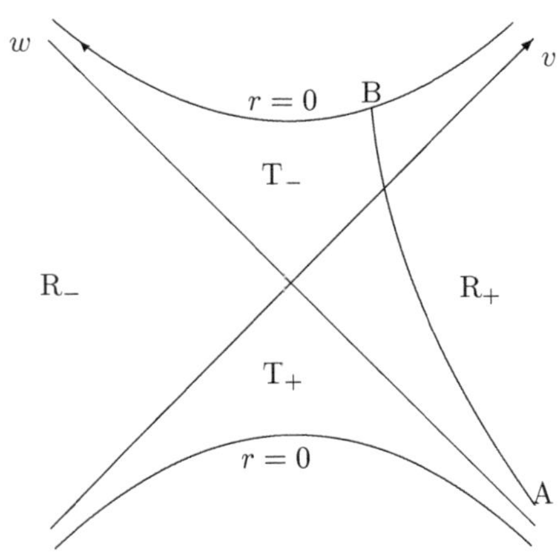

Figure 3.2 The Kruskal diagram.

The horizon $r = 2m$ is now depicted by the pair of intersecting axes $v = 0$ and $w = 0$, or, equivalently, $T = \pm R$. These axes (or these null hypersurfaces if we recall that each point on the diagram corresponds to a 2-sphere) do not belong to any of the R- or T-regions, and the same is true for their intersection point $u = v = 0$; still all points at the horizons are regular points in the Schwarzschild–Kruskal space–time, as is evident from the metric (3.82) or (3.84).

The singularity $r = 0$ is depicted by two hyperbolas: $T = -\sqrt{1 + R^2}$ (the past singularity) and $T = +\sqrt{1 + R^2}$ (the future singularity). It is the future singularity where everything gets after falling beyond the future event horizon.

Due to the null nature of the v and w coordinates, all radial null geodesics in the diagram are parallel to the v or w axes. Other causal curves (timelike as well as nonradial null trajectories) have slopes smaller than 45° from the vertical. This is true, in particular, for the lines $r =$ const in the R-regions (which depict the coordinate spheres at rest and are not geodesics).

3.4.5 *From Kruskal to Carter–Penrose diagram for the Schwarzschild metric*

In studies of the causal structure of space-time containing Killing horizons, it is often helpful to use Kruskal-like coordinate transformations which provide smooth transitions between regions separated by horizons. The results are most frequently presented with the aid of two-dimensional diagrams (Carter–Penrose diagrams) whose form contains the basic information on the global causal structure of the space–time in question. Such diagrams are more convenient than the Kruskal diagram because each R- or T-region is depicted there by a square or triangle of fixed finite size.

Let us build such a diagram for the Schwarzschild space-time using the Kruskal metric (3.82) in null coordinates:

$$ds^2 = \frac{32m^3}{r}\,\mathrm{e}^{-r/(2m)} dv\, dw - r^2 d\Omega^2, \tag{3.85}$$

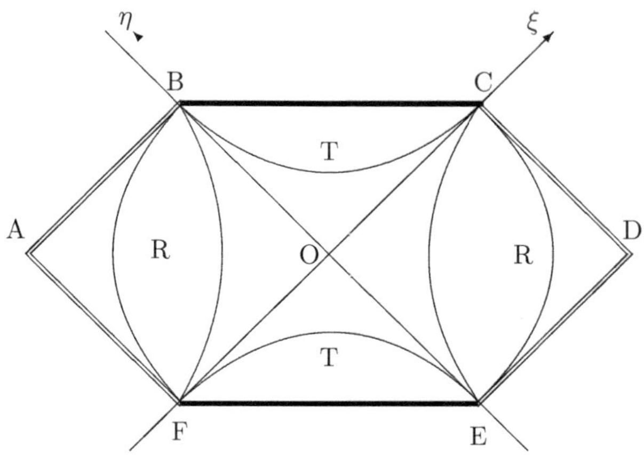

Figure 3.3 The Carter–Penrose diagram for the Schwarzschild metric.

where the radius r is related to the coordinates v and w by Eq. (3.81), or

$$-vw = \frac{r - 2m}{2m}\, \mathrm{e}^{r/(2m)}. \tag{3.86}$$

The diagram becomes compact if, instead of the coordinates $v \in \mathbb{R}$ and $w \in R$, we introduce the coordinates ξ, η, specified on the segment $(-\pi/2, \pi/2)$ by putting

$$v = \tan \xi, \qquad w = \tan \eta. \tag{3.87}$$

It is easily seen (Fig. 3.3) that the horizon is now represented by the two segments $\xi = 0$, $\eta = 0$ (BE and CF), and the singularity $r = 0$ by other two segments where $\xi + \eta = \pm\pi/2$ (BC and EF). The values $\xi = \pm\pi/2$ and $\eta = \pm\pi/2$ (the line strings FAB and CDE) correspond to an infinite radius r. The whole two-dimensional (r, t) manifold has been mapped into the interior of the hexagon ABCDEF. The null nature of the coordinates ξ and η allows for easily distinguishing spacelike, timelike and null directions at any point in precisely the same manner as it is done using the Kruskal diagram.

3.5 Global causal structure of space–times with horizons

3.5.1 *Crossing horizons in the general case*

Transformations similar to those applied to the Schwarzschild metric are of quite general nature, and, as a result, Carter–Penrose diagrams can be built in the general case of two-dimensional sections of space–time with the aid of simple correspondence rules between the behavior of metric functions and graphic images.

Consider an arbitrary static or stationary space–time where the metric depends on a spatial coordinate u, and let us restrict ourselves to its *two-dimensional section* with the metric

$$ds_2^2 = \mathrm{e}^{2\gamma(u)}dt^2 - \mathrm{e}^{2\alpha(u)}du^2. \tag{3.88}$$

(If the metric depends on other spatial coordinates, we suppose that they are fixed.) Let the surface $u = u^*$ be a horizon, so that $\mathrm{e}^{\gamma(u^*)} = 0$. Let us introduce two "radial" coordinates r_* and x by the relations

$$\begin{aligned} dr_* &= \pm\,\mathrm{e}^{\alpha-\gamma}du \\ &\Rightarrow ds_2^2 = \mathrm{e}^{2\gamma}(dt^2 - dr_*^2), \quad r_* = \text{"tortoise"}. \end{aligned} \tag{3.89}$$

$$\begin{aligned} dx &= \pm\,\mathrm{e}^{\alpha+\gamma}du, \\ &\Rightarrow ds_2^2 = \mathrm{e}^{2\gamma}dt^2 - \mathrm{e}^{-2\gamma}dx^2, \quad x = \text{quasiglobal}. \end{aligned} \tag{3.90}$$

As $u \to u^*$, at the horizon $r_* \to \pm\infty$ [it is easy to verify that a finite $r_*(u^*)$ would mean a singularity, see (3.7)], while the coordinate x can tend both to a finite and to an infinite limit. As we have already seen in Sec. 3.3, if $x \to \pm\infty$ as $u \to u^*$, the space–time under consideration is geodesically complete at $u \to u^*$ even without an extension beyond the horizon, and such a "remote horizon" is a natural boundary of the space–time. No extension is necessary in this case.

Let us address the more general case $|x(u^*)| < \infty$. Without losing generality we suppose that $x(u^*) = 0$, $r_* \to -\infty$ and $x \to +0$ as $u \to u^*$. Besides, we assume that in some finite neighborhood of $u = u^*$

$$\mathrm{e}^{2\gamma} = x^q\, F(x), \tag{3.91}$$

where $q = \text{const} \geq 1$, and $F(x)$ is an analytic function with a finite value $F(0)$. It is precisely this behavior of the metric that takes place in most situations of physical interest.

By analogy with the method of obtaining the Kruskal picture, we first introduce the null coordinates

$$V = t + r_*, \quad W = t - r_*. \tag{3.92}$$

The limit $V \to -\infty$ at fixed finite W corresponds to the past horizon (since $t \to -\infty$) while the limit $W \to \infty$ at fixed finite V to a future horizon ($t \to +\infty$). Further, if one introduces new null coordinates v and w related to V and W by

$$V = V(v), \quad W = W(w), \tag{3.93}$$

then, under some simple requirements on these functions, the mixed coordinate patch (v, W) will cover the past horizon $V \to -\infty$ while the (V, w) patch will cover the future horizon $W = +\infty$.

Indeed, consider the future horizon. It is easy to verify that a finite value of the metric coefficient g_{Vw} at $w = 0$ is achieved if, as $W \to \infty$,

$$w \sim \begin{cases} \mathrm{e}^{-WF(0)/2}, & (q = 1), \\ W^{-q/(q-1)}, & (q > 1). \end{cases} \tag{3.94}$$

A transition across the future horizon corresponds to a smooth transition of the w coordinate through zero.

On the other hand, it is also easy to verify that, under the assumptions made, as $x \to -\infty$, the following relationship between x and r_* is valid:

$$x \sim \begin{cases} \mathrm{e}^{r_* F(0)/2} & (q = 1), \\ |r_*|^{-1/(q-1)} & (q > 1). \end{cases} \tag{3.95}$$

Consequently, according to (3.91), at transitions through horizons *the coordinate x behaves in precisely the same way as w*, i.e., smoothly crosses the zero value. Thus precisely this choice of the static spatial coordinate makes it possible to describe the region beyond the horizon, despite the impossibility of describing the whole space–time in the framework of the static reference frame. It is for this reason that the coordinate x is called quasiglobal.

Depending on the parity of q, the region beyond the horizon will be an R- or T-region.

The extension through the past horizon is carried out in absolutely the same way with the aid of the (v, W) patch and leads to similar results.

Remark. It often happens in specific problems that it is hard to perform a transition to the quasiglobal coordinate in the whole region. For instance, it can be necessary to solve a transcendental equation. Let us note in this connection that for the above reasoning it is only significant that the coordinate x satisfies the relation (3.90) *near the horizon*, remaining arbitrary in the remaining space. This makes the choice much easier.

We have seen that an extension of the metric across a horizon can be described in terms of a certain distinguished coordinate (w or x), which changes its sign at such a transition. If, however, the number q is fractional in the representation (3.91), then at the values $x < 0$ the metric coefficients lose their meaning. Hence the candidate horizon is actually a space–time singularity. Such a singularity is connected with *a loss of analyticity of the metric*, and it does not necessarily imply an infinite value of any algebraic invariant of the curvature tensor. In such cases, evidently, some *differential* invariants, containing derivatives of the form $(\nabla_\alpha \nabla_\beta \ldots) R_{\mu\nu\rho\sigma}$, are singular.

We will sometimes call such surfaces *singular horizons*.

This "quantization" condition $q \in \mathbb{N}$ can turn out to be not unique if fractional powers of x appear in other metric coefficients besides g_{xx} and g_{tt}. Such examples are really sometimes considered in the literature.

3.5.2 *Construction of Carter–Penrose diagrams*

As follows from the above, for an analysis of the global causal structure of two-dimensional Lorentzian surfaces with the metric (3.88), it is convenient to use the coordinate x, specified by (3.90) and connected with the null coordinates providing a smooth extension of the metric. Thus, suppose that under some choice of the coordinates, the metric of a two-dimensional section of space–time has the form

$$ds^2 = A(x)dt^2 - \frac{dx^2}{A(x)}, \tag{3.96}$$

where the coordinate x belongs to some finite or infinite range $a < x < b$ on the real axis. According to (3.89), $r_* = \pm \int dx/A(x)$. Then, using the same reasoning as was used above in the discussion of the global properties of the Schwarzschild space–time, it is not difficult to verify that in some coordinate system one can juxtapose the end points a and b and zeros ($x = x_i$) of the function $A(x)$ to the following graphic images (assuming that there are zeros of $A(x)$ not coinciding with a or b):

$x \to a, \quad x \to b$:

$A(x) > 0, \ |r_*| < \infty$

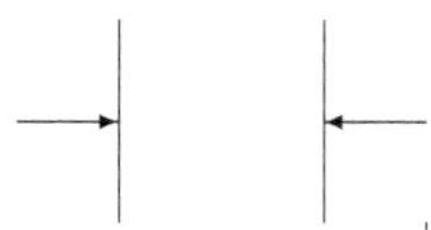

$A(x) < 0, \ |r_*| < \infty$

$x \to a, \ x \to b, \ x \to x_i$:

$A(x) \to +0, \ |r_*| \to \infty$

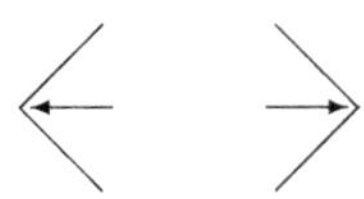

$A(x) \to -0, \ |r_*| \to \infty$

(the arrows mark the direction along which we approach the point a, b or x_i). The construction process begins from any of the end points. To each interval where the function $f(x)$ has a constant sign there corresponds the interior of a triangle or a square in the diagram. A horizon $x = x_i$ is depicted by a pair of tilted segments forming a right angle. To carry out a transition across this horizon, one should add two more tilted segments of the same length as shown in Fig. 3.4. The way of adding new segments is chosen depending on the sign of

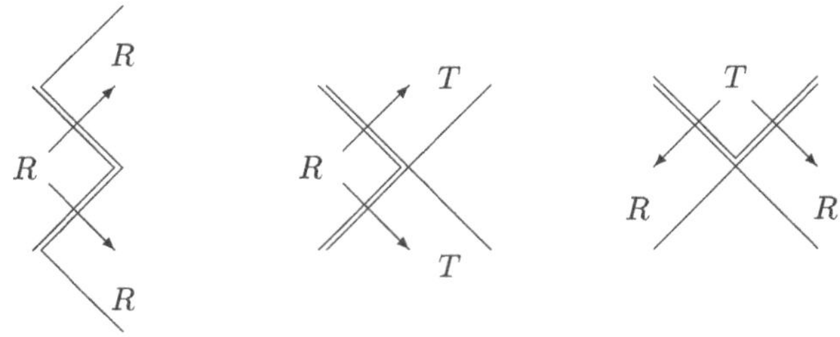

Figure 3.4 Examples of transitions across horizons. The double tilted segments show the horizon of the initial region, single segments show new segments added at the transition, and the arrows show the direction of a transition. The symbol R designates a region where $A(x) > 0$ (an R-region), the symbol T a region where $A(x) < 0$ (a T-region). In a few subsequent figures the notations are the same.

$A(x)$ at $x < x_i$ and $x > x_i$. The newly added pair of segments corresponds to the same value $x = x_i$.

One can formulate a convenient rule for transitions through horizons: **at RT and TR transitions** (i.e., from a R-region to a T-region or vice versa) **the old and new segments form an X-shaped figure while at RR and TT transitions the figure is like a zigzag**. An explanation of the rule is that a horizon with a given value of x_i is located in the diagram to the right or to the left of a R-region (and splits into a past horizon and a future horizon) whereas with respect to a T-region a horizon is located either above or below it.

The construction can continue endlessly or terminate if the diagram covers a finite area whose boundary points correspond to the boundary values $x = a$ and $x = b$, which (recall) can be finite or infinite.

For each pair of regions separated by a tilted segment (i.e., a horizon) there exists a common system of null coordinates like (3.92). The rules formulated above are sufficient for building any two-dimensional Carter–Penrose diagram as soon as the metric is written in the form (3.96). To illustrate the method, let us give examples of diagrams for the RNdS metric, for which, in the notations of (3.96),

$$A(r) = 1 - 2m/r + q^2/r^2 - \Lambda r^2/3, \tag{3.97}$$

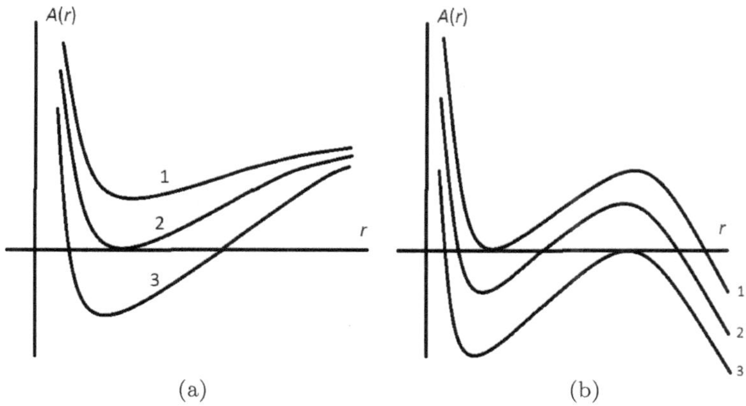

Figure 3.5 Qualitative behavior of the metric function $A(r)$ (3.97): (a) with $\Lambda = 0$ (the Reissner–Nordström metric) in the cases $m < |q|$, $m = |q|$ and $m > |q|$, curves 1, 2, 3, respectively; (b) with $0 < \Lambda q^2 < 1/4$, containing two or three horizons.

where m and q are the mass and the charge, respectively, in geometrized units. The coordinate x introduced in (3.90) coincides here with the (Schwarzschild) radial coordinate r.

The behavior of the function $A(r)$ at $r \in (0, \infty)$ is rather diverse and depends on the values of the parameters m, q, Λ. Figure 3.5 shows the qualitative form of $A(r)$ in the cases $\Lambda = 0$ (the Reissner–Nordström asymptotically flat metric) and $\Lambda > 0$ where it has either three simple roots or one simple root and one double root. Figures 3.6–3.9 show the corresponding Carter–Penrose diagrams. In the simpler cases, where $A(r)$ has a single root or two simple roots, the space–time structure is well known, see, e.g., the books [285, 409]. With the aid of the rules given above, the reader can easily build the corresponding diagrams in these and other cases.

In all Figs. 3.7–3.9, the tilted segments depict the Killing horizons whose corresponding roots r_i of the function $A(r)$ are enumerated in increasing order. The squares excluded from the diagram in Fig. 3.8 as well as the segments drawn by thick lines in Figs. 3.7 and 3.9 can play the role of branching points (a trip around any of then does not necessarily lead to the same initial region), therefore any

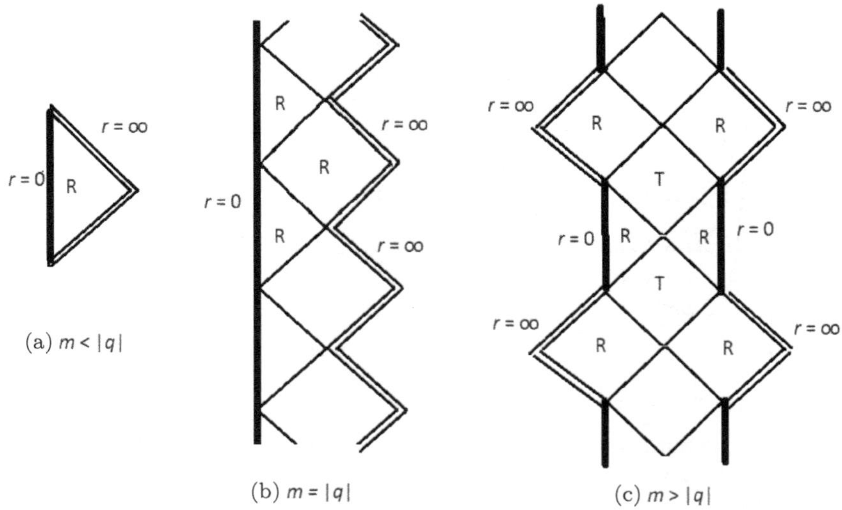

Figure 3.6 The Carter–Penrose diagrams of the Reissner–Nordström space–times in the cases (a) $m < |q|$, no horizons, (b) $m = |q|$, a single exremal horizon, and (c) $m > |q|$, two simple horizons. Both diagrams (b) and (c) are infinitely extended up and down.

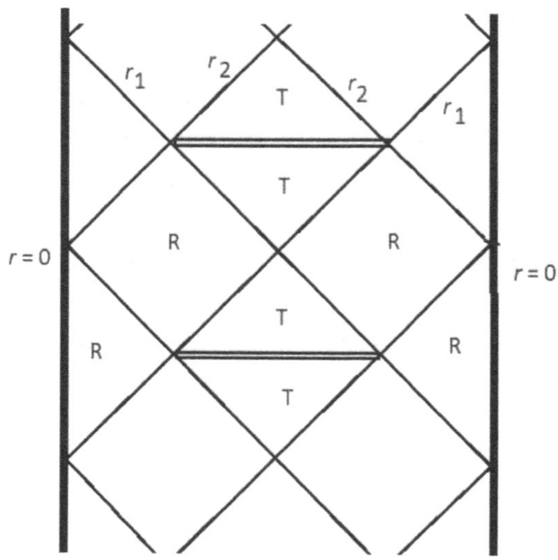

Figure 3.7 The Carter–Penrose diagram corresponding to curve 1 for $A(r)$ in Fig. 3.5. The diagram continues indefinitely both up and down. The double lines correspond to $r = \infty$.

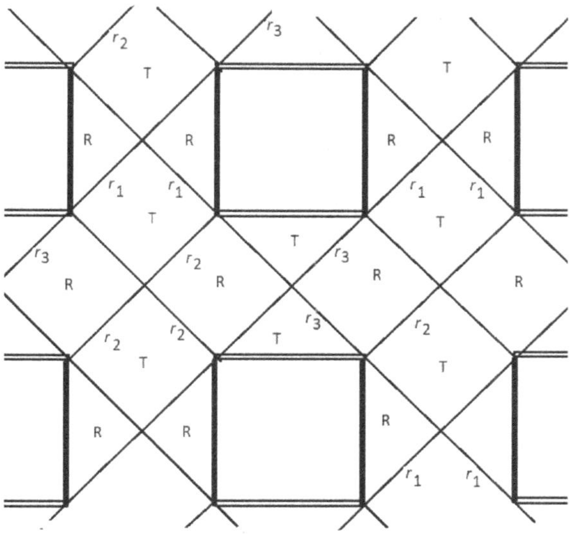

Figure 3.8 The Carter–Penrose diagram corresponding to curve 2 for $f(r)$ in Fig. 3.5. The diagram continues indefinitely in all directions and occupies the whole plane except the squares bounded by double and thick lines. The double lines correspond to $r = \infty$, the thick ones to $r = 0$.

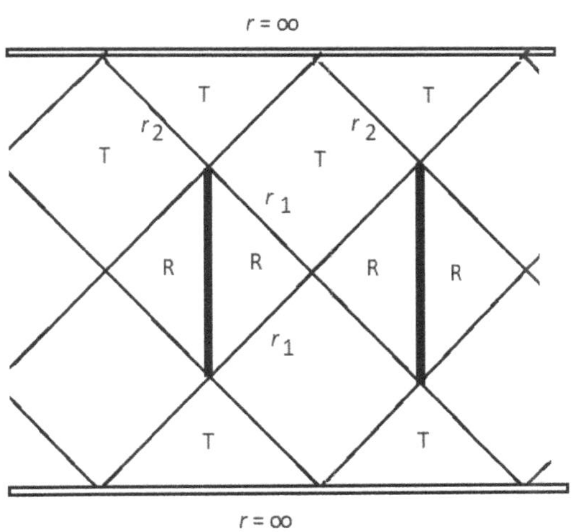

Figure 3.9 The Carter–Penrose diagram corresponding to curve 3 for $f(r)$ in Fig. 3.5. The diagram continues indefinitely to the left and to the right. The thick lines correspond to $r = 0$.

multisheet extensions of these diagrams are possible. On the other hand, identifications of isometric hypersurfaces are also possible.

3.6 A black hole as a result of gravitational collapse

As is known from astrophysics, as soon as the nuclear fuel of a star is exhausted, it experiences a catastrophic contraction (collapse) due to gravitational forces which can no longer be balanced by hydrodynamic and radiative pressure. The result of a gravitational collapse depends on which mass is taking part in it (generally speaking, it is not the star as a whole but only its inner part, its core). At masses smaller than the Chandrasekhar limit (about 1.4 $M_\odot$, where $M_\odot$ is the solar mass), the stellar inner core becomes a white dwarf with a density of the order of $10^6\,\mathrm{g{\cdot}cm^{-3}}$. At $M > 1.4\ M_\odot$, after gravitational collapse there emerges a neutron star whose density is of the order of nuclear matter density, (10^{13}–$10^{15}\,\mathrm{g{\cdot}cm^{-3}}$), while at still larger masses the mass of the collapsing core exceeds the stability limit of neutron stars (about 3–4 $M_\odot$, its precise value depends on the equation of state of neutron matter, on the structure details of the neutron star and on the choice of the theory of gravity), and the contraction leads to BH emergence. As shown by calculations, a BH should form if the initial stellar mass is $\gtrsim 10\ M_\odot$; in this process, more than half of the stellar mass is thrown away into ambient space, which appears to a distant observer as a supernova explosion.

BHs with a wide range of mass values could form from collapsing density fluctuations at very early stages of the Universe evolution, beginning with very small ones. On the other hand, observations have shown that BHs in galactic cores can be as heavy as up to a few billion solar masses [300, 332]. Some GW events seem to indicate the existence of BHs of $M \gtrsim 80\ M_\odot$ [541]. In addition to gravitational collapse, the BH mass growth mechanisms are accretion and capture of ambient bodies as well as BH mergers.

Calculations of the gravitational collapse are very complicated, there are only a few exactly solvable models (such as, for instance, the collapse of a spherically symmetric dust cloud [531]; see, e.g., [366]), while in more complex cases the problem is solved only numerically.

Still there are some general, model-independent features of gravitational collapse ending with BH formation.

3.6.1 *Internal and external regions — Birkhoff's theorem*

The main feature of gravitational collapse is that outside the contracting body the space-time remains vacuum, hence all the results obtainable for the properties of vacuum solutions to the equations of gravity are valid. Moreover, if a spherically symmetric body is collapsing, the whole process occurs with spherical symmetry, and then Birkhoff's famous theorem holds [50],[6] whose simplified formulation is "in GR, a spherically symmetric gravitational field in vacuum is necessarily static". It then reduces to the Schwarzschild solution (or Schwarzschild–de Sitter if the cosmological constant is included). From a more general and correct viewpoint, the theorem indicates the cases where the field equations, under certain conditions, induce an additional space–time symmetry that was not postulated initially. For example, for a T-region of the same Schwarzschild space–time, what follows from the Birghoff theorem is certainly not staticity (which is manifestly absent) but homogeneity, or independence of the metric on the spatial coordinate which is "inherited" from the temporal coordinate t in the static region.

Let us present a schematic proof of Birkhoff's theorem for the simplest case of spherical symmetry [the metric (3.1)] and vacuum,

[6] The theorem was actually proved earlier by J. Jebsen [308], but it has happened that his work is less well-known than Birkhoff's; see also [309]. Later on the theorem was extended to spherically symmetric systems with a cosmological constant Λ, systems with electromagnetic and scalar fields, some scalar–tensor theories (STT) of gravity, etc. The most general approach to the problem [122] consisted in studying the general conditions under which one could prove the static or homogeneous nature of the system. This included into consideration not only all the previously studied cases of the theorem in GR and STT but also many new ones. The theorem was generalized in two respects: inclusion of new symmetry types (planar, cylindrical, pseudoplanar) and new kinds of matter sources of gravity (scalar fields, gauge fields, perfect fluid, etc.). A multidimensional generalization of this approach is described in [130].

$T^\nu_\mu \equiv 0$. According to the Einstein equations, $R^\nu_\mu \equiv 0$. Suppose that the inequality (3.59) holds in the space–time region under consideration, that is, we can choose $r = e^\beta$ as a spatial coordinate. Then in the expressions (3.3) for the Ricci tensor components one can put $\dot\beta = 0$. The equation $R_{01} = -2\dot\alpha\beta' = 0$ then immediately leads to $\dot\alpha = 0$ [$\beta' \neq 0$ due to (3.59)], and among the three metric functions only γ can still be time-dependent. The Einstein equation $G^1_1 = R^1_1 - \frac{1}{2}R = 0$ contains γ only in the combination $e^{-2\alpha}\beta'\gamma'$, whence it follows that $\gamma' \equiv d\gamma/dr = \gamma'(r)$, and, $\gamma = \gamma_1(r) + \gamma_2(t)$. But the function $\gamma_2(t)$ can be excluded by a proper choice of the time coordinate, which leads to a completely static metric. The theorem is proved.[7]

The validity of Birkhoff's theorem is closely related to the fact that in GR the gravitational field has spin 2 and consequently it does not have dynamic monopole and dipole degrees of freedom. In other words, there are no monopole and dipole GWs.

As follows from Birkhoff's theorem, in the collapse of spherical bodies, the exterior space–time is unique and is described by the Schwarzschild metric at all times. From the properties of the metric it follows that, from the viewpoint of a distant observer, the stellar surface contracts to the horizon for an infinitely long time — it seems to be frozen at the Schwarzschild radius $r = 2m$. However, as follows from calculations, deflections from $r = 2m$ decay exponentially in the distant observer's time, with a characteristic time of about $2m$, which is about 10μs (the time for which light passes $2m = 3\,\text{km}$) in the case of $m = M_\odot$ [463]. Very quickly it happens that signals from the collapsing stellar surface stop coming, and it disappears from the sight of a remote observer almost instantaneously.

[7]If, in the conditions of the theorem, we replace the inequality (3.59) with its opposite, i.e., assume that we are in a T-region, then, by the same reasoning, we shall find out that the metric is again t-independent, but t is now a spatial coordinate, and the independence from it proves the space–time homogeneity. However, if one assumes that the gradient of r is null, the situation becomes slightly ambiguous and is discussed in detail in [114]. Let us also mention geometric proofs of some extensions of Birkhoff's theorem under certain assumptions on the algebraic class of the Ricci tensor [66, 240].

On the surface itself, only a finite proper time passes at contraction from any finite radius, not only to the horizon but even to the singularity $r = 0$ (recall that the singularity is cosmological in nature).

The global space–time structure for a collapsing star drastically differs from the structure of an "eternal" Schwarzschild BH depicted in Kruskal (Fig. 3.2) and Carter–Penrose (Fig. 3.3) diagrams. Thus, for instance, in Fig. 3.2, from the whole empty space–time there only remains the region to the right of the curve AB that depicts the world line of a point on the stellar surface, while everything to the left of this curve remains uncertain until a dynamic solution to the Einstein equations is known for the stellar interior. There remain only one R-region (the right one) and one T-region (the upper, contracting one).

3.6.2 *Gravitational collapse of a spherical dust cloud*

As an example of a solution describing the interior of a collapsing body, let us consider Tolman's solution [531] describing the evolution of a spherical cloud of dustlike matter in a comoving reference frame. Since for dust particles this is also a geodesic reference frame, the metric can be taken in the synchronous form

$$ds^2 = d\tau^2 - e^{2\lambda(R,\tau)}dR^2 - r^2(R,\tau)d\Omega^2, \tag{3.98}$$

where τ is the proper time along the particle trajectories labelled by different values of the radial coordinate R. The only nonzero component of the SET is $T_0^0 = \rho$, and the Einstein equations read

$$2r\ddot{r} + \dot{r}^2 + 1 - e^{-2\lambda}r'^2 = 0, \tag{3.99}$$

$$\frac{1}{r^2}(1 + \dot{r}^2 + 2r\dot{r}\dot{\lambda}) - \frac{e^{-2\lambda}}{r^2}(2rr'' + r'^2 - 2rr'\lambda') = 8\pi G\rho, \tag{3.100}$$

$$\dot{r}' - \dot{\lambda}r' = 0. \tag{3.101}$$

Equation (3.101) is easily integrated in τ giving

$$e^{2\lambda} = \frac{r'^2}{1 + f(R)}, \tag{3.102}$$

where $f(R)$ is an arbitrary function satisfying the condition $1 + f(R) > 0$. Substituting (3.102) into (3.99), we obtain the equation

$$2r\ddot{r} + \dot{r}^2 = f(R), \tag{3.103}$$

whose first integral is

$$\dot{r}^2 = f(R) + \frac{F(R)}{r}. \tag{3.104}$$

This expression makes clear the physical meaning of the function $f(R)$: since $\dot{r}$ can be loosely understood as the radial velocity of a dust particle, $f(R)$ specifies the initial dust velocity distribution: for $f \geq 0$ it is the particle velocity squared at large r ($f > 0$ and $f = 0$ correspond to hyperbolic and parabolic motion, respectively). If $f(R) < 0$ but $F(R) > 0$ (elliptic motion), the particle cannot reach infinity, and the quantity $F(R)/|f(R)|$ shows the maximum value of r accessible to it.

The meaning of the other arbitrary function, $F(R)$, becomes clear if we substitute (3.102) and (3.104) into (3.100). We obtain

$$\rho = \frac{1}{8\pi G}\frac{F'(R)}{r^2 r'}, \tag{3.105}$$

or

$$F(R) = 8\pi G \int \rho r^2 r' dR. \tag{3.106}$$

Assuming that in the initial configuration (before collapse) there was a regular center, so that $F = 0$ at $r = 0$, we can write

$$F(R) = 2GM(R), \quad M(R) = 4\pi \int_0^r \rho r^2 r' \, dR, \tag{3.107}$$

so that $M(R)$ is the mass function equal to the mass of a spherical body including all matter inside the sphere of given radius r, and $F(R)$ is the corresponding Schwarzschild radius.

Equation (3.104) can be further integrated, and the solution can be written in a parametric form for $f \neq 0$ and explicitly for $f = 0$ [366]:

$$f > 0: \quad r = \frac{F}{2f}(\cosh\eta - 1), \quad \tau_0 - \tau = \frac{F}{2f^{3/2}}(\sinh\eta - \eta); \tag{3.108}$$

$$f < 0: \quad r = \frac{F}{-2f}(1 - \cos\eta), \quad \tau_0 - \tau = \frac{F}{2(-f)^{3/2}}(\eta - \sin\eta); \tag{3.109}$$

$$f = 0: \quad r = \left(\frac{9F}{4}\right)^{1/3}(\tau_0 - \tau)^{2/3}. \tag{3.110}$$

The new arbitrary function $\tau_0(R)$ here indicates the instant of τ at which the particle with a given R reaches the singularity $r = 0$. To find out whether this singularity occurs in a R-region (and can then be called a singular center) or in a T-region (and is then cosmological in nature), we must check whether the gradient of $r(R, \tau)$ is spacelike or timelike at small r. Evidently,

$$r^{,\alpha} r_{,\alpha} = \dot{r}^2 - e^{-2\lambda} r'^2 = -1 - 2r\ddot{r}, \tag{3.111}$$

where the second equality sign follows from (3.99).

In Eq. (3.104), the second term is dominant at small r, therefore, with any $f(R)$ the function $r(R, \tau)$ near the singularity is given by (3.110), whence it follows that $r\ddot{r} \sim -(\tau_0 - \tau)^{-2/3} \to -\infty$, so that $r^{,\alpha} r_{,\alpha} \to +\infty$, and the singularity occurs in a T-region for any choice of the arbitrary functions.

Then, of interest is the question: when does a given dust particle cross the horizon thus getting from R- to T-region? The answer again follows from (3.111): since $\ddot{r} = F(R)/(2r^2)$, the gradient of r is null where $r = F(R) = 2GM(R)$, that is, a particle crosses the horizon precisely at the time instant when $r(R, \tau)$ reaches the value of the Schwarzschild gravitational radius corresponding to the interior of the sphere specified by a given R. Note that such a crossing is quite regular, and this instant is actually unnoticed in the comoving reference frame.

To obtain a global solution including both a dust distribution and an external vacuum region, it is necessary to describe such a vacuum (Schwarzschild) region in a reference frame suitable for smooth matching to the internal solution. This problem is easily solved since the vacuum solution in a proper form is obtained from the Tolman solution by putting $F(R) = \text{const}$ whence $\rho = 0$. The resulting solution reproduces the Schwarzschild metric in a geodesic reference frame as described above in Sec. 3.3.4. A global solution is then represented by the Tolman solution in which $F(R)$ is variable in a certain range of R, say, $R < R_0$, and constant at larger R, so that $f(R) = f(R_0)$.

Thus the spherical collapse of a dust cloud always leads to a BH. One of the most fundamental questions discussed in BH physics is whether or not there are matter configurations whose gravitational collapse can lead to a naked singularity instead of a BH. The so-called cosmic censorship conjecture (Penrose) requires that such unwanted entities as space–time singularities, if any, can only appear separated from external observers by BH horizons. There are counterexamples to censorship, processes where naked singularities do occur as a result of collapse, but it is still not certain whether or not such situations are generic. Moreover, an answer can be different in GR and its extensions; see, in addition to textbooks on gravity and BH physics, the reviews [191, 272, 297, 307, 550].

Chapter 4

Spherical space–times with scalar fields

In the previous chapter, we have been dealing with solutions of general relativity where the possible sources of gravity were a cosmological constant, electromagnetic fields or dustlike matter. We have seen, in particular, that the presence of a massless vector (electric or magnetic) field does not prevent the existence of BHs. In other words, a BH can possess an electric or/and magnetic charge and be a source of the corresponding fields. In this chapter we will discuss the space–time geometries that can emerge due to the possible existence of scalar fields, the simplest possible physical fields, represented by a single invariant function of the space–time coordinates. This circumstance leads to their widest use in physical theories, and it would also be quite strange if Nature avoided their real existence. And indeed, the recent discovery of the Higgs field [177] has proved their reality at least in particle physics. In gravitational physics, scalar fields play crucial roles in the formulation of numerous extensions of GR, in the construction of cosmological models as well as models of local sources of gravity.

The properties of scalar field solutions in GR are of large interest by themselves and in comparison with vacuum and electrovacuum solutions. It turns out to be possible to prove a number of general statements about scalar field configurations and to obtain many exact solutions with scalar fields, to be discussed in this book. We will begin the chapter with a discussion of solutions with a massless scalar field in GR (with a possible presence of electric or magnetic fields) and continue with those with arbitrary self-interaction potentials, for which it is hard to analytically solve the field equations but some important features of the solutions can be established. We will also briefly discuss such well-known extensions of GR as scalar-tensor theories (STT) of gravity, $f(R)$ gravity and hybrid metric-Palatini gravity (HMPG) whose structure is closely related to that of GR with scalar fields, and many results known in GR can be at least partly reformulated in these theories.

4.1 Massless minimally coupled scalar fields in GR

Let us recall that minimally coupled fields in curved space are those with the same form of the action, Lagrangian and equations as in Minkowski space with partial derivatives replaced by covariant ones. Accordingly, we consider the scalar field $\phi = \phi^{\rm min}$ with the Lagrangian

$$L_s = \frac{1}{2}\varepsilon\phi^{,\alpha}\phi_{,\alpha} - V(\phi), \quad \varepsilon = \pm 1, \tag{4.1}$$

where $V(\phi)$ is a self-interaction potential, $\varepsilon = +1$ corresponds to a usual, canonical scalar field while $\varepsilon = -1$ leads to a wrong (negative) sign of kinetic energy, such fields are called *phantom* or *ghost* scalar fields. Let us say a few words about their meaning and emergence in the theory.

Being quite unusual, such fields are often considered as purely artificial entities. However, such fields have recently attracted much interest in connection with the Dark Energy (DE) problem, a search for models able to account for the observed growing acceleration of the Universe expansion (for more details see Part II of this book, in particular, the footnote on page 322).

The main objection against a phantom field is that a field with a negative kinetic term can have an arbitrarily large negative energy of high-frequency oscillations, which, according to quantum field theory, can lead to runaway production of particles and antiparticles accompanied by addition of equal negative energy of the phantom field itself. Nothing of this kind is observed, which casts serious doubt on possible existence of phantom fields.

On the other hand, there are theoretical reasons for considering phantom fields since they naturally appear in STT and HMPG (to be discussed later in this chapter), in numerous models of string theory [495], supergravities [423] and theories in more than 11 dimensions like F-theory [330] and in some Kaluza–Klein theories (e.g., those discussed in Part III of this book). They can be effective fields, following in some reduction procedure from an underlying theory; they may not be subject to quantization (at least directly) and possibly do not interact with other fields. This, at least partly, counters the above objections.

Even not being a steady theoretical object, phantom fields seem useful for a kind of *gedanken experiments*, they can lead to models or predictions of interest which can afterwards be obtained by other, less exotic means. It is the main reason for considering them in this book.

4.1.1 *Solutions with a massless scalar field*

Consider configurations with the static metric (3.1)

$$ds^2 = e^{2\gamma(u)}dt^2 - e^{2\alpha(u)}du^2 - e^{2\beta(u)}d\Omega^2 \tag{4.2}$$

in the presence of a massless scalar field with the Lagrangian (4.1), $V \equiv 0$, and an electromagnetic field $F_{\mu\nu}$ with the Lagrangian

$$L_{\rm em} = -\frac{1}{4}F^{\mu\nu}F_{\mu\nu}. \tag{4.3}$$

The most general form of the electromagnetic potential compatible with spherical symmetry is

$$A_\mu = \delta^0_\mu A(u) + \delta^3_\mu q_m \cos\theta + \partial\Phi/\partial x^\mu, \tag{4.4}$$

where $\Phi(x^\mu)$ is an arbitrary function and q_m = const is the magnetic charge. Accordingly, the SETs of the fields ψ and $F_{\mu\nu}$ have the form

$$T^\nu_\mu[\phi] = \frac{1}{2}\varepsilon\, e^{-2\alpha}\phi'^2 \,{\rm diag}(1,-1,1,1); \tag{4.5}$$

$$T^\nu_\mu[F] = 2(\mathbf{E}^2 + \mathbf{B}^2)\,{\rm diag}(1,1,-1,-1), \tag{4.6}$$

where

$$\mathbf{E}^2 = F^{01}F_{10} = q_e^2/r^4; \quad \mathbf{B}^2 = F^{23}F_{23} = q_m^2/r^4, \tag{4.7}$$

E are **B** are the electric field strength and the magnetic induction, and

$$F^{01} = q_e\,\mathrm{e}^{-\alpha-\gamma}/r^2, \quad F_{10} = q_e\,\mathrm{e}^{\alpha+\gamma}/r^2, \tag{4.8}$$

where q_e is the electric charge. Let us denote $Q^2 = \frac{1}{2}\varkappa(q_e^2 + q_m^2)$.

The equation for ϕ and its first integral read

$$(\mathrm{e}^{\gamma+2\beta-\alpha}\phi')' = 0, \quad \phi' = \bar{C}\,\mathrm{e}^{\alpha-2\beta-\gamma}, \tag{4.9}$$

where the integration constant $\bar{C}$ is interpreted as a scalar charge.

The tensors (4.5) and (4.6) have the property $T_1^1 + T_2^2 = 0$. In this case it is helpful to use the harmonic ($\Box u = 0$) radial coordinate u [75], such that the metric coefficients in (3.1) satisfy the condition (see Sec. 3.1)

$$\alpha = 2\beta + \gamma. \tag{4.10}$$

Then the combination $\binom{1}{1} + \binom{2}{2}$ of the Einstein equations reduces to the Liouville equation

$$(\beta + \gamma)'' = \mathrm{e}^{2\beta+2\gamma}, \tag{4.11}$$

which is easily integrated giving

$$\mathrm{e}^{-\beta-\gamma} = s(k,u) := \begin{cases} k^{-1}\sinh ku, & k > 0; \\ u, & k = 0; \\ k^{-1}\sin ku, & k < 0; \end{cases} \tag{4.12}$$

where k is an integration constant; and one more integration constant has been removed by choosing the zero point of u. As a result, we arrive at the metric

$$ds^2 = \mathrm{e}^{2\gamma}dt^2 - \frac{\mathrm{e}^{-2\gamma}}{s^2(k,u)}\left[\frac{du^2}{s^2(k,u)} + d\Omega^2\right]. \tag{4.13}$$

Without loss of generality we assume $u \geq 0$, so that $u = 0$ corresponds to spatial infinity at which the metric is asymptotically flat, and we can put in addition $\gamma(0) = 0$ under a proper choice of scale along the time axis t.

At the asymptotic ($u \to 0$), the coordinate u behaves as $1/r$, so that

$$r = \mathrm{e}^{\beta} \approx \frac{1}{u}; \quad \mathrm{e}^{\alpha} \approx \frac{1}{u^2}; \quad \gamma' \to \gamma'(0) = -GM = -m, \tag{4.14}$$

where M is the Schwarzschild mass, as follows from comparing the limit of large r (4.14) with the Schwarzschild metric.

It should be noted that many exact static, spherically symmetric solutions obtained by now refer to systems with $T_1^1 + T_2^2 = 0$ (or can be reduced to them) and can therefore be found using the method described here.

The field equation for ϕ in the metric (3.1) with the coordinate (4.10) gives

$$\phi = \bar{C}u + \phi_0, \quad \phi_0 = \text{const}. \tag{4.15}$$

Let us pass on, in addition to Q, to geometrized mass and scalar charge: $m = GM$ and $C = \sqrt{\varkappa}\bar{C}$ having the dimension of length.

Among the remaining Einstein equations only one is independent (corresponding to one remaining unknown function $\gamma(u)$), but it makes sense to write down two equations: the component $\binom{0}{0}$ from (3.6), having the Liouville form, and the component $\binom{1}{1}$ from (3.5), which does not contain second-order derivatives and is a first integral of the other equations:

$$\mathrm{e}^{2\alpha} R_0^0 = -\gamma'' = -Q^2\,\mathrm{e}^{-2\gamma}, \tag{4.16}$$

$$\mathrm{e}^{2\alpha} G_1^1 = \beta'(\beta' + 2\gamma') = \frac{1}{2}\varepsilon\phi'^2 - Q^2\,\mathrm{e}^{2\gamma}. \tag{4.17}$$

Integrating (4.16), we arrive at the following metrics:

(a) for $Q \neq 0$:

$$ds^2 = \frac{dt^2}{Q^2 s^2(h, u + u_1)} - \frac{Q^2 s^2(h, u + u_1)}{s^2(k, u)}\left[\frac{du^2}{s^2(k, u)} + d\Omega^2\right]; \tag{4.18}$$

with the constants k, h, u_1, C, Q connected with each other and the mass m by the relations

$$\begin{aligned} &s^2(h,u_1) = 1/Q^2; \\ &m^2 - Q^2 = h^2 \operatorname{sign} h = k^2 \operatorname{sign} k - \varepsilon C^2/2, \end{aligned} \tag{4.19}$$

where the first equality follows from the boundary condition $\gamma(0) = 0$, and the second one from substituting the expressions (4.12) for $\beta - \gamma$ and the solution to (4.16) into (4.17);

(b) for $Q = 0$:

$$\begin{aligned} ds^2 &= \mathrm{e}^{-2hu} dt^2 - \frac{\mathrm{e}^{2hu}}{s^2(k,u)} \left[\frac{du^2}{s^2(k,u)} + d\Omega^2 \right]; \\ Gm &= h; \quad k^2 \operatorname{sign} k = \varepsilon C^2/2 + h^2. \end{aligned} \tag{4.20}$$

The function s is defined by Eq. (4.12) in all cases.

In the limit $C \to 0$, the solution (4.15)–(4.19) passes on to the Reissner–Nordström electrovacuum solution and as $Q \to 0$ to the solution (4.20); in turn, (4.20) passes on, as $C \to 0$, to the Schwarzschild vacuum solution. In other coordinates the solution (4.20) (naturally, only for the normal sign of scalar field energy, $\varepsilon = 1$) was found for the first time by I.Z. Fisher in 1948 [243] and was repeatedly rediscovered later; the solution (4.15)–(4.19) was found by R. Penney in 1969 [444]. For $\varepsilon = -1$, the solution (4.20) was obtained in [47], and (4.15)–(4.19) in [75].

Let us briefly discuss the properties of solutions for $\varepsilon = +1$.

In the electrically neutral case (4.20) (Fisher's solution), the harmonic coordinate u varies in the range $u > 0$, and $u \to \infty$ corresponds to a naked singularity, which is attractive (since $g_{00} \to 0$) for $m > 0$. The metric (4.18) behaves in a similar way at $h \geq 0$, i.e., at small charges as compared to the mass in proper units, $Q^2 \leq m^2$. The singularity may be characterized as scalar field dominated because there $\phi \to \infty$, and the total scalar field energy E_s is infinite whereas the electromagnetic field energy E_e is finite. The corresponding Reissner–Nordström electrovacuum metric possesses horizons (two for $Q^2 < m^2$ and one for $Q^2 = m^2$), which, as we see, disappear if there emerges a scalar field, however small the scalar charge C is.

At $h < 0$ (a large charge, $Q^2 > m^2$), the coordinate u is changing in the range $0 < u < u_{\max}$, where $u_{\max}$ is the closest zero of the function $\sin[|h|(u + u_1)]$. The value $u = u_{\max}$ corresponds to the central naked repulsive ($e^\gamma \to \infty$) singularity of Reissner–Nordström type, dominated by the electromagnetic field ($\phi < \infty$, $E_s < \infty$, $E_e = \infty$).

We see that in all cases the solution with a scalar field contains a naked singularity. Thus at least in spherically symmetric space–times the massless field $\phi^{\min}$ cannot be an external field of a BH. This statement, as we will see below, is one of the special cases of the so-called no-hair theorems for BHs. Such a terminology traces back to J.A. Wheeler whose famous phrase "Black holes have no hair" meant that BHs have no external parameters other than mass, electric (and/or magnetic) charge and angular momentum.

4.1.2 *Phantom scalar fields — the anti-Fisher solution*

As we saw, Fisher's solution consists of a single branch and has an attracting naked singularity. Its electrically charged counterpart consists of three branches depending on the mass-to-charge ratio.

In the case of phantom fields, the solution behavior is more diversse, and in particular, there emerge wormhole solutions. From a formal viewpoint, this is connected with the fact that in the solutions (4.18)–(4.20) one can obtain either $k < 0$ at $h \geq 0$ or $|k| > |h|$ if both $h < 0$ and $k < 0$. In these cases, at appropriate values of the constant u_1, the solution is defined on the interval $0 < u < \pi/|k|$, and at its both ends there is a flat spatial infinity. From a physical viewpoint, the emergence of wormholes is connected with violation of the Null Energy Condition $T_{\mu\nu}\xi^\mu\xi^\nu \geq 0$, $\forall\xi^\mu : \xi^\nu\xi_\mu = 0$ — see Chapters 2 and 6.

As an example, let us consider in more detail the properties of the scalar-vacuum solution (4.20) for $\varepsilon = -1$, the phantom analogue of Fisher's solution, which, by analogy with "anti-de Sitter", we suggest to call "the anti-Fisher solution."

Branch A: $k > 0$. In this case it is helpful to pass on to the quasiglobal coordinate x by the transformation

$$e^{-2ku} = 1 - \frac{2k}{x} \equiv P(x), \tag{4.21}$$

and the solution takes the form

$$\begin{aligned} ds^2 &= P^a dt^2 - P^{-a} dx^2 - P^{1-a} x^2 d\Omega^2, \\ \phi &= -\frac{C}{2k} \ln P, \end{aligned} \tag{4.22}$$

with the constants related by

$$a = h/k, \quad a^2 = 1 + C^2/(2k^2). \tag{4.23}$$

In the case $h < 0$, i.e., $m < 0$, we have $a < -1$, and, just as in the Fisher solution, a repulsive central singularity at $x = 2k$ corresponding to $u = \infty$.

(It is easy to verify that the same substitution (4.21) applied to Fisher's solution leads to the same form (4.22) of the metric and the scalar field, but with $|a| < 1$.)

The situation is, however, drastically different for $m > 0$, or $a > 1$. Indeed, the spherical radius r then has a finite minimum at $x = x_{\rm th} = (a+1)k$, corresponding to a throat of the size

$$r(x_{\rm th}) = r_{\rm th} = k(a+1)^{(a+1)/2}(a-1)^{(1-a)/2}, \tag{4.24}$$

and tends to infinity as $x \to 2k$. Moreover, for $a = 2, 3, \ldots$ the metric exhibits a horizon of order a at $x = 2k$ and admits a continuation to smaller x. A peculiar feature of such horizons is their infinite area. Such configurations have been named "cold black holes" (CBHs) [95, 97] since all of them have zero Hawking temperature.

Furthermore, all Kretschmann scalar constituents K_i [see (3.7)] behave as P^{a-2} as $x \to 2k$ and $P \to 0$. An exception is the value $a = 1$, in which case $C = 0$, $\phi = 0$, and the Schwarzschild solution is reproduced. Hence, at $x = 2k$ the metric has a curvature singularity if $a < 2$ (except for $a = 1$), a finite curvature if $a = 1$ or $a = 2$ and zero curvature if $a > 2$.

For noninteger $a > 2$, the qualitative behavior of the metric as $x \to 2k$ is the same as that near a horizon of infinite area, but a continuation beyond it is impossible due to nonanalyticity of the function $(1 - 2k/x)^a$ at $x = 2k$. Since geodesics terminate there at a finite value of the affine parameter, this is a space-time singularity (a singular horizon, as it has been called in Chapter 3) even though the curvature invariants tend there to zero.

Branch B: $k = 0$: the solution is defined in the range $u \in \mathbb{R}_+$ and is rewritten in terms of the quasiglobal coordinate $x = 1/u$ as follows:

$$ds^2 = \mathrm{e}^{-2h/x} dt^2 - \mathrm{e}^{2h/x}[dx^2 + x^2 d\Omega^2], \quad \phi = C/x. \tag{4.25}$$

As before, $x = \infty$ is a flat infinity, while at the other extreme $x \to 0$, the behavior is different for positive and negative mass $m = h/G$. Thus, for $m < 0$, $x = 0$ is a singular center ($r = 0$ and all K_i are infinite). On the contrary, for $m > 0$, $r \to \infty$, and all $K_i \to 0$ as $x \to 0$. It is again a singular horizon: despite the vanishing curvature, the nonanalyticity of the metric in terms of x makes its continuation impossible.

Branch C: $k < 0$: the solution describes a wormhole with two flat asymptotes at $u = 0$ and $u = \pi/|k|$. The metric has the form

$$\begin{aligned} ds^2 &= \mathrm{e}^{-2hu} dt^2 - \frac{k^2\, \mathrm{e}^{2hu}}{\sin^2(ku)} \left[\frac{k^2\, du^2}{\sin^2(ku)} + d\Omega^2\right] \\ &= \mathrm{e}^{-2hu} dt^2 - \mathrm{e}^{2hu}[dx^2 + (k^2 + x^2) d\Omega^2], \end{aligned} \tag{4.26}$$

where u is expressed in terms of the quasiglobal coordinate x, defined on the whole real axis $\mathbb{R}$, by $|k|u = \cot^{-1}(x/|k|)$. This metric is further discussed in Chapter 6.

4.1.3 *Cold black holes in the anti-Fisher solution*

Among different branches of the anti-Fisher solution, of considerable interest is the case of cold BHs. Let us briefly discuss their structure and properties.

For odd a, the principal geometric and causal properties, including the Carter–Penrose diagram, coincide with those of the Schwarzschild metric. Thus, at $x < 2k$, x is a temporal coordinate, t spatial, and the space–time is homogeneous and anisotropic, corresponding to the Kantowski–Sachs type of anisotropic cosmologies. The singularity at $x = 0$ ($r = 0$) is spacelike (cosmological) and is reached by all timelike geodesics in a finite time interval after crossing the horizon.

For even a, the Penrose diagram is the same as that of the extreme Reissner–Nordström space-time; however, the physical meaning of the regions where $x < 2k$ is quite different. Since g_{22} and g_{33} change their signs at the horizon, the metric at $x < 2k$ has the signature $(- + ++)$ instead of $(+ - --)$ at large x. The Lorentzian nature of space-time is still preserved, and one can verify that all geodesics are continued smoothly from one region to the other (the geodesic equations depend only on the Christoffel symbols and are invariant under the anti-isometry $g_{\mu\nu} \to -g_{\mu\nu}$). The time coordinate in that region is x since $g_{xx} < 0$ while the other diagonal components of $g_{\mu\nu}$ are positive. Thus, just as for odd a, we have a Kantowski–Sachs type cosmology with a spacelike singularity at $x = 0$ ($r = 0$). The direction of the arrow of time can be arbitrary there since timelike geodesics that penetrate from the static region become spacelike there (one cannot say for them where is the past and where is the future) and can even avoid the singularity.

The properties of the scalar field are not less exotic. According to (4.22), $\phi \to \infty$ as $x \to 2k$; this, however, does not contradict the regularity of the surface $x = 2k$ for $a \geq 2$ since the energy density

$$T_0^0 = -\frac{1}{2}A\phi'^2 = -\frac{C^2}{2}\frac{(x-2k)^{a-2}}{x^{a+2}}, \tag{4.27}$$

as well as the other components of T_μ^ν are finite there (recall that for $a < 2$ the curvature invariants also diverge, together with T_μ^ν). Thus, the infinite value of ϕ does not prevent the continuation of the space-time manifold to smaller x, where the solution is valid with $\phi = -(C \ln |P|)/(2k)$. On the other hand, the total scalar field energy, calculated as the conserved quantity corresponding to the timelike Killing vector, turns out to be infinite in the static region independently of a,

$$\mathcal{E} = \int T_0^0 \sqrt{g}\, d^3x = -2\pi C^2 \int \frac{dx}{x(x-2k)}, \tag{4.28}$$

and the integral logarithmically diverges at $x = 2k$. The divergence is related to the infinite spatial volume: the integral $\int \sqrt{{}^3g}\, d^3x$ diverges near $x = 2k$ even stronger than (4.28).

4.2 Scalar fields in some extensions of GR

4.2.1 *The general STT, Jordan and Einstein frames*

Among the most popular and widely discussed extensions of GR are scalar-tensor theories (STTs) of gravity, whose formulation rests on different versions of nonminimal couplings between scalar fields and the space–time geometry. This trend probably began with the papers by P. Jordan [310] and was actively continued in the 1960s by C. Brans and R. Dicke [73, 74]. A more general formulation of the theory, called the general (Bergmann–Wagoner–Nordtvedt) STT of gravity [48, 428, 548], to be discussed here, is described by the Lagrangian

$$L_{\rm STT} = f(\phi)\tilde{R} + h(\phi)\tilde{g}^{\mu\nu}\phi_{,\mu}\phi_{,\nu} - 2U(\phi) + L_m, \tag{4.29}$$

specified in a space–time manifold $\mathbb{M}_{\rm J}$ with the metric $\tilde{g}_{\mu\nu}$ and scalar curvature $\tilde{R}$; $U(\phi)$ is a potential, in some cases generalizing a cosmological constant; L_m is the Lagrangian of all other kinds of matter which can in general depend on $\tilde{g}_{\mu\nu}$ and ϕ. It is usually supposed that $f(\phi) > 0$ in order to have a positive effective gravitational constant. The Lagrangian (4.29) contains three arbitrary functions $f(\phi)$, $h(\phi)$ and $U(\phi)$, changing from one particular theory to another, but the arbitrariness may be reduced by choosing a parametrization of the scalar field. Thus, a popular parametrization is $f(\phi) = \phi$, $h(\phi) = \omega(\phi)/\phi$.

Let us present the well-known conformal mapping [548] which formally reduces the broad class of systems with scalar fields in the framework of the Bergmann–Wagoner–Nordtvedt STT of gravity to systems with the scalar field $\phi^{\rm min}$. Consider the Lagrangian of the general STT (4.29) specified in the space-time $\mathbb{M}_{\rm J}$ with the metric $\tilde{g}_{\mu\nu}$ and the scalar curvature $\tilde{R}$ under the assumption that the Lagrangian of the rest of matter L_m does not depend on ϕ, though in general depends on $\tilde{g}_{\mu\nu}$. This formulation, corresponding to the *Jordan conformal frame* (or *picture*) is of fundamental nature in STT since it is in this picture that the SET of matter T^μ_ν satisfies the usual conservation law $\nabla_\nu T^\nu_\mu = 0$, thus providing the standard equations of motion in the atomic system of measurement. In particular, free particles move along geodesics.

The field equations are in general easier studied and solved in the so-called *Einstein conformal frame* (or *picture*) in which the scalar field is minimally coupled to gravity. Namely, the conformal transformation [548]

$$\tilde{g}_{\mu\nu} = \frac{g_{\mu\nu}}{f(\phi)}, \qquad \frac{d\phi}{d\chi} = f\left|fh + \frac{3}{2}\left(\frac{df}{d\phi}\right)^2\right|^{-1/2} \tag{4.30}$$

leads the Lagrangian (4.29) to a Lagrangian typical of GR (up to a full divergence) with $\chi = \phi^{\rm min}$:

$$\begin{aligned} L_E &= R + \varepsilon g^{\mu\nu}\chi^{,\mu}\chi_{,\nu} + \frac{-2U(\phi) + L_m}{f^2(\phi(\chi))}, \\ \varepsilon &:= \text{sign}\left[fh + \frac{3}{2}\left(\frac{df}{d\phi}\right)^2\right] \end{aligned} \tag{4.31}$$

for a system with the scalar field χ in a space-time with the metric $g_{\mu\nu}$. It is usually supposed that the function $f(\phi)$ involved in the transformation (4.30) is finite and positive in the whole region of interest. Since in (4.31) the matter Lagrangian is divided by f^2, the SET of matter is in general not conserved, which means that matter fields or particles in $\mathbb{M}_{\rm E}$ are interacting with the scalar field.

Depending on $\varepsilon = \pm 1$, the STT splits into two classes: normal, or canonical ones, in which $\varepsilon = +1$, i.e., the kinetic energy of the scalar field in $\mathbb{M}_{\rm E}$ is positive, and anomalous or phantom (ghost), $\varepsilon = -1$, where this energy is negative.

Special examples of the theory (4.29) are GR with a minimally coupled scalar field ($f \equiv 1$, $h = \varepsilon = \pm 1$) and a conformally coupled scalar field $\phi = \phi^{\rm conf}$:

$$\begin{aligned} & f = 1 - \varepsilon\phi^2/6, \quad h = \varepsilon, \\ & \sqrt{6}\phi = \tanh\frac{\chi - \chi_0}{\sqrt{6}} \quad \text{for } \varepsilon = 1, \\ & \sqrt{6}\phi = \tan\frac{\chi - \chi_0}{\sqrt{6}} \quad \text{for } \varepsilon = -1, \end{aligned} \tag{4.32}$$

where χ_0 is an arbitrary constant.

The equation of a massless ($U(\phi) \equiv 0$) conformally coupled scalar field

$$\Box\phi + \frac{1}{6}R\phi = 0 \tag{4.33}$$

is invariant under the conformal transformation $\tilde{g}_{\mu\nu} = \Omega^{-2}(x)g_{\mu\nu}$, accompanied by the scalar field transformation $\tilde{\phi} = \Omega\phi$, as one can verify directly using the corresponding transformation formulas for the scalar curvature and the d'Alembert operator:

$$\begin{aligned} \tilde{R} &= \Omega^2 R - 6\Omega\,\Box\,\Omega + 12\Omega^{,\alpha}\Omega_{,\alpha}, \\ \tilde{\Box}\phi &= \Omega^2\,\Box\,\phi - 2\Omega\Omega^{,\alpha}\phi_{,\alpha}. \end{aligned} \tag{4.34}$$

It is easy to see that the transformation (4.30) is different from the one with respect to which Eq. (4.33) is invariant. Therefore, it becomes possible to pass on from Eq. (4.33) to the equation $\Box\chi = 0$.

Conformally coupled scalar fields have been considered in a great number of studies, beginning with those of Penrose [443] (a massless conformally invariant field) and Chernikov and Tagirov [174] (massive conformally coupled fields). The theory (4.29) with a phantom scalar was also discussed in [573] as a possible alternative to GR in astrophysical and cosmological applications.

Another, the most well-known example of an STT is the Brans–Dicke (BD) theory [74] for which

$$f(\phi) = \phi, \qquad h(\phi) = \frac{\omega}{\phi}, \qquad \phi = \exp\frac{\chi - \chi_0}{\sqrt{|\omega + 3/2|}}, \tag{4.35}$$

where $\omega = \text{const}$. This theory is normal for $\omega > -3/2$, phantom for $\omega < -3/2$ and has no dynamics at $\omega = -3/2$. This theory has also been used in a great number of studies in celestial mechanics, in astrophysics and cosmology.

The transition (4.30) is well known as a method of finding exact or approximate solutions to the field equations due to (4.29) since the equations due to (4.31) are simpler than those due to (4.29). An Einstein-frame solution having been found, its Jordan-frame counterpart is easily produced by a transformation inverse to (4.30).

There is, however, an important subtle point: if the function $f(\phi)$ in (4.29) turns to zero or infinity at some value of ϕ, it may happen that a singularity in the Einstein-frame manifold $\mathbb{M}_\text{E}$ with the metric $\bar{g}_{\mu\nu}$ transforms into a regular surface in the Jordan-frame manifold $\mathbb{M}_\text{J}$ with the metric $g_{\mu\nu}$ (or vice versa), and $\mathbb{M}_\text{J}$ should then be continued beyond this surface. Such a phenomenon, termed *conformal continuation* [80, 82], is observed in special cases of a number of scalar-vacuum and scalar-electrovacuum solutions, in particular, those of GR with conformally coupled scalar fields [56, 75] (to be discussed below) and in the BD theory [95, 97] (the so-called cold black holes).

Which conformal frame (CF) is physical? This question is rather widely discussed in non-Einsteinian theories of gravity, most frequently in STT and in multidimensional theories, see, e.g., [237, 397, 457] and numerous references therein. The review [237] even classified the authors of published papers according to their attitude to the problem: those (i) neglecting the issue; (ii) supporting the view that all frames are equivalent; (iii) recognizing the problem but giving no conclusive arguments; (iv) claiming that a Jordan frame is physical; (v) asserting that the Einstein frame is physical. Each group included tens of names, and some names even got into more than one group. Refs. [237, 397] have presented arguments in favor of the Einstein frame, and the most important ones, applicable to STT and higher-order theories (as well as multiscalar–tensor theories obtainable from multidimensional gravity) are connected with the positivity of scalar field energy and the existence of a classically stable ground state.

In our opinion [132, 133] [which turns out to be outside all the groups (i)–(v)], the above arguments could be convincing if we dealt with an "absolute", or "ultimate" theory of gravity. If, however, the gravitational action is obtained in a certain limit of a more fundamental unified or underlying theory, theoretical requirements like the existence of a stable ground state should be addressed to this underlying theory rather than its visible manifestation. In the latter, the notion of a physical CF should be only related to the properties of instruments used for measuring masses, lengths and time intervals. Moreover, different sets of instruments (different measurement systems) are described, in general, by different CFs. We thus suppose that there can be at least two different physical CFs: the fundamental one, in which the underlying field theory (or a field limit of a more fundamental theory) is specified, and the observational one, corresponding to a given set of instruments. One can say that the first CF describes what is happening "as a matter of fact", the second one — what we see.

It is also necessary to mention a large variety of STT more complex than (4.29), called Horndeski [296] and beyond-Horndeski [339] theories, which will be beyond our consideration here. The Horndeski theory is the most general four-dimensional theory where the Lagrangian, constructed from a scalar field and the metric tensor, leads to second-order equations of motion. The theory contains nonminimal coupling between different combinations built from the curvature tensor and scalar field derivatives. The beyond-Horndeski theories [339] are still more general and admit higher derivatives but have some advantages concerning instabilities of the solutions.

4.2.2 $F(R)$ *gravity*

One more very popular class of extensions of GR is called $F(R)$ theories of gravity,[1] in which the total Lagrangian reads

$$L = \frac{1}{2}F(R) + L_m, \tag{4.36}$$

[1]More frequently they are called $f(R)$ gravity, but we are using here a capital F to avoid confusion with small f from the previous subsection.

where $F(R)$ is an arbitrary function of the scalar curvature R. These theories are in turn divided into two subclasses, depending on which variables are treated as dynamic variables in the theory of gravity, the metric $\tilde{g}_{\mu\nu}$ or the connection $\Gamma^{\rho}_{\mu\nu}$, and accordingly the action integral with the Lagrangian (4.36) is varied with respect to these variables to obtain the equations of motion. In the special case $F(R) = R$, in which the theory reduces to GR, the resulting equations are the same in both cases [439] (and the whole formallism is then called the metric or Palatini formulation of GR). However, in the general case of (4.36) the equations are different, and the corresponding subclasses of theories are called "metric $F(R)$ gravity" and "Palatini $F(R)$ gravity," respectively.

It turns out that the metric $F(R)$ theories of gravity are equivalent to the BD theory with $\omega = 0$ and potentials $U(\phi)$ whose form depends on the choice of the function $F(R)$ (see, e.g., the reviews [238, 508] and references therein). Indeed, the field equations due to (4.36)

$$-\frac{1}{2}F\delta^{\nu}_{\mu} + (R^{\nu}_{\mu} + \nabla_{\mu}\nabla^{\nu} - \delta^{\nu}_{\mu}\Box)F_R - T^{\nu}_{\mu} = 0 \qquad (4.37)$$

(where $F_R := df/dR$, and it is assumed that $F_R > 0$ and $F_{RR} := d^2F/dR^2$ is not identically zero) are equivalent to those due to the scalar-tensor Lagrangian (4.29) with the functions

$$f(\phi) = F_R, \quad h(\phi) = 0, \quad 2U(\phi) = RF_R - F(R), \qquad (4.38)$$

which is nothing else but the BD theory with $\omega = 0$, under the parametrization $f(\phi) = \phi$. It should be stressed that the two equivalent forms of the theory correspond to the same (Jordan) conformal frame $\mathbb{M}_J$ and employ the same metric $g_{\mu\nu}$. Thus, the condition $f(\phi) > 0$ that provides a positive effective gravitational constant takes the form $F_R > 0$ in $F(R)$ gravity. A transition to the Einstein frame $\mathbb{M}_E$ is thus carried out according to

$$\tilde{g}_{\mu\nu} = \frac{g_{\mu\nu}}{F_R}, \quad \chi = \sqrt{\frac{3}{2}}\ln F_R, \qquad (4.39)$$

and results in a Lagrangian in $\mathbb{M}_{\rm E}$ similar to (4.31)

$$L_E = R + \varepsilon g^{\mu\nu}\chi^{,\mu}\chi_{,\nu} - 2V(\chi) + \frac{L_m}{F_R^2}, \quad V(\chi) = \frac{RF_R - F}{2F_R^2}. \quad (4.40)$$

4.2.3 *Hybrid metric-Palatini gravity*

The recently proposed [280] hybrid metric-Palatini gravity (HMPG) theory separately introduces the Riemannian metric $g_{\mu\nu}$ and the independent connection $\hat{\Gamma}^{\alpha}_{\mu\nu}$ considering the total action [280]

$$S = \frac{1}{2\varkappa}\int d^4x\sqrt{-g}[R + F(\mathcal{R})] + S_m, \quad (4.41)$$

where $R = R[g]$ is the Ricci scalar corresponding to $g_{\mu\nu}$, $F(\mathcal{R})$ is a function of the scalar $\mathcal{R} = g^{\mu\nu}\mathcal{R}_{\mu\nu}$, with the Ricci tensor $\mathcal{R}_{\mu\nu}$ built in the standard way from the connection $\hat{\Gamma}^{\alpha}_{\mu\nu}$; moreover, as usual, $g = \det(g_{\mu\nu})$, $\varkappa$ is the gravitational constant, and S_m is the action of nongravitational matter.

Since the action (4.41) is varied in both $g^{\mu\nu}$ and $\hat{\Gamma}^{\alpha}_{\mu\nu}$, this theory combines the metric and Palatini approaches to gravity and thus extends the $F(R)$ theories. It has been shown that this theory does not contradict the local tests of gravity in the Solar system [155], fairly well describes the dynamics of galaxies and galaxy clusters thus addressing the dark matter problem [153], and is able to explain the modern cosmological observations that tell us about the properties of an accelerating Universe. A detailed exposition of HMPG and its achievements may be found in the reviews [154, 282]. An even more general theory, containing an arbitrary function of two variables, R and $\mathcal{R}$, is presented in [57], see a recent further work on this theory in [415].

Varying the action separately in $g^{\mu\nu}$ and $\hat{\Gamma}^{\alpha}_{\mu\nu}$, one obtains the field equations according to which [154, 193] $\hat{\Gamma}^{\alpha}_{\mu\nu}$ is a Riemannian connection corresponding to the metric $h_{\mu\nu} = \phi g_{\mu\nu}$, conformal to $g_{\mu\nu}$, and, as a result, there is only one dynamic degree of freedom in addition to $g_{\mu\nu}$, expressed in the scalar field $\phi = F_{\mathcal{R}} \equiv dF/d\mathcal{R}$.

The whole theory can then be reformulated in terms of $g_{\mu\nu}$ and ϕ as a scalar-tensor theory with the gravitational part of the action [154, 280]

$$S_g = \int d^4x\sqrt{-g}\left[(1+\phi)R - \frac{3}{2\phi}(\partial\phi)^2 - V(\phi)\right], \tag{4.42}$$

where[2] the scalar field potential is related to $F(\mathcal{R})$ by

$$V(\phi) = \mathcal{R}F_{\mathcal{R}} - F(\mathcal{R}). \tag{4.43}$$

The action (4.42) corresponds to the original formulation of the HMPG in [280]. In the review paper [154], the authors introduce one more constant Ω_A into the action, so that in (4.42) one has $(\Omega_A + \phi)$ before R instead of $(1+\phi)$. They further remark that in the limit $\Omega_A \to 0$ the theory (4.41) turns into pure Palatini $F(R)$ gravity in which the scalar field has no dynamics (precisely as in the BD theory with $\omega = -3/2$, to which (4.42) is converted if one replaces $(1+\phi \mapsto \phi)$, while in the limit $\Omega_A \to \infty$ the theory (4.41) turns into metric $F(R)$ gravity. In what follows we adhere to the original STT formulation (4.42).

It is easy to notice that Eq. (4.42) represents a special case of the Bergmann–Wagoner–Nordtvedt STT (4.29) in which

$$f(\phi) = 1+\phi, \quad h(\phi) = -\frac{3}{2\phi}. \tag{4.44}$$

The conformal transformation (4.30) then leads to the Einstein-frame action

$$S_g = \int d^4x\sqrt{-\bar{g}}\left[\overline{R} + \varepsilon\bar{g}^{\mu\nu}\bar{\phi}_{,\mu}\bar{\phi}_{,\nu} - \frac{V(\phi)}{f^2(\phi)}\right], \tag{4.45}$$

[2]Unlike [154, 193, 280] and many other papers, we are using here the metric signature $(+---)$, therefore, the sign before $(\partial\phi)^2 = g^{\mu\nu}\phi_\mu\phi_\nu$ is different from that in the cited papers (the minus here corresponds to a phantom), while the meaning of (4.42) remains the same as there. We will also safely omit the factor $1/(2\varkappa)$ before the gravitational part of the action integral because we will only deal with vacuum configurations where $S_m = 0$.

where bars mark quantities obtained using the transformed metric $\overline{g}_{\mu\nu}$, and $\varepsilon = -\operatorname{sign}\phi$, so that

$$\phi = -\tanh^2 \frac{\bar{\phi}}{\sqrt{6}} \quad (n = +1, \quad -1 < \phi < 0), \tag{4.46}$$

$$\phi = \tan^2 \frac{\bar{\phi}}{\sqrt{6}} \quad (n = -1, \quad \phi > 0). \tag{4.47}$$

Thus, depending on the sign of ϕ, the theory splits into canonical and phantom sectors, and the emergence of the latter looks more natural since in this case all three metrics $g_{\mu\nu}$, $\overline{g}_{\mu\nu}$ and $h_{\mu\nu} = \phi g_{\mu\nu}$ have the same signature. Let us also note that values of ϕ smaller than -1 lead to a negative effective gravitational constant and are thus manifestly nonphysical.

The substitution $\phi = -\varepsilon\chi^2/6$ converts the action (4.42) to the form

$$S_g = \int d^4x\sqrt{-g}\Big[(1-\varepsilon\chi^2/6)R + \varepsilon(\partial\chi)^2 - W(\chi)\Big], \tag{4.48}$$

with $W(\chi) = V(\phi)$. The action (4.48) describes GR where the source of gravity is a conformally coupled scalar field (as mentioned in [154]), which has the usual sign of kinetic energy if $\phi < 0$ ($\varepsilon = 1$) and is of phantom nature if $\phi > 0$ ($\varepsilon = -1$).

In the massless case, $V(\phi) \equiv W(\chi) = 0$, the field equations due to (4.42) imply that all vacuum solutions (such that $S_m = 0$) have both zero Ricci scalars, $R = \mathcal{R} = 0$ (see [193]). A general inverse result is also valid [85]:

Let there be a vacuum solution with $R \equiv 0$ and a non-constant scalar field in a theory (4.42), with $V \equiv 0$, then this STT reduces either to GR with a conformally coupled scalar field (which may be canonical or phantom) or to pure conformal scalar field theory.

A question of interest is: suppose we have found a solution of STT with some $V(\phi)$, then, what is the corresponding HMPG? In other words, given $V(\phi)$, can we determine $F(\mathcal{R})$?

For the case $V(\phi) \equiv 0$, Eq. (4.43) gives simply $F(\mathcal{R}) = \text{const}\cdot\mathcal{R}$. For $V(\phi) \not\equiv 0$, since $\phi = F_{\mathcal{R}}$, the relation (4.43) is a Clairaut equation (see, e.g., [314]) whose solution consists of a regular family that contains only linear functions,

$$F(\mathcal{R}) = H\mathcal{R} - V(H), \quad H = \text{const}, \tag{4.49}$$

and the so-called singular solution which is an envelope of the regular family and may be presented in a parametric form:

$$\begin{aligned} F(\mathcal{R}) &= \phi\mathcal{R} - V(\phi), \\ \mathcal{R} &= dV/d\phi. \end{aligned} \tag{4.50}$$

This issue is discussed in more detail in [193].

As we remember, in STT there is some controversy on the physical interpretations of different conformal frames. Unlike that, in HMPG it is assumed from the very beginning that it is the Jordan frame, used for the original formulation of the theory, that is to be interpreted as the physical one and should be confronted to observations and experiments. On the other hand, the geometric foundation of this theory can be treated as one of the possible justifications of the emergence of phantom fields in the theory and their employment in further model construction.

In Sec. 4.3, we will discuss static, spherically symmetric solutions of GR with a massless conformal scalar field, coinciding with those of HMPG under the condition $V(\phi) = 0$.

4.3 Geometries with conformally coupled scalar fields

The scalar–vacuum and scalar–electrovacuum gravitational fields with ϕ^{conf} are obtained from (4.15)–(4.20) with the aid of the transformation (4.30), (4.32); as a result, the metric acquires a conformal factor substantially changing the nature of the geometry. In what follows we will only discuss scalar–vacuum space–times, and scalar–electrovacuum ones may be easily analyzed in the same manner, see, e.g., [75]. For certainty, we will present the Jordan-frame scalar field ϕ in the parametrization (4.42); a transition to χ according to (4.48) is carried out using the relation $\phi = -\varepsilon\chi^2/6$.

4.3.1 *Models with a canonical field ($\varepsilon = 1$)*

According to the above-said, the Jordan-frame metric and the scalar field ϕ in the theory (4.32) or (4.42) with $V(\phi) = 0$ and $\phi < 0$ ($\varepsilon = +1$) may be presented as [75, 85, 88]

$$ds_J^2 = \cosh^2\psi\left\{ e^{-2hu}dt^2 - \frac{k^2 e^{2hu}}{\sinh^2 ku}\left[\frac{k^2du^2}{\sinh^2 ku} + d\Omega^2\right]\right\}, \quad (4.51)$$

$$\phi(u) = \tanh^2\psi, \quad \psi := \bar{\phi}/\sqrt{6} = Cu + \psi_0, \quad (4.52)$$

where $\psi = \bar{\phi}/\sqrt{6}$, $C = \text{const}$ may be interpreted as a scalar charge, $\psi_0 = \text{const}$, and one more constant k is expressed via C and h:

$$k = \sqrt{h^2 + 3C^2}. \quad (4.53)$$

The radial coordinate u in (4.51), (4.52) is borrowed from $\mathbb{M}_E$ (with the metric in curly brackets in (4.51)) and is there a harmonic coordinate, but it is not harmonic in the whole metric (4.51). The solution is defined at $u > 0$ so that $u = 0$ corresponds to spatial infinity; near it, the spherical radius $r = \sqrt{-g_{22}}$ behaves as $r \sim 1/u$, and the Schwarzschild mass is (see Chapter 3, Eq. (3.10))

$$m_J = h - C\tanh\psi_0. \quad (4.54)$$

On the other end of the u range, as $u \to \infty$, there are three kinds of behavior:

- $C < h$: we have $g_{00} \to 0$, and $r \to \infty$. It is a naked attracting singularity located beyond a throat, the kind of singularity called a "space pocket" by P. Jordan [310].
- $C > h$: in this case, $g_{00} \to \infty$ and $r \to 0$, so this is a naked singularity at the center, repulsive for test particles.
- $C = h > 0$: both g_{00} and r tend to finite limits, so that $u = \infty$ is a regular sphere, and a continuation beyond it is necessary.

To extend the solution for $C = h$ beyond $u = \infty$, let us put

$$y = \coth hu, \quad u = \frac{1}{2h}\ln\frac{y+1}{y-1}. \quad (4.55)$$

The metric becomes

$$ds_J^2 = \frac{(y+y_1)^2}{1-y_1^2}\left[\frac{dt^2}{(y+1)^2} - \frac{h^2}{y^4}(y+1)^2(dy^2 + y^2 d\Omega^2)\right], \qquad (4.56)$$

where $y_1 = \coth(\psi_0/C)$. The sphere $u = \infty \leftrightarrow y = 1$ is now manifestly regular. We obtain:

- $y \to \infty$ is flat spatial infinity.
- $y_1 < 0 \Rightarrow y = -y_1 > 0$ is a *naked attracting singularity* at the center ($r \to 0$).
- $y_1 > 0 \Rightarrow y \to 0$ is one more flat infinity, and the whole configuration is a *traversable wormhole* (see Chapter 6).
- $y_1 = 0 \Rightarrow y = 0$ is a double horizon. Passing on to the coordinate $r = h(y+1)$, we obtain

$$ds_J^2 = \left(1-\frac{h}{r}\right)^2 dt^2 - \left(1-\frac{h}{r}\right)^{-2} dr^2 - r^2 d\Omega^2, \qquad (4.57)$$

  which is the well-known *BH solution with a scalar charge and a conformal scalar field* [39, 56], sometimes called the BBMB BH solution. Its geometry coincides with that of the Reissner–Nordström solution with $q = m$.

One can notice that the substitution (4.55) loses its meaning at $y < 1$. Accordingly, the relation $\phi = -\tanh^2\psi$ is also meaningless at $\phi < -1$. Instead, after the conformal continuation, we have [82] in the Einstein frame another "copy" of the Fisher solution, where, instead of $\phi = -\tanh^2\psi$, we have $\phi = -\coth^2\psi$, and the conformal factor in (4.51) is $\sinh^2\psi$. At the transition surface $y = 1$, the field ϕ crosses the critical value $\phi = -1$, and beyond it, at $\phi < -1$, there is an "antigravitational" region, with a negative effective gravitational constant, where, in other words, the graviton becomes a ghost [138].

We see that the solutions with $V \equiv 0$ and $n = +1$ generically contain naked singularities, while the only existing BH and wormhole solutions are special, emerge due to conformal continuations, and each of them contains an "antigravitational" region.

4.3.2 *Models with a phantom field* ($\varepsilon = -1$)

Assuming $\phi > 0$, $n = -1$, we obtain, quite similarly to (4.51) and (4.52),

$$ds_J^2 = \cos^2\psi \left\{ \mathrm{e}^{-2hu} dt^2 - \frac{e^{2hu}}{s^2(k,u)} \left[\frac{du^2}{s^2(k,u)} + d\Omega^2 \right] \right\}, \quad (4.58)$$

$$\phi(u) = \tan^2\psi, \quad \psi := \bar{\phi}/\sqrt{6} = Cu + \psi_0, \quad (4.59)$$

where the function $s(k,u)$ is defined in (4.12). For the integration constants k, h and C we now have

$$k^2 \operatorname{sign} k = h^2 - 3C^2. \quad (4.60)$$

We see that $\operatorname{sign} k$ is not fixed now, and accordingly the solution splits into three branches.

Let us assume, without loss of generality, $|\psi_0| < \pi/2$. Then, as before, the metric (4.58) is asymptotically flat at $u = 0$ ($r \equiv \sqrt{-g_{22}} \to \infty$ where $r \sim 1/u$,[3] and the Schwarzschild mass is

$$m = h + C\tan\psi_0. \quad (4.61)$$

Other properties of the solution depend on the sign of k, taking into account the definition of $s(k,u)$ in (4.12).

Branch A: $k > 0$. The metric reads

$$ds_J^2 = \cos^2\psi \left\{ \mathrm{e}^{-2hu} dt^2 - \frac{k^2\, \mathrm{e}^{2hu}}{\sinh^2 ku} \left[\frac{k^2 du^2}{\sinh^2 ku} + d\Omega^2 \right] \right\},$$

$$\psi = \psi_0 + Cu. \quad h^2 = 3C^2 + k^2. \quad (4.62)$$

The only difference from (4.51) is the conformal factor $\cos^2(Cu+\psi_0)$ instead of $\cosh^2(Cu + \psi_0)$, which drastically changes the metric behavior. Indeed, as u grows starting from zero, $\psi(u)$ ultimately reaches the value where $\cos\psi = 0$ and, according to Eq. (4.59), $\phi \to \infty$. This happens where $Cu + \psi_0 \to \pi/2$ if $C > 0$ and where

[3] The conformal factor $\cos^2\psi$ is not normalized to unity at $u = 0$ if $\psi_0 \neq 0$, which, however, does not affect the further description.

$Cu + \psi_0 \to -\pi/2$ if $C < 0$. Other quantities involved in the metric are there evidently finite. Thus it is a naked central (since $r \to 0$) singularity, and it is attractive for test particles due to $g_{00} \to 0$.

Branch B: $k = 0$. In this case, the solution has the form

$$ds_J^2 = \cos^2\psi\left[\mathrm{e}^{-2hu}dt^2 - \mathrm{e}^{2hu}\left(\frac{du^2}{u^4} + \frac{d\Omega^2}{u^2}\right)\right],$$
$$\psi = \psi_0 + Cu, \quad h^2 = 3C^2. \tag{4.63}$$

As in Branch A, the coordinate u ranges from zero to the value where $\cos\psi = 0$ (say, $\psi = \pi/2$) and $\phi = \infty$, and we observe again a central attractive singularity.

Branch C: $k < 0$. Now the solution reads

$$ds_J^2 = \cos^2\psi\left[\mathrm{e}^{-2hu}dt^2 - \frac{k^2\,\mathrm{e}^{2hu}}{\sin^2 ku}\left(\frac{k^2du^2}{\sin^2 ku} + d\Omega^2\right)\right],$$
$$\psi = \psi_0 + Cu, \quad h^2 = 3C^2 - k^2. \tag{4.64}$$

The solution behavior crucially depends on ψ_0 at given k, C and depends on which of the quantities $\sin|k|u$ or $\cos\psi$ will be the first to vanish as u grows starting from zero. For asymptotic flatness we should assume that at $u = 0$ the factor $\cos^2\psi$ is nonzero, hence $|\psi_0| < \pi/2$ without loss of generality. Then, three possible behaviors should be singled out, see Fig. 4.1 (we assume for certainty $C > 0$).

C1: $(\pi/2 - \psi_0)/C < \pi/|k|$. The solution terminates at $u = u_s = (\pi/2 - \psi_0)/C$, where $\cos\psi = 0$, and $u = u_s$ is a naked central singularity quite similar to those in branches A and B.

C2: $(\pi/2 - \psi_0)/C > \pi/|k|$. The solution terminates at $u_* = \pi/|k|$ where $\sin ku = 0$, corresponding to the second flat spatial infinity, where the radius r infinitely grows while g_{tt} and ϕ remain finite, and the Schwarzschild mass is there equal to

$$m_* = -\,\mathrm{e}^{hu_*}(h\cos\psi_* + C\sin\psi_*), \tag{4.65}$$

($\psi_* = \psi_0 + C\pi/|k| < \pi/2$ is the value of ψ at $u = u_*$). Such a wormhole solution is only quantitatively different from its anti-Fisher and BD analogs, see, e.g., [75, 137].

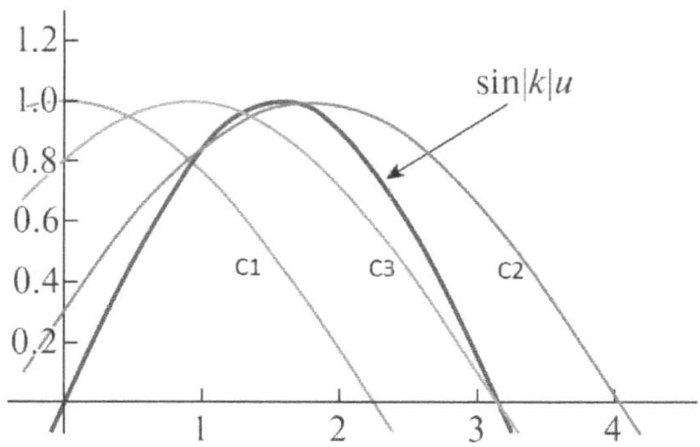

Figure 4.1 The behaviors C1, C2, C3 of the metric (4.64) is illustrated by the corresponding curves. We assume $|k| = 1$; curves C1–C3 plot $\cos\psi$ for $C = 0.7$ and different ψ_0. Curve C1 corresponds to a naked singularity, C2 to a wormhole, and C3 to a black hole.

C3: $(\pi/2 - \psi_0)/C = \pi/|k|$. In this intermediate case, at $u = u_1 = \pi/|k|$ vanish both $\sin|k|u$ and $\cos\psi$, the spherical radius $r = \sqrt{-g_{\theta\theta}}$ is finite but $\phi = \infty$. Near $u = u_1$, the metric behaves as

$$ds_J^2 = C^2\left[e^{-2hu_1}\Delta u^2 dt^2 - e^{2hu_1}\frac{du^2}{\Delta u^2} - e^{2hu_1}d\Omega^2\right], \tag{4.66}$$

where $\Delta u = u_1 - u$. Consequently, $u = u_1$ is a double (extremal) horizon, and the metric should be continued beyond it.

The condition $(\pi/2-\psi_0)/C = \pi/|k|$ leads to $\psi_0 = \pi(1/2 - C/|k|)$, thus $C < |k|$, hence the plot of $\cos\psi$ is wider than that of $\sin ku$. Therefore, as u further grows (describing the region beyond the horizon), the next zero of $\sin ku$, $(u = u_2 = 2\pi/|k|)$ is reached before a zero of $\cos\psi$. The value $u = u_2$ corresponds to the second flat spatial infinity, in full similarity with wormhole solutions. This space-time is also globally regular but is now only one-way traversable due to emergence of the horizon. Figure 4.2 shows the corresponding Carter–Penrose diagram.

This BH solution has much in common with the one with the metric (4.57) [39, 56, 75]. Both BHs are described by special solutions to the Einstein–scalar equations, both are asymptotically flat and extremal (with zero Hawking temperature), and in both cases the supporting scalar fields turn to infinity on the horizon, whereas the

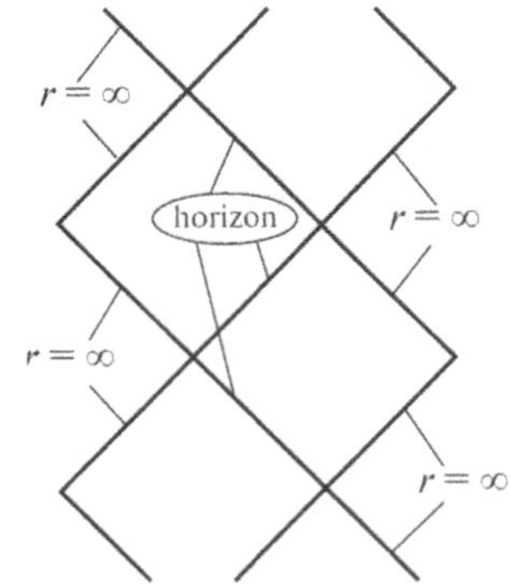

Figure 4.2 Carter–Penrose diagram for a regular black hole with the metric (4.64). The diagram infinitely extends up and down.

effective stress-energy tensors T_μ^ν are finite there (as is evident from finiteness of the Einstein tensor components G_μ^ν). Also, in both cases the scalar curvature is zero in the whole space, and the solutions are obtained from their Einstein-frame counterparts using conformal continuations [75, 82].

However, the solution (4.57) has a singular center $r = 0$ (the geometry is the same as that of the extreme Reissner–Nordström space–time), while the space–time (4.64) is globally regular and has no center at all.

It happens that none of the static, spherically symmetric solutions of GR with conformal scalar fields, being also solutions of the theory (4.42) with $V \equiv 0$, have simple horizons with finite Hawking temperature, in contrast to the results announced in [193].

A geometry with infinitely many horizons. There is a one-parameter family of solutions of interest obtained if we put

$$C = |k|, \quad \psi_0 = -\pi/2, \tag{4.67}$$

(thus abandoning the asymptotic flatness requirement). We then have $\cos^2\psi = \sin^2 ku$, and the Jordan-frame metric reads

$$ds_J^2 = \sin^2 ku\, \mathrm{e}^{-2hu} dt^2 - k^2\, \mathrm{e}^{2hu} \left(\frac{k^2 du^2}{\sin^2 ku} + d\Omega^2 \right), \tag{4.68}$$

where $h = \pm k$ according to (4.60). This metric, with $u \in \mathbb{R}$, describes a space–time unifying an infinite number of static regions (each described by a half-wave of the function $\sin ku$), separated by double horizons located at each $u = \pi n/|k|$, with any integer n. In this case, the Jordan-frame manifold $\mathbb{M}_{\rm J}$ unifies a countable number of Einstein-frame manifolds $\mathbb{M}_{\rm E}$, each of the latter representing an anti-Fisher wormhole whose both infinities turn into horizons in $\mathbb{M}_{\rm J}$. Another example of a manifold obtained by infinitely many conformal continuations was obtained in [82], using a solution for a conformally coupled scalar field ϕ with a nonzero potential $U(\phi)$ and the normal sign of kinetic energy. In that example, the transition from one region to another occurred through ordinary surfaces $S_{\rm trans}$ of finite radius, there were no horizons, and the whole $\mathbb{M}_{\rm J}$ was either completely static (shaped as an infinitely long tube with a periodically changing radius) or completely cosmological (forming a $(2+1)$ cosmology with a periodically changing scale factor). In the present case, all transitions surfaces $S_{\rm trans}$ are double horizons, and the structure is aperiodic due to the factors $e^{\pm hu}$ in (4.68). The corresponding global structure is shown in Fig. 4.3.

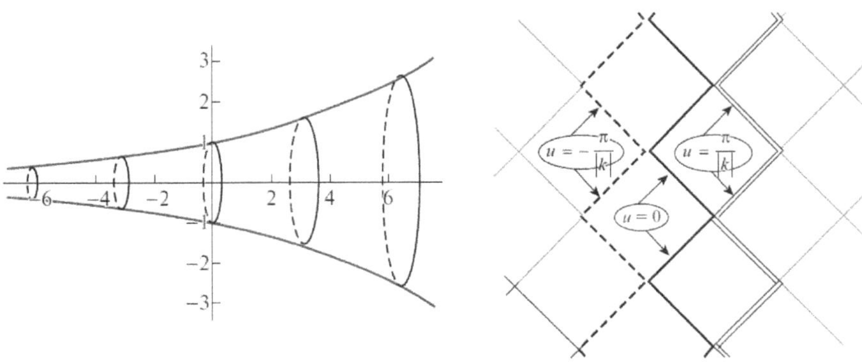

(a) (b)

Figure 4.3 The manifold $\mathbb{M}_{\rm J}$ with the metric (4.68). (a) The spatial image: a horn with cross-sections whose radius grows with growing u. For certainty, we have put $|k| = 1$; the selected sections at $u = n\pi$ present the horizon spheres. (b) The Carter–Penrose diagram, occupying the whole plane. Horizons with $u = -\pi/|k|, 0, \pi/|k|$ are shown by the dashed, thick and double broken lines, respectively.

Another example of a manifold with infinitely many horizons [173] has been obtained for a family of phantom dilaton-Einstein–Maxwell black holes.

4.4 Some other STT solutions

4.4.1 *Solutions with nonconformal coupling*

Vacuum and electrovacuum solutions have also been discussed for GR with a scalar field having a more general nonminimal coupling with gravity, corresponding to the Lagrangian (4.29) with

$$f(\phi) = 1 - \xi\phi^2, \quad \xi = \text{const} > 0;$$
$$h(\phi) = 1, \quad U(\phi) \equiv 0. \tag{4.69}$$

At $\xi = 0$ and $\xi = 1/6$ this reduces to minimal and conformal couplings, respectively. For $\xi \neq 1/6$, it has been shown that the solution properties are to a large extent similar to those for $\xi = 1/6$: they are generically singular, but under some special relation between the integration constants, the singularity of the Einstein frame corresponds to a regular sphere in the Jordan frame [30, 110–112].

If $\xi > 1/6$, all solutions continued beyond this sphere have a second flat infinity, i.e., describe wormholes, but, as always in such cases, the region beyond the transition sphere is anti-gravitational in the sense that the effective gravitational constant is negative there. Similar wormhole solutions also exist for $\xi < 1/6$, but only under an additional inequality for integration constants, otherwise the conformally continued solution possesses a naked singularity [30, 112].

All these wormhole solutions with zero or small electric charges prove to be unstable under radial perturbations [110–112], and the characteristic time of perturbation growth is of the order of the time needed for a light signal to cover a distance of the order of the throat radius. Meanwhile, numerical calculations have shown [112] that the perturbation growth increment diminishes as the charge grows. It therefore seems possible that some values of the electric charge can lead to a stable wormhole.

BH solutions like (4.57) are absent at $\xi \neq 1/6$.

4.4.2 *Vacuum and electrovacuum solutions in the Brans–Dicke theory*

All static, spherically symmetric solutions of the BD theory (4.29), (4.35) with $U(\phi) \equiv 0$ are easily obtained from those described in Sec. 4.1 using the transformation (4.30). Namely, in the general theory (4.29) one can choose the BD scalar field parametrization $f(\phi) = \phi$, $h(\phi) = \omega(\phi)/\phi$, so that

$$L_{\mathrm{STT}} = \phi\tilde{R} + \frac{\omega(\phi)}{\phi}\tilde{g}^{\mu\nu}\phi_{,\mu}\phi_{,\nu} - 2U(\phi) + L_m, \qquad (4.70)$$

and the transformation (4.30) reads

$$\tilde{g}_{\mu\nu} = \frac{1}{\phi}g_{\mu\nu}, \quad \frac{d\phi}{d\psi} = \phi\Big|\omega(\phi) + 3/2\Big|^{-1/2}. \qquad (4.71)$$

where the metrics $\tilde{g}_{\mu\nu}$ and $g_{\mu\nu}$ correspond to the Jordan and Einstein frames, respectively. In BD theory, $\omega = \mathrm{const}$, so the fields ϕ and ψ are related by

$$\phi = \mathrm{e}^{\psi/\sqrt{|\omega+3/2|}}. \qquad (4.72)$$

Recall that the BD theory is normal for $\omega > -3/2$, phantom for $\omega < -3/2$ and has no dynamics at $\omega = -3/2$; accordingly, Eq. (4.72) is meaningless at $\omega = -3/2$.

Substituting the metrics from Sec. 4.1 for $\tilde{g}_{\mu\nu}$ and the corresponding expressions for ψ into (4.71) and (4.72), we obtain scalar-(electro)vacuum solutions of the BD theory. Their properties are analyzed in detail in quite a number of papers, see, e.g., [78, 95, 97, 137] and references therein. One can note that in the scalar–vacuum case ($q = 0$), the conformal mapping (4.71) with $\omega = \mathrm{const}$ applied to Fisher's solution and the three branches of the anti-Fisher solution leads to Brans's four classes of vacuum BD solutions [73] (in other notations). The main conclusions are as follows.

1. In the case of a normal BD field, i.e., $\omega + 3/2 > 0$, all solutions possess naked singularities, with the only exception: a special wormhole solution existing at $\omega = 0$ [78] due to a conformal continuation, similar to that described in Sec. 4.3.

2. One of the generic types of BD solutions for all $\omega < -3/2$ is a wormhole with two asymptotically flat infinities, obtained from the anti-Fisher wormhole by an everywhere regular conformal transformation.
3. Other generic types of BD solutions for all $\omega < -3/2$ contain singular horizons with an infinite area; their singular nature is connected with violation of analyticity rather than an infinite curvature. Some of these singularities are remote, i.e., correspond to infinite values of the canonical parameter along geodesics that reach them. In such cases the space–time is geodesically complete.
4. There is a discrete family of solutions with $\omega < -3/2$, parametrized by two integers, where the horizon of infinite area is regular and admits an extension beyond it (see diagrams 2b and 2d in Fig. 5.3). These configurations have been termed cold BHs since they inevitably have a zero Hawking temperature. Many of these configurations do not contain any singularities.
5. The electric charge adds some branches of solutions as compared with the vacuum case but does not drastically change the situation with singularities, BHs and wormholes [97].

4.4.3 *Summary for massless scalar fields*

From our consideration of exact static solutions with massless scalar fields, one can draw the following basic conclusions, mostly concerning BHs.

1. In GR with a normal, minimally coupled scalar field $\phi^{\rm min}$ ($\varepsilon = +1$), there are no solutions with horizons, in agreement with the so-called no-hair theorems, see Sec. 4.5.
2. In GR with a phantom minimally coupled scalar field $\phi^{\rm min}$ ($\varepsilon = -1$), there are "cold BH" solutions containing horizons of infinite area under special discrete values of the parameters (integration constants). There are no horizons of finite area.
3. In GR with a conformal scalar field $\phi^{\rm conf}$, with the normal sign of kinetic energy ($\varepsilon = +1$), there is a special solution (4.57) in

the form of an extremal BH with the metric coinciding with the extremal Reissner–Nordström one.

4. In GR with a phantom conformal scalar field $\phi^{\rm conf}$, generic solutions are wormhole ones and those with naked singularities. There is also a special two-parameter family of regular BHs with the structure shown in Fig. 4.2.
5. All solutions obtained for massless conformally coupled fields in GR are simultaneously solutions of HMPG with zero potential $V(\phi)$.
6. Other kinds of nonminimally coupled scalar fields, scalar–tensor theories: at $\varepsilon = +1$ cases like item 3 are possible in principle but are yet unknown; at $\varepsilon = -1$ the situation is similar to item 2. For instance, such discrete families of electrically neutral and charged cold BHs have been found in the BD theory. Some of them are globally regular.

4.5 Scalar fields with arbitrary potentials — no-go theorems

4.5.1 *What is the use of no-go theorems?*

In most of the applications, it is necessary to consider more general scalar fields than those discussed in the previous section, namely, fields with nonzero potentials as well as various interactions, generalized or modified scalar field theories. Meanwhile the equations for such self-gravitating scalar fields can be explicitly integrated only in very rare special cases even for fields with high symmetry, such as isotropic models in cosmology and static, spherically symmetric systems. Therefore, of great value are general statements or theorems about the properties of such systems which can be obtained without completely solving the field equations.

For static, spherically symmetric scalar–vacuum configurations in GR, described by the action

$$
\begin{aligned}
S &= \int d^4x \, \sqrt{|g|}\,\big[R/\varkappa + L_{\rm sc}\big],\\
L_{\rm sc} &= g^{\mu\nu}\phi_{,\mu}\phi_{,\nu} - 2V(\phi),
\end{aligned}
\tag{4.73}
$$

where R is the scalar curvature and $\varkappa$ the gravitational constant, the following theorems are known:

A. "No-hair" theorems ([8, 40]), claiming that asymptotically flat black holes with a nontrivial scalar field cannot exist under certain conditions imposed on the potential $V(\phi)$.

B. The generalized Rosen theorem, claiming that a particle-like solution (PLS) (that is, an asymptotically flat solution with a regular center) cannot exist if $V \geq 0$ [136].

C. The nonexistence theorem for regular solutions without a center (e.g., wormhole solutions) [81].

D. The global structure theorem [81], claiming that the list of possible global causal structures (and the corresponding Carter–Penrose diagrams) for configurations with arbitrary potentials $V(\phi)$ and any asymptotic behavior is the same as the analogous list in the case $\phi = \text{const}$, namely it includes the Minkowski (or AdS), Schwarzschild, de Sitter and Schwarzschild–de Sitter structures.

A number of known exact solutions represent examples of configuration admitted by these theorems: black hole solutions with scalar fields where $V(\phi) \geq 0$ but without asymptotic flatness; asymptotically flat PLS and black holes with at least partly negative potentials $V(\phi)$, etc. All this, taken together, leads to a clear enough picture of what can and what cannot be expected from minimally coupled scalar fields in GR. A detailed description of all these results is beyond the scope of our presentation here, see, e.g., [81, 136] and references therein.

There are quite a number of generalizations of the action (4.73), for which it is of interest to know whether or not the above Statements A–D hold, and if yes, under which additional conditions. In particular:

1. Multidimensional spherically symmetric configurations in GR, with the scalar field Lagrangian given in (4.73) and the metric

$$ds^2 = \mathrm{e}^{2\gamma} dt^2 - \mathrm{e}^{2\alpha} du^2 - \mathrm{e}^{2\beta} d\Omega_{d_0}^2, \tag{4.74}$$

where α, β, γ are functions of the radial coordinate u and $d\Omega^2_{d_0}$ is a linear element on a d_0-dimensional sphere of unit radius.

2. More general scalar field Lagrangians in GR, such as $L_{\rm sc} = F(\phi,\ I)$, where $I = g^{\mu\nu}\phi_{,\mu}\phi_{,\nu}$ and F is an arbitrary function of two variables (the so-called generalized k-essence).

3. Sets of scalar fields $\vec{\phi} = \{\phi^K\}$, $K = \overline{1,N}$ of σ model type, considered as coordinates in the N-dimensional target space $\mathbb{T}_\phi$, so that

$$L_{\rm sc} = H_{KL}(\vec{\phi})(\partial\phi^K,\ \partial\phi^L) - 2V(\vec{\phi}), \tag{4.75}$$

where the target-space metric H_{KL} (usually supposed to be positive-definite) and the potential V are functions of the N variables ϕ^K, and we use the notation

$$(\partial y,\ \partial z) = g^{\mu\nu}\partial_\mu y \partial_\nu z. \tag{4.76}$$

4. Scalar-tensor theories (STT) of gravity, with the D-dimensional action

$$S_{\rm CTT} = \int d^D x\sqrt{g}[f(\phi)\mathcal{R} + h(\phi)(\partial\phi)^2 - 2U(\phi)], \tag{4.77}$$

where $(d\phi)^2 = (\partial\phi,\ \partial\phi)$ and f, h, U are arbitrary functions of the scalar field.

5. Theories of gravity with higher derivatives (curvature-nonlinear theories) where the scalar curvature R in (4.73) is replaced by a function $f(\mathcal{R})$, or, more generally, including, for example, the invariant $R_{\mu\nu}R^{\mu\nu}$.

6. More general multidimensional configurations such as products like

$$\mathbb{M}^D = \mathbb{R}_u \times \mathbb{M}_0 \times \mathbb{M}_1 \times \mathbb{M}_2 \times \cdots \times \mathbb{M}_n, \tag{4.78}$$

where $\mathbb{M}_{\rm ext} = \mathbb{R}_u \times \mathbb{M}_0 \times \mathbb{M}_1$ is the "external" manifold with the metric (4.74), $\mathbb{R}_u \subseteq \mathbb{R}$ is the range of the radial coordinate u, $\mathbb{M}_1$ is the time axis, and $\mathbb{M}_0 = \mathbb{S}^{d_0}$. Furthermore, $\mathbb{M}_2, \ldots, \mathbb{M}_n$ are the "internal" factor spaces of arbitrary dimensions d_i, $i = 2, \ldots, n$; by this notation, $\dim \mathbb{M}_0 = d_0$ and $\dim \mathbb{M}_1 = d_1 = 1$.

This list can be continued and/or its items can be combined to obtain new generalizations.

We will consider here static, spherically symmetric systems in GR, in space–times of arbitrary dimension D with a sufficiently general σ-model source. The proofs turn out to be almost the same as for similar systems in four dimensions (4.73), but the present formulation is well suited for (multi-)scalar–tensors and multidimensional generalizations.

4.5.2 *Basic equations*

Consider D-dimensional GR with the set of scalar fields (4.75). As in four dimensions, the Einstein equations can be written in two equivalent forms:

$$G_\mu^\nu \stackrel{\text{def}}{=} \mathcal{R}_\mu^\nu - \frac{1}{2}\delta_\mu^\nu \mathcal{R} = -\varkappa T_\mu^\nu, \tag{4.79}$$

or

$$\mathcal{R}_\mu^\nu = -\varkappa \widetilde{T}_\mu^\nu \stackrel{\text{def}}{=} -\varkappa^2 \left(T_\mu^\nu - \frac{\delta_\mu^\nu}{D-2} T_\alpha^\alpha \right), \tag{4.80}$$

where T_μ^ν is the stress-energy tensor (SET). For the fields (4.75) it is given by

$$T_\mu^\nu = \partial^\nu \vec{\phi} \partial_\mu \vec{\phi} - \frac{1}{2}\delta_\mu^\nu L_{\text{sc}}, \tag{4.81}$$

or in the "tilded" version,

$$\widetilde{T}_\mu^\nu = \partial^\nu \vec{\phi} \partial_\mu \vec{\phi} - \delta_\mu^\nu V(\vec{\phi}), \tag{4.82}$$

where two arrows denote a scalar product in the target space $\mathbb{T}_\varphi$: $\vec{a}\vec{b} = H_{KL} a^K b^L$.

The static, spherically symmetric metric (4.74) is written with an arbitrary radial coordinate u. Now, it is convenient for our purposes to use the quasiglobal coordinate $u = x$ corresponding to the gauge condition $\alpha + \gamma = 0$, so that the metric takes the form

$$ds^2 = A(x)dt^2 - \frac{du^2}{A(x)} - r^2(x) d\Omega_{d_0}^2, \tag{4.83}$$

where we have denoted $r(x) = \mathrm{e}^{\beta}$ and $A(x) = \mathrm{e}^{2\gamma} \equiv \mathrm{e}^{-2\alpha}$. This choice is preferable for considering Killing horizons, described as regular zeros of the function $A(x)$. The reason is that, as we saw in Chapter 3, in a close neighborhood of a horizon the coordinate x defined in this way varies (up to a positive constant factor) like manifestly well-behaved Kruskal-like coordinates used for an analytic continuation of the metric. This fact refers to the geometry of the (t, x) subspace and does not depend on the geometry of other dimensions and even on their number.

With this choice of the coordinate gauge, the nonzero components of the Ricci tensor for the metric (4.83) read

$$
\begin{aligned}
R^t_t &= -\frac{1}{2r^{d_0}}(A' r^{d_0})', \\
R^x_x &= -\frac{1}{2r^{d_0}}(A' r^{d_0})' - d_0 A \frac{r''}{r}, \\
R^\theta_\theta &= \frac{d_0 - 1}{r^2} - A\left[\frac{r''}{r} + (d_0 - 1)\frac{r'^2}{r^2} + \frac{A' r'}{Ar}\right],
\end{aligned} \tag{4.84}
$$

where θ is any of the spherical angles in (4.83). Accordingly, the scalar field equations and four different combinations of components of Eq. (4.54) can be written as follows:

$$[A r^{d_0} H_{KL} (\varphi^L)']' = r^{d_0} \partial V / \partial \varphi^K; \tag{4.85}$$

$$(A' r^{d_0})' = -(4/d_0) r^{d_0} \varkappa V; \tag{4.86}$$

$$d_0 r''/r = -\varkappa (\vec{\phi}')^2; \tag{4.87}$$

$$A(r^2)'' - r^2 A'' + (d_0 - 2) r' (2Ar' - A'r) = 2(d_0 - 1); \tag{4.88}$$

$$d_0(d_0 - 1)(1 - Ar'^2) - d_0 A' r r' = \varkappa[-Ar^2 (\vec{\phi}')^2 + 2r^2 V]. \tag{4.89}$$

Equations (4.86)–(4.88) are the components $\binom{t}{t}$, $\binom{t}{t} - \binom{x}{x}$ and $\binom{t}{t} - \binom{\theta}{\theta}$ of (4.54), respectively, and (4.89) is the $\binom{x}{x}$ component of (4.79). We have written $(N + 4)$ equations for $(N + 2)$ unknowns φ^K, A and r, but there are only two independent equations among (4.86)–(4.89); in particular, (4.89) is a first integral of the other equations. So this set of equations is determined.

4.5.3 Global structure theorems

The first theorem concerns the nonexistence of wormholes, horns and flux tubes. A *wormhole* is, by definition, a configuration with two asymptotic regions that have large $r(x)$ (see Chapter 6), hence $r(x)$ must have at least one regular minimum. A *flux tube* is a configuration with $r = \text{const} > 0$, a static $(d_0 + 1)$-dimensional cylinder. A *horn* is a geometry that tends to a flux tube at one of its asymptotics, which happens if $r(x) \to \text{const} > 0$ at one of the ends of the range of x. Such "horned particles" with a flat asymptotic region have been discussed as possible remnants of black hole evaporation [27].

Theorem 4.1 (on configurations without a center). *Equations* (4.85)–(4.89) *for* $D \geq 4$ *and positive-definite* H_{KL} *do not admit* (i) *solutions where the function* $r(x)$ *has a regular minimum,* (ii) *solutions describing a horn, and* (iii) *flux-tube solutions with* $\varphi^K \neq \text{const}$.

A proof rests on Eq. (4.87), implying $r'' \leq 0$, which actually expresses the null energy condition valid for the SET T_μ^ν as long as the matrix H_{KL} is positive-definite. As a result, not only are wormholes as global entities impossible but even wormhole throats.

Another theorem concerns the possible number and order of Killing horizons, coinciding with the number and order of zeros of $A(x)$. Recall that a simple (first-order) or any odd-order horizon separates a static region, $A > 0$ (R-region), from a nonstatic region, $A < 0$ where (4.74) is a homogeneous cosmological Kantowski–Sachs-type metric (a T-region). A horizon of even order separates regions with the same sign of $A(x)$.

The disposition of horizons unambiguously determines the global causal structure of space–time (up to identification of isometric surfaces, if any). The following theorem severely restricts such possible dispositions.

Theorem 4.2 (on the global structure). *Consider solutions to Eqs.* (4.85)–(4.88) *for* $D \geq 4$. *Let there be a static region* $a < x < b \leq \infty$. *Then:*

(i) *all horizons are simple;*
(ii) *no horizons exist at* $x < a$ *and at* $x > b$.

A proof of this theorem [81, 136] employs the properties of Eq. (4.88), which can be rewritten in the form

$$r^4 B'' + (d_0 + 2) r^3 r' B' = -2(d_0 - 1), \qquad (4.90)$$

where $B(x) = A/r^2$. At points where $B' = 0$, we inevitably have $B'' < 0$. Therefore $B(x)$ cannot have a regular minimum. Having once become negative while moving to the left or to the right along the x axis, $B(x)$ (and hence $A(x)$) cannot return to zero or positive values.

By Theorem 4.2, there can be at most two simple horizons around a static region. A second-order horizon separating two nonstatic regions can appear, but this horizon is then unique, and the model has no static region, it consists of two T-regions.

The possible dispositions of zeros of the function $A(x)$, and hence the list of possible global causal structures, are thus the same as in the case of vacuum with a cosmological constant. The latter is a solution to Eqs. (4.85)–(4.88) with $\varphi^K = \text{const}$, $\varkappa V = \Lambda = \text{const}$, and the metric

$$ds^2 = A(r)dt^2 - \frac{dr^2}{A(r)} - r^2\, d\Omega^2_{d_0}, \qquad (4.91)$$

$$A(r) = 1 - \frac{2m}{(d_0 - 1) r^{d_0 - 1}} - \frac{2\Lambda r^2}{d_0(d_0 + 1)}. \qquad (4.92)$$

This is the multidimensional Schwarzschild-de Sitter (or Tangherlini-de Sitter) solution. Its special cases correspond to the Schwarzschild ($d_0 = 2$) and Tangherlini [528] ($d_0 \geq 2$) solutions[4] when $\Lambda = 0$ and the de Sitter solution in arbitrary dimensions when $m = 0$, called anti-de Sitter (AdS) in the case $\Lambda < 0$. For $\Lambda > 0$, if m is positive

---

[4]The mass M in conventional units, say, grams, is obtained by writing $m = GM$ where G is a $(d_0 + 2)$-dimensional analogue of Newton's constant. The coefficient of m is chosen in (4.94) and accordingly in (4.92) in such a way that at large r in the case $\Lambda = 0$, when the space–time is asymptotically flat, a test particle at rest experiences a Newtonian acceleration equal to $-GM/r^{d_0}$.

but smaller than the critical value

$$m_{\rm cr} = \frac{d_0 - 1}{d_0 + 1}\left[\frac{d_0(d_0-1)}{2\Lambda}\right]^{(d_0-1)/2}, \tag{4.93}$$

there are two horizons, the one at smaller r being interpreted as a black hole horizon and the other as a cosmological horizon. At $m = m_{\rm cr}$, the two horizons merge, and there are two homogeneous nonstatic regions separated by a double horizon. The solution with $m > m_{\rm cr}$ is purely cosmological and has no Killing horizon. In the cases $m < 0$ and/or $\Lambda < 0$ there is at most one simple horizon.

Possible behaviors of $A(r)$ and the corresponding Carter–Penrose diagrams are presented in Figs. 4.4 and 4.5.

In (2+1)-dimensional gravity ($d_0 = 1$) we have a still shorter list of global structures: at most one simple horizon is possible.

Theorems 1 and 2 are independent of the form of the potential and of any assumptions about spatial asymptotics.

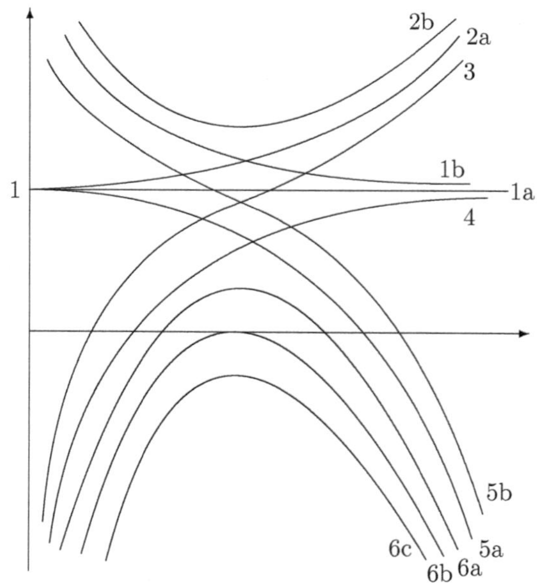

Figure 4.4 The behavior of $A(r)$, Eq. (4.92), for different values of m and Λ.

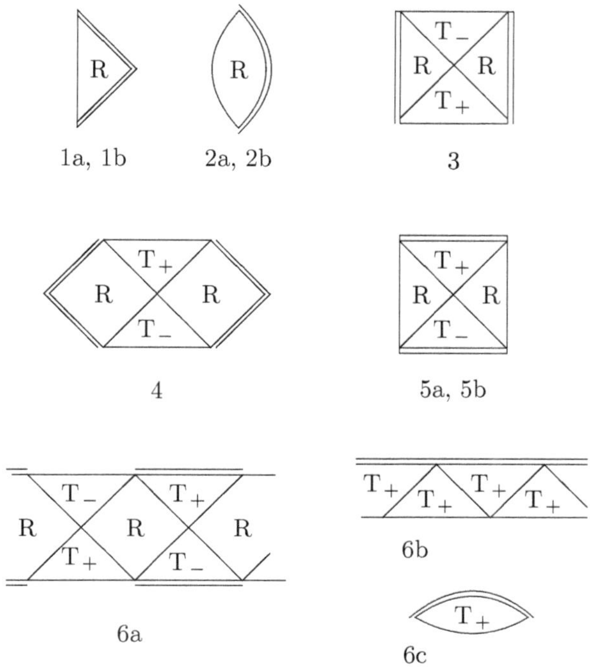

Figure 4.5 Carter–Penrose diagrams for different cases of the metric (4.91) and (4.92), labeled according to Fig. 4.4. The R and T letters correspond to R and T space–time regions while T_+ and T_- denote expanding and contracting T regions (i.e., with r increasing and decreasing with time, respectively). Single lines on the border of the diagrams denote $r = 0$, double lines $r = \infty$. Diagrams 6b and 6c are drawn for the case of expanding Kantowski-Sachs cosmologies; to obtain diagrams for contracting models, one should merely interchange $r = 0$ and $r = \infty$ and replace T_+ with T_-. Diagrams 6a and 6b admit identification of isometric timelike sections.

4.5.4 *No-hair theorem*

Let us now consider asymptotically flat space–times, which means, in terms of the metric (4.83), that without loss of generality, $r \approx x$ and the function $A(x) \approx A(r)$ has the Tangherlini form (4.92) with $\Lambda = 0$:

$$A(r) = 1 - \frac{2m}{(d_0 - 1)r^{d_0 - 1}}, \tag{4.94}$$

as $x \to \infty$. It then follows from the field equations that the SET components, and hence the quantities V and $(\vec{\phi})^2$, decay at large $x \approx r$ quicker than $r^{-(d_0+1)}$.

Let us now prove the following no-hair theorem [107], extending to our system the theorems known in four dimensions [8, 40]:

Theorem 4.3 (on external scalar fields of black holes). *Given Eqs.* (4.85)–(4.89) *for* $D \geq 4$, *with a positive-definite matrix* $H_{KL}(\vec{\phi})$ *and* $V(\vec{\phi}) \geq 0$, *the only asymptotically flat black hole solution is characterized by* $V \equiv 0$, $\vec{\phi} = \text{const}$ *and the Tangherlini metric* (4.91), (4.94) *in the whole range* $h < x < \infty$ *(the domain of outer communication of the BH), where* $x = h$ *is the event horizon.*

At the event horizon $x = h$ we have by definition $A = A(h) = 0$, and $A > 0$ at $x > h$. By Theorem 4.2, the horizon should be simple, so that $A \sim x - h$ as $x \to h$. Consider the function

$$\mathcal{F}_1(x) = \frac{r^{d_0}}{r'}[2V - A\vec{\phi}'^2]. \tag{4.95}$$

One can verify that

$$\mathcal{F}'(x) = \mathcal{F}_2(x) \stackrel{\text{def}}{=} r^{d_0-1}\left[2d_0 V + (d_0 - 1)\frac{\vec{\phi}'^2}{r'^2} + A\vec{\phi}'^2\right]. \tag{4.96}$$

To do so, when calculating $\mathcal{F}_1'$, one should substitute $\vec{\phi}''$ from (4.85), r'' from (4.87) and A' from (4.89). Let us integrate (4.96) from h to infinity:

$$\mathcal{F}_1(\infty) - \mathcal{F}_1(h) = \int_h^\infty \mathcal{F}_2(x)\,dx. \tag{4.97}$$

Since $r'(\infty) = 1$ and $r'' \leq 0$, we have $r' > 1$ in the whole range of x, but $r'(h) < \infty$. Indeed, regularity of the horizon implies a finite value of the Kretschmann scalar given by (3.7), hence finite values of all its constituents. Of interest for us is the condition $|A'r'| < \infty$. Since $A'(h) > 0$ (the horizon is simple due to the Global structure theorem), its finiteness means $r'(h) < \infty$.[5]

[5]The expressions (3.7) are written for the case $D = 4$, but for the metric (4.83) they have the same form up to numerical factors.

The quantity $\mathcal{F}_1(h)$ should be finite, since otherwise we would have either V or $A\vec{\phi}'^2$ infinite, leading to infinite SET components (see (4.81)) and, via the Einstein equations, to a curvature singularity.

If, however, we admit a nonzero value of $A\vec{\phi}'^2$ at $x = h$, the integral in (4.97) will diverge at $x = h$ due to the second term in brackets in (4.96), and this in turn leads to an infinite value of $\mathcal{F}_1(h)$. Therefore, $A\vec{\phi}'^2 \to 0$ as $x \to h$, and we conclude that

$$\mathcal{F}_1(h) = \frac{2r^{d_0}(h)}{r'(h)} V(h) \geq 0.$$

On the other hand, $\mathcal{F}_1(\infty) = 0$ due to the asymptotic flatness conditions. Thus, in Eq. (4.97) there is a nonpositive quantity in the left-hand side and a nonnegative quantity on the right. The only way to satisfy (4.97) is to put $V \equiv 0$ and $\vec{\phi}' \equiv 0$ in the whole range $x > h$, and the only solution for the metric then has the Tangherlini form. □

As follows from the scalar field equations (4.85), the equality $V = 0$ should take place where $\partial V/\partial\varphi^K = 0$, i.e., at an extremum or saddle point of the potential, and it should obviously be a minimum for stable equilibrium.

It is of interest that one of the key points of the above proof, that $A\vec{\phi}'^2 = 0$ at $x = h$, might be obtained from smoothness considerations. Indeed, since $A \sim x - h$ near $x = h$, a nonzero value of $A\vec{\phi}^2$ means that some of $(\phi^K)'$ behave as $(x-h)^{-1/2}$, violating the C^1 requirement for the scalar fields. Our proof is "more economical" since it only uses the requirement of space–time regularity at the horizon.

One can also note that our no-hair theorem is in a complementarity relation with a recent black hole uniqueness theorem [471] (see [472] for recent generalizations and [288, 400] for earlier reviews). In D-dimensional general relativity coupled to the σ-model (4.75) with $V \equiv 0$, it has been proved without assuming spherical symmetry at the outset that "the only black hole solution with a regular, nonrotating event horizon in an asymptotically flat, strictly stationary domain of outer communication is the Schwarzschild–Tangherlini solution with a constant mapping ϕ" [471]. In contrast to that, our Theorem 4.3 applies to σ-models with arbitrary $V(\vec{\phi}) \geq 0$ but selects the Tangherlini solution among spherically symmetric configurations.

There are a number of other theorems of similar nature whose common meaning is that any stationary (rather than static or spherically symmetric only) black hole is completely and uniquely characterized by solutions to the Einstein equations determined by the parameters M (mass), Q (electric [and magnetic] charge) and J (angular moment). It is this statement that is meant in Wheeler's claim [561] that "Black holes have no hair". There exist many generalizations and analogues of such theorems in extensions of GR in diverse dimensions as well as for some non-asymptotically flat configurations (see, e.g., [40, 239, 400] for reviews).

4.5.5 *Two expressions for the mass and the properties of particle-like solutions*

In this section, we will discuss particle-like solutions, i.e., solutions with a flat asymptotic behavior and a regular center. We begin by deriving two general expressions for the active gravitational (Tangherlini) mass m of a D-dimensional configuration with the metric (4.74) and an arbitrary SET compatible with the regular center and asymptotic flatness conditions.

One expression is easily obtained from the $\binom{t}{t}$ component of Eq. (4.79) which may be written in the curvature coordinates ($u = r$) in the following way:

$$\frac{d_0}{(d_0 - 1)r^{d_0}} \frac{dm}{dr} = \varkappa^2 T^t_t, \tag{4.98}$$

where $m(r)$ is the mass function,

$$m(r) \stackrel{\text{def}}{=} \frac{d_0 - 1}{2} r^{d_0 - 1}(1 - \mathrm{e}^{-2\alpha}), \tag{4.99}$$

being a generalization of the well-known four-dimensional mass function $m(r) = \frac{1}{2} r(1 - \mathrm{e}^{-2\alpha})$. For a system with a regular center ($r = 0$), the function $m(r)$, expressed from (4.98) as

$$m(r) = \frac{d_0 - 1}{d_0} \varkappa \int_0^r T_t^t r^{d_0} dr, \tag{4.100}$$

can be interpreted as the mass inside a sphere of radius r. If, in addition, the space–time is asymptotically flat, this integral converges at large r and if taken from zero to infinity, gives the full Tangherlini mass $m = m(\infty)$. The constant $\varkappa$ is expressed in terms of d_0 and the multidimensional Newtonian constant G (such that $m = GM$) if we require the validity of the usual expression for mass in terms of density, $M = \int T_t^t \, dv$ (dv being the element of volume) in the flat space limit. One thus obtains

$$\varkappa = \frac{d_0}{d_0 - 1} s(d_0) G, \quad s(d_0) = 2\pi^{(d_0+1)/2} / \Gamma\Big(\frac{d_0 + 1}{2}\Big), \tag{4.101}$$

where $s(d_0)$ is the area of a d_0-dimensional sphere of unit radius and Γ is Euler's gamma function. In case $D = 4$ we have, as usual, $\varkappa = 8\pi G$.

Equation (4.100) for the Tangherlini mass is easily rewritten in terms of an arbitrary radial coordinate, e.g., the quasiglobal coordinate x used in Eqs. (4.85)–(4.89):

$$m = \frac{d_0 - 1}{d_0} \varkappa^2 \int_{x_c}^{\infty} T_t^t(x) r^{d_0} r' dx, \tag{4.102}$$

where x_c is the value of x at the centre.

On the other hand, one can integrate the $\binom{t}{t}$ component of Eq. (4.54), which, in terms of the same x coordinate for R_t^t, assumes the form

$$\frac{1}{2r^{d_0}} (A' r^{d_0})' = \varkappa^2 \widetilde{T}_t^t. \tag{4.103}$$

For an asymptotically flat metric (4.83) with a regular center, integration of (4.103) over the whole range of x gives

$$m = \frac{\varkappa^2}{d_0} \int_{x_c}^{\infty} [(d_0 - 1) T_t^t - T_i^i] r^{d_0} dx, \tag{4.104}$$

where the index i enumerates spatial coordinates. It is a multidimensional analogue of Tolman's well-known formula [531] for the mass of a regular matter distribution in general relativity. Comparing the expressions (4.102) and (4.104), we obtain the following *universal identity valid for any particle-like static, spherically symmetric configuration in D-dimensional GR*:

$$\int_{x_c}^{\infty} [(r'-1)(d_0-1)T_t^t + T_i^i] r^{d_0} dx = 0. \tag{4.105}$$

For the σ-model (4.75), Eq. (4.104) takes the form

$$m = -\frac{2\varkappa}{d_0} \int_{x_c}^{\infty} V(\vec{\phi}) r^{d_0} dx, \tag{4.106}$$

leading to a multidimensional version of what has been previously called the generalized Rosen theorem [136]: *a static, spherically symmetric particle-like solution with positive mass cannot be obtained with scalar fields having a nonnegative potential* V.

An even stronger no-go theorem follows from the universal identity (4.105) [107]:

Theorem 4.4 (on particle-like solutions). *Equations* (4.85)–(4.89) *with* $D \geq 4$ *for the* σ*model* (4.75) *do not admit any particle-like solution if the matrix* H_{KL} *is positive-definite and* $V \geq 0$.

In other words, even negative-mass particle-like solutions can only be obtained with (at least partly) negative potentials.

To prove the theorem, it is sufficient to show that the expression in brackets in (4.105) is positive for any nontrivial solution under the conditions of the theorem — it will then follow that the integral is positive, hence it is not a particle-like configuration. This expression is

$$\frac{1}{2}(d_0-1)A(\vec{\phi}')^2 + V[2+(d_0-1)r'].$$

Its positivity is evident since, as already mentioned, $r' = 1$ at the flat asymptotic and, due to $r'' \leq 0$, we have $r' \geq 1$ in the whole range of x. □

Theorem 4.2 is the most universal: it holds for any scalar fields and, even more than that, for any matter source of gravity possessing the property $T_t^t = T_\theta^\theta$. The other theorems hold under the assumption that the matrix H_{KL} is positive-definite, or, if there is a single scalar field, it is canonical. If there is at least one phantom field, Theorems 4.1, 4.3 and 4.4 can be violated.

Particular examples of solutions with scalar fields having $V(\phi) \neq 0$ will be discussed in further chapters.

Chapter 5

More about black holes

In this chapter, we will continue the discussion of selected problems of BH physics. As before, we will deal with spherically symmetric configurations and describe two examples of "hairy" black holes in GR for which the source of gravity is a scalar field with a nonzero potential. Next, we will discuss the general problem of singularities in BH space–times and possible regular BH models both in GR and in some of its extensions: STT, HMPG and one of the versions of the brane-world concept, the so-called RS2 brane worlds. We will also describe a possible way of singularity avoidance in Schwarzschild space time due to vacuum quantum effects in strong gravitational fields. Lastly, we will very briefly discuss more general but more realistic rotating axially symmetric BHs as well as BH thermodynamics.

5.1 Examples of black holes with nonzero scalar field potentials

As we saw in Chapter 4, and in agreement with the no-hair Theorem 4.3, massless scalar fields are compatible with BHs only in special cases and in space–times exhibiting conformal continuations. Moreover, all such BHs are "cold" in the sense that they possess extremal horizons with zero Hawking temperature. The situation is in general different in the case of nonzero potentials, as will be demonstrated using some examples.

We will consider scalar–vacuum systems in GR with the Lagrangian

$$L = R + \varepsilon g^{\mu\nu}\partial_\mu\phi\partial_\nu\phi - 2V(\phi), \quad \varepsilon = \pm 1, \tag{5.1}$$

in static, spherically symmetric space–times with the metric in quasiglobal coordinates,

$$ds^2 = A(x)dt^2 - \frac{dx^2}{A(x)} - r^2(x)d\Omega^2, \tag{5.2}$$

assuming also that the scalar field is $\Phi = \Phi(x)$. The complete set of Einstein–scalar equations reads (these are Eqs. (4.85)–(4.89) for $d_0 = 2$ and a single scalar, choosing the units where $\varkappa = 1$)

$$[Ar^2\phi']' = r^2\partial V/\partial\phi, \tag{5.3}$$

$$(A'r^2)' = -2r^2V, \tag{5.4}$$

$$2r''/r = -\varepsilon\phi'^2, \tag{5.5}$$

$$A(r^2)'' - r^2A'' = 2, \tag{5.6}$$

$$2(1 - Ar'^2) - 2A'rr' = -Ar^2\phi'^2 + 2r^2V, \tag{5.7}$$

where primes denote d/dx. Among these equations, only three are independent, for example, (5.4), (5.5) and (5.6), the other two are their consequences. If $V(\phi)$ is known, there are three unknowns, the scalar field $\phi(x)$ and the two metric functions $A(x)$ and $r(x)$, for which we thus have a determined set of equations.

As already mentioned, it is in general a hard problem to solve analytically the Einstein–scalar equations when the potential $V(\phi)$ is specified. However, for many purposes, such as obtaining particular examples of the possible system behavior, it is quite helpful to use the inverse problem method: to specify one of the unknown functions and to use the equations for determining all the rest.

We will use this method by specifying the behavior of the spherical radius $r(x)$. Equation (5.6) is easily integrated giving

$$(A/r^2)' = 2(x_0 - x)/r^4, \quad x_0 = \text{const}, \tag{5.8}$$

and then, if $r(x)$ is known, $A(x)$ is found by one more integration. After that, Eq. (5.1) yields $\phi(x)$ up to an integration constant, and the potential $V(x)$ is found from (5.4). Moreover, if $r'' \neq 0$, the derivative $\phi'(x)$ does not change its sign, $\phi(x)$ is monotonic, hence $V(\varphi)$, having been found in a parametric form, is a well-determined function.

5.1.1 *Example 1:* $r^2(x) = x^2 - a^2$

Under the assumption $r^2(x) = x^2 - a^2$, the constant a plays the role of a length scale. Let us assume $a = 1$, thus expressing all quantities in terms of this arbitrary length. Since now $r''/r = -(x^2-1)^{-2} < 0$, according to Eq. (5.5), we are dealing with a canonical scalar field, $\varepsilon = +1$. Denoting $\psi = \phi/\sqrt{6}$, we obtain the asymptotically flat (as $x \to \infty$) solution in the form [136]

$$A(x) = 1 - 3mx + \frac{3}{2}(x^2 - 1)\ln\frac{x+1}{x-1}, \tag{5.9}$$

$$\psi(x) = \frac{1}{2\sqrt{3}}\ln\frac{x+1}{x-1} + \psi_0, \qquad \psi_0 = \text{const}, \tag{5.10}$$

$$V(\psi) = \frac{3m}{x^2-1}\left[6x + (3x-1)\ln\frac{x+1}{x-1}\right], \tag{5.11}$$

(in (5.11) V is a function of x, but its ψ dependence is easily restored by substituting $x = x(\psi)$ determined from (5.10)). In this solution, $x \to \infty$ is flat spatial infinity, m is the Schwarzschild mass, and the value $x = 1$ corresponds to the spherical radius $r = 0$, which is a naked central singularity if $m \leq 1/3$ and a singularity hidden under an event horizon if $m > 1/3$, see Fig. 5.1. In the "massless" case, $m = 0$, we have $A \equiv 1$, $U \equiv 0$, and the present solution coincides with the case $h = 0$ of Fisher's solution (4.20), though expressed using another radial coordinate.

The potential V is proportional to m, it is everywhere negative, it is singular at $x = 1$ and rapidly vanishes at infinity:

$$V(x) \sim \frac{\ln(x-1)}{x-1} \qquad \text{as } x \to 1,$$

$$V(x) \approx -\frac{8m}{5x^5} \qquad \text{as } x \to \infty. \tag{5.12}$$

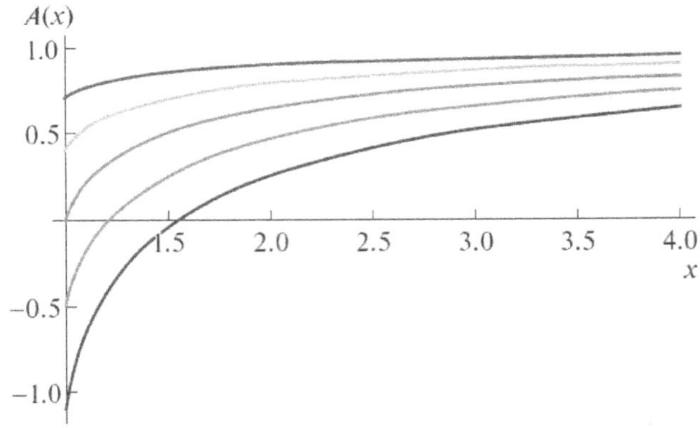

Figure 5.1 The function $A(x)$ according to (5.9) for $m = 0.1, 0.2, 1/3, 0.5, 0.7$ (upside down). It has a simple zero if $m > 1/3$, the solution then describes a black hole. All solutions have a singularity at $x = -1$, where $r = 0$.

This example shows that generic scalar–vacuum solutions with nonzero potentials (which cannot be nonnegative according to the no-hair theorem) contain naked singularities or simple horizons. In the latter case there is also a singularity at $r = 0$, and the global structure is the same as that of Schwarzschild space–time.

5.1.2 *Example 2:* $r^2(x) = x^2 + a^2$

Another analytic solution is obtained under the assumption $r(x) = \sqrt{x^2 + a^2}$ [105], and, as before, we here put $a = 1$ as an arbitrary length scale. Then the inequality $r''/r = (x^2 + 1)^{-2} > 0$ reveals the phantom nature of $\phi = \sqrt{6}\psi$, and we have the solution [105]

$$A(x) = 1 + 3mx - 3m(x^2 + 1)\,\text{arccot}\, x, \tag{5.13}$$

$$\psi(x) = \frac{1}{\sqrt{3}} \arctan x + \psi_0, \tag{5.14}$$

$$V(\psi) = \frac{6m}{(x^2 + 1)}\big[-3x + (3x^2 + 1)\,\text{arccot}\, x\big]. \tag{5.15}$$

In this solution, $x \in \mathbb{R}$, $x \to \infty$ is flat spatial infinity, and m has the meaning of the Schwarzschild mass. The behavior of the solution as $x \to -\infty$ is different, depending on the sign of m:

- $m < 0 : A \sim x^2$, $U \to \text{const} < 0$ — the solution describes a wormhole with an AdS limit at the "far end".
- $m = 0 :\ A \to 1$, $U \equiv 0$ — it is the simplest (Ellis) twice asymptotically flat wormhole with zero mass [75, 229] ($A \equiv 1$, $U \equiv 0$).
- $m > 0 :\ A \sim -x^2$, $U \to \text{const} > 0$ — we obtain a regular black hole with a de Sitter expansion far beyond the horizon instead of a singularity (a *black universe* [93, 105]), to be discussed in more detail and from a more general viewpoint in Sec. 5.2.

The behavior of $A(x)$ and $V(x)$ is shown in Fig. 5.2.

We see that the solutions with a phantom scalar and $r^2(x) = x^2 + a^2$ are nonsingular, and the spherical radius $r(x)$ nowhere turns to zero. Meanwhile, as in the previous example with a canonical scalar, we obtain BHs with simple horizons as generic configurations. The Carter–Penrose diagram of such a regular BH space–time is again the same as that of the Schwarzschild space–time, but the segments that depicted the past and future singularities in the Schwarzschild

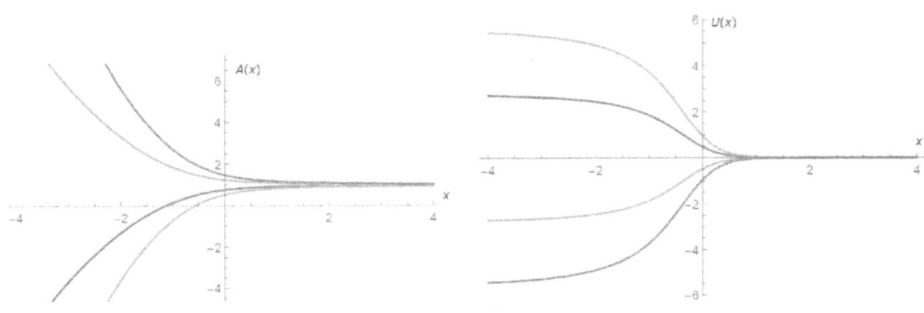

(a) (b)

Figure 5.2 The metric function $A(x)$ (a) and the potential $V(\psi(x))$ (b) for the solution (5.2), (5.13)–(5.15), $m = -0.1, -0.05, 0.05, 0.1$ [bottom-up for $A(x)$, upside down for $U(x)$]. As $x \to -\infty$, the metric is asymptotically de Sitter if $m > 0$ and AdS if $m < 0$.

case now correspond to the past and future de Sitter cosmological infinities.

5.1.3 *Extensions*

As we discussed in Sec. 4.2, the scalar-vacuum solutions of GR may be also considered as the Einstein-frame solution of such extensions of GR as STTs, $F(R)$ gravity and HMPG. The greatest freedom is observed concerning STT: their general Lagrangian (4.29) contains three arbitrary functions, of which one can be absorbed by choosing the scalar field parametrization. In $F(R)$ gravity there is a single arbitrary function: this theory is identical to the Brans–Dicke STT with the coupling constant $\omega = 0$ and an arbitrary potential function $V(\phi)$. Lastly, HMPG is equivalent to another STT, coinciding with GR with a conformally coupled scalar field, also having an arbitrary potential. In all these cases, any particular solution of GR containing a minimally coupled scalar field can be easily transformed to the corresponding solution of any of the above-mentioned alternatives of GR, formulated in their Jordan frames as well as in any other conformal frames. An evident exception is $F(R)$ gravity whose Einstein frame cannot rest on a massless scalar and must necessarily contain a nontrivial potential function. One should also bear in mind the emergence of conformal continuations for many special solutions.

Let us here discuss, similarly to Sec. 4.3, how the above examples of Einstein-frame solutions with $r^2(x) = x^2 \pm a^2$ transform to solutions with conformal coupling which, as we know, coincide with HMPG solutions.

For the solution (5.9)–(5.11) with a **canonical field**, the Jordan-frame metric is

$$ds_J^2 = \cosh^2\psi\left[A(x)dt^2 - \frac{dx^2}{A(x)} - r^2(x)d\Omega^2\right]. \tag{5.16}$$

The conformal factor $\cosh^2\psi$ is well-behaved at $x > 1$, tends to a constant at large x (so that the metric ds_J^2 is also asymptotically flat), and blows up as $x \to 1$:

$$\cosh^2 \psi \approx \cosh^2 \psi_0 + \frac{\sinh(2\psi_0)}{\sqrt{3}x}, \qquad x \to \infty,$$
$$\cosh^2 \psi \sim (x-1)^{-1/\sqrt{3}}, \qquad x \to 1. \tag{5.17}$$

Meanwhile, in the same limit $x \to 1$,

$$A(x) \approx 1 - 3m + 3m(x-1)[\ln 2 - 1 - \ln(x-1)].$$

Thus, the conformal factor cannot regularize the metric at $x = 1$: it enhances the singularity if $m \neq 1/3$ (for example, g_{tt} remains finite in the Einstein frame but blows up in Jordan's) and only modifies it if $m = 1/3$. As a whole, the conformal factor only deforms the metric at $x > 1$ but does not change it qualitatively. We conclude that the HMPG solution with the potential according to (4.45), that is,

$$U(\phi) = V(\psi) f^2(\phi) = V(\psi)/\cosh^4 \psi \tag{5.18}$$

describes a black hole with a simple horizon in the case $m > 1/3$. The black hole mass is equal to m in the Einstein frame, while in Jordan's we have

$$m_J = m \cosh \psi_0 - \frac{1}{\sqrt{3}} \sinh \psi_0. \tag{5.19}$$

They coincide if $\psi_0 = 0$.

For the solution (5.13)–(5.15) with a **phantom field**, in Jordan's frame we have the metric

$$ds_J^2 = \cos^2 \psi \left[A(x) dt^2 - \frac{dx^2}{A(x)} - r^2(x) d\Omega^2 \right], \tag{5.20}$$

and the geometry crucially depends on the value of ψ_0. The range of ψ is

$$\text{Range } (\psi) = \left(\psi_0 - \frac{\pi}{2\sqrt{3}},\ \psi_0 + \frac{\pi}{2\sqrt{3}} \right), \tag{5.21}$$

its length is $\pi/\sqrt{3} < \pi$, smaller than π, length of the segment where $\cos \psi > 0$. The spatial asymptotic value $x \to \infty$ corresponds to

$\psi = \psi_1 = \psi_0 + \pi/(2\sqrt{3})$, and the Schwarzschild mass is there

$$m_J = m \cos\psi_1 - \frac{1}{\sqrt{3}} \sin\psi_1. \tag{5.22}$$

If $\cos\psi \neq 0$ in the whole range (5.21) (for example, if $|\psi_0| < \pi(\sqrt{3} - 1)/(2\sqrt{3}) \approx 0.663$), then the conformal factor $\cos^2\psi$ is everywhere positive and regular, it then only deforms the metric ds_E^2 but does not change it qualitatively.

Otherwise, the conformal factor $\cos^2\psi$, in general, creates a singularity at some finite $x = x_s$. It destroys wormhole solutions, producing a central attracting singularity instead of their regular far end; a similar singularity is created instead of a horizon in black universe solutions if $x = x_s$ belongs to a static region or is precisely a horizon; it is like a big bang (or crunch) if it happens to be in a nonstatic region beyond the horizon. Lastly, if $\psi_0 = \pi/2 - \pi/(2\sqrt{3})$, so that $\psi_1 = \pi/2$, then the conformal factor destroys the asymptotic flatness of the solution.

We conclude that in the phantom sector all three kinds of solutions are generic: black hole ones, wormhole ones and those with naked singularities. The black hole solutions can be regular (black-universe type) or singular beyond the horizon, depending on the value of ψ_0.

The addition of an electromagnetic field substantially widens the diversity of possible black hole structures.

5.2 Regular black holes and black universes

5.2.1 *Different kinds of regular black holes*

One of the long-standing problems in (BH) physics, both in GR and its alternatives, is the existence of curvature singularities beyond the event horizon in the BH solutions obtained under the simplest and the most natural physical conditions (the Schwarzschild, Reissner–Nordström, Kerr and Kerr–Newman solutions of general relativity and their counterparts in many alternative theories). Singularities are places where GR or another classical theory of gravity does not work since they do not belong to a Riemannian space–time. Therefore, a full understanding of BH physics requires avoidance of singularities or/and modification of the corresponding classical theory and addressing quantum effects. There have been numerous attempts on this trend, many of them being connected with the hope that the singularities of classical gravity will be cured by effects of quantum gravity, to be briefly discussed below in this section. However, of great interest are also the opportunities of singularity avoidance in the framework of classical gravity.

Let us discuss the possible geometry of classical nonsingular BHs, restricting ourselves, as before, to asymptotically flat static, spherically symmetric configurations. We begin with the general static, spherically symmetric metric (5.2), written in terms of the quasiglobal coordinate x, which is, as we already know, particularly convenient for dealing with Killing horizons. At a flat asymptotic ($x \to \infty$) we have, without loss of generality, $A(x) \to 1$ and $r(x) \approx x$. A center $x = x_c$ (if any) corresponds to $r = 0$ in a static region; horizons (if any) are described by regular zeros of the function $A(x)$, and their number, order and disposition determine the global causal structure of space–time.

The four simplest types of regular four-dimensional, asymptotically flat, static, spherically symmetric BHs known in the literature can be presented as follows (see Fig. 5.3):

1. BHs with a regular center ($r \approx \text{const}\,(x - x_c)$, $A \to A_c > 0$, $A(dr/dx)^2 \approx 1 + O(r^2)$ as $x \to x_c$). Since a regular center can only be located in a R-region, such a BH must have at least two simple horizons or one double horizon, and its causal structure is then represented by the same Carter–Penrose diagram as that of the nonextreme or extreme Reissner–Nordström BH (diagrams 1b and 1c in Fig. 5.3), respectively. A larger number of horizons is not excluded, leading to more complex causal structures.
2. BHs without a center, having second-order horizons of infinite area (so-called cold BHs because such horizons are always characterized by zero Hawking temperature) [95, 97, 150]. They may have different causal structures. In one case, the Carter–Penrose diagram coincides with that of Kerr's extreme BH (see below),

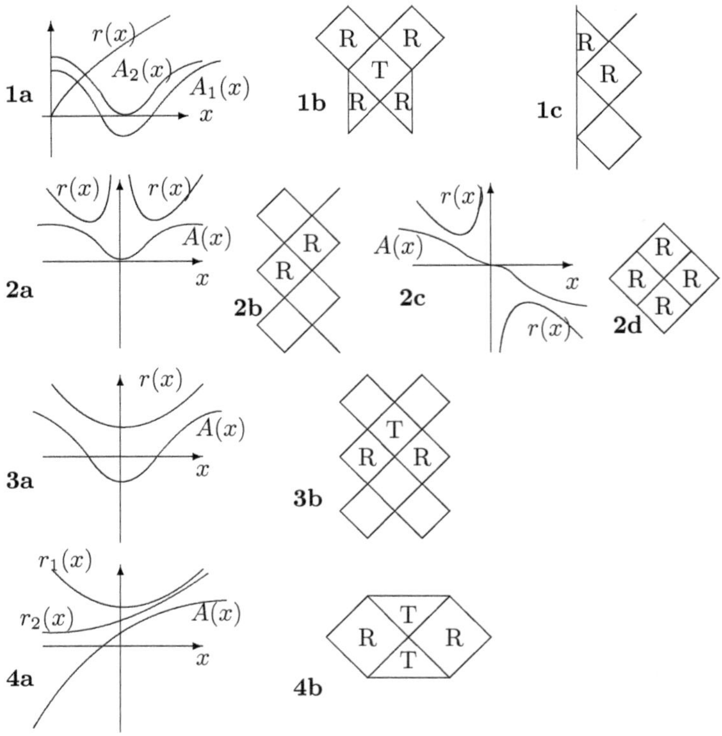

Figure 5.3 Plots showing the qualitative behavior of the metric functions and Carter–Penrose diagrams for four different types of regular static, spherically symmetric BHs. Diagrams 1b and 1d (like those for the nonextreme and extreme Reissner–Nordström metrics) refer to curves A_1 and A_2 in plot 1a, respectively. Diagram 2b, like that for the extreme Kerr metric, refers to plot 2a, diagram 2d to plot 2c, 3b to 3a and 4b to 4a. The R and T letters in the diagrams designate the R and T space–time regions. Diagrams 1b, 1c, 2b and 3b are infinitely extendible upward and downward. In all diagrams, all inner slanting lines depict horizons while all boundaries correspond to $r = \infty$, with the following exceptions: the verticals in diagrams 1b and 1d describe a regular center, $r = 0$; the horizontal lines in diagram 4b correspond either to $r = \infty$ or to $r = r_0 > 0$, according to the curves $r_1(x)$ or $r_2(x)$ at large negative x.

consisting of an infinite tower of R-regions but certainly without a ring singularity which is present in Kerr's solution (plots 2a and diagram 2b in Fig. 5.3); we have obtained this structure in Sec. 4.3. In another case, there are only four R-regions (plots 2c and diagram 2d).

3. BHs whose causal structure coincides with that of a nonextreme Kerr BH, again without a singular ring [98, 164] (diagram 3b).
4. Regular BHs with a Schwarzschild-like causal structure [105] (diagram 4b) but with cosmological expansion instead of a singularity. Such BHs have been termed *black universes* [99].

Type 1 of regular BHs traces back to Bardeen's work [32] which puts forward the very idea of regular BHs instead of singular ones and suggested, as an example, a particular BH configuration with

$$r \equiv x, \quad A(x) = 1 - \frac{Mx^2}{(x^2+q^2)^{3/2}}, \tag{5.23}$$

where M, q = const, and two horizons exist provided $q^2 < (16/27)M^2$. Later on there appeared numerous examples ([25, 79, 224] and others) of regular BH solutions where, just as in Eq. (5.23), $r \equiv x$. This, by virtue of the Einstein equation, implies that the stress-energy tensor of the matter source satisfies the condition

$$T_0^0 \equiv T_1^1 \tag{5.24}$$

or, in other words, $\rho = -p_r$ ($\rho = T_0^0$ is the energy density and $p_r = -T_1^1$ is the radial pressure). The condition (5.24) is invariant under radial boosts, making it possible to ascribe the source to vacuum matter [224]. Indeed, it is a definitive property of vacuum that there is no comoving reference frame: this kind of matter looks the same whatever is our motion in the radial direction. At the regular center in this case the matter equation of state has necessarily the form of a cosmological constant [100], $T_\mu^\nu \propto \delta_\mu^\nu$.

It has also been shown [79] that regular BHs with $r \equiv x$ and any $A(x)$ satisfying the regular center conditions may be obtained as magnetic monopole solutions of general relativity coupled to gauge-invariant nonlinear electrodynamics with the Lagrangian $L(F)$, $F := F_{\mu\nu}F^{\mu\nu}$ ($F_{\mu\nu}$ is the electromagnetic field tensor): the arbitrariness in $A(x)$ corresponds to the freedom of choosing the function $L(F)$. Solutions with an electric charge were shown [79] to be impossible whatever be the choice of $L(F)$ if $L(F)$ is the same in the whole space; this theorem, however, may be circumvented by assuming different forms of $L(F)$ near the center and at large r [147], i.e., by requiring a sort of phase transition(s) at some value(s) of the radial coordinate.

Other examples of type 1 regular BHs have also been found and discussed, see, e.g., [21, 44, 287, 371, 401, 413, 497, 509] and references therein. Among them, the BH models [21, 413, 509] are related to different ideas of quantum gravity (the so-called UV-complete quantum gravity and noncommutative geometry) realized in a semiclassical approximation.

Type 2 regular BHs, with and without an electric charge, have been obtained [95, 97, 150] in the framework of the Brans–Dicke scalar–tensor theory (STT) with the coupling constant $\omega < -3/2$, when the theory is of phantom nature. Their existence requires fine tuning in the form of specific relations between ω and the integration constants of the corresponding exact solutions. It is of interest to note that these regular cold Brans–Dicke BHs have singular counterparts in the Einstein frame, in other words, in general relativity with a minimally coupled phantom scalar field [see the anti-Fisher solution (4.22)] [93].

Type 3 regular BHs have been found [98, 164] as static, spherically symmetric solutions to the effective equations [500] describing 4D gravity in an RS2 type brane world. It has been shown that such regular solutions are generic in a certain range of the integration constants [98] and that many of them are stable, at least under some kinds of perturbations [4]. It should be stressed, however, that these 4D equations do not form a closed set; to study the full 5D geometry of the bulk one should solve the corresponding 5D equations, and only tentative results in this direction are known [165].

Type 4 configurations have been obtained [105] as generic solutions to the Einstein–scalar equations for the case of minimally coupled phantom scalar fields with certain potentials, see the previous section. Let us recall that such scalar fields have recently become popular in the cosmological context since they are able to provide an equation of state with the pressure to density ratio $p/\varepsilon = w < -1$. It is this kind of equation of state that is probably required for the dark energy (DE) component of the material content of the Universe to account for its accelerated expansion (see, e.g., the reviews [187, 489] and references therein). Type 4 configurations, termed black universes [99], are, in our view, of particular interest, and in what follows we will discuss them in some more detail.

Let us note that the above list of structures is certainly incomplete: thus, one could easily imagine regular configurations with a larger number of horizons (e.g., draw in a proper place a smooth peak twice crossing the x axis in any plot of $A(x)$ in Fig. 5.3).

An interesting example of a structure somewhat similar to a black universe has appeared in studies on the basis of a semiclassical limit of loop quantum gravity [412, 442], see Fig. 5.4. This "quantum-corrected Schwarzschild black hole" also has a Kantowski–Sachs temporal infinity instead of the singularity $r = 0$, but there are at least three important points in which it differs from our black universes: (1) the minimum radius $r_{\min}$ is always reached in the T-region; (2) $r_{\min}$ is of the order of the Planck length whereas in our models it is arbitrary, and (3) this configuration does not become isotropic at late times like ours.

Other conclusions on a quantum-corrected Schwarzschild space–time have been obtained in [58] (approaching at large times a Nariai-type metric with a constant spherical radius) and [21, 413, 509]

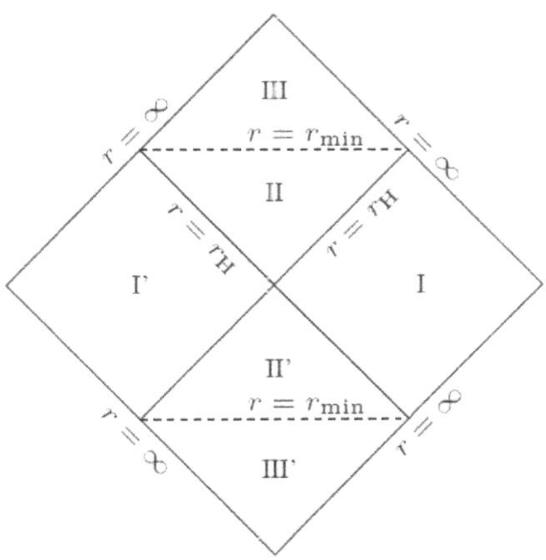

Figure 5.4 A regular BH structure obtained in [442]. I and I' are R-regions like those in the Schwarzschild space–time, II and III are Kantowski–Sachs regions (II' and III expanding, II and III' contracting).

(acquiring a regular center). This diversity is one of manifestations of an as yet uncertain choice of a correct theory of quantum gravity.

5.2.2 *Black universes with minimally coupled scalar fields*

Let us return to considering scalar–vacuum space–times with the Lagrangian (5.1), the metric (5.2) and $\phi = \phi(x)$. The sustem is governed by three independent combinations of the Einstein equations,

$$(A'r^2)' = -2r^2 V, \tag{5.25}$$

$$2r''/r = -\varepsilon\phi'^2, \tag{5.26}$$

$$A(r^2)'' - r^2 A'' = 2. \tag{5.27}$$

As we know from Sec. 4.5.3, Eq. (5.27) severely restricts the possible dispositions of zeros of the function $A(x)$ and hence the global causal structure of space–time [81], leading to the Global Structure Theorem (Theorem 4.2), according to which the choice of possible types of global causal structure is precisely the same as for the general Schwarzschild–de Sitter solution with arbitrary values of the mass and cosmological constant. This result equally applies to normal and phantom fields since Eq. (5.27) does not contain ε. It holds for any sign and shape of $V(\phi)$ and under any assumptions on the asymptotics. BHs with scalar hair (respecting the existing no-hair theorems) are not excluded. Examples of (singular) BHs with both normal (e.g., [136, 168, 432]) and phantom [196] scalar hair are known. However, BHs with a regular center (type 1 according to the previous subsection) are ruled out for our system since their existence requires a regular minimum of $B(x) = A(x)/r^2$.

As shown in [105], the system (5.25)–(5.27) with $\varepsilon = -1$ has as many as 16 types of regular solutions with flat, de Sitter and AdS asymptotic behaviors. Let us discuss asymptotically flat configurations, for which $A(x) \to 1$ and $r(x) \approx x$ as $x \to \infty$. Then, as x decreases from infinity, the derivative r' also decreases according to Eq. (5.26), the decrease rate depending on the details of the system. If the decrease is slow enough, $r(x)$ will reach zero at some finite x.

It can be a regular center only if horizons are absent (otherwise there would be a type 1 regular BH which is ruled out here), and a particlelike solution is then obtained instead of a BH one.

Assuming the absence of singularities, other opportunities are $r(x) \to r_0 = \text{const}$ and $r(x) \to \infty$ as $x \to -\infty$. (Note that any kind of oscillatory behavior of $r(x)$ is ruled out by (5.26) according to which $r'' \geq 0$.)

In the first case, according to (5.6) (that leads to (5.8)),

$$A' \approx -2x/r_0^2 \to +\infty, \quad A \approx -x^2/r_0^2 \quad \text{as } x \to -\infty.$$

So, this "r_0 asymptotic" is located in a T-region. The radius r_0 is related to the limiting value of the potential: $V \to -1/r_0^2$. Changing the notations $-x \to T$ and $t \to y$ (since x here becomes a temporal coordinate and the former time t a spatial one), we can write the asymptotic form of the resulting Kantowski–Sachs metric as

$$ds^2 \approx \frac{r_0^2}{T^2}\, dT^2 - \frac{T^2}{r_0^2} dy^2 - r_0^2 d\Omega^2 \quad \text{as } T \to \infty. \tag{5.28}$$

It is a highly anisotropic universe, exhibiting no expansion in the two angular directions and an exponential (in terms of the physical time $\tau \sim \log T$) expansion in the third direction x. From the viewpoint of an observer at large positive x, this universe is located beyond the event horizon of a BH.

In the case $r \to \infty$, the most interesting opportunity is that $r \sim |x|$ as $x \to -\infty$. Assuming $r \approx -ax$, $a = \text{const} > 0$, we obtain from (5.8) and (5.25):

$$A \approx 1/a^2 - Ca^2x^2, \quad V \approx 3Ca^2, \quad C = \text{const}. \tag{5.29}$$

It is easy to verify that $C = 0$ leads to a Minkowski metric at large negative x (though the time rate will be different from that at large positive x if $a \neq 1$), an anti-de Sitter metric if $C < 0$ and a de Sitter metric if $C > 0$. In the cases $C \leq 0$ horizons are absent since otherwise the function $B(x) = A/r^2$ would have a minimum at some finite x, which cannot happen due to (5.8). Thus possible solutions with $C \leq 0$ describe traversable wormholes (see Chapter 6), and the details of their geometry depend on the particular choice of $V(\phi)$.

Lastly, with $C > 0$ we obtain a de Sitter asymptotic behavior of the solution, describing (since we are now in a T-region) isotropic expansion or contraction. From the viewpoint of an external observer, located at large positive x, it is a BH, but a possible BH explorer, instead of hitting a singularity, has now a chance to survive for a new life in an expanding, gradually isotropizing Kantowski–Sachs universe. The specific BH profile and the isotropization mode after crossing the horizon depend on the choice of $V(\phi)$.

Since, according to the Global Structure Theorem (see above), an asymptotically flat configuration can have only one simple horizon, such a BH has a Schwarzschild-like causal structure, but the singularity $r = 0$ in the Carter–Penrose diagram is now replaced by the de Sitter infinity $r = \infty$.

A simple example of such a black universe has already been obtained in Sec. 5.1.2 using the ansatz [105]

$$r = (x^2 + a^2)^{1/2}, \quad a = \text{const} > 0, \tag{5.30}$$

and the inverse problem scheme. The solution can be rewritten in the form

$$B(x) = \frac{A(x)}{r^2(x)} = \frac{c}{a^2} + \frac{1}{a^2 + x^2} + \frac{x_0}{a^3}\left(\frac{ax}{a^2 + x^2} + \arctan\frac{x}{a}\right), \tag{5.31}$$

$$\phi = \pm\sqrt{2}\arctan(x/a) + \phi_0, \tag{5.32}$$

$$V = -\frac{c}{b^2}\frac{r^2 + 2x^2}{r^2} - \frac{x_0}{a^3}\left(\frac{3ax}{r^2} + \frac{r^2 + 2x^2}{r^2}\arctan\frac{x}{a}\right), \tag{5.33}$$

with $c, \phi_0, x_0 = \text{const}$, and $r = r(x)$ given by (5.30). In particular,

$$B(\pm\infty) = -\frac{1}{3}V(\pm\infty) = \frac{2ac \pm \pi x_0}{2a^3}. \tag{5.34}$$

Choosing in (5.32), without loss of generality, the plus sign and $\phi_0 = 0$, we obtain for $V(\phi)$ $(\psi := \phi/\sqrt{2})$:

$$V(\phi) = -\frac{c}{a^2}(3 - 2\cos^2\psi) - \frac{x_0}{a^3}[3\sin\psi\cos\psi + \psi(3 - 2\cos^2\psi)]. \tag{5.35}$$

The solution behavior is controlled by two integration constants: c that moves the plot of $B(x)$ up and down, and x_0 showing the maximum of $B(x)$. Both $r(x)$ and $B(x)$ are even functions if $x_0 = 0$, otherwise $B(x)$ loses this symmetry. Asymptotic flatness at $x = +\infty$ implies $2ac = -\pi x_0$ while the Schwarzschild mass, defined in the usual way, is $m = x_0/3$.

Under this asymptotic flatness assumption, for $x_0 = m = 0$ we obtain the simplest symmetric configuration, the Ellis wormhole [229]: $A \equiv 1$, $V \equiv 0$. For $x_0 < 0$, according to (5.34), we obtain a wormhole with $m < 0$ and an AdS metric at the far end, corresponding to the cosmological constant $V_- < 0$. For $x_0 > 0$, when $V_- > 0$, there is a black universe with $m > 0$ and a de Sitter asymptotic far beyond the horizon.

The horizon radius $r(x_h)$ can be obtained by solving the transcendental equation $A(x_h) = 0$, where $A(x)$ is given by Eq. (5.31). It depends on the parameters m and $a = \min r(x)$ and cannot be smaller than the parameter a, which plays the role of a scalar charge since $\psi \approx \pi/2 - a/x$ at large x. Since $A(0) = 1 + c$, the throat $x = 0$ is situated in the R-region if $c > -1$, i.e., if $3\pi m < 2a$, precisely at the horizon if $3\pi m = 2a$ and in the T-region beyond the horizon if $3\pi m > 2a$. These relations between the constants m and b prompt (and this is probably the case in more general situations) that if the BH mass dominates over the scalar charge, there is no throat in the static region, and for a distant observer the BH looks approximately the same as usual in general relativity.

It should be mentioned that spherically symmetric geometries in which a minimum of the radius r takes place in a T-region and thus represents a bounce in one of the Kantowski–Sachs scale factors have attracted some attention and have received the name of "black bounce" geometries [387, 503].

As follows from the scalar field equation (5.3) and from (5.25), the potential V tends to a constant and, moreover, $dV/d\phi \to 0$ at each end of the x range. It is a general property of all classes of regular solutions indicated in [105]. More precisely, *a regular scalar field configuration requires a potential with at least two zero-slope points (not necessarily extrema) at different values of* ϕ.

Suitable potentials are, e.g., $V = V_0 \cos^2(\phi/\phi_0)$ and the Mexican hat potential $V = (\lambda/4)(\phi^2 - \eta^2)^2$ where V_0, ϕ_0, λ, η are constants. A flat infinity at $x = +\infty$ certainly requires $V_+ = 0$, while a de Sitter asymptotic can correspond to a maximum of V since phantom fields tend to climb up the slope of the potential rather than roll down, as is evident from Eq. (5.3). Accordingly, Faraoni [236], considering spatially flat isotropic phantom cosmologies, has shown that if $V(\phi)$ is bounded above by $V_0 = \text{const} > 0$, the de Sitter solution is a global attractor. Very probably this conclusion extends to Kantowski–Sachs cosmologies after isotropization.

One more point may be stressed: the late-time de Sitter expansion rate is entirely determined by the corresponding potential value $V_- > 0$ (which, in our notation according to (5.1), coincides with the effective cosmological constant Λ at late times) rather than by the details of the solution such as the Schwarzschild mass defined at the flat asymptotic.

We can conclude that black universes are a generic class of solutions to the Einstein–scalar equations in the case of phantom scalars with proper potentials.

Thus we have classified the possible geometries of static, spherically symmetric, and asymptotically flat BHs and discussed in more detail a particular type, the black universes. Such hypothetical configurations combine the properties of a wormhole (absence of a center, a regular minimum of the area function) and a BH (a horizon separating R- and T-regions).

Quite naturally, such unusual objects require unusual matter for their existence. We saw how they can be obtained using a phantom scalar field. As shown in [99], such solutions also exist in much more general frameworks for the description of phantom matter, namely, STT of gravity and the so-called k-essence. A conclusion is that a great number of models of phantom matter produce such solutions.

The latter circumstance leads to the idea that our Universe could appear from phantom-dominated collapse in another, "mother" universe and undergo isotropization (e.g., due to particle creation) soon after crossing the horizon. It is known that a Kantowski–Sachs nature of our Universe is not excluded observationally [11] if its isotropization had happened early enough, before the last scattering epoch (at redshifts $z \gtrsim 1000$). One can notice that we are thus facing one more mechanism of universes multiplication, in addition to the well-known mechanism existing in the chaotic inflation scenario [378].

As we saw in Sec. 4.2.1, there are arguments both *pro et contra* phantom fields, and the latter still seem somewhat stronger. It is reasonable to try to avoid such fields in modeling real or hypothetic objects and phenomena. In what follows, we describe one such attempt, leading to regular BH models model in brane-world scenarios of RS2 type [98]. We can also mention two other versions of black universe models with scalar fields: one also in an RS2 type brane world, with a canonical scalar field and without any phantom matter [101]; another [102] uses the notion of a "trapped ghost" [140], i.e., a scalar field which has phantom properties only in a restricted region of space, a strong-field region, whereas far away from it all standard energy conditions are observed. This can explain why ghosts, or phantoms, are not observed under usual physical conditions but certainly does not remove the basic shortcomings of phantom fields, see Sec. 4.2.1.

5.2.3 *Regular black holes in a brane world*

The brane world concept describes our world as a 4D surface (brane) supporting all or almost all matter fields and embedded in a higher-dimensional space–time (called the bulk). This concept traces back to the early 80s and leads to a variety of consequences in cosmology, gravitational and particle physics, see the reviews [186, 393, 427, 474]. In particular, brane worlds turn out to be a natural framework for wormholes (see Sec. 6.3.4) since there the modified Einstein equations [500] (5.36) contain a source term E_μ^ν of geometric origin which need not observe the usual energy conditions. As we shall see now, it is this source term that can replace phantom fields in constructing regular BH models.

The modified Einstein equations (5.36) used here correspond to the so-called RS2 scenario: a single brane in a $\mathbb{Z}_2$-symmetric five-dimensional bulk, with all fields except gravity confined to the brane. It generalizes the second Randall–Sundrum model comprising a single Minkowski brane in an AdS bulk [460]. In other brane-world scenarios, the effective 4D Einstein equations also contain terms similar to E_μ^ν, e.g., on codimension-1 branes without Z_2 symmetry [570] and in Dvali–Gabadadze–Porrati brane worlds [223] with different kinds of induced gravity terms [486]. Thus we can anticipate that regular BH and black-universe solutions similar to ours exist in such brane worlds as well, though probably under some other conditions.

The gravitational field on the brane is described by the modified Einstein equations [500]

$$G_\mu^\nu = -\Lambda_4\delta_\mu^\nu - \varkappa_4^2 T_\mu^\nu - \varkappa_5^4 \Pi_\mu^\nu - E_\mu^\nu, \tag{5.36}$$

where $G_\mu^\nu = R_\mu^\nu - \frac{1}{2}\delta_\mu^\nu R$ is the 4D Einstein tensor, Λ_4 is the 4D cosmological constant expressed in terms of Λ_5 and the brane tension λ:

$$\Lambda_4 = \frac{1}{2}\left(\Lambda_5 + \frac{1}{6}\varkappa_5^4\lambda^2\right); \tag{5.37}$$

$\varkappa_4^2 = 8\pi G_N = \varkappa_5^4\lambda/(6\pi) = m_4^{-2}$ is the 4D gravitational constant; G_N is Newton's constant of gravity, and m_4 is the 4D Planck mass;

$\varkappa_5 = m_5^{-3/2}$, m_5 being the 5D Planck energy scale;

T_μ^ν is the SET of matter trapped on the brane;

Π_μ^ν is a tensor quadratic in T_μ^ν, obtained from matching the 5D metric across the brane:

$$\Pi_\mu^\nu = \frac{1}{4}T_\mu^\alpha T_\alpha^\nu - \frac{1}{2}TT_\mu^\nu - \frac{1}{8}\delta_\mu^\nu\left(T_{\alpha\beta}T^{\alpha\beta} - \frac{1}{3}T^2\right), \tag{5.38}$$

where $T = T_\alpha^\alpha$; and E_μ^ν is the "electric" part of the 5D Weyl tensor projected onto the brane. In proper 5D coordinates,

$$E_{\mu\nu} = \delta_\mu^A\delta_\nu^C {}^{(5)}C_{ABCD}n^B n^D, \tag{5.39}$$

where $A, B, \ldots$ are 5D indices and n^A is the unit normal to the brane. By construction, E_μ^ν is traceless, such that $E_\mu^\mu = 0$.

Other characteristics of E_μ^ν are unknown without specifying the 5D metric, hence the set of equations (5.36) is not closed. In isotropic cosmology this leads to an additional arbitrary constant in the field equations, connected with the density of "dark radiation". For static, spherically symmetric systems to be discussed here, this freedom is expressed in a single arbitrary function of the radial coordinate.

Reasons for neglecting Π_μ^ν. Let us show that under quite reasonable conditions we can neglect the tensor Π_μ^ν in (5.36).

We put $\Lambda_4 = 0$, so that

$$|\Lambda_5| = \frac{1}{6}\varkappa_5^4\lambda^2 = 6\pi^2(\varkappa_4/\varkappa_5)^4, \tag{5.40}$$

and use the observational restriction on the bulk length scale ℓ which follows from the recent short-range Newtonian gravity tests [388], showing that Newton's inverse-square law holds at length scales greater than about 0.1 mm. This means that if we live on an RS2-like brane, the bulk length scale can be estimated as

$$\ell = (6/|\Lambda_5|)^{1/2} \lesssim 10^{-2} \text{ cm}. \tag{5.41}$$

Note that the 4D Planck scale in our notation is

$$m_4 = \varkappa_4^{-1} = 8\pi G_{\rm N} \approx 2.4{\cdot}10^{18}\,\text{GeV},$$
$$l_4 = 1/m_4 = \varkappa_4 \approx 8{\cdot}10^{-33}\,\text{cm}.$$

Combining (5.40) and (5.41), we obtain

$$m_5/m_4 = (\pi\ell/l_4)^{-1/3} \gtrsim 10^{-10}, \tag{5.42}$$

so that the 5D Planck energy scale in this scenario is at least about 10^8 GeV.

Now, we can assert that the term with Π_μ^ν is negligible in (5.36) as compared with the T_μ^ν term as long as

$$\varkappa_5^4 W^2 \ll \varkappa_4^2 W,$$

where W characterizes the magnitude of T_μ^ν, say, the absolute value of the largest component of T_μ^ν or

$$W \ll m_5^6/m_4^2 = m_4^4(m_5/m_4)^6, \tag{5.43}$$

where $m_4^4 \approx 3.5{\cdot}10^{73}\text{GeV}^4 \approx 8.4{\cdot}10^{90}\ \text{g}{\cdot}\text{cm}^{-3}$ is the Planck density while the second factor is, according to the experimental bound (5.42), about 10^{-60} or larger. As a result, for the "density" W we have

$$W \ll 10^{30}\ \text{g}{\cdot}\text{cm}^{-3}.$$

Recalling that the density of nuclear matter is about $10^{13}\ \text{g}{\cdot}\text{cm}^{-3}$, it is clear that this bound certainly holds for any thinkable matter.

Solution of the trace equation. Our interest here will be in selecting a general class of static, spherically symmetric BH solutions to Eqs. (5.36) without specifying E_μ^ν. The BH construction procedure will be formulated in the general case when both matter and the cosmological constant Λ_4 are present and the space–time asymptotic properties are not specified.

We will not discuss the possible bulk properties of models in question and only note that the existence of the corresponding solutions to the higher-dimensional equations of gravity (in our case, the 5D vacuum Einstein equations with a cosmological term) is guaranteed by the Campbell–Magaard type embedding theorems [151, 396]; see also a more recent exposition and more references in [494].

Due to the arbitrariness of E_μ^ν, the only unambiguous equation among (5.36) is the trace equation. To solve it, it turns out to be helpful to write the static, spherically symmetric metric in 4 dimensions in the curvature coordinates,

$$ds^2 = A(r)dt^2 - \frac{dr^2}{B(r)} - r^2 d\Omega^2, \tag{5.44}$$

so that the Ricci scalar is

$$R = \frac{2}{r^2}(1-B) - B\left[\frac{A_{rr}}{A} - \frac{A_r^2}{2A^2} + \frac{A_r B_r}{2AB} + \frac{2}{r}\left(\frac{A_r}{A} + \frac{B_r}{B}\right)\right], \tag{5.45}$$

where the subscript r means d/dr.

If we treat Eqs. (5.36) as the conventional Einstein equations with an effective SET $T^\nu_{\mu\,\text{eff}}$, i.e., $G_\mu^\nu = -\varkappa_4^2 T^\nu_{\mu\,\text{eff}}$, then the effective density

$\rho_{\rm eff}$, radial pressure $p_{r\,\rm eff}$ and tangential pressure $p_{\perp\rm eff}$ are expressed in terms of A and B as follows:

$$G_t^t = -\varkappa_4^2\rho_{\rm eff} = \frac{B-1}{r^2} + \frac{B_r}{r},$$
$$G_r^r = \varkappa_4^2 p_{r\,\rm eff} = \frac{B-1}{r^2} + \frac{BA_r}{Ar},$$
$$G_\theta^\theta = G_\phi^\phi = \varkappa_4^2 p_{\perp\rm eff} = \frac{B}{4}\left[\frac{2A_{rr}}{A} - \frac{A_r^2}{A^2} - \frac{A_rB_r}{AB} + \frac{2}{r}\left(\frac{A_r}{A} + \frac{B_r}{B}\right)\right]. \tag{5.46}$$

The trace of Eqs. (5.36) has the form

$$R = 4\Lambda_4 + \varkappa_4^2 T_\alpha^\alpha + \varkappa_5^4 \Pi_\alpha^\alpha. \tag{5.47}$$

Assuming that the right-hand side is a known function of r, i.e., that $R = R(r)$ is known, Eq. (5.47) may be written as a linear first-order equation with respect to $f(r) := rB(r)$ [98, 139, 540]:

$$A(rA_r + 4A)f_r + [r(2AA_{rr} - A_r^2) + 3AA_r]f = 2A^2[2 - r^2R(r)]. \tag{5.48}$$

Its general solution is

$$f(r) = \frac{2A\,{\rm e}^{3\Gamma}}{(4A + rA_r)^2}\int (4A + rA_r)[2 - r^2R(r)]\,{\rm e}^{-3\Gamma}\,dr \tag{5.49}$$

where

$$\Gamma(r) = \int \frac{A_r dr}{4A + rA_r}. \tag{5.50}$$

Thus, choosing any smooth function $A(r)$, we obtain $f(r)$ from (5.49), and, after fixing the integration constant, the metric is known completely.

If the function $R(r)$ is not specified, Eq. (5.49) is simply another form of the trace of the Einstein equations. It is valid for *any* static, spherically symmetric metric, at least in ranges of r where the quantity $A(4A + rA_r)$ is finite and nonzero and where $r = \sqrt{-g_{\theta\theta}}$ is an admissible coordinate. The latter means, in particular, that Eq. (5.49) is applicable to a wormhole metric only on one (but either) side of a wormhole throat (see [139] for details) and is evidenly invalid for flux-tube metrics, characterized by $g_{\theta\theta} = \text{const.}$

For a given SET T_μ^ν, the dependence $R(r)$ is not always known, but in the vacuum case $T_\mu^\nu = 0$, so that $R(r) = 4\Lambda_4$, the solution to the Einstein equations can always be written in the form (5.49) under the above evident restrictions.

Conditions at horizons. Before singling out BH solutions on the basis of Eq. (5.49), let us first formulate the conditions under which the generating function $A(r)$ leads to a metric with a Killing horizon. As we remember, for this purpose preferential is the quasiglobal coordinate x specified by the condition $g_{tt}g_{xx} = -1$, so that the metric (5.44) is then rewritten in the form

$$ds^2 = \mathcal{A}(x)dt^2 - \frac{dx^2}{\mathcal{A}(x)} - r^2(x)d\Omega^2, \tag{5.51}$$

where the variables are connected with those in Eq. (5.44) as follows:

$$\mathcal{A}(x) = A(r), \quad r(x) = r, \quad \mathcal{A}(x)(dr/dx)^2 = B(r). \tag{5.52}$$

Both $\mathcal{A}$ and r must be smooth functions of x near the horizon $x = h$, so that

$$\mathcal{A}(x) \sim (x-h)^k, \quad r(x) \approx r_h + \text{const} \cdot (x-h)^s, \tag{5.53}$$

where $k = 1, 2, \ldots$ is the order of the horizon (the horizon is simple if $k = 1$, double if $k = 2$, etc.) while $s = 1, 2, \ldots$ characterizes the behavior of $r(x)$; $r_h > 0$ is the horizon radius (leaving aside possible horizons of infinite radius as in "cold BHs" [93, 95]). Generically but not necessarily one has $s = 1$. If $s = 1$ (i.e., dr/dx is finite at the horizon), the coordinate r can be used for continuing the metric through the horizon on equal grounds with x. The continuation will be analytic if both $\mathcal{A}(x)$ and $r(x)$ are analytic at $x = h$.

Assuming (5.53) and directly employing the relations (5.52), we find that

$$f(r) \sim B(r) \sim (r - r_h)^{(k+2s-2)/s} \quad \text{as} \quad x \to h. \tag{5.54}$$

On the other hand, substituting $A = \mathcal{A}(x) \sim (r - r_h)^{k/s}$ into the solution (5.49) and assuming that the quantity $Q(r) = 2 - r^2 R(r)$ is finite at $r = r_h$ (it is generically so), it is easy to obtain that near r_h

$$f(r) \sim B(r) \sim (r - r_h)^{2-k/s}[(r - r_h)^{k/s} + C], \tag{5.55}$$

where C is an integration constant and $C = 0$ corresponds to the case where integration in (5.49) is performed from r_h to r. Comparing the exponents in (5.54) and (5.55), we see that

- in case $C = 0$: $k = 2$, s is not restricted;
- in case $C \neq 0$: $k = 1$, s is not restricted.

Thus, to obtain a solution with a horizon at $r = r_h$, we should take $A(r)$ behaving as $(r - r_h)^{k/s}$ with $k = 1$ or 2 and $s \in \mathbb{N}$.

An important point is that, under the condition $R(r_h) \neq 2/r_h^2$, a horizon can be either simple ($k = 1$) or double ($k = 2$); horizons of higher orders k do not appear. This is true, in particular, for all static, spherically symmetric metrics with $R = 0$.

Let us now look what changes when the function $2 - r^2R(r)$ vanishes at $r = r_h$. One can write

$$Q(r) := 2 - r^2R(r) \sim (r - r_h)^p, \quad p = 0, 1, 2, \ldots, \quad \text{as} \quad r \to r_h, \tag{5.56}$$

preserving the assumptions (5.53). So $p = 0$ corresponds to the above generic case $Q(r_h) \neq 0$ and $p > 0$ means that $Q(r)$ has a zero of order p. Then the expression (5.54) for $f(r)$ remains the same but (5.55) must be replaced with

$$f(r) \sim B(r) \sim (r - r_h)^{2-k/s}[(r - r_h)^{k/s+p} + C]. \tag{5.57}$$

Thus, if $C \neq 0$, the metric behaves as before, while for $C = 0$ we obtain

$$\begin{aligned} &A(r) \sim (r - r_h)^{p+2/s}, \qquad B(r) \sim (r - r_h)^{p+2}, \\ &\mathcal{A}(u) \sim (u - h)^{ps+2}, \end{aligned} \tag{5.58}$$

that is, a horizon of order $ps + 2$.

Assuming $A = 0$ at $r = r_h > 0$, one can rewrite Eq. (5.49) in the form

$$f(r) \equiv rB(r) = \frac{2A\,\mathrm{e}^{3\Gamma}}{(4A + rA_r)^2}\left\{\int_{r_h}^{r}(4A + rA_r)[2 - r^2 R(r)]\,\mathrm{e}^{-3\Gamma}\,dr + C\right\}. \tag{5.59}$$

The above analysis shows that we obtain a metric with a horizon at $r = r_h$ in two cases:

(i) $A \sim (r - r_h)^{1/s}$, as $r \to r_h$, $s \in \mathbb{N}$. Then Eq. (5.59) leads to a metric with a simple horizon if $C \neq 0$ and with a horizon of order $2 + ps$ if $C = 0$.
(ii) $A \sim (r - r_h)^{2/s}$ as $r \to r_h$, s odd. Then Eq. (5.59) leads to a metric with a horizon of order $2 + ps$ in case $C = 0$.

Here, as before, the parameter p characterizes the behavior of $Q(r)$ according to Eq. (5.56).

Item (ii) does not include solutions with $C \neq 0$. The point is that in case $A \sim (r - r_h)^2$, $C \neq 0$, the metric is singular at $r = r_h$, as is confirmed by calculating the Kretschmann scalar. For odd numbers $s > 1$, the metric with $A \sim (r - r_h)^{2/s}$, $C \neq 0$ has a finite Kretschmann scalar but loses analyticity at $r = r_h$ and cannot be analytically continued through this sphere.

Definition and algorithms. We have been so far discussing local conditions at possible Killing horizons. Let us now turn to space-time properties at large and try to select BH metrics. We do not need a general rigorous definition of a BH [549] which in turn needs such notions as strong asymptotic predictability, trapped regions etc. The following working definition will be appropriate for our purposes.

Definition. *The metric (5.51) is said to describe a BH if (a) the functions* $\mathcal{A}(x)$ *and* $r(x)$ *are analytic in the range* $\mathcal{R}[x] : \ h \le x < x_{\max}$ *where* $x_{\max}$ *may be finite or infinite; (b)* $r(x) > 0$ *in* $\mathcal{R}[x]$*, and* $r(x_{\max}) > r(h) = r_h$*; (c)* $\mathcal{A}(x) > 0$ *at* $x > h$*, and* $\mathcal{A}(h) \sim (x - h)^k$*,* $k \in \mathbb{N}$ *as* $r \to r_h$.

Item (c) means that $x > h$ is a static region (R-region) of a static, spherically symmetric space-time while the sphere $x = h$, a boundary of this region, is a Killing horizon of a certain order k. So, in usual terms, our working definition describes the domain of outer communication of a BH, and $x = h$ is its event horizon.

The definition uses the x coordinate rather than r, due to its advantage at horizon description. The difference is really essential: there are metrics which behave non-analytically in terms of r at $r = r(h) = r_h$ but analytically in terms of x at $x = h$.

The analyticity requirement rejects possible cases of restricted smoothness. In our view, if we are dealing with a field configuration, its non-analyticity at a certain surface must have a physical reason, e.g., a phase transition, and it seems too artificial, a kind of perfect fine tuning, to assume that the phase transition occurs precisely at a horizon.

We do not require $r(x_{\max}) = \infty$ since we do not want to rule out metrics with cosmological horizons like the Schwarzschild–de Sitter space–time where an R-region is situated between a BH horizon and a cosmological horizon. We, however, adopt the requirement $r(x_{\max}) > r_h$ to exclude configurations with only cosmological horizons.

Let us return to the solution (5.49), or (5.59). Due to its generality, it certainly describes all BH metrics, at least piecewise. We can, however, formulate explicit requirements to the generating function $A(r)$ under which Eq. (5.49) leads *algorithmically* to a BH metric. Namely, let there be a range

$$\mathcal{R}[r]: \quad r_{\max} > r > r_h, \quad r_h > 0, \tag{5.60}$$

in which the r.h.s. of Eq. (5.48) is positive:

$$Q(r) = 2 - r^2 R(r) > 0. \tag{5.61}$$

Then the above items (i) and (ii) lead to the following BH construction algorithms.

(BH1) *Specify a function $A(r)$, positive and analytical in $\mathcal{R}[r]$, in such a way that $g(r) = 4A + rA_r > 0$ in $\mathcal{R}[r]$ and $A \sim (r - r_h)^{1/s}$, $s \in \mathbb{N}$, as $r \to r_h$. Then the functions $A(r)$ and*

$B(r)$ given by Eq. (5.59) *with $C \geq 0$ determine a BH metric* (5.44) *with a horizon at $r = r_h$. The horizon is simple if $C > 0$; in case $C = 0$ it is of the order $2+ps$ if $Q(r)$ behaves according to Eq.* (5.56).

(BH2) *Specify a function $A(r)$, positive and analytical in $\mathcal{R}[r]$, in such a way that $g(r) = 4A + rA_r > 0$ in $\mathcal{R}[r]$ and $A \sim (r - r_h)^{2/s}$ as $r \to r_h$, s being an odd positive integer. Then the functions $A(r)$ and $B(r)$ given by Eq.* (5.59) *with $C = 0$ determine a BH metric* (5.44) *with a horizon at $r = r_h$ of the order $2 + ps$ if $Q(r)$ behaves according to Eq.* (5.56).

Both algorithms (BH1) and (BH2) lead to double horizons in case $C = 0$ if $Q(r_h) > 0$.

To obtain **asymptotically flat** BHs, one should assume $r_{\max} = \infty$ and restrict the choice of $A(r)$ to functions compatible with asymptotic flatness. Properly choosing the time scale, we can require a Schwarzschild behavior of A: $A = 1 - 2m/r + o(1/r)$ as $r \to \infty$, where m is the Schwarzschild mass. Equation (5.59) then leads to the proper behavior of $B(r)$, i.e., $B \to 1$ as $r \to \infty$, provided the curvature R decays quickly enough: $R(r) = o(r^{-3})$ as $r \to \infty$.

One can notice that both algorithms use the r coordinate whereas the BH definition uses x. A transition to x is carried out with Eq. (5.52). The conditions of (BH1) and (BH2) guarantee that $B > 0$ at $r > r_h$, therefore, choosing $dr/dx > 0$ in (5.52), we evidently satisfy the BH definition.

The condition (5.61) can be weakened: what is actually required is that the integral in (5.59) should remain positive in $\mathcal{R}[r]$.

Another condition of both algorithms, $g(r) = 4A + rA_r > 0$ in $\mathcal{R}[r]$, allows one to avoid a singular point of Eq. (5.48), where the coefficient of the derivative f_r vanishes. This coefficient also vanishes at horizons, which have been already discussed. The points where $A \neq 0$ but $g = 0$, if any, also deserve special attention. Though, the equality $g = 0$ itself has no evident geometric (and hence physical) meaning, it only represents some technical difficulty in our description with the aid of Eq. (5.48).

Points where $g = 0$ are avoided by most of asymptotically flat BH metrics. Indeed, in this case $g \geq 0$ at the horizon and $g = 4$ at infinity;

the condition $g(r) > 0$ means that $r^4 A$ is a strictly increasing function of r, which is the case, e.g., for all functions $A(r)$ monotonically growing from 0 at the horizon to 1 at infinity.

However, one can easily verify that such points inevitably appear in other important situations, e.g., between a regular center and a cosmological horizon or between two simple horizons.

It can be shown that if the choice of the generating function $A(r)$ leads to $g(r)$ with a simple zero at some $r = r_s \in \mathcal{R}[r]$, then there is a unique nonzero value of the constant C (namely, such that integration in Eq. (5.49) starts from r_s) making it possible to avoid a singularity of the metric at $r = r_s$. So Algorithm (BH1) survives, but it now gives a single solution with a horizon instead of a family parametrized by C. Algorithm (BH2) requires $C = 0$ and therefore does not work.

Generic behavior of the solutions: wormholes and regular BHs. Let us discuss the properties of the metric in the generic situation that leads to a BH by Algorithm (BH1): let $A(r)$ be a well-behaved positive function at $r > r_h$ and have a simple zero at $r = r_h$, and let also $Q(r_h) > 0$. Then, in a small neighborhood of $r = r_h$, we can write

$$\begin{aligned} A(r) &= A_1(r - r_h) + o(r - r_h), \\ B(r) &= B_1 C\,(r - r_h) + B_2(r - r_h)^2 + o\left((r - r_h)^2\right), \end{aligned} \tag{5.62}$$

with fixed positive constants A_1, B_1, B_2. The integration constant C is a family parameter, and $C = 0$ is its critical value at which the solution drastically changes its properties.

If $C < 0$, B turns to zero at $r = r_{\min} = r_h + |C|B_1/B_2 > r_h$. (Here, $|C|$ should be small enough for the solution to remain in a range where Eq. (5.62) is still approximately valid.) One obtains $B(r) \approx B_2(r - r_h)(r - r_{\min})$, so that $B(r)$ has a simple zero at $r = r_{\min}$, whereas $A(r_{\min}) > 0$. Such a behavior of the metric functions corresponds to a symmetric wormhole throat at $r = r_{\min}$ [139]. The substitution $r = r_{\min} + x^2$, $x \in \mathbb{R}$, makes the metric (5.44) regular at $r = r_{\min}$ ($x = 0$), and all metric coefficients are even functions of x. Thus, our solution does not reach the anticipated horizon $r = r_h$ and describes a symmetric wormhole.

In case $C = 0$, as already described, we obtain a double horizon at $r = r_h$, and the geometry is smoothly continued to smaller r, where further properties of the metric depend on the specific choice of $A(r)$.

If $C > 0$, then $B > 0$ at $r > r_h$, turns to zero at the horizon $r = r_h$ and again turns to zero in the T-region at $r = r_{\min} = r_h - C B_1/B_2 < r_h$. This, as before, bounds the range of r from below in a way similar to a wormhole throat. The coordinate singularity at $r = r_0$ is again removed by the transformation $r = r_{\min} + x^2$, but now x is a temporal coordinate in a T-region, where the metric (5.44) describes a Kantowski–Sachs cosmology with two scale factors $r(x)$ and $A(r(x))$ and the $\mathbb{R} \times \mathbb{S}^2$ topology of spatial sections. Hence, $x = 0$ is the time instant at which $r(x)$ experiences a bounce. It is quite the same "black bounce" [503] that occurs in black universe models we discussed above.

Assuming asymptotic flatness at large r, one can use the standard methodology to obtain the corresponding Carter–Penrose diagram describing the global causal structure: it coincides with that of the Kerr or Kerr–Newman non-extremal BH [549] (but without a ring singularity) and contains an infinite sequence of R and T regions.

By continuity, this behavior is preserved in a finite range of C values. We conclude that each generic choice of $A(r)$ with a simple zero at $r = r_h$ leads to a family of solutions unifying symmetric wormholes($C < 0$), extremal BHs ($C = 0$) and regular nonextremal BHs ($C > 0$), see Fig. 5.5.

Example. Let us assume for the generating function $A(r)$ the Schwarzschild form, $A(r) = 1 - 2m/r$, $m = \text{const} > 0$ and consider the solution (5.49) in the vacuum case $R \equiv 0$. We obtain

$$f(r) = \frac{(r - 2m)(r - r_0)}{r - 3m/2}, \tag{5.63}$$

where r_0 is an integration constant. The metric takes the form

$$ds^2 = \left(1 - \frac{2m}{r}\right) dt^2 - \frac{1 - 3m/(2r)}{(1 - 2m/r)(1 - r_0/r)} dr^2 - r^2 d\Omega^2. \tag{5.64}$$

The Schwarzschild metric is restored in the special case $r_0 = 3m/2$. The metric (5.64) was obtained by Casadio, Fabbri and Mazzacurati [164] in search for new brane-world BHs and by Germani and Maartens [255] as a possible external metric of a homogeneous star on the brane. Without repeating their more detailed descriptions, let us outline the main points in our notations.

BH metrics appear according to Algorithm (BH1), where $C = 0$ corresponds to $r_0 = 2m$ and $C > 0$ to $r_0 < 2m$. In the case $r_0 > 2m$, the metric (5.64) describes a symmetric traversable wormhole [139].

In the case $r_0 = 2m$ we have a double horizon at $r = 2m$: near $r = 2m$, the coordinate r is connected with the quasiglobal coordinate x by $r - 2m \sim (x - x_h)^2$, x_h being the value of x at the horizon, and $\mathcal{A}(x) \sim (x - x_h)^2$. The Carter–Penrose diagram coincides with that of the extreme Reissner–Nordström metric [Fig. 5.5(b)] with the only difference that the timelike curvature singularity occurs at $r = 3m/2$ instead of $r = 0$.

In the case $r_0 < 2m$, as in the Schwarzschild metric, $r = 2m$ is a simple horizon, and, as described in [164], the space–time structure depends on the sign of $\eta = r_0 - 3m/2$. If $\eta < 0$, the structure

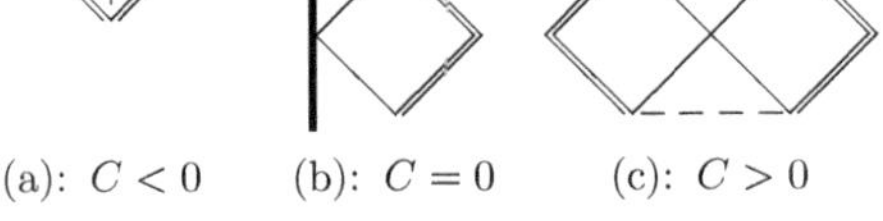

Figure 5.5 Carter–Penrose diagrams for a generic family of asymptotically flat solutions: (a) a symmetric wormhole, (b) an extremal BH in case there is a singularity inside the inner R region, and (c) a regular BH. Diagrams (b) and (c) can be infinitely continued upward and downward. The letters R and T mark static (R) and cosmological (T) regions, respectively. Spatial infinity ($r = \infty$) is shown by double lines, horizons ($r = r_h$) by single thin lines and the singularity in (b) by a thick line. Dashed lines show the wormhole throat in diagram (a) and the bouncing time instants in diagram (c).

is that of a Schwarzschild BH, but the spacelike curvature singularity is located at $r = 3m/2$ instead of $r = 0$. If $\eta > 0$, the solution describes a nonsingular BH with a bounce at $r = r_0$ inside the horizon. The corresponding global structure [164] is the same as that of a nonextremal Kerr BH [Fig. 5.5(c)].

Thus, the metric properties in the whole range $r_0 > 3m/2$ of the integration constant r_0 entirely conform to our description for both positive and negative C.

The components of the effective SET (5.46) have the form

$$\varkappa_4^2 T^t_{t\,\text{eff}} = \frac{m(2r_0 - 3m)}{r^2(2r - 3m)^2},$$

$$\varkappa_4^2 T^r_{r\,\text{eff}} = \frac{2r_0 - 3m}{r^2(2r - 3m)},$$

$$\varkappa_4^2 T^\theta_{\theta\,\text{eff}} = -\frac{(r - m)(2r_0 - 3m)}{r^2(2r - 3m)^2}. \tag{5.65}$$

To analyze NEC violations, we find

$$\varkappa_4^2 \left(T^t_{t\,\text{eff}} - T^r_{r\,\text{eff}}\right) = \frac{2(2r_0 - 3m)(2m - r)}{r^2(2r - 3m)^2}. \tag{5.66}$$

In an R-region the quantity (5.66) is equal to $\rho_{\text{eff}} + p_{r\text{eff}}$, and in the wormhole case ($r \geq r_0 > 2m$) it is negative, exhibiting NEC violation. In the "black bounce" case ($r_0 < 2m$ and $r > r_0 > 3m/2$) the quantity (5.66) is positive in the T-region $r < 2m$, but since there the coordinates r and t exchange their roles, the SET components also change their meanings: T^t_t is there "minus pressure" while T^r_r is the density, therefore, again, the NEC is violated. However, all that happens to the effective SET due to tidal influence of the bulk, while ordinary matter is absent.

More examples of brane BH solutions and their analysis can be found, e.g., in [98, 164].

5.3 Rotating black holes

Until now we have been discussing only spherically symmetric BHs, whose angular momentum is zero. Now, we will very briefly describe the basic properties of rotating BHs. We will assume them to be electrically neutral both for simplicity and in connection with the small role of possible BH charges in astrophysical conditions.

Rotating BHs form as a result of the gravitational collapse of rotating bodies (but practically all celestial bodies are rotating) and at BH mergers. Therefore, rotating BHs are much more realistic than spherically symmetric ones. Their structure and properties are, however, much more complicated.

Stationary rotating BHs in GR are described by the Kerr metric [321]

$$ds^2 = \left(1 - \frac{2mr}{\rho^2}\right) dt^2 + \frac{4m\,r\,a^2 \sin^2\theta}{\rho^2} d\phi\,dt - \frac{\rho^2}{\Delta} dr^2 - \rho^2 d\theta^2$$
$$-\left(r^2 + a^2 + \frac{2m\,r\,a^2}{\rho^2}\right) \sin^2\theta\,d\phi^2, \tag{5.67}$$

where

$$a = J/m, \qquad \rho^2 = r^2 + a^2\cos^2\theta, \qquad \Delta = r^2 - 2m\,r + a^2. \tag{5.68}$$

Here, $(t,\ r,\ \theta,\ \phi)$ are the so-called Boyer–Lindquist coordinates [68] (initially, Kerr obtained his metric in less convenient notations). The metric (5.67) represents an asymptotically flat (as $r \to \infty$) stationary, axially symmetric solution to the Einstein equations in vacuum ($T^\nu_\mu \equiv 0$). The stationarity and axial symmetry are determined by the existence of the Killing vectors $\xi^\mu = (1,0,0,0)$ and $\eta^\mu = (0,0,0,1)$, respectively (with the usual order of enumerating the coordinates: $t,\ r,\ \theta,\ \phi$). The stationarity, instead of staticity, of the metric manifests itself in its noninvariance under the substitution $t \mapsto -t$.

The solution is parametrized by two constants, the mass m and the angular momentum J; at $J = 0$ it turns into the Schwarzschild solution.

A true curvature singularity occurs at $\rho = 0$, $r = 0$, $\theta = \pi/2$, and an inspection shows that this singularity has the shape of a ring (this fact is not evident in the Boyer–Lindquist coordinates and is established using another parametrization of the metric, see, e.g., [250, 409]). The surfaces where $\Delta = 0$, i.e.,

$$r_\pm = m \pm \sqrt{m^2 - a^2}, \tag{5.69}$$

are horizons. Evidently, the horizons exist if $m^2 \geq a^2$, and in the case of inequality there are two simple horizons while at equality there is only one extremal (double) horizon.

Consider the metric (5.67) with horizons. In this case, the norm of the Killing vector ξ^μ, equal to

$$\xi_\mu \xi^\mu = g_{tt} = \frac{\Delta - a^2 \sin^2\theta}{\rho^2}, \tag{5.70}$$

vanishes on the surface where

$$\Delta = a^2 \sin^2\theta, \tag{5.71}$$

i.e., outside the horizon (with the surface $\Delta = 0$, i.e., the horizon $r = r_+$, it has only two common points, namely, the poles $\theta = 0$ and $\theta = \pi$, see Fig. 5.6). The surface (5.71) is called the stationarity limit because stationary observers cannot exist on and inside this surface. Besides, the surface (5.71) is a surface of infinite redshift for photons emitted from it to spatial infinity.

The region of space between the horizon $r = r_+$ and the surface (5.71) has received the name of the ergosphere. This name is connected with the opportunity to extract the rotational energy of the

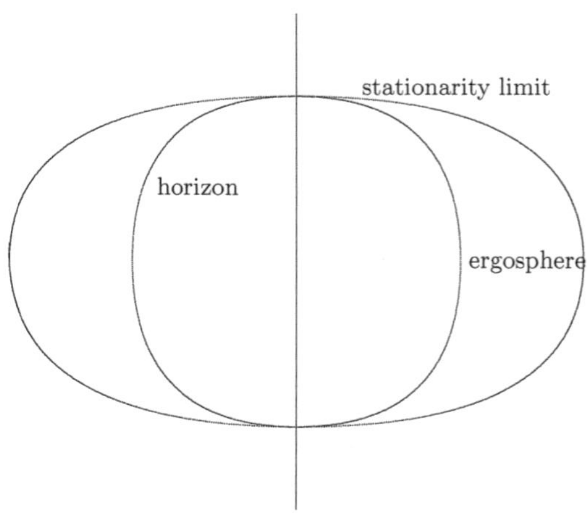

Figure 5.6 The horizon and the ergosphere of a Kerr BH, a meridional section.

Kerr BH with the aid of the Penrose process [445]. The point is that the ergosphere contains orbits of massive particles with a negative total energy (this means that the absolute value of the binding energy exceeds the particle mass). Then one can launch, say, a spacecraft into the ergosphere, where it will emit particles with negative energy and return to the usual spatial region with a larger energy than it initially had. As to particles with a negative energy, their motion is directed against the BH rotation and, being absorbed by the BH, they diminish its rotation rate and total rotational energy.

It has been shown [26, 268, 269, 575] that particle collisions close to horizons of rotating black holes can lead to extremely high energies of the order of the Grand Unification scale in the center-of-mass reference frame of colliding particles. According to [26], close to the horizon of an extremal Kerr black hole, there is a resonance for the center-of-mass (CM) energy of two scattering particles: a pole for some special value of the angular momentum of the scattered particle was found, showing that the center-of-mass energy can be arbitrarily large. The papers [268, 269] show that the same effect can occur in nonextremal black holes if one takes into account the possibility of multiple scattering of a particle: at the first scattering close to the horizon the particle gets an angular momentum close to its critical value, and at the second scattering, due to the resonance effect, it can acquire an energy at the level of high energy physics, Grand Unification or even Planck values.

Assuming that a rotating black hole is located at the center of an active galactic nucleus, the time needed for a particle to come from the accretion disk to the first scattering point was estimated [269] and turned out to be of the order of a week. Then, the time needed for a high-energy particle to leave the ergosphere to be observed on Earth as a particle of ultra-high energy cosmic rays (UHECR) was evaluated. This time was also found to be quite reasonable: of the order of days, plus certainly the time needed to travel to the observer.

Electrically charged stationary rotating BHs are described by the Kerr–Newman solution to the Einstein–Maxwell equations, characterized by the metric in the Boyer–Lindquist coordinates similar to those in (5.67)

$$ds^2 = \left(1 - \frac{2mr - q^2}{\rho^2}\right) dt^2 + \frac{(2m\,r - q^2)2a^2 \sin^2\theta}{\rho^2} d\phi\, dt - \frac{\rho^2}{\Delta} dr^2$$
$$-\rho^2 d\theta^2 - \left(r^2 + a^2 + \frac{(2m\,r - q^2)a^2}{\rho^2}\right) \sin^2\theta\, d\phi^2, \qquad (5.72)$$

with

$$\rho^2 = r^2 + a^2\cos^2\theta, \qquad \Delta = r^2 - 2m\,r + a^2 + q^2, \qquad (5.73)$$

where q is the electric charge; m and a have the same meaning as in Kerr's solution. Evidently, ρ^2 is the same as in (5.68) while Δ has the new contribution q^2. The electromagnetic field of the BH is determined by the vector potential A_μ given by

$$A_\mu dx^\mu = -\frac{qr}{\rho^2}(dt - a\sin^2\theta\, d\phi). \qquad (5.74)$$

This solution reduces to the Reissner–Nordström solution if $a = 0$, and to the Kerr solution if $q = 0$. A detailed analysis of the properties of Kerr–Newman BHs can be found in the book [250].

At large charges and/or angular momenta ($q^2 + a^2 > m^2$), the Kerr and Kerr–Newman solutions are no longer BHs but describe the gravitational fields of naked singularities. It is of interest that if we juxtapose these parameters with masses, spins and charges of known elementary particles, they correspond to such naked singularities. Surprisingly, it turns out that the above solutions with rotation do really describe the particle properties fairly well, even including some of their quantum features, see [145, 146] and references therein.

5.4 Black hole thermodynamics and evaporation

5.4.1 *Four laws of BH thermodynamics*

By definition, a BH can emit nothing (in the framework of the classical theory) because neither photons nor massive particles are able to leave it. Nevertheless, as found for the first time by Bekenstein [38],

BHs satisfy some relations identical to the well-known relations of thermodynamics, and we can ascribe to them such thermodynamic characteristics as temperature (which is in general nonzero) and entropy.

Hawking [284] discovered, while considering quantum processes near horizons, that BHs emit radiation like black bodies with a certain temperature, and he derived an exact relationship between the BH entropy and the BH horizon area. A little later, four laws of BH thermodynamics were formulated [256]. As a result, there emerged a consistent description of thermodynamic systems which include both conventional matter and BHs.

The so-called *zeroth law of BH thermodynamics* is formulated as follows:

The Zeroth Law. *The surface gravity of a BH does not change from point to point of the event horizon.*

The surface gravity is, by definition, the value of the free-fall acceleration, calculated at the event horizon in terms of the particle acceleration. For a Schwarzschild BH, the surface gravity $\varkappa$ is

$$\varkappa = 1/(4m), \tag{5.75}$$

while for a Kerr BH rather a bulky calculation gives (see, e.g., [270])

$$\varkappa = \frac{r_+ - m}{2mr_+} = \frac{\sqrt{m^2 - a^2}}{2m\left(m + \sqrt{m^2 - a^2}\right)}. \tag{5.76}$$

Equations (5.75) and (5.76) contain only the BH parameters, without any angular dependence, thus confirming the Zeroth Law.

The First Law of BH thermodynamics reads

$$dm = \frac{\varkappa}{8\pi} dA + \Omega\, dJ, \tag{5.77}$$

where A is the horizon area,

$$A = 4\pi(r_+^2 + a^2) \tag{5.78}$$

(at $a = 0$ we obtain the expression for the Schwarzschild BH, $A = 4\pi r_+^2 = 16\pi m^2$), and Ω is the angular velocity at the horizon,

$$\Omega = \frac{a}{r_+^2 + a^2}. \tag{5.79}$$

The equality (5.77) is similar to the first law of conventional thermodynamics $dU = TdS + dW$ (U is the internal energy, S the entropy, W the work). Therefore, Eq. (5.77) may be pronounced as follows: $\Omega\, dJ$ is the work executed over the BH by adding the angular momentum dJ. Further reasoning leads to the following identifications for the BH temperature and entropy [284]:

$$T_{\rm BH} = \varkappa/(2\pi), \quad S_{\rm BH} = A/4. \tag{5.80}$$

In particular, from the expression (5.76) it follows that the temperature is zero only for an extremal BH with a double horizon ($|a| = m$).

The Second Law of BH thermodynamics is similar to the statement that the entropy cannot decrease:

The Second Law. *There is no physical process that could decrease the horizon area of a BH.*

The quantum process of BH evaporation due to Hawking radiation violates this classical law.

Lastly, the Third Law asserts the nonaccessibility of an absolute zero of temperature, or, as is sometimes formulated, the nonexistence of negative temperatures. For a BH, a similar statement reads [33]:

The Third Law. *There is no procedure able to bring the BH temperature to zero by a finite sequence of operations.*

In other words, it is impossible to make an extremal BH from a non-extremal one, to say nothing of exceeding the extremality limit (to reach $|a| > m$) and to obtain a naked singularity from a BH.

5.4.2 *Black hole evaporation*

As already mentioned, according to [284], BHs evaporate, emitting like black bodies with the temperature $T_{\rm BH} = \varkappa/(2\pi)$, where the surface gravity at the horizon $\varkappa$ is determined by Eqs. (5.75) and (5.76).

A qualitative physical picture of BH evaporation consists in that virtual particle-antiparticle pairs, permanently appearing and disappearing according to quantum field theory, are subject to a sufficiently strong gravitational field in the neighborhood of the horizon. Then, with a certain probability (which is larger for larger space–time curvatures), the gravitational field can drag apart such virtual pairs, so that one of the particles falls to the horizon while the other escapes to infinity. As particles with nonzero energy leave the BH, the latter loses its mass.

A substitution of the corresponding constants leads to the following approximate expression for the temperature of a Schwarzschild BH in kelvins:

$$T_{\rm BH} \approx 10^{-7} m_\odot/m \ {\rm K}, \qquad (5.81)$$

where $m_\odot$ is the solar mass. Thus the temperatures of BHs of solar (*a fortiori*, galactic) masses is very small, and their evaporation occurs very slowly: in fact, they increase their masses much more rapidly due to accretion of ambient matter. The smaller a BH, the larger is the curvature of space near its horizon and higher its temperature. In particular, the temperature becomes smaller if mass is added, thus BHs have a negative heat capacity.

The loss of mass (energy) of a BH can be calculated according to the Stefan–Boltzmann law, $-\dot{E}/A = \sigma T_{\rm BH}^4$, where σ is the Stefan–Boltzmann constant. Integration of this expression leads to the following BH mass as a function of time:

$$m(t) = (m_0^3 - 3Kt)^{1/3}, \quad K = \frac{1}{15360\pi} = \frac{\hbar c^4}{15360\pi G^2}, \qquad (5.82)$$

where m_0 is the BH mass at $t = 0$; after the last equality sign, the expression is given in conventional units. At the time instant $t_f = m_0^3/(3K)$, the evaporation terminates, but it is so far not clear whether the BH must completely disappear or if there remains a certain object with a mass of Planckian order of magnitude, $m \sim 10^{-5}$ g. The point is that as the mass approaches its Planck value (hence also the horizon radius approaches the Planck length $\sim 10^{-33}$ cm), the so far unknown laws of quantum gravity must come into force.

A second before the end of the process, as one can easily find, the BH mass is of the order 10^9 g $\sim 10^{33}$ GeV. It is the energy released in the last second of BH evaporation.

On the other hand, if one equates the lifetime $t_f = m_0^3/3K$ of a BH with a given mass m_0 to the age of the Universe, $t_U \sim 13 \cdot 10^9$ years, we obtain the mass $m_0 = m_U \sim 10^{15}$ g. Thus all primordial BHs of mass smaller than 10^{15} g, having formed in the first seconds (or millennia) of the Universe existence, have already evaporated and contributed to the chemical composition of the modern Universe, while the remaining BHs of masses close to m_U must now and then explode in the modern epoch. Heavier BHs will have a much longer life.

Chapter 6

Wormholes

6.1 The notion of a wormhole

In this chapter, we will discuss one more type of configurations involving a significant space–time curvature, other than BHs: the wormholes. Above all, let us fix the terminology.

The term "wormhole" was introduced by J.A. Wheeler [559] in 1955. Wormholes are generally understood as comparatively narrow "bridges", or "tunnels" between comparatively large or infinitely extended space–time regions. Needless to say that such an opportunity is widely used by authors of science fiction whose heroes easily and without much emotion leave behind them interstellar distances or thousands of years.

The possible appearance of a wormhole geometry is illustrated by Fig. 6.1, which presents two-dimensional images instead of three-dimensional ones. Each of the surfaces depicted contains a part having the shape of a thin tube that joins either flat "universes" (a) or spherical ones (b) (a dumbbell-like configuration), or a flat one with a spherical one (c) (a hanging-drop configuration), or distant regions of the same "universe" (d and e).

We see that the notion of a wormhole has in essence an intuitive nature since it rests on such words as "comparatively narrow", etc., but it contains an element that admits a rigorous definition. It is a throat, defined as a two-dimensional surface of minimal area, which does not admit a further contraction. In Fig. 6.1, the throats are one-dimensional circles of minimum radius, which conditionally depict

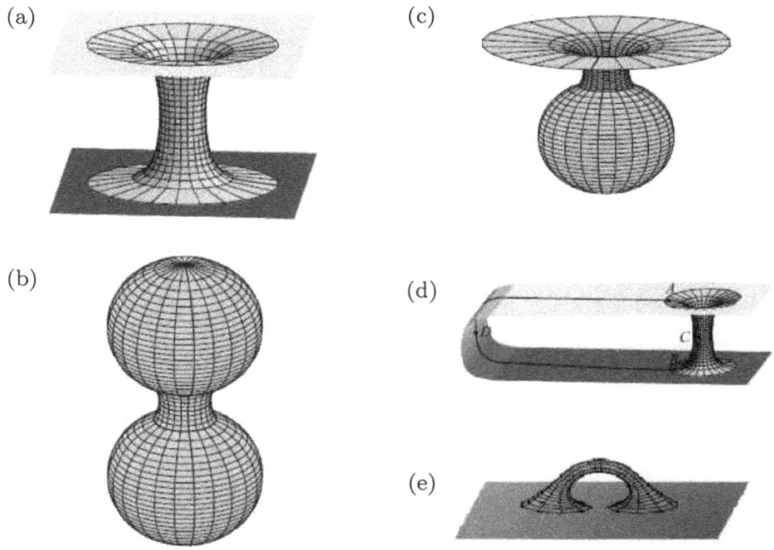

Figure 6.1 Diverse wormhole configurations, two-dimensional analogues.

two-dimensional spheres of minimum area. It is obvious, however, that in general a wormhole need not be spherically symmetric, and its throat is not necessarily a sphere.

One should also add that all these shapes correspond to spatial sections of the four-dimensional space–time and, as usual, these spatial sections are not only able to evolve with time but can in general also depend on the choice of reference frames (recall the discussion of relativity of the spatial section topology in Sec. 2.2).

In the case of static, spherically symmetric space–times described by the general metric (3.1), it is not difficult to formulate the common requirements to be imposed on the metric coefficients which, if satisfied, enable one to say that this metric describes a wormhole geometry.

The condition that there are two large regions separated by a throat means that the function $r \equiv e^{\beta}(u)$, which is equal to the curvature radius of the coordinate sphere $u = \text{const}$, $t = \text{const}$ at each value of u, should have a regular minimum $r = r_{\rm min} > 0$ at some value of u and assume values $r \gg r_{\rm min}$ far from this minimum.

Another condition is that the wormhole must be freely traversable in both directions. So the existence of a proper three-geometry is insufficient: one must also require the absence of horizons. It is a fundamental distinction from BHs. For the metric (3.1), this requirement reads $g_{tt} \equiv e^{2\gamma} > 0$ in the whole range of the radial coordinate.

The simplest example of a wormhole metric is

$$ds^2 = dt^2 - dx^2 - (x^2 + a^2)d\Omega^2, \quad a > 0, \tag{6.1}$$

it differs from the Minkowski flat-space metric only by the emergence of the term a^2. But this drastically changes the geometry: at $a = 0$ we have the Minkowski space, and the coordinate x changes from zero to infinity, $x = 0$ is a regular center, and an extension to negative values of x is meaningless. However, if $a > 0$, the value $x = 0$ is a regular minimum of the function $r(x) = \sqrt{u^2 + a^2}$, i.e., a throat, nothing prevents consideration of $x < 0$, and $x \to -\infty$ is again a spatial infinity of the same kind as that at $x \to +\infty$. The 3-geometry corresponds to the image (a) in Fig. 5.1 and is time-independent.

To illustrate the importance of the second condition, let us make a small historical digression.

Right after the advent of general relativity, the researchers began to reflect on what could be expected if space–time is not simply curved but is strongly curved. As early as in 1916, the birth year of GR, Schwarzschild found his famous solution, Eq. (3.20), which describes, in modern terms, the simplest BH. Less well known is the work of Ludwig Flamm [242] also dated 1916, where he noticed that the Schwarzschild solution describes something like a bridge or shortcut between two worlds or two parts of the same world, i.e., just what modern authors call a wormhole.

Let us explain this observation by passing on from the curvature coordinates in the metric (3.20) to the isotropic coordinates (3.2) by the transformation

$$r = u\left(1 + \frac{m}{2u}\right)^2. \tag{6.2}$$

As a result, the Schwarzschild metric acquires the form

$$ds^2 = \frac{(2 - m/u)^2}{(2 + m/u)^2} dt^2 - \left(1 + \frac{m}{2u}\right)^4 (du^2 + u^2 d\Omega^2), \tag{6.3}$$

where the coordinate $u \approx r$ as $r \to \infty$ and has the value $u = m/2$ at the horizon $r = 2m$. Here, it is convenient to introduce the dimensionless variable $y = 2u/m$. Then

$$ds^2 = \frac{(y - 1)^2}{(y + 1)^2} dt^2 - \frac{m^2}{4} \frac{(y + 1)^4}{y^4} (dy^2 + y^2 d\Omega^2), \tag{6.4}$$

and the horizon corresponds to $y = 1$. The R-region $r > 2m$ is now described as the range $y > 1$. It is, however, easy to verify that the metric (6.4) does not change at an inversion, $y \mapsto 1/y$. It means that the range $0 < y < 1$ describes one more R-region, with quite the same geometry as the region $y > 1$, and the value $y = 0$ corresponds to another spatial infinity quite similar to the one at ($y = \infty$); moreover, the value $y = 1$, where the coordinate spheres have a minimum area equal to $16\pi m^2$, corresponds to a throat in the spatial section of the 4-manifold under study. If the geometry were restricted to the spatial section only, we would have all reasons to say that there is a wormhole. However, everything changes due to $g_{tt} = 0$ at $y = 1$: as we know from the previous chapters, it is a horizon, whose existence means that the 4-geometry is inextendible in the static reference frame. For a more complete description one can introduce, e.g., the Kruskal coordinates, in which the complete geometry of the Schwarzschild space–time can be described. The resulting complete picture implies that the metric (6.4) describes only the two R-regions represented by the left and right quadrants in the Kruskal diagram. (Compare: the metric (3.20) at $r > 2m$ describes only one of the R-regions.) The sphere $y = 1$, which seems to be a boundary between the two R-regions when we look at Eq. (6.4), turns out to be a pair of different surfaces in the full description, and there is a whole T-region between them. The paradox is explained, in particular, by the fact that the horizon $r = 2m$ (or $y = 1$) *does not belong* to any of the R-regions, hence the unity of their description by the metric (6.4) is only seeming, and the coordinate regions $y > 1$ and $y < 1$, corresponding to different R-regions of the Schwarzschild space–time, should be considered quite separately.

Metrics like (6.4), containing horizons, are sometimes said to describe "nontraversable wormholes". In our view, in such cases, to avoid confusion, one should speak of quite a different class of relativistic objects, namely, black holes. Even more than that, as we have seen while building the Carter–Penrose diagrams, *all* black holes with simple horizons (and any other horizons of odd order) contain quite the same pairs of R-regions as the Schwarzschild black hole, and all of them could be called "nontraversable wormholes" on the same grounds (more precisely, at the same level of uselessness).

It should be noted, though, that some families of solutions to the gravitational field equations can contain metrics describing objects of both classes at different values of the parameters, see, e.g., Sec. 5.2.3 and [98].

6.2 A wormhole as a time machine

The works on wormhole physics are certainly only theoretical; others are impossible right now. They can be divided into two trends, supplementary to each other. In one of them, the researchers assume the existence of wormholes from the very beginning and try to infer what can come out of that. An important part of this trend is what can be called wormhole astrophysics: assuming that there are wormholes of astrophysically relevant size, it is reasonable to seek their observational signatures. In the other trend, the researchers (quite frequently, the same authors) try to find out under which conditions, natural or artificial, wormholes can emerge, and what will be their properties and peculiarities. Among such works are certainly numerous attempts to obtain particular wormhole solutions to the gravitational field equations.

Let us here dwell upon some questions belonging to the first trend, leaving aside for a while the question of "what the wormholes can be made of". We will discuss the (probably) most interesting version of a wormhole assuming that it connects different regions of the same space–time. To begin with, let us suppose that there is a stable wormhole somewhere in space, and that the space–time outside it is flat or almost flat, where, as usual, the laws of special relativity hold. Then, simply from the symmetries of Minkowski space it is clear that a wormhole can be not only an intergalactic tunnel but also an accelerator and even a time machine.

Indeed, the wormhole has two flat asymptotic regions. In each of them one can choose a reference frame (RF) (for simplicity, an inertial one) in which the wormhole mouth is at rest. But it follows from nowhere that it is the same RF for both mouths! And even if it is the same RF, and in the regions where the mouths are located, the clocks are synchronized in some reasonable way, it follows from nowhere that after passing through the wormhole the traveler will get into the same epoch by this common time, since the Minkowski metric is invariant under time translations. In other words, even if we know where a given wormhole (say, found in space by chance) leads, we in general cannot know "to when" it leads.

The invariance of the Minkowski metric under boosts, i.e., transformations from one inertial RF to another, leads to an uncertainty in the velocity of the mouth of exit with respect to the RF of the mouth of entry. Thus, a traveler which has passed through the wormhole can find himself in motion with an arbitrary relativistic velocity with respect to the initial RF and (which is more important if the mouths are far from each other) with respect to the ambient bodies.

Thus, the appearance, role and function of a wormhole depends on the positions and motion of its mouths, not simply in space but in space–time.

Another statement of the problem is also of interest. Suppose we are able to create a wormhole here and now, with quite macroscopic parameters, so that one can pass through a wormhole from one mouth to the other for a few seconds, and the mouths themselves are placed relatively close to each other somewhere in circumterrestrial space. Is it possible to convert such a wormhole into a time machine? It is shown by Morris, Thorne and Yurtsever [417] how to do that: the idea is to leave one of the mouths, to be denoted A, at rest, while the other, B (which should in principle behave as an ordinary massive body) should be accelerated to velocities comparable with the speed of light, then returned back and stopped near A. Then, due to the familiar special-relativistic effect, the Lorentzian time slowing-down on the moving body as compared to the one at rest, the time elapsed at B is smaller than at A. The particular difference depends on the velocity reached and the travel duration. It is actually just the well-known twin paradox: the twin returning from an interstellar trip at relativistic speed turns out to be younger than his brother sitting at home. If the difference between the mouths is, say, six months, then, being located near mouth A in January 2021, we would be able to see the green leaves of summer 2020 through the wormhole — and will really return there if we pass it through! If we again approach mouth A (as we agreed, it is not too far from B) and passing through the wormhole once more, we would get into last year's snow of January 2020. And so on, until we get tired... Naturally, under the assumption that the wormhole is stable or we ourselves are able to maintain it in operation.

Let us quote the beginning of the article [417], which looks like a slogan of wormhole physics: "Normally theoretical physicists ask: 'What are the laws of physics?' and/or 'What do these laws predict about our Universe?' In this letter we ask, instead, 'What constraints do the laws of physics place on the activities of an arbitrarily advanced civilization?' This will lead to some intriguing queries about the laws themselves."

Soon after the works of Thorne's group (1988) appeared the paper by Frolov and Novikov [249] who showed that even without special effort a wormhole in an external gravitational field, as time passes, inevitably turns into a time machine. Indeed, suppose that one of the mouths (A) of a stable wormhole is located near the surface of a neutron star and the other (B) far from it. Then the time rate is slower near A and faster near B, this difference is stored, and eventually there should appear closed timelike trajectories threading the wormhole. The time machine is ready.

The possible existence of wormholes would substantially affect black hole physics as well, see a discussion in the same paper [249]. Thus, nothing prevents us from assuming that mouth A of a wormhole is located outside a black hole while the other, B, is inside the horizon. Such a wormhole may be used to rescue an astronaut who fell into the black hole as well as for exploring the inner properties of the hole itself. Strictly speaking, the black hole horizon is then not an event horizon any more. This issue is actively discussed in modern literature in connection with the phenomenon of quantum entanglement at BH evaporation [172, 527].

We have seen that if there is a stable wormhole at our disposal, we can rather easily obtain a time machine. The above reasoning makes it possible to conclude that the laws of physics (among which we include the notion of gravity as space–time curvature), in principle, do not forbid the existence of wormholes and consequently closed timelike and null trajectories that violate the causality principle. As a result, in papers on theoretical physics one can meet discussions of paradoxes whose place could be only expected in science fiction. What about meeting another (third, tenth) copy of oneself? What about the well-known "grandmother paradox" — the opportunity to penetrate into the past and kill one's grandmother in her childhood (or, in a milder version, simply to marry the young grandmother, then a girl, instead of the young grandfather)? But then the hero himself — where could he come from?

One of the well-known ways to surmount such and other paradoxes involves the hypothesis of parallel worlds, also called the Multiverse hypothesis, which is closely related to Everett's many-worlds interpretation of quantum theory [234]. Quantum processes are known to be of probabilistic nature, and any of them can have a few outcomes. By Everett, all possible outcomes are realized, but each of them in its own universe. Thus, any quantum event (including, by the way, all particle interactions in any part of the Universe) increases the number of parallel worlds. Being so exotic and evidently cumbersome, such an interpretation is free of logical contradictions and rather naturally explains not only the quantum paradoxes but also the above-mentioned "grandmother paradox": getting to the past, the traveler actually creates a new universe, where a girl looking like his grandmother in youth can exist, but she is not at all the mother of his mother...

Another possible solution (more economical but not so easy to understand) was suggested by Novikov as the "self-consistency principle" [430], according to which "something can be realized locally only if it is globally self-consistent". It means [277] that events in a time loop affect each other along a closed cycle, and not only the past determines the future, but the future takes part in forming the past, which, whatever strange it can seem, does not lead to a violation of the causality principle.

There is, though, the radically conservative viewpoint that nature does not admit and cannot admit time loops — something like a chronological censorship. As any censorship, it cannot cause sympathy with the authors and many of their compatriots. But to put it seriously, this hypothesis, like any other, needs proof on the basis of the first principles. (As is easy to understand, any hypothesis claiming that something cannot exist can be only rejected by experiment but not proved.)

6.3 Wormholes as solutions to gravitational field equations

The works belonging to the second trend are more numerous and are also of much interest: among them there are searches for specific wormhole models, discussions of their properties and peculiarities that determine whether or not they might be realized, what could be done with them and how to use them.

Above all, if wormholes or, wider, space–times containing time loops are not forbidden completely, they should be quite rare in Nature, otherwise they would be observed long ago. Thus, if a theory pretends to describe Nature, it should contain a mechanism forbidding or at least strongly preventing the formation of macroscopic wormholes.

No doubt, GR pretends to describe reality. However, there are many solutions describing wormholes and other spaces with closed timelike loops. But all of them are as a rule considered to be unrealistic or, in a sense, not dangerous. Thus, a very interesting solution to the Einstein equations has been found by Gödel [259]: it is a homogeneous stationary universe, rotating as a whole. It contains closed timelike trajectories, but a calculation shows that the smallest total duration of such a loop is much larger that the age of our Universe.

Important constraints follow from the very structure of the Einstein equations. The idea is that if we substitute a metric, possessing some properties of interest, to the left-hand side of the equations, it becomes possible to find explicit constraints on the possible form of the right-hand side (that is, the SET) compatible with these properties. Let us look how it is done using, as a simple but important example, static, spherically symmetric space–times.

6.3.1 *Spherically symmetric wormholes — general properties*

Let us, as before, begin with the general form (3.1) of a static, spherically symmetric metric

$$ds^2 = e^{2\gamma(u)} dt^2 - e^{2\alpha(u)} du^2 - e^{2\beta(u)} d\Omega^2, \tag{6.5}$$

and consider some general properties of static, spherically symmetric wormholes, assuming the SET in the general form compatible with this kind of symmetry, viz.,

$$T_\mu^\nu = \mathrm{diag}(\rho,\ -p_r,\ -p_\perp,\ -p_\perp), \tag{6.6}$$

where ρ, p_r, $p_\perp$ are the density, radial pressure and transversal pressure, respectively, which can in general be arbitrary functions of the radial coordinate u.

We will make use of two different radial coordinates, namely, the quasiglobal ($u = x$) and curvature ($u = r$) coordinates.

With the quasiglobal coordinate x ($\alpha + \gamma = 0$), the metric has the form

$$ds^2 = A(x)dt^2 - \frac{dx^2}{A(x)} - r^2(x)d\Omega^2, \tag{6.7}$$

and the conditions on a wormhole throat $x = x_{\rm th}$, expressing the absence of a horizon and a finite regular minimum of r, have the form

$A > 0$, $r > 0$, $r' = 0$, $r'' > 0$. Then two of the Einstein equations (see (3.5), (3.6))

$$R^0_0 - R^1_1 = 2A\frac{r''}{r} = -\varkappa(\rho + p_r), \tag{6.8}$$

$$G^1_1 = \frac{1}{r^2}[-1 + A'rr' + Ar'^2] = \varkappa p_r, \tag{6.9}$$

lead to the following conditions on the throat:

$$\rho + p_r < 0, \quad p_r < 0. \tag{6.10}$$

Note that there is no restriction on the sign of the density. The conditions (6.10) are of local nature and do not depend on any assumptions on the space–time or matter properties far from the throat. It is clear that both inequalities are quite unusual for macroscopic matter, though for field matter the second condition (6.10) can be easily satisfied, e.g., for a radial electric field we have, according to (3.14), $p_r = -\rho$.

It is easy to verify that the first inequality (6.10) violates one of the most important conditions, satisfied by the majority of known kinds of matter. It is the Null Energy Condition (NEC). In the general case, this condition is expressed by the inequality

$$T_{\mu\nu}\xi^\mu\xi^\nu \geq 0, \quad \forall\, \xi^\mu : \quad \xi^\nu\xi_\mu = 0. \tag{6.11}$$

In the metric (6.5), for the null vector $\xi^\mu = (\mathrm{e}^{-\gamma},\ \mathrm{e}^{-\alpha},\ 0,\ 0)$, this condition takes the form

$$T^0_0 - T^1_1 \geq 0. \tag{6.12}$$

For the tensor (6.6) it reads $\rho + p_r \geq 0$, contrary to (6.10). Thus, the wormhole existence requires NEC violation.

Matter that violates the NEC is commonly called exotic. The conclusion that it is this kind of matter that is necessary for wormhole existence is not restricted to cases that are static or spherically symmetric and has quite a general nature: it holds true for throats of any spatial configuration, both stationary and time-dependent [291, 292] (though for the latter even the definition of a throat is rather complicated and admits a few versions [53, 292]).

But let us return to static spherical symmetry and prove a no-go theorem of interest [87]:

A static, spherically symmetric traversable wormhole with $r \to \infty$ and $A(x)r^2(x) \to \infty$ on both sides of the throat cannot be supported by any isotropic matter source with $p_r = p_\perp$.

Indeed, if $p_r = p_\perp$, it follows $R^x_x = R^\theta_\theta$, and the difference of the corresponding Einstein equations gives

$$r^2 A'' + 2Arr'' - 2Ar'^2 + 2 = 0. \tag{6.13}$$

The substitution $A(x) = D(x)/r^2(x)$ converts it to

$$D'' - \frac{4D'r'}{r} + \frac{4Dr'^2}{r^2} + 2 = 0. \tag{6.14}$$

A possible minimum of $D(x)$ at some $x = x_0$ requires $D' = 0$ and $D'' \geq 0$. Meanwhile, if $D' = 0$, Eq. (6.14) gives $D'' \leq -2$, hence it is necessarily a maximum.

However, an asymptotically flat traversable wormhole requires $r \to \infty$ and $A \to 1$ as $x \to \pm\infty$, in an asymptotically AdS wormhole there must be $A \sim r^2$ at large r, etc. In all such cases (more precisely, in all cases where $D = Ar^2 \to \infty$ on both sides far from the throat) $D(x)$ should have a minimum, which, as we have seen, is impossible. This completes the proof.

This theorem excludes twice asymptotically flat and twice asymptotically AdS wormholes as well as those asymptotically flat on one end and AdS on the other. What is not excluded, is that one or both asymptotic regions are de Sitter: in this case, $r \to \infty$ but $A \sim -r^2$ at large r, and it is not necessary to have a minimum of $D(x)$.

The existing wormhole solutions where an isotropic fluid occupies a finite region of space also do not contradict the theorem since they inevitably require "heavy" thin shells on the boundary between the fluid and vacuum regions [384, 521], and these shells are highly anisotropic in the sense that a tangential pressure is nonzero while the radial one is not defined (the radial direction is orthogonal to the shell).

One more quite general result in static spherical symmetry can be obtained if we write the metric in terms of the curvature coordinates:

$$ds^2 = e^{2\gamma(r)}dt^2 - e^{2\alpha(r)}dr^2 - r^2 d\Omega^2. \tag{6.15}$$

Identifying the line elements (6.7) and (6.15), it is not hard to verify that on a throat, if any, the following conditions hold: $|\gamma| < \infty$, $e^{-2\alpha} \equiv B(r) = 0$, $dB/dr \neq 0$. One of the Einstein equations (3.5) has the form

$$-G_0^0 = \frac{1}{r^2}(1 - e^{-2\alpha}) + \frac{2\alpha' e^{-2\alpha}}{r} = \varkappa\rho \tag{6.16}$$

and can be integrated: putting $e^{-2\alpha} = 1 - 2m(r)/r$, we bring it to the form $2m'(r) = \varkappa\rho r^2$ and consequently,

$$m(r) = 4\pi G \int_{r_0}^{r} \rho r^2 dr, \tag{6.17}$$

where r_0 is an integration constant. The function $m(r)$ is called the mass function; in the case of an asymptotically flat configuration with a regular center it really gives the Schwarzschild mass if one integrates from zero to large radii where the metric is approximately Schwarzschild. However, the integral (6.17) is quite general for the metric (6.15).

In the case of an asymptotically flat wormhole, integrating in (6.17) from the throat $r = r_{\rm th}$ to large values of the radius, we obtain

$$r_{\rm th} = 2m - \varkappa \int_{r_{\rm th}}^{\infty} \rho r^2 dr, \tag{6.18}$$

since on the throat $B(r) \equiv 1 - 2m(r)/r = 0$. Recalling that $2m(\infty) = 2m$ is the Schwarzschild radius corresponding to the mass m, we come to the following conclusion:

The radius of a wormhole throat is larger than the Schwarzschild radius corresponding to the wormhole mass at its flat asymptotic only if the density of matter supporting the wormhole is negative; the throat radius is smaller than the Schwarzschild one if the density is positive.

It is really an important constraint: indeed, if we wish to have a wormhole with a throat of a few meters or kilometers rather than something of planetary or stellar size, we must take into account that meters and kilometers are just the scale of planetary and stellar Schwarzschild radii (for the Earth it is of the order of 1 cm, for the Sun about 3 km). To avoid huge gravity near the wormhole mouth and/or throat corresponding to such a mass, we must invoke not simply exotic matter, but matter with negative density.

In the next subsections, we will consider a few examples of exact static, spherically symmetric wormhole solutions to the gravitational field equations.

6.3.2 *Wormholes in the anti-Fisher solution*

The simplest example is given by the anti-Fisher solution described by Eq. (4.26) with $\varepsilon = -1$ and $k < 0$: it describes a wormhole with two spatial asymptotics at $u = 0$ and $u = \pi/|k|$. The metric has the form

$$\begin{aligned} ds^2 &= \mathrm{e}^{-2hu} dt^2 - \frac{k^2\, \mathrm{e}^{2hu}}{\sin^2(ku)} \left[\frac{k^2\, du^2}{\sin^2(ku)} + d\Omega^2 \right] \\ &= \mathrm{e}^{-2hu} dt^2 - \mathrm{e}^{2hu} [dx^2 + (k^2 + x^2) d\Omega^2], \end{aligned} \tag{6.19}$$

where the harmonic coordinate u is expressed in terms of the quasiglobal coordinate x, specified on the whole axis $\mathbb{R}$, by $|k|u = \operatorname{arccot}(x/|k|)$.

At $m > 0$, the wormhole attracts the ambient test matter on one asymptotic ($\rho \to \infty$) and repels it on the other ($\rho \to -\infty$), and vice versa in the case $m < 0$. At $m = 0$ we obtain the simplest solution called the Ellis wormhole (although H. Ellis's paper [229] discussed anti-Fisher wormholes of any mass).

The wormhole throat is located at $x = h$ and has the size

$$r_{\mathrm{th}} = (h^2 + k^2)^{1/2} \exp\left(\frac{h}{k} \cot^{-1} \frac{h}{k} \right). \tag{6.20}$$

It has been shown in [260] that the anti-Fisher wormholes are unstable under small radial perturbations, and the characteristic time of perturbation growth is of the order $r_{\rm th}/c$, i.e., the time needed for a light beam to cover a distance of the order of the throat size.

Analogues of the metric (6.19) with a nonzero electric charge are also known [75]; they can be obtained from (4.24), (4.25) at $\varepsilon = -1$: these are counterparts of Penney's solution [444] with a phantom scalar field. The stability of static, spherically symmetric configurations will be discussed in the next chapter.

Quite naturally, wormholes can also be supported by minimally coupled phantom scalar fields with nonzero potentials, see, e.g., [105]; a more unusual observation is that wormhole solutions can be obtained with the "trapped ghosts" [140], scalar fields which exhibit phantom properties only in a strong-field restricted region of space whereas in the rest of space all standard energy conditions are observed. Such wormholes can only exist with nonzero scalar field potentials. An example of such a solution will be considered below.

6.3.3 *Wormholes with nonminimally coupled scalar fields*

In GR with a massless conformal scalar field having the Lagrangian (4.45), ($W = 0$), and an electric field there exists a family of solutions to the field equations having the metric [75] (slightly more general than (4.51))

$$ds^2 = (y + y_0)^2 \left[\frac{dt^2}{(y + y_1)^2} - h^2 \frac{(y + y_1)^2}{y^4} (dy^2 + y^2 d\Omega^2) \right], \quad (6.21)$$

$$y_0 = \tanh hu_0, \quad y_1 = \coth hu_1, \quad (6.22)$$

and the scalar field

$$\phi = \sqrt{6}\frac{1 + yy_0}{y + y_0}. \quad (6.23)$$

In the electrically neutral case we have $y_1 = 1$; spatial infinity corresponds to $y = \infty$, where the scalar field tends to $\phi = \phi_\infty = \sqrt{6}y_0$.

The metric (6.22) describes a wormhole in the case $y_0 > 0$; the coordinate y then has the range $y > 0$. It is easy to verify that as

$y \to 0$, as well as for $y \to \infty$, the metric becomes flat. Thus, $y = 0$ is the second spatial infinity, where $e^{2\gamma} \to (y_0/y_1)^2$ and $\phi \to (1/y_0)\sqrt{6}$. Thus, we have a wormhole with a throat at $y = \sqrt{y_0 y_1}$.

The metric (6.22) (in other notations) and its generalizations to other nonminimally coupled scalar fields (the Lagrangian (4.29), $f(\phi) = 1 - \xi\phi^2$, $h(\phi) = 1$, where $\xi > 0$ is an arbitrary constant) were discussed by Barcelo and Visser [29, 30] as examples of wormholes existing even with a scalar field with the normal sign of kinetic energy.

However, all wormholes supported by nonminimally coupled scalar fields and, more generally, wormholes existing due to conformal continuations in scalar–tensor theories [82] have a common peculiarity: at the transition sphere, at which the nonminimal coupling function $f(\phi)$ in the Lagrangian (4.1) passes through zero, the effective gravitational constant also changes its sign. As a result, the second spatial infinity of such wormholes is situated in a region with a negative effective gravitational constant (this observation is sometimes expressed by the words "the graviton is there a ghost"). The plausibility of such configurations is thus questionable; also, evidently, they cannot connect distant regions of the same universe, and their remote mouth should be situated in a world where the physical laws are very unusual from our point of view. Besides, it has been established [111, 112] that electrically neutral (or weakly charged) wormholes of this kind are unstable under spherically symmetric perturbations (see also Chapter 7), and there are reasons to believe that such an instability has a common nature since it is related to the field behavior near the transition surface with $f = 0$.

(For more details on conformal continuations in static, spherically symmetric configurations in scalar–tensor and $f(R)$ gravity, the necessary and sufficient conditions for their existence and the possible full geometries of the continued space–times see [82, 91, 92].)

More generally, in the framework of STT and $f(R)$ theories of gravity (for instance, in the Brans–Dicke theory), wormhole solutions evidently do exist and can be obtained from those of GR with a phantom scalar field with the aid of the conformal mapping (4.3) provided the conformal factor $f(\phi)$ is regular everywhere. Thus, all wormhole solutions with a massless Brans–Dicke scalar are transforms of the anti-Fisher wormholes discussed above, except for the special case $\omega = 0$ [78] where a special wormhole solution exists due to a conformal continuation (see Sec. 6.3.5).

6.3.4 *Wormhole construction by solving the trace of the Einstein equations*

In this section, we demonstrate the significant level of arbitrariness in wormhole solutions of the Einstein equations if we do not specify the form of exotic matter from the beginning. We will show that even if, for simplicity, we restrict ourselves to matter with zero trace of the SET, the wormhole solutions form a set parametrized by one arbitrary function plus an integration constant [139]. To some extent, this presentation is similar to Sec. 5.2.3.

Let us start with the general static, spherically symmetric metric in 4 dimensions written in the curvature coordinates

$$ds^2 = e^{2\gamma(r)}dt^2 - e^{2\alpha(r)}dr^2 - r^2 d\Omega^2. \qquad (6.24)$$

If the SET of matter has a zero trace, the equation $R = 0$ is valid, and it can be written as a linear first-order equation with respect to the functon $F(r) := r\,e^{-2\alpha}$

$$F_r(2 + r\gamma_r) + F(2r\gamma_{rr} + 2r\gamma_r^2 + 3\gamma_r) = 2, \qquad (6.25)$$

where the subscript r means d/dr.

Let us note that Eq. (6.25) is of particular interest because the trace equation is the only uniquely determined equation describing the 4D gravitational field in an RS type brane world since the set of equations for gravity on the brane [500] contains a contribution $E_{\mu\nu}$ depending on the 5D Weyl tensor, having zero trace. (The basic concepts of the brane-world scenario are briefly discussed in Sec. 5.2.3.) In vacuum, where matter on the brane is absent and the 4D cosmological constant is zero (a natural assumption for scales much smaller than the size of the Universe), these equations reduce to

$$G_{\mu\nu} = -E_{\mu\nu}, \qquad (6.26)$$

where $G_{\mu\nu}$ is the four-dimensional Einstein tensor corresponding to the brane metric $g_{\mu\nu}$ while $E_{\mu\nu}$ is the projection of the five-dimensional Weyl tensor onto the brane. The traceless tensor $E_{\mu\nu}$ connects gravity on the brane with the bulk geometry (and is sometimes called the tidal SET), so that the set of equations (6.26) is not closed. Due to its geometric origin, $E_{\mu\nu}$ does not necessarily satisfy the energy conditions applicable to ordinary matter. Thus, examples are known [544] when negative energies on the brane are induced by gravitational waves or black strings in the bulk. Therefore, if the brane world concept is taken seriously, $E_{\mu\nu}$ can be the most natural "matter" supporting wormholes.

Thus, the wormhole metrics obtained with the algorithms to be described here may be considered as vacuum wormhole metrics in a brane world [139].

The general solution of Eq. (6.25) is

$$F(r) = \frac{2\,\mathrm{e}^{-2\gamma+3\Gamma}}{(2+r\gamma_r)^2}\int (2+r\gamma_r)\,\mathrm{e}^{2\gamma-3\Gamma}\,dr,$$

$$\Gamma(r) = \int \frac{\gamma_r dr}{2+r\gamma_r}. \tag{6.27}$$

(it is another formulation of (5.49), (5.50), more convenient for wormhole description). Choosing $\gamma(r)$ arbitrarily, we obtain $F(r)$ from (6.27), and, after fixing the integration constant, the metric is known completely at least in the region where e^{γ} and e^{α} are smooth and nonzero.

From the solution (6.27), wormhole metrics (as black hole metrics in Sec. 5.2.3) are built algorithmically by specifying the generating function $\gamma(r)$ with desirable properties.

Let us note for reference purposes that in many articles on wormholes, beginning with [417], the function $\mathrm{e}^{2\alpha}$ in the curvature coordinates is written in the form $[1 - b(r)/r]^{-1}$, where $b(r)$ is called "the shape function". This name is, in our view, not quite appropriate since $b(r)$ bears information on the wormhole profile in a very indirect form; one could more reasonably call $r(l)$ in the metric (6.28) a shape function. In fact (see (6.17)) $b(r) = 2m(r)$, where

$m(r)$ is the mass function. The function $F(r)$ introduced here is $F(r) = r - b(r) = r - 2m(r)$.

Let us now make clear how to choose the function $\gamma(r)$ (the so-called redshift function) and the integration constant in Eq. (6.27) in order to obtain a wormhole solution.

The coordinate r, which proves to be convenient for solving Eq. (6.25), is not an admissible coordinate in the whole space for wormhole solutions since in this case r has at least one minimum, and the solution in terms of r therefore splits into at least two branches. As an admissible coordinate one can take, e.g., the Gaussian coordinate l (proper length along the radial direction) connected with r by the relation $l = \int e^{\alpha} dr$, and the metric is rewritten as

$$ds^2 = e^{2\gamma(l)} dt^2 - dl^2 - r^2(l) d\Omega^2. \tag{6.28}$$

We seek static, traversable, twice asymptotically flat wormhole solutions. So we require:

1. There should be two flat asymptotics: $l \in \mathbb{R}$; $r \approx |l| \to \infty$ and $\gamma = \text{const} + O(r^{-1})$ as $l \to \pm\infty$;
2. Both functions $r(l) > 0$ and $\gamma(l)$ should be smooth (at least C^2) in the whole range $l \in \mathbb{R}$.

This guarantees the absence of curvature singularities and horizons (the latter correspond to $\gamma \to -\infty$ which is ruled out). This also means that $r(l)$ should have at least one regular minimum, $r_{\min} > 0$ (throat), at some value of l. Moreover, returning to functions of r, we see that at a flat asymptotic $e^{\alpha} \to 1$ and $F(r) \approx r$.

Suppose, without loss of generality, that a minimum of $r(l)$, that is, a wormhole throat, is located at $l = 0$. Then $r(0) = r_0 > 0$, $r_l(0) = 0$ and (generically) $r_{ll}(0) > 0$, where the subscript l denotes d/dl. Near $l = 0$ one has $r - r_0 \sim l^2$, hence the metric function $e^{2\alpha(r)}$ behaves as $(r - r_0)^{-1}$, and $F(r) = r\, e^{-2\alpha} \sim r - r_0$. In other words, a simple zero of $F(r)$ is *an indicator of a wormhole throat* provided $\gamma(r)$ is smooth and finite at the same r.

On the other hand, the derivative $\gamma_l(0)$ may be zero (which is always the case if the wormhole is symmetric with respect to the throat) or nonzero. If $\gamma_l(0) = 0$, we shall have $\gamma_r(r_0) < \infty$. If, on the contrary, $\gamma_l(0) \neq 0$, then near r_0 we have $\gamma_r \sim 1/|l| \sim \sqrt{r-r_0}$, so that

$$\gamma(r) \approx \gamma(r_0) + k\sqrt{r-r_0}, \quad k > 0. \tag{6.29}$$

We cannot put $k < 0$ since then we would obtain the expression $2 + r\gamma_r$ ranging from 2 (at spatial infinity) to $-\infty$ at $r = r_0$, so that $2 + r\gamma_r$ would vanish at some $r > r_0$ causing a singularity in (6.27).

We are now ready to single out a class of symmetric wormhole metrics (WH1) and a class of potentially asymmetric wormhole metrics (WH2) on the basis of the solution (6.27).

WH1. Specify the function $\gamma(r)$, smooth in the range $r_0 \le r < \infty$, $r_0 > 0$, in such a way that $\gamma(\infty) = 0$, $\gamma_r(r_0) < \infty$, and $2 + r\gamma_r > 0$ in the whole range. Fix the integration constant in (6.27) by performing integration from r_0 to r. Then these $\gamma(r)$ and $F(r)$ determine a wormhole which has a throat at $r = r_0$ and is symmetric with respect to the throat.

Indeed, by construction, $F(r) \sim r - r_0$ near r_0. Introducing the new coordinate x by the relation $r = r_0 + x^2$, we have $e^{2\alpha}dr^2 \sim (r-r_0)^{-1}dr^2 = 4dx^2$, which leads to a perfectly regular metric whose coefficients are all even functions of $x \in \mathbb{R}$. Both $x \to +\infty$ and $x \to -\infty$ are flat asymptotics.

Each $\gamma(r)$ chosen as prescribed creates a family of symmetric wormholes with zero scalar curvature. The family is parametrized by the throat radius r_0, taking arbitrary values in the range where $\gamma(r)$ is regular and $2 + r\gamma_r > 0$.

Another procedure is applicable to functions $\gamma(r)$ behaving according to Eq. (6.29).

WH2a. Specify the function $\gamma(r)$, smooth in the range $r_0 \le r < \infty$, $r_0 > 0$, such that $\gamma(\infty) = 0$, $2 + r\gamma_r > 0$ in the whole range, and Eq. (6.29) holds near r_0. Then, for proper values of the integration constant in (6.27), the sphere $r = r_0$ is a wormhole throat, and the solution is smoothly continued beyond it.

Indeed, the solution (6.27) may be rewritten as follows:

$$F(r) = \frac{e^{-2\gamma+3\Gamma}}{\left(1+\frac{1}{2}r\gamma_r\right)^2}\left[\int_{r_0}^{r}\left(1+\tfrac{1}{2}r\gamma_r\right)e^{2\gamma-3\Gamma}\,dr + C\right]. \qquad (6.30)$$

Suppose $C > 0$. Then $F(r)$ behaves near r_0 as $r - r_0 =: x^2$, while $\gamma = \gamma(r_0) + kx + O(x^2)$. The metric behaves smoothly at $r = r_0$ $(x = 0)$ in terms of the new coordinate x and can be continued through this sphere. One cannot, however, guarantee that this continuation will lead to another flat spatial infinity to yield an asymmetric wormhole, since the further behavior of $\gamma(x)$ and $F(x)$ may lead to anything, including a horizon or a singularity.

If we choose $C \leq 0$ in (6.30), we obtain two other situations:

WH2b. If $C < 0$, then $f(r_0) < 0$; recalling that $f \sim r$ at large r, we see that $F(r) = 0$ at some value $r = r_1 > r_0$, where γ_r is finite, and we return to the circumstances described as WH1, obtaining a symmetric wormhole with $r \geq r_1$, and the sphere $r = r_1$ is its throat.

WH2c. If $C = 0$, then near r_0 we obtain $F(r) \sim (r - r_0)^{3/2}$, and the metric is regularized at $r = r_0$ by another substitution: $r - r_0 = \xi^4$. As a result, Eq. (6.29) yields

$$\gamma = \gamma(r_0) + k\xi^2 + \text{further even powers of } \xi,$$

and we again obtain a symmetric wormhole, but now with a quartic behavior of r near its minimum as a function of the admissible coordinate $\xi \in \mathbb{R}$.

Now we will give some examples, presenting expressions for the metric functions γ and F, the effective "tidal" energy density ρ and the sum $\rho + p_r$, which characterizes violation of the NEC.

We use the time scale of a distant observer at rest and so always assume that $e^\gamma \to 1$ as $r \to \infty$.

Example 1. The simplest example is obtained for $\gamma \equiv 0$. Choosing any $r_0 > 0$ and applying the W1 algorithm, we simply obtain $F(r) = r - r_0$. This is a symmetric wormhole solution known as the spatial

Schwarzschild geometry [190][1]:

$$ds^2 = dt^2 - \left(1 - \frac{r_0}{r}\right)^{-1} dr^2 - r^2 d\Omega^2$$
$$= dt^2 - 4(r_0 + x^2)dx^2 - (r_0 + x^2)^2 d\Omega^2. \tag{6.31}$$

The effective SET E_μ^ν has the form $E_\mu^\nu = \mathrm{diag}(0, -p_r, p_r/2, p_r/2)$ with zero density and the radial pressure

$$p_r = -r_0/r^3. \tag{6.32}$$

Example 2. Our next example uses the Schwarzschild form of γ and has actually been discussed in Sec. 5.2.3:

$$\mathrm{e}^{2\gamma} = 1 - \frac{2m}{r}, \qquad m > 0. \tag{6.33}$$

Choosing any $r_0 > 2m$, we obtain according to the W1 prescription:

$$F(r) = \frac{(r - 2m)(r - r_0)}{r - 3m/2}, \tag{6.34}$$

$$ds^2 = \left(1 - \frac{2m}{r}\right)dt^2 - \frac{1 - 3m/(2r)}{(1 - 2m/r)(1 - r_0/r)}dr^2 - r^2 d\Omega^2$$
$$= \frac{x^2 + r_0 - 2m}{x^2 + r_0}dt^2 - \frac{4(x^2 + r_0)(x^2 + r_0 - 3m/2)}{x^2 + r_0 - 2m}dx^2$$
$$- (x^2 + r_0)^2 d\Omega^2. \tag{6.35}$$

This is evidently a symmetric wormhole geometry for any $r_0 > 2m \geq 0$, or for any $r_0 > 0$ in case $m < 0$. The Schwarzschild metric is restored from (6.35) in the special case $r_0 = 3m/2$.

The SET components of interest are

$$\rho = \frac{m(2r_0 - 3m)}{r^2(2r - 3m)^2}; \quad \rho + p_r = -\frac{2(r - 2m)(2r_0 - 3m)}{r^2(2r - 3m)^2}. \tag{6.36}$$

---

[1] This example, in which the spatial metric is the same as in the Schwarzschild space–time, clearly shows that it is absolutely meaningless to try to distinguish BH and wormhole space–times on the basis of their spatial sections, as is done by some authors: the difference between them is hidden in the temporal part of the metric.

Other examples are discussed in [139]. We can conclude that even the equation $R = 0$ leads to a great number of wormhole solutions. Symmetric solutions of class WH1 can be obtained from any $\gamma(r)$ providing asymptotic flatness; asymmetric wormhole solutions belonging to class WH2a require somewhat more special conditions and are more difficult to obtain from the general solution (6.27). One such example is analyzed in [139], with

$$e^{2\gamma} = \left(1 - b + b\sqrt{1 - 2m/r}\right)^2, \tag{6.37}$$

with constants $m > 0$, $b \neq 0$, 1. It has been shown that an asymmetric wormhole is obtained from (6.37) only if $b \in (0, 1/2)$, otherwise there is a singularity or a horizon beyond the throat ($r_{\rm th} = 2m$).

As follows from the analyzed examples, wormholes are not always connected with negative (effective) energy densities ρ. They can appear with $\rho > 0$, but only with comparatively large negative pressures maintaining violation of the NEC; recall in this respect our general reasoning in Sec. 6.3.1.

All the above refers to metrics satisfying the condition $R=0$ in four dimensions, which admit an interpretation as vacuum brane metrics. We did not discuss the possible bulk properties of models in question; it can be noted that the existence of the corresponding solutions to the 5D equations of gravity (in our case, the 5D vacuum Einstein equations with a cosmological constant) is guaranteed by the Campbell–Magaard type embedding theorems [151, 396]. A recent discussion of these theorems, applied in particular to brane-world scenarios, and more references can be found in [494]. It has also been claimed that "any 4D space–time with $R = 0$ gives rise to a 3-brane world without surface stresses embedded in a five-dimensional space–time" [545] since the embedding contains a very significant arbitrariness. Nevertheless, a complete model requires knowledge of the full 5D space–time. In other words, one should "evolve" the 4-metric into the bulk, using this 4-matric as initial data for the five-dimensional equations. It is rather a difficult task, as was demonstrated in a study of particular black hole solutions in [164, 167]. There are, however, two favorable circumstances. One is the wealth of wormhole solutions: there is actually an arbitrary function $\gamma(r)$ leading to wormholes on the brane, which must in turn lead to a wide choice of suitable bulk functions. The other is the global regularity of wormhole space–times, and one can expect that the bulk incorporating them will also be regular. (It may be recalled that it was the singular nature of black hole solutions that caused some technical difficulties in [167].)

6.3.5 *Alternative gravity and vacuum as wormhole supporters*

The algorithms presented in the previous subsection demonstrate the wealth and diversity of wormhole metrics even in the narrow class of spherically symmetric space–times. Still this does not remove the question of possible real material sources of wormhole geometry satisfying, under the same symmetry, the conditions (6.10).

As we have seen, usual classical (nonphantom) matter, including fields, does not lead to wormholes since it respects the NEC This result is valid in the Einstein theory, but it extends to a large class of alternative theories of gravity where the gravitational equations are reduced to those of GR with an additional nonphantom scalar field. These are the scalar-tensor theories (STT) under the condition $\varepsilon = +1$ (see Sec. 4.1, Eqs. (4.1)–(4.3)), $F(R)$ theories, which have recently become very popular and where, instead of the GR Lagrangian (const $\cdot\, R$), the gravitational Lagrangian is a certain function $F(R)$, and the class of theories that unifies the STT and $F(R)$ theories, where the Lagrangian contains a function of two variables $F(\phi, R)$. The transformation from the initial equations involving the physical metric $g_{\mu\nu}$ (the so-called Jordan picture) to the equations of GR with minimally coupled scalar fields (the Einstein picture) have been obtained in a general form in the papers [548] for STT, [34, 35, 220, 320, 562] for $F(R)$ theories and [394] for $F(\phi, R)$ theories.

Indeed, let the metric $g_{\mu\nu}$ in the initial manifold $\mathbb{M}_J$ (the Jordan picture) and the metric $\overline{g}_{\mu\nu}$ in the manifold $\mathbb{M}_E$ (the Einstein picture) be related by the conformal mapping

$$g_{\mu\nu} = f(x)\overline{g}_{\mu\nu}, \tag{6.38}$$

where $f(x)$ is an everywhere regular function of the space–time coordinates, bounded both above and below by some positive constants. According to [291], in GR, for the existence of a static throat (defined as a minimal 2-surface in 3-space) it is necessary that the material source in the Einstein equations violate the NEC. Thus, if the NEC is respected, wormholes and even throats cannot exist. Furthermore, under the assumptions made, the transformation (6.38) always transfers a flat asymptotic in $\mathbb{M}_E$ to a flat asymptotic in $\mathbb{M}_J$ and vice versa (though the corresponding Schwarzschild masses may differ due to nonconstancy of f). If we suppose that there is a twice asymptotically flat wormhole in $\mathbb{M}_J$, then each of its flat asymptotics has a counterpart in $\mathbb{M}_E$, and due to smoothness of the transformation we obtain a wormhole in $\mathbb{M}_E$, contrary to what was said above. Therefore, static asymptotically flat wormholes are absent in $\mathbb{M}_J$ as well.

This simple reasoning [107, 138] does not require any assumption on the spatial symmetry and works under very general assumptions on the original theory. It admits extensions to dynamic wormholes [138], since in GR they also require NEC violation, although introducing the notion of a throat faces some difficulties in the dynamic case [53, 292]. The asymptotic flatness condition can also be weakened in an evident manner.

It should be stressed that throats in Jordan's picture are not forbidden since local violations of the energy conditions can occur. However, as we have seen, there are no wormholes as global entities.

If the requirements on the function $f(x)$ are weakened, one can obtain wormholes in Jordan's picture in some theories (as we have seen above for a conformal scalar field); however, such wormholes possess some pathological features, for instance, the effective gravitational constant either tends to infinity at the second flat asymptotic [92] or has the wrong sign (see Sec. 5.3.2).

Thus it is rather difficult to obtain more or less realistic wormholes in classical field theory without explicitly invoking phantom fields, at least in GR and its simplest extensions in four dimensions. One may, though, recall the above-discussed case of hybrid metric-Palatini gravity (HMPG) in which the emergence of a phantom scalar field is justified by its purely geometric origin from an independent connection.

Some phantom-free wormhole-like solutions in GR can still be recalled, beginning with Kerr's metric with a large angular momentum and Zipoy's static vacuum solution [580] as well as their electromagnetic and scalar extensions [104, 402], but all of them possess a naked ring singularity that bounds a disk playing the role of a throat. Therefore, despite their asymptotic flatness, they can hardly be welcomed. Others [24, 152], being phantom-free, lack asymptotic flatness. All such shortcomings may probably be interpreted as signs of the topological censorship [246, 247].

More opportunities appear in multidimensional theories. As we saw in the previous subsection, in the brane-world scenario the 4D gravitational field equations contain a geometric source which has no reason to respect the usual energy conditions and can thus create wormholes (but there remains the difficulty of solving the full 5D equations). In more general theories of gravity in four dimensions, the terms effectively added to the matter SET as compared with GR can also lead to wormhole formation: an example can be found in the recent paper [318] where a family of static, spherically symmetric wormhole solutions was obtained numerically in dilatonic Einstein–Gauss–Bonnet theory with the Lagrangian

$$L = \text{const} \cdot \left[R + \frac{1}{2}(\partial\phi)^2 + \alpha\, \mathrm{e}^{-\lambda\phi} R^2_{\text{GB}} \right], \tag{6.39}$$

where $R^2_{\text{GB}} = R_{\mu\nu\rho\sigma}R^{\mu\nu\rho\sigma} - 4R_{\mu\nu}R^{\mu\nu} + R^2$ is the Gauss–Bonnet invariant, α and λ are constants. In these solutions, the scalar field is canonical, nonphantom, and there is no other matter violating the NEC, so only the Gauss–Bonnet contribution is responsible for the wormhole's existence. Curiously, among these solutions there are no force-free ones, i.e., those with $g_{00} = \mathrm{e}^{2\gamma} = \text{const}$. Nonexistence of such wormholes has been proved [103] for a class of Lagrangians more general that (6.39), with an arbitrary potential $V(\phi)$ added and an arbitrary function $h(\phi)$ instead of $\mathrm{e}^{-\lambda\phi}$.

More general geometries, such as those involving torsion, can also lead to phantom-free wormhole solutions, as has been demonstrated in [108, 109] in the framework of the Einstein–Cartan theory of gravity.

Another promising opportunity is connected with quantum phenomena. We have already mentioned the concept of space–time foam existing at Planck scales due to large chaotic quantum fluctuations of the geometry and topology [559]. Among other formations, wormholes are elements of space–time foam. Under usual conditions, the existence of the foam cannot be noticed at macroscopic and even atomic and subatomic scales, but at the earliest stages of the Universe evolution, during its rapid expansion (inflation, see the future chapters), some of the tiny wormholes get a chance to expand to a large size. There exists an idea (see, e.g., [333, 334] and references therein) that such wormholes of kiloparsec size (or like) form a whole chaotic network in the Universe, and the Friedmann homogeneous and isotropic space–time is only obtained as a result of averaging. It has been shown that the inhomogeneities of this foam-like cosmic structure may be observed as manifestations of dark matter (thus solving the well-known problem of mass deficit in galaxies and clusters of galaxies for explaining their dynamic properties [323, 438]). However, the authors themselves admit [335] that by now there are quite a number of unsolved questions in this scenario: what is the wormhole distribution by size and mass? Why are they not yet observed astronomically? Do they take part in the present expansion of the Universe? And so forth. And the same question is probably the most important: which material source supports the existence of large wormholes after they left the Planck scale?

A possible answer is: vacuum polarization due to quantum fields. By quantum field theory, as is well known, the vacuum is the field state with minimum energy and without real particles. This state is not an absolute rest but rather the chaotic creation and annihilation of pairs of all possible virtual particles which all together form some energy density, which is in general not small and can be of any sign. Both depend on the space configuration where the vacuum is considered and on the matter present there. This phenomenon, called vacuum polarization, does really exist and has been repeatedly confirmed by experiment. In particular, the famous Casimir effect, attraction of two closely located plane metal plates in empty space, is explained by a difference in the properties of the electromagnetic field vacuum in the region between the plates and in the whole remaining space.

Calculation of the physical vacuum properties is a difficult task, but some known results look promising from the viewpoint of wormhole existence. Thus, unlike macroscopic matter, the vacuum may not necessarily obey the energy conditions, and it has been found [290, 359] that vacuum polarization can in principle support static wormholes of any size. However, as shown by Hsu [144], the so-called semiclassical matter, suitable for wormhole construction, possesses a special kind of instability due to quantum fluctuations: a wormhole, being left to itself, will be distorted and will be eventually destroyed (but what will then form is not clear). This instability is certainly not universal and not inherent to all wormhole models. Furthermore: if we discuss on the opportunities of principle, from a physical viewpoint, of an "arbitrarily advanced civilization," we can surely admit that, having an interest in interstellar transport, it will launch some negative feedback mechanism able to damp the perturbations.

6.3.6 *Wormholes with a trapped ghost*

We have seen that some non-Einsteinian models can lead to phantom-free wormhole models. There is nevertheless a good reason to adhere to GR and macroscopic matter and fields if our interest is in obtaining (potentially) realistic and manageable wormholes since it is this theory that is quite well verified by experiment at the macroscopic level and even serves as a tool in a number of engineering applications such as, for instance, GPS navigation.

Meanwhile, macroscopic phantom matter has never been observed, and this puts to doubt whether it can exist in principle and whether it is possible to obtain realistic wormholes, suitable, for example, for an interstellar travel, even in a remote future and even by a highly advanced civilization.

An interesting opportunity to circumvent these problems and to obtain wormhole configurations in GR is to find such a kind of matter that possesses phantom properties only in a restricted region of space, somewhere close to the throat, whereas far away from it all standard energy conditions are observed [140]. As an example of such matter, we can consider a minimally coupled scalar field with the Lagrangian

$$L_s = h(\phi) g^{\mu\nu} \partial_\mu \phi \partial_\nu \phi - V(\phi), \tag{6.40}$$

as a source of gravity in static, spherically symmetric configurations, with arbitrary functions $h(\phi)$ and $V(\phi)$. If $h(\phi)$ has a variable sign, it cannot be absorbed by redefinition of ϕ in its whole range. A case of interest is that $h > 0$ (then is, the scalar field is canonical, with positive kinetic energy) in a weak field region and $h < 0$ (the scalar field is phantom, or ghost) in some restricted region where a wormhole throat can be expected. In this sense it can be said that the ghost is trapped. A possible transition between $h > 0$ and $h < 0$ in cosmology was considered in [363].

Phantom fields are known to produce not only wormhole configurations but also different kinds of regular black holes, including black universes, see Chapter 5. The same can be expected from trapped ghost models, and such models have been really found [84, 102].

In this subsection, we will show how both wormhole and black universe solutions with a trapped-ghost scalar field can be obtained. We will first explain why the "trapped-ghost" strategy cannot be realized for a massless scalar field ($V(\phi) \equiv 0$), then describe some general properties of the system and present explicit examples of "trapped-ghost" wormhole and black-universe solutions, obtained using the inverse-problem method. The stability problem for such spherically symmetric scalar field configurations will be discussed in Chapter 7.

Massless fields. If we again consider a static, spherically symmetric metric in GR and a massless scalar field $\phi(u)$ as a source of gravity [$V(\phi) = 0$ in (6.40)], we can solve the field equation in quite the same manner as in Sec. 4.1.1, but instead of the relation between the constants (4.20), that is, $k^2 \operatorname{sign} k = h^2 = \varepsilon C^2/2$, we now obtain

$$k^2 \operatorname{sign} k = m^2 + h(\phi)\phi'^2, \tag{6.41}$$

where the prime means d/du, and u is the harmonic radial coordinate. It implies that $h(\phi)\phi'^2 = \text{const}$, hence $h(\phi)$ cannot change its sign within a particular solution characterized by certain fixed values of the integration constants m and k. The situation does not change if, instead of a single scalar field, there is a nonlinear sigma model with multiple scalar fields ϕ^a and the Lagrangian

$$L_\sigma = h_{ab} g^{\mu\nu} \partial_\mu \phi^a \partial_\nu \phi^b, \tag{6.42}$$

where h_{ab} are functions of ϕ^a. The metric then remains the same (that is, Fisher or anti-Fisher), and a relation similar to (6.41) reads [94]

$$k^2 \operatorname{sign} k = m^2 + h_{ab}(\phi^a)'(\phi^b)'.$$

Therefore, the quantity that determines the canonical or phantom nature of the scalars, $h_{ab}(\phi^a)'(\phi^b)'$, is constant. If the matrix h_{ab} is not positive-definite, some of the scalars can be phantom, but in each particular solution the scalar field system preserves its canonical or phantom nature in the whole space.

Fields with nonzero potentials $V(\phi)$. Let us consider a static, spherically symmetric metric (6.7) written in terms of the quasiglobal coordinate x,

$$ds^2 = A(x)dt^2 - \frac{dx^2}{A(x)} - r^2(x)d\Omega^2. \tag{6.43}$$

and write the Einstein–scalar equations for a scalar with the Lagrangian (6.40):

$$2(Ar^2 h\phi')' - Ar^2 h'\phi' = r^2 dV/d\phi, \tag{6.44}$$

$$(A'r^2)' = -2r^2 V, \tag{6.45}$$

$$r''/r = -h(\phi)\phi'^2, \tag{6.46}$$

$$A(r^2)'' - r^2 A'' = 2, \tag{6.47}$$

$$-1 + A'rr' + Ar'^2 = r^2(hA\phi'^2 - V), \tag{6.48}$$

where the prime again denotes d/dx. It is easy to verify that Eqs. (6.44) and (6.48) follow from (6.45)–(6.47), which, given the potential $V(\phi)$ and the kinetic function $h(\phi)$, form a determined set of equations for the unknowns $r(x)$, $A(x)$, $\phi(x)$. Moreover, Eq. (6.47) can be integrated giving

$$B'(x) \equiv (A/r^2)' = 2(3m - x)/r^4, \qquad (6.49)$$

where $B(x) \equiv A/r^2$ and m is an integration constant equal to the Schwarzschild mass if the metric (6.43) is asymptotically flat as $x \to \infty$ ($r \approx x$, $A = 1 - 2m/x + o(1/x)$). If there is a flat asymptotic as $u \to -\infty$, the Schwarzschild mass there is equal to $-m$ ($r \approx |u|$, $A = 1 + 2m/|u| + o(1/u)$.

Thus we have a general result for *any* solution with two flat asymptotic regions in the presence of *any* potential $V(\phi)$ compatible with such behavior (such solutions can represent wormholes or regular black holes): we inevitably have masses of opposite signs, just as is the case in the well-known special solution — the anti-Fisher solution discussed in Sec. 6.3.2.

It is also clear that $m = 0$ in all symmetric solutions to Eqs. (6.44)–(6.48), such that $r(x)$ and $A(x)$ are even functions. Indeed, in this case $B'(x)$ is odd, hence $m = 0$ in (6.49).

Models with a trapped ghost. If one specifies the functions $V(\phi)$ and $h(\phi)$ in the Lagrangian (6.40), it is, in general, hard to solve the above equations. Alternatively, to find examples of solutions possessing some particular properties, one employs the inverse problem method, choosing some of the functions $r(x)$, $A(x)$ or $\phi(x)$ and then reconstructing the form of $V(\phi)$ and/or $h(\phi)$. We will do so, choosing a function $r(x)$ suitable for both wormhole and black-universe models. Then $A(x)$ is found from (6.49) and $V(x)$ from (6.45). The function $\phi(x)$ is found from (6.46) provided $h(\phi)$ is known; alternatively, using the scalar field parametrization freedom, we can, vice versa, choose a monotonic function $\phi(x)$ (which will yield an unambiguous function $V(\phi)$) and find $h(x)$ from Eq. (6.46).

Let us discuss which kind of function $r(x)$ should be chosen for our purposes.

1. In both wormhole and black universe solutions there must be a minimum of $r(x)$ ($x = 0$ without loss of generality), so that

$$r(0) = a, \quad r'(0) = 0, \quad r''(0) > 0, \quad a = \text{const} > 0. \qquad (6.50)$$

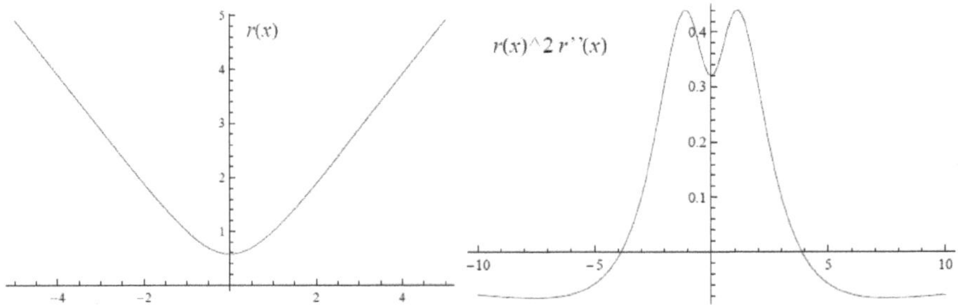

Figure 6.2 Plots of $r(x)$ and r^2r'' for $n = 3$.

2. In a trapped-ghost configuration, by definition, the kinetic coupling function $h(\phi)$ is negative near the minimum of r and positive far from it. According to (6.46), this means that r'' is positive at small $|x|$ and negative at sufficiently large $|x|$.
3. If our model is asymptotically flat or asymptotically (anti-) de Sitter at large $|x|$, we should have

$$r(x) \approx |x| \quad \text{as} \quad x \to \pm\infty.$$

A simple example of the function $r(x)$ satisfying the requirements 1–3 is [84] (see Fig. 6.2)

$$r(u) = a\frac{(x/a)^2 + 1}{\sqrt{(x/a)^2 + n^2}}, \qquad n = \text{const} > 2, \tag{6.51}$$

where a is an arbitrary constant that has the meaning of the throat radius and can be used as the length scale. It differs from the function $r(x)$ used in [140] and leads to slightly simpler resulting expressions.

In what follows, we put $a = 1$, which actually means that the length scale is arbitrary but the quantities r and m (the Schwarzschild mass in our geometrized units), etc., with the dimension of length, are expressed in units of a, the quantities B, V and others with the dimension (length)$^{-2}$ in units of a^{-2}, etc.; the quantities A, ϕ, h are dimensionless. Since

$$r''(x) = \frac{1}{a}\frac{x^2(2-n) + n(2n-1)}{(x^2+n)^{5/2}},$$

we have $r'' > 0$ at $x^2 < n(2n-1)/(n-2)$ and $r'' < 0$ at larger $|x|$, as required; it is also clear that $r \approx a|x|$ at large $|x|$. It guarantees $h < 0$ at small $|x|$ and $h > 0$ at large $|x|$ (see Fig. 6.1, right panel).

Further integration can be performed analytically but leads to rather cumbersome expressions for $B(x)$ and other quantities, therefore we will restrict ourselves to the choice $n = 3$. An inspection shows that a particular choice of the parameter $n > 2$ does not change the qualitative features of the solutions though certainly affects their numerical characteristics.

Integrating (6.49) for $n = 3$, we obtain

$$B(x) \equiv \frac{A(x)}{r^2(x)} = B_0 + \frac{1}{6(1+x^2)^3}[26 + 24x^2 + 6x^4 + 3mx(69 + 100x^2 + 39x^4)] + \frac{39m}{2} \arctan x, \tag{6.52}$$

where B_0 is an integration constant.

Now suppose that our system is asymptotically flat at large positive x. Since $B = A/r^2$ and $A \to 1$ at infinity, we require $B \to 0$ as $x \to \infty$ and thus fix B_0 as

$$B_0 = -\frac{39\pi m}{4}.$$

The form of $B(x)$ (and consequently $A(x) = Br^2$) now depends on the mass m, see Fig. 6.3. We see that with a negative mass we obtain $B(x)$ tending to a positive constant as $x \to -\infty$, hence $A \sim r^2$, and we obtain a wormhole with an anti-de Sitter (AdS) asymptotic behavior at the "far end" (an M-AdS wormhole for short, where M stands for Minkowski). With zero mass there is a twice asymptotically flat (M-M) wormhole, and lastly in the case $m > 0$ $B(x)$ changes its sign and tends to a negative constant: we thus obtain a black universe with a de Sitter (dS) asymptotic at large negative x.

Now we know the metric completely, and the remaining quantities $\phi(x)$ and $V(\phi(x))$ are easily found from Eqs. (6.46) and (6.45), respectively (Fig. 6.4). To construct V as an unambiguous function

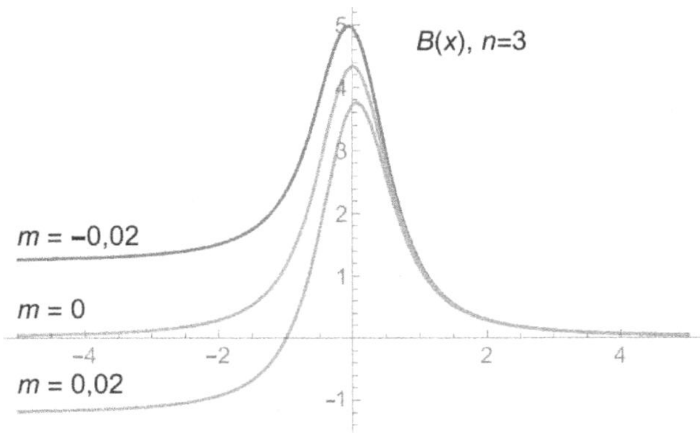

Figure 6.3 Plots of $B(x)$ for different values of m. With $m < 0$ we obtain an asymmetric M-AdS wormhole, with $m = 0$ a symmetric M-M wormhole, and with $m > 0$ a black universe with a dS behavior as $x \to -\infty$.

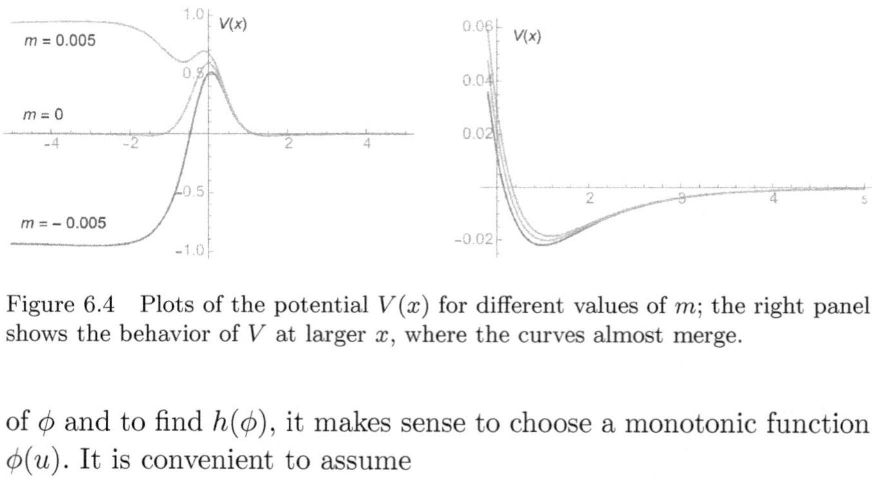

Figure 6.4 Plots of the potential $V(x)$ for different values of m; the right panel shows the behavior of V at larger x, where the curves almost merge.

of ϕ and to find $h(\phi)$, it makes sense to choose a monotonic function $\phi(u)$. It is convenient to assume

$$\phi(x) = \frac{1}{\sqrt{3}} \arctan \frac{x}{\sqrt{3}}, \tag{6.53}$$

so that ϕ has a finite range: $\phi \in (-\phi_0, \phi_0)$, $\phi_0 = \pi/(2\sqrt{3})$, which is common to kink configurations. Thus, we have $x = \sqrt{3}\tan(\sqrt{3}\phi)$, whose substitution into the expression for $V(x)$, found from (6.45), gives $V(\phi)$ defined in this finite range.

The expression for the kinetic coupling function $h(\phi)$ is then found from (6.46) as follows:

$$h(\phi) = \frac{x^2 - 15}{x^2 + 1} = \frac{3\tan^2(\sqrt{3}\phi) - 15}{3\tan^2(\sqrt{3}\phi) + 1}. \tag{6.54}$$

The function $h(\phi)$ produced by Eq. (6.54) is also defined in the interval $(-\phi_0, \phi_0)$ and can be extended to $\mathbb{R}$ by supposing $h(\phi) \equiv 1$ at $\phi| \geq \phi_0$. Evidently, the NEC is violated where and only where $h(\phi) < 0$.

For $V(x)$ we obtain [84]

$$\begin{aligned} V(x) = {} & \frac{1}{12(1+x^2)^2(3+x^2)^3}\{32(-6 + x^2 + 3x^4) \\ & + 6mx(2655 + 6930x^2 + 5420x^4 + 1326x^6 + 117x^8) \\ & + 117m(1+x^2)^2(15 + 89x^2 + 29x^4 + 3x^6) \\ & \times (-\pi + 2\arctan x)\}. \end{aligned} \tag{6.55}$$

The function $V(\phi)$ can also be extended to the whole real axis, $\phi \in \mathbb{R}$, by supposing $V(\phi) \equiv 0$ at $\phi \geq \phi_0$ (since $V(x = +\infty) = 0$) and $V(\phi) = V(-\phi_0) > 0$ at $\phi < -\phi_0$.

It is easy to verify that the asymptotic values of the function $B(x)$ as $x \to -\infty$ are directly related to those of the potential V which in this case plays the role of an effective cosmological constant:

$$V(-\infty) = 117m\pi/2 = -3B(-\infty), \tag{6.56}$$

and thus, as already mentioned, a negative $B(-\infty)$, corresponding to $m > 0$, leads to a dS asymptotic, with $m = 0$ and $B(-\infty) = 0$ it is flat, and with $B(-\infty) > 0$ $(m < 0)$ it is AdS (Fig. 6.3).

More complicated models with various global structures emerge if, in addition to a scalar field, there is an electromagnetic field. Examples of such solutions, which include M-M and M-AdS wormholes and regular black holes with up to three horizons, are presented in [61]: the structures obtained are similar to those found earlier with a scalar field which is everywhere phantom [62].

6.4 Cylindrical wormholes

As we already know, one of the main problems in wormhole physics in GR is the necessity of exotic, or phantom matter for wormhole existence, violating the Weak Energy Condition (WEC) and its part, the NEC, at least near a throat [246, 386, 417, 551] and at least for wormholes with compact throats of finite area and for asymptotically flat spacetimes. Meanwhile, asymptotic flatness is a necessary requirement if a wormhole is expected to be observable from distant regions with small curvature. A mathematical manifestation of this fundamental issue is known as topological censorship [246, 247, 539].

Cylindrical symmetry seems to be a way to avoid topological censorship because, even assuming asymptotic flatness in the radial (r) direction, we have no such flatness in the longitudinal one, z, since at large $|z|$ spacetime at given r remains equally curved as it was at small $|z|$. And, provided the system is t-independent (stationary), the same is true for the null (z, t) direction. So, there is hope to obtain phantom-free wormholes, asymptotically flat in the radial direction.

Among numerous known cylindrical wormhole solutions in GR (see, among others, [65, 83, 123–125, 127] and references therein) many are phantom-free but none are asymptotically flat, for the same reason for which the Levi-Civita vacuum solution is only asymptotically flat if it is completely flat, i.e., only when the entire spacetime is Minkowiski. To solve the problem, the following construction was suggested [125]: a central wormhole region with a throat, surrounded on both sides by Minkowski regions, matched through thin cylindrical shells Σ_- and Σ_+. These shells will possess nonzero densities and pressures, their SETs being determined by jumps of the extrinsic curvature according to the well-known Darmois–Israel formalism. A phantom-free model will be obtained if in the internal region and on the two surfaces all SETs respect the WEC.

In this section, we will describe the conditions on the geometry and SETs of matter for obtaining static or stationary (rotating) cylindrically symmetric wormhole configurations and present some particular configurations. Among them are two examples of rotating phantom-free wormhole models, asymptotically flat on both sides of the throat.

6.4.1 *Wormhole definitions and existence conditions*

We begin with the stationary cylindrically symmetric metric

$$ds^2 = e^{2\gamma(x)}[dt - E(x)\,e^{-2\gamma(x)}\,d\varphi]^2 - e^{2\alpha(x)}dx^2 \\ - e^{2\mu(x)}dz^2 - e^{2\beta(x)}d\varphi^2, \qquad (6.57)$$

where x, $z \in \mathbb{R}$ and $\varphi \in [0, 2\pi)$ are the radial, longitudinal and angular coordinates. The coordinate x is specified up to a reparametrization $x \to f(x)$, so its range depends on both its choice (the "gauge") and the geometry itself. The quantity $r = e^{\beta}$ has the meaning of the circular radius. The off-diagonal component E corresponds to rotation, unless $E\,e^{-2\gamma} = \text{const}$, in which case one can make $E = 0$ by redefining (resynchronizing) the time coordinate. If $E = 0$, the metric (6.57) is static and coincides with (A.1).

The metric (6.57) is said to describe a *wormhole* if either (i) the circular radius $r(x) = e^{\beta(x)}$ has a regular minimum (the corresponding surface $x = \text{const}$ is called an *r-throat*) and is large or infinite far from this minimum or (ii) the same is true for the area function $a(x) = e^{\mu+\beta}$ (its minimum defines an a-throat) [125, 127]. For certainty, we choose the definition connected with the radius $r(x)$, looking more evident: while moving to smaller r, we approach a would-be symmetry axis $r = 0$, but, instead, we reach a minimum of r and observe its subsequent growth, paying no attention to what happens to the z scale.

Asymptotic flatness in some limit of x implies $r \to \infty$, finite limits of γ, μ and E (thus no rotation) and $e^{-\alpha}r' \to 1$ to provide a Euclidean shape of large circles. It is clear that if a wormhole is asymptotically flat on both extremes of the x range, it contains both r- and a-throats.

Let us first consider the static case, putting $E(x) \equiv 0$ in (6.57). Then the nonzero components of the Ricci tensor read

$$\begin{aligned}
R_0^0 &= \mathrm{e}^{-2\alpha}\left[\gamma'' + \gamma'(\gamma' - \alpha' + \beta' + \mu')\right], \\
R_1^1 &= \mathrm{e}^{-2\alpha}\left[\gamma'' + \mu'' + \beta'' + \gamma'^2 + \mu'^2 + \beta'^2 - \alpha'(\gamma' + \mu' + \beta')\right], \\
R_2^2 &= \mathrm{e}^{-2\alpha}\left[\mu'' + \mu'(\gamma' - \alpha' + \beta' + \mu')\right], \\
R_3^3 &= \mathrm{e}^{-2\alpha}\left[\beta'' + \beta'(\gamma' - \alpha' + \beta' + \mu')\right],
\end{aligned} \tag{6.58}$$

where the prime denotes d/dx, and the coordinates are numbered according to the scheme $(0, 1, 2, 3) = (t, x, z, \phi)$. It is also useful to write the component G_1^1 of the Einstein tensor $G_\mu^\nu = R_\mu^\nu - \frac{1}{2}\delta_\mu^\nu R$ that contains only first-order derivatives:

$$G_1^1 = -\,\mathrm{e}^{-2\alpha}\left(\gamma'\mu' + \beta'\gamma' + \beta'\mu'\right). \tag{6.59}$$

The coordinate x can be chosen in different ways, depending on convenience and taste, and is usually fixed by some relation between the functions $\alpha, \beta, \gamma, \mu$. For example, many authors use the condition $\alpha = \mu$. However, the components R_μ^ν take the simplest form if we choose the harmonic radial coordinate defined by the condition [76],

$$\alpha = \beta + \gamma + \mu. \tag{6.60}$$

Let us here, for illustration, show how the Levi-Civita vacuum solution [135, 374] is obtained using (6.60). The equations $R_\mu^\nu = 0$ then immediately lead to $\beta'' = \gamma'' = \mu'' = 0$, whence

$$\beta(u) = bu + b_0, \quad \gamma(u) = cu + c_0, \quad \mu(u) = hu + h_0, \tag{6.61}$$

with six integration constants, among which $b_0,\ c_0,\ h_0$ can be turned to zero by changing scales along the t and z axes and by choosing the zero point of the u coordinate. Moreover, the first-order equation $G_1^1 = 0$ leads to a relation between the remaining constants $b,\ c,\ h$:

$$bc + bh + ch = 0, \tag{6.62}$$

hence there are precisely two significant constants. The metric takes the form

$$ds^2 = e^{2cu}dt^2 - e^{2(b+c+h)u}du^2 - e^{2hu}dz^2 - e^{2bu}d\phi^2. \qquad (6.63)$$

In its traditional form with the gauge $\alpha = \mu$ it looks more complex.

In the general static case, assuming the most general SET of matter compatible with the symmetry,

$$T_\mu^\nu = \mathrm{diag}(\rho, -p_x, -p_z, -p_\phi) \qquad (6.64)$$

and using the expressions (6.58) for the Ricci tensor components, it is easy to show [127] that at an a-throat near a minimum of $e^{\beta+\mu}$ one should inevitably have

$$\rho < p_x \leq 0, \qquad (6.65)$$

which means the necessity of a negative density, violating the WEC. Thus static, twice asymptotically flat wormholes are impossible. On the other hand, an r-throat only implies

$$\rho - p_x - p_z + p_\phi < 0, \qquad (6.66)$$

which does not necessarily violate the WEC. Though, if $p_z = p_\phi$ (as, for example, in isotriopic fluids), (6.66) reads $\rho - p_x < 0$, violating the dominant energy condition if $\rho > 0$.

Thus in the static case we can have phantom-free wormhole solutions in the above sense but without two-side asymptotic flatness. Other restrictions also follow from combinations of (6.70), for example, the following no-go theorem [127]:

A static, cylindrically symmetric wormhole, mirror-symmetric with respect to its throat, cannot exist in GR with matter whose SET satisfies the conditions $T_2^2 = T_3^3$ and $T_0^0 \geq 0$.

This concerns, in particular, isotropic fluids with any equations of state and scalar fields $\Phi(x)$ with a variety of Lagrangians, such as $L_s = (\partial\Phi)^2 - 2V(\Phi)$ with any potentials V and more general k-essences with $L_s = f(X, \Phi)$ where $X = (\partial\Phi)^2$ and f is an arbitrary function.

What changes if there is rotation? To make it clear, we note that rotation in spacetime with the metric (6.57) can be characterized by the angular velocity $\omega(x)$ of a congruence of timelike curves defined as [125, 361, 362]

$$\omega = \tfrac{1}{2}(E\,\mathrm{e}^{-2\gamma})'\,\mathrm{e}^{\gamma-\beta-\alpha}. \tag{6.67}$$

Furthermore, if our reference frame is comoving with matter in its rotation, then the SET component $T_0^3 = 0$, hence (due to Einstein's equations) the Ricci tensor component $R_0^3 \sim (\omega\,\mathrm{e}^{2\gamma+\mu})' = 0$, and [125]

$$\omega = \omega_0\,\mathrm{e}^{-\mu-2\gamma}, \quad \omega_0 = \mathrm{const}. \tag{6.68}$$

Consequently, from (6.67) we find

$$E(x) = 2\omega_0\,\mathrm{e}^{2\gamma(x)} \int \mathrm{e}^{\alpha+\beta-\mu-3\gamma} dx. \tag{6.69}$$

The nontrivial components of the Einstein equations in the form $R_\mu^\nu = \varkappa \widetilde{T}_\mu^\nu = \varkappa(T_\mu^\nu - \frac{1}{2}\delta_\mu^\nu T)$ are

$$\begin{aligned}
R_0^0 &= \mathrm{e}^{-2\alpha}\left[\gamma'' + \gamma'(\sigma' - \alpha')\right] + 2\omega^2 = \varkappa\widetilde{T}_0^0,\\
R_1^1 &= \mathrm{e}^{-2\alpha}\left[\sigma'' + \sigma'^2 - 2U - \alpha'\sigma'\right] - 2\omega^2 = \varkappa\widetilde{T}_1^1,\\
R_2^2 &= \mathrm{e}^{-2\alpha}\left[\mu'' + \mu'(\sigma' - \alpha')\right] = \varkappa\widetilde{T}_2^2,\\
R_3^3 &= \mathrm{e}^{-2\alpha}\left[\beta'' + \beta'(\sigma' - \alpha')\right] - 2\omega^2 = \varkappa\widetilde{T}_3^3,\\
R_3^0 &= G_3^0 = E\,\mathrm{e}^{-2\gamma}\left(R_3^3 - R_0^0\right) = \varkappa\widetilde{T}_3^0,
\end{aligned} \tag{6.70}$$

where we use the notations

$$\sigma = \beta + \gamma + \mu, \quad U = \beta'\gamma' + \beta'\mu' + \gamma'\mu'. \tag{6.71}$$

From (6.70) it follows [125] that the diagonal components of the Ricci (R_μ^ν) and Einstein (G_μ^ν) tensors split into those for the static metric (6.57) with $E = 0$ plus contributions involving ω:

$$\begin{aligned}
R_\mu^\nu &= {}_sR_\mu^\nu + {}_\omega R_\mu^\nu, \quad {}_\omega R_\mu^\nu = \omega^2\,\mathrm{diag}(2, -2, 0, -2),\\
G_\mu^\nu &= {}_sG_\mu^\nu + {}_\omega G_\mu^\nu, \quad {}_\omega G_\mu^\nu = \omega^2\,\mathrm{diag}(3, -1, 1, -1),
\end{aligned} \tag{6.72}$$

where ${}_sR_\mu^\nu$ and ${}_sG_\mu^\nu$ are the static parts (see (6.58)). The tensors ${}_sG_\mu^\nu$ and ${}_\omega G_\mu^\nu$ (each separately) obey the conservation law $\nabla_\alpha G_\mu^\alpha = 0$ in terms of the static metric. Therefore, ${}_\omega G_\mu^\nu/\varkappa$ behaves as one more SET with exotic properties (the effective energy density is $-3\omega^2/\varkappa < 0$) that favors the existence of wormholes, as confirmed by a number of examples [65, 123, 125].

By (6.70), if the diagonal components of Einstein's equations are solved, their only off-diagonal component is automatically fulfilled.

6.4.2 *Examples of static wormholes*

A number of such examples are presented in [127]. Let us here mention one of them, with a massless scalar field Φ having the Lagrangian $L_s = \varepsilon g^{\mu\nu}\Phi_{,\mu}\Phi_{,\nu}$ in the presence of a cosmological constant $\Lambda = -3\lambda^2 < 0$ ($\lambda > 0$), where, as before, $\varepsilon = 1$ corresponds to a canonical scalar and $\varepsilon = -1$ to a phantom one. Then the field equations for the metric (6.57) with $E = 0$, in the harmonic gauge $\alpha = \beta + \gamma + \mu$, have the family of solutions [127]

$$
\begin{aligned}
&\Phi = Cx, \quad \gamma = \frac{1}{3}\alpha - Ax,\\
&\beta = \frac{1}{3}\alpha + Ax + Bx, \qquad \mu = \frac{1}{3}\alpha - Bx,\\
&\mathrm{e}^{-\alpha} = \begin{cases} (\lambda/h)\sinh(hx), & h > 0,\\ \lambda x, & h = 0,\\ (\lambda/h)\sin(hx), & h < 0, \end{cases}
\end{aligned}
\tag{6.73}
$$

where A, B, C, h are integration constants related by

$$
h^2 \operatorname{sign} h = 3(A^2 + AB + B^2) + 3\varepsilon\varkappa C^2. \tag{6.74}
$$

Among them, wormhole solutions are those where $\beta(x) \to \infty$ on both ends of the range of x. Assuming $x > 0$, we see that as $x \to 0$ we have $\beta \to \infty$ for any values of the constants. Let us look when $\beta \to \infty$ at large x.

With a normal scalar field, $\varepsilon = 1$, we have $h > 0, C \neq 0, x \in \mathbb{R}_+$, and $\alpha \approx -hx \ \Rightarrow\ 3\beta \approx (3A + 3B - h)x$, and we obtain a wormhole if $A + B > h/3$. The metric can be written as

$$ds^2 = Y(\mathrm{e}^{-2Bx}dt^2 - Y^2dx^2 - \mathrm{e}^{-2Ax}dz^2 - \mathrm{e}^{2Ax+2Bx}d\phi^2),$$
$$Y := \left(\frac{h}{\lambda \sinh(hx)}\right)^{2/3}. \tag{6.75}$$

Other wormhole solutions are obtained with $h \leq 0$. Thus, if $h < 0$, then $x \in (0, \pi/|h|)$, and all metric functions $\alpha, \beta, \gamma, \mu$, being governed by $\sin(|h|x)$, tend to infinity on both ends of this range, where repulsive singularities occur.

Among these solutions, there is a symmetric subfamily: it corresponds to $\varepsilon = -1$, $C \neq 0$, $A = B = 0$, $h < 0$. In accord with the above said, it has negative energy density, $T^0_0 = -\Phi'^2 \mathrm{e}^{-2\alpha} + \Lambda/\varkappa$. In terms of the Gaussian coordinate $l = (1/\lambda)\log\tan(hx/2)$ ($l \in \mathbb{R}$ is a length in the radial direction), the metric in this case can be written as

$$ds^2 = -dl^2 + \left(\frac{|h|}{\lambda}\cosh\frac{l}{\lambda}\right)^{2/3}(dt^2 - dz^2 - d\phi^2). \tag{6.76}$$

6.4.3 *Examples of rotating wormholes*

Assuming $E \neq 0$ and $\omega \neq 0$, we return to (6.68)–(6.70) and see, in particular, that (6.68) is incompatible with asymptotic flatness since finite values of $\gamma(x)$ and $\mu(x)$ imply finite ω whereas ω should vanish at flat infinity. Accordingly, among numerous known rotating cylindrical wormhole solutions, none are asymptotically flat, and to obtain configurations potentially observable by astronomers, one has to use matching with external regions on some surfaces $\Sigma_\pm$ as described above.

Furthermore, in a phantom-free model, all involved SETs should respect the WEC, and it is a separate challenge to find such solutions. For example, it has been shown [123, 125] that a construction with two external Minkowski regions around a wormhole region does not lead to phantom-free models: the SET on at least one of the shells $\Sigma_\pm$ violates the WEC, if the SET of matter filling the internal wormhole region has the property $T^0_0 = T^3_3$. This no-go theorem [83] was proved by analyzing the matching conditions using the solution of the equation $R^0_0 = R^3_3$. Therefore neither of wormhole solutions known for scalar fields (including those with exponential and other potentials [123]) are suitable for our purpose.

WEC on a junction surface. In what follows, we will present two examples of models that solve the problem designated above. But before that let us formulate a convenient criterion of WEC fulfillment on a junction cylinder Σ ($x = x_s$) separating two regions $\mathbb{D}_-$ ($x \le x_s$) and $\mathbb{D}_+$ ($x \ge x_s$) with two different metrics of the form (6.57). The metric on Σ should be the same when looking from $\mathbb{D}_-$ or $\mathbb{D}_+$, therefore

$$[\beta] = 0, \quad [\mu] = 0, \quad [\gamma] = 0, \quad [E] = 0, \tag{6.77}$$

where for any $f(x)$, $[f] = f(x_s + 0) - f(x_s - 0)$. The coordinates t, z, ϕ are identified in the whole space, while the radial coordinates in $\mathbb{D}_+$ and $\mathbb{D}_-$ may be different since all quantities involved in the matching conditions are insensitive to this choice.

The material content of Σ is found using the Darmois–Israel formalism [46, 303] in terms of discontinuities of the extrinsic curvature of Σ.

$$S_a^b = (8\pi G)^{-1}[\widetilde{K}_a^b], \quad \widetilde{K}_a^b := K_a^b - \delta_a^b K_c^c, \tag{6.78}$$

where $a, b, c = 0, 2, 3$, and $K_{ab} = \frac{1}{2}\,\mathrm{e}^{-\alpha} g'_{ab}$. The question is whether S_a^b satisfies the WEC requirements

$$S_{00}/g_{00} = \sigma_s \ge 0, \qquad S_{ab}\xi^a\xi^b \ge 0, \tag{6.79}$$

where σ_s is the surface energy density and ξ^a an arbitrary null vector on Σ. The second inequality makes the content of the NEC as a part of the WEC, and taken together, they provide $\sigma_s \ge 0$ in any reference frame on Σ.

The condition $\sigma_s \ge 0$ is easily found and reads

$$[\mathrm{e}^{-\alpha}(\beta' + \mu')] \le 0. \tag{6.80}$$

The radial coordinates x can be chosen in different ways in $\mathbb{D}_+$ and $\mathbb{D}_-$, and the prime denotes a coordinate derivative defined for each region separately.

The NEC should hold for *any* null vector ξ^a on Σ. However, instead of directly verifying this, we can note that the WEC holds if in a comoving reference frame the density σ_s and the pressures p_i in all directions satisfy the inequalities

$$\sigma_s \geq 0, \quad \sigma_s + p_i \geq 0. \tag{6.81}$$

A difficulty is that Σ is in general not described in a comoving frame, but the needed quantities can be found as eigenvalues of S_{ab} written in an orthonormal frame. In this way we obtain the following WEC criterion [65, 124]:

$$\begin{aligned} &a \geq 0, \quad a + c \geq 0, \\ &a + c + \sqrt{(a-c)^2 + 4d^2} + 2b \geq 0, \end{aligned} \tag{6.82}$$

where

$$\begin{aligned} a &= -[\mathrm{e}^{-\alpha}(\beta' + \mu')], \quad b = [\mathrm{e}^{-\alpha}(\beta' + \gamma')], \\ c &= [\mathrm{e}^{-\alpha}(\gamma' + \mu')], \quad d = -[\omega]. \end{aligned} \tag{6.83}$$

Minkowski regions. If we place Minkowski regions $\mathbb{M}_\pm$ around "internal" rotating wormhole regions, we must take the Minkowski metric in a rotating reference frame. To do so, in the inertial-frame metric $ds^2_{\mathrm{M}} = dt^2 - dX^2 - dz^2 - X^2 d\varphi^2$ we substitute $\varphi \to \varphi + \Omega t$ ($\Omega = \mathrm{const}$ is the angular velocity) to obtain

$$ds^2_{\mathrm{M}} = dt^2 - dX^2 - dz^2 - X^2(d\varphi + \Omega dt)^2. \tag{6.84}$$

The relevant quantities in the notations of (6.57) are

$$\begin{aligned} &\mathrm{e}^\alpha = \mathrm{e}^\mu = 1, \quad \mathrm{e}^{2\gamma} = 1 - \Omega^2 X^2, \quad \mathrm{e}^{2\beta} = \frac{X^2}{1 - \Omega^2 X^2}, \\ &E = \Omega X^2, \quad \omega = \frac{\Omega}{1 - \Omega^2 X^2}. \\ &\gamma' = -\frac{\Omega^X}{1 - \Omega^2 X^2}, \quad \beta' = \frac{1}{X} + \frac{\Omega^2 X}{1 - \Omega^2 X^2}. \end{aligned} \tag{6.85}$$

This metric is stationary and suitable for matching if $|X| < 1/|\Omega|$, so that the linear rotational velocity is smaller than the speed of light.

A model with stiff matter, $p = \rho$. The perfect fluid with $p = \rho$ is a maximally stiff matter compatible with causality, where the velocities of sound and light coincide. This circumstance makes easier finding exact solutions, including wave and inhomogeneous ones [77]; stiff matter also finds various applications in cosmology, see, e.g., [72, 434] and references therein.

A rotating wormhole solution with such a fluid reads [65]

$$ds^2 = \left(Pdt - \frac{y}{\omega_0}d\varphi\right)^2 - \frac{dy^2}{2\omega_0^2(k^2+y^2)} - dz^2 - (k^2+y^2)\frac{d\varphi^2}{2\omega_0^2},$$
$$\varkappa\rho_0 = \omega_0^2 = \text{const.} \tag{6.86}$$

The solution contains an arbitrary time scale factor $P > 0$ and integration constants k and ω_0. It is regular in the whole range $y \in \mathbb{R}$, but at $y^2 > k^2$ it has $g_{33} > 0$, so that φ-circles are closed timelike curves.

Consider matching with the Minkowski metric (6.84) on Σ_+ ($y = y_0 > 0$, $X = X_0 > 0$). The matching conditions yield

$$P^2 = 1 - \Omega^2 X^2, \quad \Omega X^2 = \frac{Py}{\omega_0}, \quad \frac{k^2+y^2}{2\omega_0^2} = \frac{X^2}{P^2}, \tag{6.87}$$

where the index "zero" near X and y is omitted without risk of confusion. We have six system parameters: four (P, k, ω_0, y) from the internal metric (6.86) and two (Ω, X) from the external one. Choosing three of them as independent ones: $X = X_0$ (having the dimension of length), $y = y_0$ and P, we obtain from (6.87)

$$\Omega = \frac{\sqrt{1-P^2}}{X}, \quad \omega_0 = \frac{Py}{\sqrt{1-P^2}X}, \quad k^2 = y^2\frac{1+P^2}{1-P^2}. \tag{6.88}$$

As a result, the quantities a, b, c, d from (6.83) are expressed as

$$a = \frac{P^3y-1}{P^2X}, \quad b = \frac{1-Py}{X},$$
$$c = \frac{P^2-1}{P^2X}, \quad d = \frac{Py-1+P^2}{P^2X\sqrt{1-P^2}}. \tag{6.89}$$

The factor $1/X$ is common and irrelevant for the inequalities of interest. An inspection shows that under the condition

$$y \geq \frac{2-P^2}{P^3} \tag{6.90}$$

the inequalities (6.82) are satisfied.

What changes on Σ_- specified by $X = -X_0 < 0$ and $y = -y_0 < 0$? The common factor $1/X$ becomes negative. But the signs of all discontinuities also change: while on Σ_+ we had $[f] = f_{\rm out} - f_{\rm in}$ for any f, on Σ_- we have the opposite. Therefore, the WEC requirements do not change and hold with (6.90) applied to $|y|$ instead of y. Thus we have obtained a completely phantom-free wormhole model.

We also notice that (6.88) implies $y_0^2 < k^2$, hence $y^2 < k^2$ in the whole internal region, and there are no closed timelike curves.

A model with similar properties but with a z-directed magnetic field and a fluid with $p = w\rho$, with a constant $w \in (0,\ 1)$, has been obtained and discussed in [126]. It passes on to the presently described model in the limit $w \to 1$, in which the magnetic field vanishes.

A model with an anisotropic fluid. One more model was obtained in [124] with the SET

$$T_\mu^\nu = \rho\,{\rm diag}(1,-1,1,-1) \ \oplus\ T_3^0 = -2\rho E\,{\rm e}^{-2\gamma}, \tag{6.91}$$

describing an anisotropic fluid chosen by analogy with that of a longitudinal magnetic field in the static case, which cannot be directly extended to involve rotation. (A full description of the anisotropic fluid formalism in the metric (6.57) may be found, e.g., in [194]; Ref. [64] describes a general framework for finding solutions with anisotropic fluids in the same class of space–times.)

The solution, in the harmonic gauge $\alpha = \beta + \gamma + \mu$, is written as

$$\begin{aligned}
\rho &= \rho_0\,{\rm e}^{-2\gamma-2\mu}, \qquad \rho_0 = {\rm const} > 0,\\
r^2 &\equiv {\rm e}^{2\beta} = \frac{r_0^2}{Q^2(x_0^2-x^2)}, \qquad {\rm e}^{2\gamma} = Q^2(x_0^2-x^2),\\
{\rm e}^{2\mu} &= (x_0-x)^{1-x/x_0}(x_0+x)^{1+x/x_0},
\end{aligned}$$

$$E = \frac{r_0}{2x_0^2}\left[2x_0 x + (x_0^2 - x^2)\ln\frac{x_0 + x}{x_0 - x}\right],$$

$$x_0 := \frac{|\omega_0|}{\varkappa\rho_0 r_0}, \qquad Q^2 := \varkappa\rho_0 r_0^2. \tag{6.92}$$

The solution contains integration constants ω_0, ρ_0 and r_0, and dimensionless constants x_0 and Q introduced for convenience. The coordinate x ranges from $-x_0$ to x_0. Since $r \to \infty$ as $x \to \pm x_0$, it is a wormhole solution, and it is symmetric with respect to the throat $x = 0$. However, the extremes $x = \pm x_0$ are curvature singularities, with the Kretschmann invariant $\sim |x_0 - x|^{-4}$.

To construct an asymptotically flat configuration, we join it at some $x = \pm x_s < x_0$ to Minkowski regions at $X = \pm X_s$ and obtain

$$X = r_0, \quad Q^2(x_0^2 - x^2) = 1 - \Omega^2 X^2 =: P^2,$$

$$2x_0^2\sqrt{1 - P^2} = 2xx_0 + (x_0^2 - x^2)\ln\frac{x_0 + x}{x_0 - x}, \tag{6.93}$$

where the index s near x is omitted. To satisfy $[\mu] = 0$, we change the z scale in $\mathbb{M}_\pm$, obtaining $-g_{zz} = M^2 := e^{2\mu(x_s)}$ taken from the internal metric.

It is convenient to use the ratio $y = x_s/x_0$ and to introduce the notation $L = L(y) = \ln[(1 + y)/(1 - y)]$. So we have

$$M = M(y) = \bigl(1 - y\bigr)^{-(1-y)/2}\bigl(1 + y\bigr)^{-(1+y)/2},$$

$$P^2(y) = (1 - y^2)\left[1 - yL(y) - \frac{1}{4}(1 - y^2)L^2(y)\right]. \tag{6.94}$$

Calculation of the quantities a, b, c, d used in the WEC requirements (6.82) on $\Sigma_\pm$ gives

$$a = -\frac{1}{P^2(y)} + \frac{M(y)}{x_0^2}\left(\frac{y}{1 - y^2} + \frac{1}{2}L(y)\right),$$

$$b = 1,$$

$$c = -\frac{1}{P^2(y)} + 1 + \frac{M(y)}{x_0^2}\left(\frac{y}{1-y^2} - \frac{1}{2}L(y)\right),$$

$$\pm d = -\frac{\sqrt{1-P^2(y)}}{P^2(y)} + \frac{M(y)}{x_0^2(1-y^2)}, \tag{6.95}$$

where the insignificant factor $1/r_0$ at each of them has been ignored.

The expressions for a, b, c and d^2 depend on two parameters, x_0 and y, and coincide on Σ_+ and Σ_-, hence the conditions (6.82) for the two surfaces also coincide. A numerical analysis shows that [124]:

- The condition $0 < P^2(y) < 1$, required by construction, holds for $0 < y \lesssim 0.564$;
- The conditions $a > 0$ and $a > c$ hold in a significant range of x_0 and y, for example, for $x_0 = 0.5,\ y \in (0.15, 0.47)$ and for $x_0 = 0.3,\ y \in (0.05, 0.53)$;
- The condition $a + c > 0$ also holds in a certain range of the parameters, e.g, for $x_0 = 0.5,\ y \in (0.15, 0.38)$ and for $x_0 = 0.3$, $y \in (0.05, 0.51)$.

Thus, in the parameter space (x_0, y) there is a significant range in which this asymptotically flat wormhole model completely satisfies the WEC. One can also verify that $g_{33} < 0$ in the internal region between Σ_- and Σ_+, hence the model does not contain closed timelike curves.

This section demonstrates the possible existence of cylindrical wormholes asymptotically flat on both sides of the throat without WEC violation. It should be noted that the study of such systems is at an early stage, and many questions are yet to be answered, including, above all, possible observational signatures of such objects if they can really exist.

To conclude this section, we would like to mention that cylindrical thin shells have been used in many studies not as auxiliary objects needed to reach asymptotic regions but as the main sources of gravity leading to wormhole solutions, see [228, 244, 496] and references therein. This trend is beyond the scope of our book.

6.5 Observational effects — wormhole astrophysics

In the recent years, the question of possible wormhole existence in our Universe became a subject of serious analysis, which is evidently related to the discovery of the Universe's accelerated expansion, repeatedly mentioned in this book. Let us recall the main facts. Acceleration replaced deceleration rather recently from a cosmological viewpoint, about 3–4 billion years ago, according to the growing observational evidence, which continues to grow more and more. As is well known, if the evolution of an isotropic universe is dominated by a cosmological constant (such that the corresponding density and pressure are related by $\rho + p = 0$), then it develops according to the de Sitter law, with a constant acceleration. For the acceleration to grow with time, it is necessary to suppose $\rho + p < 0$, which violates the weak and null energy conditions. (Though, the cosmological constant is also within the observational bounds.) At present, there must be about 70 per cent of this exotic type of matter, called phantom dark energy (DE), and this fraction is increasing. Indeed, unlike ordinary matter, whose density decreases with increasing volume, phantom matter behaves in just the opposite way: its density grows with growing volume.

But it is just such exotic matter, whatever be its origin and physical nature, that is required as a construction material for wormholes. Though, there is an important subtle point if we try to use a phantom DE suitable for cosmology, that is, an isotropic fluid with $\rho + p < 0$, to make a wormhole. If matter is isotropic, a simple no-go theorem [87] excludes twice asymptotically flat and twice asymptotically AdS wormholes as well as those asymptotically flat on one end and AdS on the other (see Sec. 6.3.1). What is not excluded, is that one or both asymptotic regions are de Sitter, and it can be of interest, in particular, because such wormholes may exist in an inflationary universe and provide causal connections between otherwise distant regions [87].

A number of papers discuss the hypothesis that some active galactic nuclei, quasars and/or other compact astrophysical objects can be nothing else than wormhole entrances, and one seeks specific features able to distinguish wormholes from BHs or other, less familiar objects predicted by theorists (such as boson stars [407, 490], gravastars [403, 418], quasi-black holes [369, 370], etc.) sometimes designated as "black hole mimickers" [522]. Various potentially observable effects of wormholes in astronomy are discussed, among other papers, in [86, 216, 217, 281, 336], and in more general reviews [385, 386, 539]. see also numerous references therein.

Many effects, such as gravitational lensing and the existence of accretion disks, are common to different classes of compact objects, so it is in general hard to expect that they can reveal specific wormhole features. Moreover, many calculations of these effects have very little predictive power, especially if they deal with particular wormhole models. A reason is that there is too great an uncertainty in the nature of the exotic matter supporting the wormhole geometry: as we have seen in Sec. 6.3.4, even under the condition that the scalar curvature is zero, wormhole metrics are obtained with an arbitrary function, and in the more general static case there are two arbitrary functions. All this arbitrariness refers to static, spherically symmetric geometries, to say nothing about more general and realistic cases. More than that, if the exotic matter supporting a wormhole occupies a finite volume in space, then the regions outside it are, as always, described by usual, most probably vacuum solutions to the equations of gravity (e.g., Schwarzschild and Kerr) with their usual characteristics.

However, there are some effects able to reveal wormhole peculiarities. An interesting point is that a wormhole entrance may be a source of a monopole magnetic field [62]. In such a case, there is no necessity of isolated magnetic poles (monopoles): the magnetic lines of force, as well as electric ones, can simply thread the wormhole, realizing Wheeler's idea of electric or magnetic "charge without charge" [559]. Other distinguishing effects are (see, e.g., [319, 431]) (1) a possible observation of objects in remote parts of the Universe (or another universe) through the throat (assuming that the throat is wide enough and filled with a transparent substance); there is a certain probability to see the same object beside the wormhole and through it with different redshifts [216, 217, 431]; (2) specific modes of particle acceleration and jet shapes due to possible radial magnetic fields; (3) test particle orbits impossible in other settings, such as oscillations across the throat, if there is a minimum of the gravitational potential; (4) gravitational lensing in a strong-field regime [86, 188, 200, 456, 487, 530], etc.

According to Harko *et al.* [281], a wormhole signature can still be obtained from the properties of accretion disks: it has been obtained that the intensity of a flux emerging from the disk surface is greater for wormholes than for rotating black holes with the same Schwarzschild mass and accretion rate. The conversion efficiency of the accreting mass into radiation was also calculated [281] shownning that rotating wormholes provide a much more efficient engine for transformation of the accreting mass into radiation than Kerr black holes.

Curiously, according to [15], it turns out that similar wormholes in the Brans–Dicke theory are "quasi-Schwarzschild objects", and their accretion energy fluxes are smaller by about an order of magnitude as compared to their GR counterparts considered in [281].

Let us give as an example a brief sketch of one of such studies. Pozanenko and Shatskiy [454] considered a promising effect connected with observation through a wormhole throat. They remark that if radiation from another universe comes in this way, then a distant observer in our Universe will perceive a wormhole throat like a point source. Furthermore, the brightest known sources are gamma-ray bursts (GRBs). Their brightness in the gamma range may exceed that of the entire sky. If GRB phenomena are present in the other universe (or anywhere beyond the wormhole), an observer in our Universe will detect repeated aperiodic gamma-ray flashes coming from a single point spatially coinciding with the wormhole.

Objects like that do exist: these are sources of soft repeated gamma-ray bursts called soft gamma repeaters. Although most of them can be reliably associated with magnetars in our Galaxy, it still makes sense to check whether or not some of those repeaters can be observational signatures from GRBs of another universe. A calculation of power spectra from GRBs transmitted by a wormhole has led to a conclusion [454] that known soft gamma repeaters are very unlikely to be wormhole candidates.

Thus, for the moment, there are no clear observational traces of natural wormholes in the Universe, but the rapidly growing wealth of observational data leaves a hope that such evidence will sooner or later appear. If, certainly, they really do exist.

Chapter 7

Stability of spherically symmetric configurations

7.1 Preliminaries

In the previous chapters, we have described a number of self-gravitating static and stationary configurations described by solutions to the field equations. To see whether or not such solutions can lead to viable models of some real objects in Nature, one needs to check their stability against various perturbations. This guarantees that the system under consideration will not be immediately destroyed after its formation. The perturbations can be large or small, and can appear as fluctuations in the system itself or be caused by external influence. In all such cases we are dealing with various stability problems.

Perturbations and stability of vacuum and electrovacuum solutions to the Einstein equations (Schwarzschild, Reissner–Nordström, Kerr, Kerr–Newman) have been studied by many authors; a comprehensive review and bibliography can be found in the books [169, 250]. The most important conclusion of these studies is that all these solutions are stable under any small perturbation and have certain spectra of quasinormal modes, i.e., oscillations characterizing their late-time response to any disturbances.

Many papers are devoted to various more general and multidimensional BHs (most often string theory-motivated ones) and their stability, see, e.g., [131, 262] and the recent reviews [302, 356].

Less attention has been paid to solutions with scalar fields although they have some peculiarities of interest. Their main feature is that, unlike gravitational (spin-2) and electromagnetic (spin-1) perturbations, scalar (spin-0) time-dependent perturbations can preserve spherical symmetry. This circumstance is directly related to the Birkhoff theorem which holds for gravitational and electromagnetic fields and in a very rough formulation sounds like that: "spherically symmetric vacuum or electrovacuum space-times are necessarily static" (see more details in Chapter 3), but a scalar field inserts dynamics, so that both the scalar itself and the metric tensor become essentially time-dependent.

Accordingly, in pure vacuum and electrovacuum space–times the multipole expansion of small (linear) perturbations begins with dipole (if there is an electromagnetic field) and quadrupole terms, whereas a scalar field inserts a monopole term. Moreover, it turns out that monopole (spherically symmetric) perturbations are the most "dangerous" in the sense of instability. The point is that in many important cases linear perturbation equations can be reduced to single Schrödinger-like equations with certain effective potentials with respect to some "wave function", while the role of an energy level is played by a frequency squared. Therefore negative energy levels correspond to imaginary frequencies and hence a possible exponential growth of the perturbations, i.e., instability. The experience shows that higher multipoles lead to positive contributions into the effective potentials, like centrifugal barriers in quantum mechanics. Therefore, if a system to be studied is unstable, this instability will most probably manifest itself under monopole perturbations.

In this chapter, we will discuss in some detail the stability of static, spherically symmetric solutions to the Einstein equations with scalar fields as sources. Following [106], we will describe a general methodology of studying small spherically symmetric perturbations of scalar–vacuum configurations with any potential $V(\phi)$ (see (7.1)) and pay special attention to space–times with throats. The difficulty with the latter lies in the fact that the effective potentials $V_{\rm eff}(x)$ for perturbations (not to be confused with the self-interaction potential $V(\phi)$) possess a singularity at the throat, which prevents a complete perturbation analysis. It is for this reason that in some previous stability studies of anti-Fisher solutions [19, 95] no unstable modes were noticed whereas a numerical perturbation analysis of Shinkai and Hayward [499] revealed an instability in the simplest representative of this family of solutions, the Ellis massless wormhole [75, 229].

Gonzalez *et al.* [260], analyzing the stability of anti-Fisher wormholes, made a proper substitution in the perturbation equation (a special case of Darboux transformations [257, 572], also called S-deformations [301, 302]) and regularized the effective potential $V_{\rm eff}$. As a result, they were able to consider a previously missed perturbation mode with a nonzero perturbation of the throat radius and showed that it can exponentially grow, which means that the anti-Fisher wormholes are unstable. We will verify that a similar methodology can be applied to much more general self-gravitating scalar field configurations including those with arbitrary self-interaction potentials $V(\phi)$ and electromagnetic fields. To this end, we will prove that, generically, (i) the effective potential $V_{\rm eff}$ for such configurations has precisely the form required for regularization by S-deformation and (ii) any solution of the transformed wave equation with a regularized potential leads to a regular perturbation of the background static configuration.

As a particular example, we study the stability of all anti-Fisher solutions (see Sec. 4.1) under spherically symmetric perturbations. We prove the instability of Branch A and B solutions and confirm the conclusions of [260] for Branch C solutions (wormholes). A study of particular solutions with nonzero potentials $V(\phi)$ (black universes) is also carried out.

7.2 Perturbation equations

7.2.1 *General relations*

Consider an action with a single scalar field, but more general than described previously: we include an electromagnetic field and admit a dilatonic-type interaction between the scalar and electromagnetic fields:

$$\mathcal{S} = \frac{1}{16\pi}\int \sqrt{-g}d^4x[R + 2h(\phi)g^{\alpha\beta}\phi_{;\alpha}\phi_{;\beta} - 2V(\phi) - S(\phi)F_{\mu\nu}F^{\mu\nu}], \tag{7.1}$$

where $h > 0$ for a normal scalar field with positive kinetic energy and $h < 0$ for a phantom one; the function $S(\phi) > 0$, characterizing the scalar–electromagnetic interaction, is arbitrary. The field equations are

$$4h\nabla^{\mu}\nabla_{\mu}\phi + 2h_{\phi}\phi^{\mu}\phi_{\mu} + 2V_{\phi} + S_{\phi}F_{\mu\nu}F^{\mu\nu} = 0, \tag{7.2}$$

$$\nabla_{\mu}(S(\phi)F^{\mu\nu}) = 0, \tag{7.3}$$

$$R^{\nu}_{\mu} - \frac{1}{2}\delta^{\nu}_{\mu}R = -T^{\nu}_{\mu}, \tag{7.4}$$

where $S_{\phi} \equiv dS/d\phi$ and T^{ν}_{μ} is the SET:

$$\begin{aligned} T^{\nu}_{\mu} &= T^{\nu}_{\mu}[\phi] + T^{\nu}_{\mu}[F], \\ T^{\nu}_{\mu}[\phi] &= h(\phi)[2\phi_{\mu}\phi^{\nu} - \delta^{\nu}_{\mu}\phi^{\alpha}\phi_{\alpha}] + \delta^{\nu}_{\mu}V(\phi), \\ T^{\nu}_{\mu}[F] &= S(\phi)\Big[- 2F_{\mu\alpha}F^{\nu\alpha} + \frac{1}{2}\delta^{\nu}_{\mu}F_{\alpha\beta}F^{\alpha\beta}\Big]. \end{aligned} \tag{7.5}$$

The general static, spherically symmetric metric may be written, as before, in the form

$$ds^2 = \mathrm{e}^{2\gamma}dt^2 - \mathrm{e}^{2\alpha}du^2 - \mathrm{e}^{2\beta}d\Omega^2, \tag{7.6}$$

where γ, α and β are functions of the radial coordinate u, and there remains a coordinate freedom of choosing u. We also use the notation $r(u) \equiv \mathrm{e}^{\beta}$.

We are going to study linear spherically symmetric perturbations of solutions to the field equations due to (7.1). So, the metric has the form (7.6) but now $\gamma(u,t) = \gamma(u) + \delta\gamma$ and similarly for other quantities, with small "deltas"; for the scalar field we have $\phi(u,t) = \phi(u) + \delta\phi(u,t)$. The most general electromagnetic field compatible with spherical symmetry is described by the 4-potential

$$A_\mu = \delta^0_\mu A_0 + \delta^3_\mu q_m \cos\theta + \partial_\mu \Phi, \tag{7.7}$$

where q_m is a magnetic charge and Φ an arbitrary function. The electromagnetic field equations give

$$S(\phi)\, \mathrm{e}^{\alpha+2\beta+\gamma} F^{10} = q_e, \tag{7.8}$$

where q_e is an electric charge. We thus have

$$F_{\mu\nu}F^{\mu\nu} = 2\,\mathrm{e}^{-4\beta}(-q_e^2/S^2 + q_m^2). \tag{7.9}$$

For the SETs we obtain

$$T^\nu_\mu[\phi] = \varepsilon\, \mathrm{e}^{-2\alpha}\phi'^2\, \mathrm{diag}(1,\ -1,\ 1,\ 1) + \delta^\nu_\mu V, \tag{7.10}$$

$$T^\nu_\mu[F] = \eta\, \mathrm{e}^{-4\beta} Q(\phi)\, \mathrm{diag}(1,\ 1,\ -1,\ -1), \tag{7.11}$$

where $Q = Q(\phi) = q_e^2/S(\phi) + q_m^2 S(\phi)$. We will consider the equations governing linear radial perturbations of a static, spherically symmetric solution to the field equations (7.2)–(7.4), following the lines of [106, 121].

Preserving only linear terms with respect to time derivatives, we can write all nonzero components of the Ricci tensor as

$$R^0_0 = \mathrm{e}^{-2\gamma}(\ddot\alpha + 2\ddot\beta) - \mathrm{e}^{-2\alpha}[\gamma'' + \gamma'(\gamma' - \alpha' + 2\beta')], \tag{7.12}$$

$$R^1_1 = \mathrm{e}^{-2\gamma}\ddot\alpha - \mathrm{e}^{-2\alpha}[\gamma'' + 2\beta'' + \gamma'^2 + 2\beta'^2 - \alpha'(\gamma' + 2\beta')], \tag{7.13}$$

$$R^2_2 = R^3_3 = \mathrm{e}^{-2\beta} + \mathrm{e}^{-2\gamma}\ddot\beta - \mathrm{e}^{-2\alpha}[\beta'' + \beta'(\gamma' - \alpha' + 2\beta')], \tag{7.14}$$

$$R_{01} = 2[\dot\beta' + \dot\beta\beta' - \dot\alpha\beta' - \dot\beta\gamma'], \tag{7.15}$$

where dots and primes denote $\partial/\partial t$ and $\partial/\partial u$, respectively. The zero-order (static) scalar, $\binom{0}{0}$, $\binom{1}{1}$, $\binom{2}{2}$ components of Eq. (7.4) are

$$2h[\phi'' + \phi'(\gamma' + 2\beta' - \alpha')] + h'\phi' = \mathrm{e}^{2\alpha} P_\phi; \tag{7.16}$$

$$\gamma'' + \gamma'(\gamma' + 2\beta' - \alpha') = \mathrm{e}^{2\alpha}(-V + Q\,\mathrm{e}^{-4\beta}); \tag{7.17}$$

$$\gamma'' + 2\beta'' + \gamma'^2 + 2\beta'^2 - \alpha'(\gamma' + 2\beta') = -2h\phi'^2 + \mathrm{e}^{2\alpha}(-V + Q\,\mathrm{e}^{-4\beta}); \tag{7.18}$$

$$-\,\mathrm{e}^{2\alpha-2\beta} + \beta'' + \beta'(\gamma' + 2\beta' - \alpha') = -P\,\mathrm{e}^{2\alpha}, \tag{7.19}$$

where the subscript ϕ denotes $\partial/\partial\phi$ and

$$P := V(\phi) + Q\,\mathrm{e}^{-4\beta}, \tag{7.20}$$

The first-order perturbed equations (scalar, $R_{01} = \ldots$, and $R_2^2 = \ldots$) read

$$\begin{aligned}
&2\,\mathrm{e}^{2\alpha-2\gamma}h\delta\ddot{\phi} - 2h[\delta\phi'' + \delta\phi'(\gamma' + 2\beta' - \alpha') \\
&\quad + \phi'(\delta\gamma' + 2\delta\beta' - \delta\alpha')] - 2\delta h[\phi'' + \phi'(2\beta' + \gamma' - \alpha')] \\
&\quad - h'\delta\phi' - \phi'\delta h' + \delta(\mathrm{e}^{2\alpha}P_\phi) = 0, \qquad (7.21) \\
&\delta\dot{\beta}' + \beta'\delta\dot{\beta} - \beta'\delta\dot{\alpha} - \gamma'\delta\dot{\beta} = -h\phi'\delta\dot{\phi}, \qquad (7.22) \\
&\delta(\mathrm{e}^{2\alpha-2\beta}) + \mathrm{e}^{2\alpha-2\gamma}\delta\ddot{\beta} - \delta\beta'' - \delta\beta'(\gamma' + 2\beta' - \alpha') \\
&\quad - \beta'(\delta\gamma' + 2\delta\beta' - \delta\alpha') = \delta(\mathrm{e}^{2\alpha}P). \qquad (7.23)
\end{aligned}$$

Equation (7.22) may be integrated in t, and since we are interested in time-dependent perturbations, we omit the appearing arbitrary function of u describing static perturbations and obtain

$$\delta\beta' + \delta\beta(\beta' - \gamma') - \beta'\delta\alpha = -h\phi'\delta\phi. \tag{7.24}$$

Let us note that we are now having two independent forms of arbitrariness: one is the freedom of choosing a *radial coordinate* u, the other is the choice of a *perturbation gauge*, or, in other words, a reference frame in the perturbed space–time, which can be expressed in imposing a certain relation for $\delta\alpha$, $\delta\beta$, etc. In what follows, we will make use of both kinds of freedom. All the above equations have so far been written in the most universal form, without coordinate or gauge fixing.

7.2.2 *Master equation*

Preserving the coordinate arbitrariness, we will now choose the simplest possible gauge $\delta\beta \equiv 0$. Then Eq. (7.24) expresses $\delta\alpha$ in terms of $\delta\phi$:

$$\beta'\delta\alpha = h(\phi)\phi'\delta\phi. \tag{7.25}$$

Equation (7.23) expresses $\delta\gamma' - \delta\alpha'$ in terms of $\delta\alpha$ and $\delta\phi$:

$$\beta'(\delta\gamma' - \delta\alpha') = 2\,e^{2\alpha-2\beta}\delta\alpha - \delta(e^{2\alpha}P). \tag{7.26}$$

Substituting all this into (7.21), we finally obtain the wave equation

$$e^{2\alpha-2\gamma}\delta\ddot{\phi} - \delta\phi'' - \delta\phi'(\gamma' + 2\beta' - \alpha' + h'/h) + U\delta\phi = 0, \tag{7.27}$$

$$U \equiv \frac{1}{2}\phi'^2\left(\frac{h_\phi^2}{h^2} - \frac{h_{\phi\phi}}{h}\right) + e^{2\alpha}\left[\frac{2h\phi'^2}{\beta'^2}(P - e^{-2\beta}) + \frac{2\phi'}{\beta'}P_\phi + \frac{h_\phi}{2h^2}P_\phi - \frac{P_{\phi\phi}}{2h}\right]. \tag{7.28}$$

This expression for U directly generalizes the one obtained in [106] for scalar–vacuum configurations. The latter is restored if we assume $h = \epsilon = \pm 1$ and $q_e = q_m = 0$.

Let us pass on to the "tortoise" coordinate z defined by the condition $\alpha = \gamma$ and connected with an arbitrary coordinate u by

$$du/dz = e^{\gamma-\alpha}. \tag{7.29}$$

Let us also change the unknown function $\delta\phi \mapsto \psi$ according to

$$\delta\phi = \psi(z,t)\,e^{-\eta}, \quad \eta' = \beta' + \frac{h'}{2h}. \tag{7.30}$$

Thus, we reduce the wave equation to its canonical form, also called the master equation for radial perturbations:

$$\ddot{\psi} - \psi_{zz} + V_{\text{eff}}(z)\psi = 0 \tag{7.31}$$

(the index z denotes d/dz), with the effective potential

$$V_{\text{eff}}(z) = e^{2\gamma-2\alpha}[U + \eta'' + \eta'(\eta' + \gamma' - \alpha')] \tag{7.32}$$

(the prime denotes $\partial/\partial u$, as before). More explicitly,

$$V_{\text{eff}}(z) = e^{2\gamma}\left[\frac{2h\phi'^2}{\beta'^2}(P - e^{-2\beta}) + \frac{2P'}{\beta'} - \frac{h_\phi P_\phi}{4h^2} + \frac{P_{\phi\phi}}{2h}\right] + e^{2\gamma-2\alpha}\Big[\beta'' + \beta'^2 + (\gamma' - \alpha')(\beta' + h'/h)\Big]. \tag{7.33}$$

A further substitution, possible since the background is static,

$$\psi(z,t) = Y(z)\,e^{i\omega t}, \quad \omega = \text{const}, \tag{7.34}$$

leads to the Schrödinger-like equation

$$Y'' + [\omega^2 - V_{\text{eff}}(z)]Y = 0. \tag{7.35}$$

Now, if there is a nontrivial solution to (7.35) with $\operatorname{Im}\omega < 0$ satisfying some physically reasonable conditions at the ends of the range of z (in particular, the absence of ingoing waves), then the static system is unstable since $\delta\phi$ can exponentially grow with t. Otherwise our static system is stable in the linear approximation. Thus, as usual in such studies, the stability problem is reduced to a boundary-value problem for Eq. (7.35) — see, e.g., [96, 106, 111–113, 121, 128, 260].

Note that all the above relations are written without fixing the background radial coordinate u.

The gauge $\delta\beta = 0$ is technically the simplest one, but apparently causes some problems when applied to wormholes and other configurations with throats. The reason is that the assumption $\delta\beta = 0$ leaves invariable the throat radius, while perturbations must in general admit its time dependence [121, 260]. It may even seem that the emergence of a pole in V_{eff} due to $\delta\beta$ in the denominator in (7.28) is an artifact of the gauge. It can be shown, however, by analogy with [106, 121, 260], that Eq. (7.35) is in fact gauge-invariant, while $\delta\phi$ is a representation of a gauge-invariant quantity in the gauge $\delta\beta = 0$. In the next subsection, we will discuss this issue in some detail.

7.2.3 *Gauge-invariant perturbations*

As in all numerous problems with perturbation dynamics in gravitation theory, it is necessary to make sure that we are dealing with real perturbations rather than pure coordinate effects, therefore it is necessary to construct gauge-invariant quantities from perturbations.

Consider the general metric (7.6) and small coordinate transformations $x^a \mapsto x^a + \xi^a$ in the (t, u) subspace:

$$t = \bar{t} + \Delta t(t, u), \quad u = \bar{u} + \Delta u(t, u), \tag{7.36}$$

where Δt and Δu are supposed to be small. This transformation adds an increment to any scalar with respect to transitions to new coordinates in the (t, u) subspace (2-scalar, for short), such as, e.g., $\phi(t, u)$ and $r(t, u)$:

$$\Delta\phi = \dot{\phi}\Delta t + \phi'\Delta u \approx \phi'\Delta u \tag{7.37}$$

in the linear approximation since both $\dot{\phi}$ and Δt are small, and similarly for $r(t, u)$ and other 2-scalars, if any. If there are perturbations $\delta\phi$ and δr, the transformation (7.36) changes them as follows:

$$\begin{aligned} \delta\phi &\mapsto \overline{\delta\phi} = \delta\phi + \Delta\phi = \delta\phi + \phi'\Delta u, \\ \delta r &\mapsto \overline{\delta r} = \delta r + \Delta r = \delta r + r'\Delta u, \end{aligned} \tag{7.38}$$

since in our static background solution $\dot{\phi} = \dot{r} = 0$. It then follows that the combination

$$\Psi \equiv r'\delta\phi - \phi'\delta r, \tag{7.39}$$

does not change under the transformation (7.36), or — in other words — is gauge-invariant. Note that the prime here denotes d/du in the background static configuration, and u is an arbitrary admissible radial coordinate, in particular, it may be thought of as our "tortoise" coordinate z used to formulate the boundary-value problem for perturbation modes.

One can notice that combinations constructed like (7.39) from *any* 2-scalars (for example, such as e^{ϕ} and $\beta = \ln r$, or two different linear combinations of ϕ and r) are also gauge-invariant. Moreover, gauge-invariant is Ψ multiplied by any 2-scalar or any combination of background quantities which are known and fixed functions of u and are certainly also gauge-independent.

The physical properties of perturbations must not depend on which gauge-invariant quantity Ψ is chosen to describe them. Meanwhile, with different Ψ, the effective potentials can, in general, also be different. However, given a specific background configuration, in order that the theory be consistent, these different potentials should lead to the same perturbation spectrum.

Due to gauge invariance of Ψ, equations that govern it may be written in any admissible gauge, in particular, $\delta\beta = 0$, and Eq. (7.31) for ψ then may be considered as a result of substituting the invariant $\psi = (r/r')\Psi = r\delta\phi$ in a manifestly correct equation for $\Psi = r'\delta\phi$.

We conclude that Eqs. (7.31) and (7.35) are themselves gauge-invariant since their unknown functions are gauge invariants (though written down in a particular gauge) while other quantities involved are taken from the background solution.

A similar problem for cosmological perturbations is discussed, e.g., in the review [71].

7.3 Throats, trapped ghosts and stability

7.3.1 *Perturbations near a throat*

We have verified that Eq. (7.35) is gauge-invariant, and it really makes sense to pose boundary-value problems to find out whether or not there are unstable perturbation modes. However, the stability problem is complicated by the singular nature of the effective potential V_{eff}. There are two kinds of singularities: one is related to throats, if any, since there are terms proportional to $1/\beta'^2$ and $1/\beta'$ in V_{eff} (and $\beta' = 0$ on a throat); the other exists at possible transitions from canonical to phantom scalar since V_{eff} contains terms with h^{-2} and h^{-1} while $h = 0$ at such a transition. In this subsection we discuss singularities on a throat, partly following [106, 260].

Suppose that a throat occurs at $z = 0$ (without loss of generality) and that the spherical radius $r(z) \equiv \mathrm{e}^{\beta}$ has a generic minimum at $z = 0$, so that we have a Taylor expansion of the form $r(z) = r_0 + \frac{1}{2}r_2 z^2 + o(z^2)$ with $r_0 > 0,\ r_2 > 0$. Also, without loss of generality, in a certain neighborhood of the throat we can put $h(\phi) = -1$ since a throat definitely requires $h < 0$, and $h(\phi) = -1$ can be achieved by a regular redefinition of ϕ. Then, using the zero-order (background) equations, it is not difficult to show that near the throat

$$V_{\text{eff}}(\phi) = 2/z^2 + O(1), \tag{7.40}$$

and there is no $O(z^{-1})$ contribution. We see that there is an infinitely high potential wall separating perturbations on the left of the throat ($z < 0$) from those on the right ($z > 0$) because a natural boundary condition at $z = 0$ is $\psi(0) = 0$. It immediately implies that all other perturbations vanish at $z = 0$. We are thus unable to consider a mode of obvious physical significance, the one corresponding to dynamic changes of the throat radius.

This difficulty can be overcome using the so-called S-deformation method applied to the potential V_{eff}. It has been used in [301, 302] for transforming a partly negative potential to a positive-definite one for perturbations of higher-dimensional black holes. Later, with its aid, Gonzalez *et al.* [260] transformed a singular potential to a nonsingular one for perturbations of the anti-Fisher wormholes and found that an exponentially growing mode, making such wormholes unstable, is just the one connected with evolving throat radius. The same scheme in a more general formulation was used in [106, 121] for revealing instabilities in other solutions with throats, including some black universe models. Let us briefly describe the method.

Consider a wave equation of the type (7.31)

$$\ddot{\psi} - \psi'' + W(z)\psi = 0, \tag{7.41}$$

with an arbitrary potential $W(z)$ (whose specific example is the above potential V_{eff}). Let us present $W(z)$ in the form

$$W(z) = S^2(z) + S', \tag{7.42}$$

which is actually a Riccati equation with respect to $S(z)$, so such a function can in general be found. Then Eq. (7.41) can be rewritten as

$$\ddot{\psi} + (\partial_z + S)(-\partial_z + S)\psi = 0. \tag{7.43}$$

Now, if we introduce the new function

$$\chi = (-\partial_z + S)\psi, \tag{7.44}$$

then, applying the operator $-\partial_z + S$ to the left-hand side of Eq. (7.43), we obtain the following wave equation for χ:

$$\ddot{\chi} - \chi'' + W_{\text{reg}}(z)\chi = 0, \tag{7.45}$$

with the new effective potential

$$W_{\text{reg}}(x) = -S' + S^2 = -W(z) + 2S^2. \tag{7.46}$$

If a static solution $\psi_s(z)$ of Eq. (7.41) is known, so that $\psi_s'' = W(z)\psi_s$, then we can choose

$$S(z) = \psi_s'/\psi_s \tag{7.47}$$

to carry out the above transformation. Furthermore, if we assume that $W(x) \approx az^{-n}$ at small z, then, requiring that $W_{\rm reg}$ should be finite at $z = 0$ and using (7.46), we easily find that necessary conditions for such a singularity in $W(z)$ to be removable by the above transformation is $n = 2$, $a = 2$, that is, $W \approx 2/z^2 + W_0 + O(z)$, where W_0 is a constant. Fortunately, by Eq. (7.40), the potential $V_{\rm eff}$ behaves precisely in this way. A point of interest is that such removal is only possible for $W \to +\infty$ as $z \to 0$, so that a potential wall in $W(z)$ can be removed but a potential well cannot.

From (7.46) with finite $W_{\rm reg}$ it also follows that at small z

$$S = -1/z - W_0 z + O(z^2),$$
$$W_{\rm reg}(z) = 3W_0 + O(z). \tag{7.48}$$

To summarize, we have seen that singularities of the effective potential $V_{\rm eff}$ at a throat can be regularized in the generic case [106, 260]. What is of equal importance, it has also been shown that regular solutions to the regularized wave equation really describe regular perturbations of both the scalar field and the metric. It was this procedure that made it possible to prove the instability of anti-Fisher (Ellis type [75, 229]) wormholes [260] and other scalar field configurations in GR [106, 121].

7.3.2 *Regular perturbations near a throat*

Suppose we have found a solution $\chi(x,t)$ to Eq. (7.45), satisfying the appropriate boundary conditions. The function χ is regular at $z = 0$ since the potential $W_{\rm reg}$ is regular there. If χ is a growing function of t, it probably indicates an instability of the initial static configuration; but this is indeed the case only if $\chi(z,t)$ creates regular perturbations of the metric functions α, β, γ and the scalar field ϕ.

Let us look at how it happens. Given $\chi(z,t)$, a solution to (7.41), or (7.31), is found as $\psi = (\partial_z + S)\chi$. Generically, χ is finite at $z = 0$

while $S \approx 1/z$ at small z, hence $\psi \sim 1/z$, and according to (7.30) we obtain $\delta\phi \to \infty$ at the throat. This result is in fact quite natural since the relation (7.30) corresponds to the gauge $\delta r = \mathrm{e}^{\beta}\delta\beta = 0$, in which the throat radius is fixed, whereas we are seeking perturbations with nonzero δr on the throat. So, it is necessary to pass on to another gauge, which is easily done due to gauge invariance of the quantity ψ given by

$$\psi = r\delta\phi - \frac{r\phi_z}{r_z}\delta r. \tag{7.49}$$

Namely, a finite expression for δr is obtained in the gauge $\delta\phi = 0$ provided $\phi_z(0) \neq 0$ since then

$$\delta r = -\frac{r_z}{r\phi_z}\psi, \tag{7.50}$$

while the product $r_z\psi$ is finite. It remains to find $\delta\alpha$ and $\delta\gamma$ from the perturbation equations in the gauge $\delta\phi = 0$. From Eqs. (7.22) and (7.21) we find

$$\beta_x\delta\alpha = \delta\beta_z + \delta\beta(\beta_z - \gamma_z), \tag{7.51}$$

$$\delta\gamma_x = \delta\alpha_z - 2\delta\beta_z - \frac{2\epsilon}{\phi_z}V_\phi\,\mathrm{e}^{2\alpha}\delta\alpha, \tag{7.52}$$

but here we are again facing a problem: according to (7.51), in general $\delta\alpha$ diverges at the throat where $\beta_z = 0$. This divergence is only avoided if the right-hand side of (7.51) behaves like $\beta_z \sim u - u_{\text{th}} \sim z$.

Surprisingly, this is the case in a generic situation, as can be verified in a general form using near-throat expansions. Indeed, let us preserve the previous assumptions, that the throat is a generic minimum of $\beta(z)$ and that the background quantity $\phi'(z) \neq 0$. Let us assume, in addition, that

(i) The function $\chi(z,t)$ that solves Eq. (7.45) is finite and nonzero at $z = 0$.

(ii) the function $S(z)$ behaves at small z according to (7.48).

Our task is to estimate the right-hand side of Eq. (7.51) in the order $O(z^0)$: if it is zero, it means that $\beta_z\delta\alpha \sim z \sim \beta_z$ and thus

$\delta\alpha(0)$ is finite; the remaining metric perturbation $\delta\gamma$ is then found from (7.52) and is also finite.

Taking $\delta\beta = \delta r/r$ from (7.50) and substituting ψ as $\psi = \chi_z + S\chi \sim -\chi_0/z$, where $\chi_0 = \chi(0)$, we find

$$\delta\beta = \frac{\chi_0\,\mathrm{e}^{-\beta}}{\phi_z}\Big(\beta_2 + \frac{1}{2}\beta_3 z + \cdots\Big).$$

Then we substitute this expression to (7.51) to obtain

$$\beta_z\delta\alpha \approx \frac{\chi_0\,\mathrm{e}^{-\beta}}{\phi_z^2}\Big[-\phi_{zz}\beta_2 + \frac{1}{2}\phi_z(\beta_3 - 2\gamma_z\beta_2)\Big],$$

where all quantities are taken at $z = 0$. Now, ϕ_{zz} can be expressed from the background equation (7.16), β_{zz} from (7.19); we can use the fact that $\beta_2 = \beta_{zz}(0)$, etc., and we can also ignore all terms proportional to β_z. After these substitutions we finally obtain

$$\beta_z\delta\alpha \propto \mathrm{e}^{2\gamma}V_\phi\Big[\epsilon\beta_{zz} + \frac{1}{2}\phi_z^2\Big]_{z=0}.$$

But the expression in the square brackets vanishes due to the difference of Eqs. (7.17) and (7.18), which proves that $\beta_x\delta\alpha = O(x)$ and thus $\delta\alpha(0)$ is finite.

We conclude that *under generic assumptions on the system behavior, regularization of the potential V_{eff} always leads to finite perturbations of the background static solution.*

7.3.3 *Perturbations near a transition surface* $\mathbb{S}_{\text{trans}}$ *(where $h(\phi) = 0$)*

The effective potential V_{eff} also possesses singularities at the values of the radial coordinate where $h = 0$, which exist in cases where the function $h(\phi)$ in (7.1) changes its sign. This happens in the framework of the trapped ghost concept. Such singularities are quite different in nature than those occurring on a throat: they are rather similar to those found at transition surfaces in systems with conformal continuations [82, 111, 112]. The latter phenomenon is observed, for instance, in scalar–tensor theories (STTs) of gravity where it sometimes happens that the whole manifold in the Einstein frame maps to only a part of the Jordan frame manifold (or vice versa). It was found that if such configurations are static and spherically symmetric, then their monopole perturbations are described by wave equations like (7.31) with effective potentials that have a singularity of the form

$$V_{\text{eff}} \approx -\frac{1}{4\Delta z^2} \quad \text{as} \quad \Delta z \to 0 \tag{7.53}$$

if $\Delta z = z - z_0$ and $z = z_0$ is a transition surface where the Einstein frame metric has a singularity while the Jordan frame metric is regular [111].

Indeed, let us assume that near the value $z = z_0$ where $h = 0$ the metric functions α, β, γ as well as the ϕ field are regular, so that the following Taylor expansions are valid:

$$\begin{aligned}
\alpha(z) &= \alpha_0 + \alpha_1 \Delta z + \cdots, \quad \text{the same for } \beta(z),\ \gamma(z),\\
\phi(z) &= \phi_0 + \phi_1 \Delta z + \frac{1}{2}\phi_2 \Delta z^2 + \cdots,\\
h(z) &= h_1 \Delta z + \frac{1}{2} h_2 \Delta z^2 + \cdots, \qquad \phi_1 \neq 0, \quad h_1 \neq 0.
\end{aligned} \tag{7.54}$$

It can be shown that in this generic situation the effective potential (7.33) is represented near $z = z_0$ by the Laurent series expansion

$$V_{\text{eff}}(z) = -\frac{1}{4\Delta z^2} + \frac{1}{2\Delta z}\left[\frac{h_2}{h_1} + 2\left(\beta_1 - \alpha_1 + \frac{\phi_2}{\phi_1}\right)\right] + O(1). \tag{7.55}$$

Thus, the surface $\mathbb{S}_{\text{trans}}$ creates an infinitely deep potential well for perturbations. In quantum mechanics such a potential in the stationary Schrödinger equation would mean that there are negative energy levels of unlimited depth. Returning to a stability study, it is tempting to conclude that there are perturbation modes with $\omega^2 < 0$ and an arbitrarily large $|\omega|$, and the corresponding growth ($\delta\phi \sim e^{|\omega|t}$) of the perturbations then immediately leads them out of the linear regime and necessitates a nonlinear or nonperturbative analysis.

However, the stability study implies other physical requirements on ψ in Eq. (7.31) or Y in (7.35): $\delta\phi$ should be finite in the whole space and should vanish at infinities or horizons, whereas quantum mechanics only requires quadratic integrability of ψ.

With the potential (7.55), the general solution of (7.35) at small z has, independently of ω, the leading terms

$$Y(z) = \sqrt{|\Delta z|}(C_1 + C_2 \ln|\Delta z|) + O(\Delta z^{3/2}),$$
$$C_1,\ C_2 = \text{const.} \tag{7.56}$$

On the other hand, according to (7.30), $\delta\phi \sim Y/\sqrt{h} \sim Y/\sqrt{|\Delta z|}$, therefore,

$$\delta\phi \sim C_1 + C_2 \ln|\Delta z| + O(z), \tag{7.57}$$

and for physically meaningful perturbations we must require $C_2 = 0$. Thus, the negative pole of the potential (7.55) leads to a constraint which should be added to the ordinary boundary conditions, which, combined with the equation itself, in more conventional cases completely determine the solution to a boundary-value problem. Due to this constraint it may happen that the boundary-value problem with Eq. (7.35) has no discrete spectrum. Indeed, suppose we have a background configuration with a surface $z = z_0$ where $h = 0$, and there is a solution to (7.35) satisfying the boundary conditions at infinities and/or a horizon; then there is in general zero probability that this solution will be finite at $z = z_0$. There could be an exception if the background were $\mathbb{Z}_2$-symmetric relative to the surface $z = 0$, but here it is manifestly not the case because $h > 0$ on one side of $z = 0$ and $h < 0$ on the other. A similar situation was discussed in [404] concerning the stability of black holes with a conformal scalar field [39, 56], and after all, with some more reasoning, it was concluded that such black holes are stable under monopole perturbations.[1]

[1]Let us recall that, contrary to this situation, the quantum-mechanical requirement of quadratic integrability of $Y(z)$ does hold since the integral $\int z^n \ln^2 z\, dz$ converges at $z = 0$ for any $n \geq 0$. The quantum particle in a field with such a potential well, as is sometimes said, "falls onto the center" [365] because the wave function, as we unlimitedly descend to deeper and deeper negative energy levels, more and more concentrates near $z = z_0$.

We conclude that if there is a transition from a canonical scalar to a phantom one, it plays a strong stabilizing role for any such configuration. One cannot, however, say whether or not a particular trapped-ghost solution is stable or not without its investigation because this evidently depends on the details of each model.

7.3.4 *Transition surfaces and throats*

If a configuration contains a throat, then, as we saw, the master equation for perturbations should be regularized in order to include all perturbation modes. The regularized effective potential $W_{\rm reg}$, obtained by the S-transformation (7.44), (7.46) from $W = V_{\rm eff}$, strongly differs from $V_{\rm eff}$. If, in addition, there is a transition surface $\mathbb{S}_{\rm trans}$ ($z = z_0$) at which $h = 0$, it is of interest to know how $W_{\rm reg}$ looks like at and near such a surface and to which consequences it leads.

Using (7.42) and (7.46) for $W(z) \approx 1/(4\Delta z)$, it is easy to obtain that near $z = z_0$

$$S(z) \approx \frac{1}{2\Delta z}, \qquad W_{\rm reg}(z) \approx \frac{3}{4\Delta z^2}. \tag{7.58}$$

The transformed Schrödinger equation (7.45) with this form of $W_{\rm reg}$ has the following solution in the leading approximation near $z = z_0$:

$$\chi(z) = C_3 \Delta z^{3/2} + C_4 \Delta z^{-1/2}. \tag{7.59}$$

To find an expression for $\delta\phi(z)$, it is necessary to convert χ to ψ by the transformation inverse to (7.44), that is,

$$\psi(z) = (\partial_z + S)\chi(z), \tag{7.60}$$

and then to obtain $\delta\phi$ using Eq. (7.30) (actually, $\delta\phi \sim \psi/\sqrt{|\Delta z|}$). It turns out that $(\partial_z + S)\Delta z^{3/2} = 2\sqrt{\Delta z}$, so that the term with C_3 in (7.59) which vanishes at $z = z_0$ precisely transforms to the term with C_1 in (7.57). However, being applied to $\Delta z^{-1/2}$, the transformation (7.30) gives zero, which actually means $o(|\Delta z|^{-3/2}$, i.e., the second solution to (7.41) may only be obtained using a more precise approximation of χ than (7.59). However, the direct operator $-\partial_z+S$ applied to $\psi = \sqrt{\Delta z}\ln|\Delta z|$ (see (7.57)) leads to $\chi = |\Delta z|^{-1/2}$ in accord with (7.59), therefore we can be sure that a regular expression for $\delta\phi$ at $z = z_0$ is obtained on equal grounds from (7.57) with $C_2 = 0$ or from (7.59) with $C_4 = 0$. These requirements lead to boundary conditions for Eq. (7.41) or (7.45) on $\mathbb{S}_{\rm trans}$.

The stability problem for configurations containing $\mathbb{S}_{\rm trans}$ then reduces to a set of boundary-value problems posed on each range of z between an infinity and $\mathbb{S}_{\rm trans}$or between a horizon and $\mathbb{S}_{\rm trans}$ (the latter in the case of a black universe). If the system contains a throat, one should apply these boundary conditions to Eq. (7.45) with a regularized potential $W_{\rm reg}$, and if there is no throat, one can use Eq. (7.41) with the potential $W = V_{\rm eff}$.

7.4 The stability problem for some particular solutions

7.4.1 *Instabilities of the Fisher and anti-Fisher solutions*

The static solutions. Let us briefly recall these solutions to the Einstein–scalar equations with $V(\phi) = 0$, described in detail in Chapter 4, partly using slightly different notations.

These solutions can be written jointly for $\varepsilon = \pm 1$ using the harmonic coordinate u as

$$ds^2 = e^{-2mu}dt^2 - \frac{e^{2mu}}{s^2(k,u)}\left[\frac{du^2}{s^2(k,u)} + d\Omega^2\right], \quad \phi = Cu, \qquad (7.61)$$

where the integration constants m (the Schwarzschild mass), C (the scalar charge) and k are related by the equality

$$2k^2 \operatorname{sign} k = 2m^2 + \epsilon C^2. \qquad (7.62)$$

The function $s(k,u)$ is defined according to (4.12). The coordinate u is defined in the whole range $u > 0$ for $k \geq 0$ and in the range $0 < u < \pi/|k|$ for $k < 0$. The value $u = 0$ in all cases corresponds to flat spatial infinity: at small u, the spherical radius is $r(u) \approx 1/u$, and the metric becomes approximately Schwarzschild with mass m.

In the case $k > 0$, we pass on to the quasiglobal coordinate x by the transformation

$$e^{-2ku} = 1 - 2k/x =: P(x), \tag{7.63}$$

and the solution takes the form

$$ds^2 = P^a dt^2 - P^{-a} dx^2 - P^{1-a} x^2 d\Omega^2, \quad \phi = -\frac{C}{2k} \ln P(u), \tag{7.64}$$

with the constants related by

$$a = m/k, \quad a^2 = 1 - \varepsilon C^2/(2k^2). \tag{7.65}$$

Fisher's solution [243] corresponds to $\epsilon = +1$, it consists of a single branch $k > 0$, and $|a| < 1$ in (7.64). It is defined in the range $x > 2k$, and $u = 2k$ is a naked central ($r = 0$) singularity which is attractive for $m > 0$ and repulsive for $m < 0$. The Schwarzschild solution is restored at $C = 0$, $a = 1$ for $m > 0$ and at $C = 0$, $a = -1$ for $m < 0$.

The *anti-Fisher solution* ($\epsilon = -1$) [47, 75] splits into three branches with the following basic properties.

- **Branch A, $k > 0$:** we have again the solution (7.64), but now $|a| > 1$. For $m < 0$, that is, $a < -1$, as in Fisher's solution, there is a naked central singularity at $x = 2k$. A drastically different situation is observed at $m > 0$, $a > 1$: there is a throat at $x = x_{\rm th} = (a+1)k$, and $r \to \infty$ as $x \to 2k$. Moreover, for $a = 2, 3, \ldots$ the metric exhibits a horizon of order a at $u = 2k$ and admits a continuation to smaller x [93]. These horizons have an infinite area and zero Hawking temperature (the so-called cold black holes [95]). The throat radius does not coincide with the (geometrized) Schwarzschild mass $m = ak$ but is of the same order of magnitude.

  The metric (7.64) has a curvature singularity at $x = 2k$ if $a < 2$ (except for the Schwarzschild case $a = 1$), a finite curvature if $a = 2$ and zero curvature if $a > 2$. At noninteger $a > 2$, we are dealing with weaker singularities termed "singular horizons" (see Sec. 4.1 and [95]).
- **Branch B, $k = 0$:** the solution is defined in the range $x \in \mathbb{R}_+$ and is rewritten in terms of the quasiglobal coordinate $x = 1/u$ as

follows:

$$ds^2 = \mathrm{e}^{-2m/x}dt^2 - \mathrm{e}^{2m/x}[dx^2 + x^2 d\Omega^2], \quad \phi = C/x. \tag{7.66}$$

As before, $x = \infty$ is a flat infinity, while at the other extreme, $x \to 0$, the behavior is different for positive and negative mass. For $m < 0$, $x = 0$ is a singular center ($r = 0$), while for $m > 0$, $r \to \infty$, and all $K_i \to 0$ as $x \to 0$. This is again a singular horizon: the nonanalyticity of the metric in terms of x makes its continuation impossible. The throat occurs at $x = m$ and has the size $\mathrm{e} \cdot m$, e being the base of natural logarithms.

- **Branch C, $k < 0$:** the solution describes a wormhole with two flat asymptotics at $u = 0$ and $u = \pi/|k|$. The metric has the form [75, 229]

$$\begin{aligned} ds^2 &= \mathrm{e}^{-2mu}dt^2 - \frac{k^2\,\mathrm{e}^{2mu}}{\sin^2(ku)}\left[\frac{k^2\,du^2}{\sin^2(ku)} + d\Omega^2\right] \\ &= \mathrm{e}^{-2mu}dt^2 - \mathrm{e}^{2mu}[dx^2 + (k^2 + x^2)d\Omega^2], \end{aligned} \tag{7.67}$$

where u is expressed in terms of the quasiglobal coordinate $x \in \mathbb{R}$ by $\bar{k}u = \operatorname{arccot}(x/\bar{k})$, and we denote $-k = \bar{k} > 0$. If $m > 0$, the wormhole is attractive for ambient test matter at the first asymptotic ($x \to \infty$) and repulsive at the second one ($x \to -\infty$), and vice versa in the case $m < 0$. The wormhole throat occurs at $x = m$.

Perturbations: Fisher's solution

For a massless scalar field [$V \equiv 0$ in (7.1)], the effective potential in (7.31) takes the following form in terms of the quasiglobal coordinate x:

$$V_{\text{eff}} = -\epsilon\frac{A\phi'^2}{r'^2} + \frac{Ar''}{r} + \frac{A'r'}{r} = -\epsilon\frac{A\phi'^2}{r'^2} + \frac{A}{r^2} - \frac{A^2r'^2}{r^2}, \tag{7.68}$$

where $A(x) := \mathrm{e}^{2\gamma(x)} = \mathrm{e}^{-2\alpha(x)}$, and the second equality in (7.68) follows from Eq. (7.19).

Calculating V_{eff} for the solution (7.64), we find a common expression for both $\varepsilon = +1$ and $\varepsilon = -1$:

$$V_{\text{eff}}(u) = kP^{2a} \frac{S(x)}{x^2(x - k(1+a))^2(x - 2k)^2}.$$

$$S(x) = 2ax^3 - 3(1+a)^2kx^2 + 2(3 + 4a + 3a^2 + 2a^3)k^2x - (1+a)^4k^3. \tag{7.69}$$

Since in Fisher's solution $a < 1$, the binomial $x - k(1+a)$ is positive at all $x \geq 2k$, and the only singularity in V_{eff} is $x \to 2k$, coinciding with the singularity of the background solution. Near the singularity, at which according to (7.29) we can put $z = 0$, $V_{\text{eff}}(z) \sim -1/(4z^2)$, a negative pole in agreement with [113] and many subsequent papers.

The boundary condition at spatial infinity ($x \to \infty$, $z \to \infty$) is natural: $\delta\phi \to 0$, or $\psi \to 0$. For $x \to 2k$, where the background field ϕ tends to infinity, the boundary condition is not so evident. In [113] and other papers, dealing with minimally coupled or dilatonic scalar fields, the minimal requirement was used providing the validity of the perturbation scheme:

$$|\delta\phi/\phi| < \infty. \tag{7.70}$$

(The requirement of absence of ingoing waves then does not lead to further restrictions.) Under this boundary condition it is easy to conclude that there are solutions to the Schrödinger-like equation (7.35) with any $\omega^2 < 0$, which means that the static field configuration is unstable, in agreement with the previous work [113] (see also [93] for details).

Perturbations: The anti-Fisher solution

- **Branch A.** The effective potential has the same form (7.69), but now, since $a > 1$ (we restrict ourselves to the case $m > 0$), the potential has a positive pole at $x = x_{\text{th}} = k(a+1) > 2k$, the throat in the background configuration.

To remove this singularity, we find the following simple static solution to Eq. (7.31):

$$\psi_{s+}(x) \propto r(x)\frac{ax - x_{\text{th}}}{x - x_{\text{th}}}. \tag{7.71}$$

Applying the technique described in Sec. 7.3.1 with these $\psi_s(x)$, we calculate the new effective potential $W_{\text{reg}} = W_A(x)$:

$$\begin{aligned} W_A(x) &= \left(1 - \frac{2k}{x}\right)^{2a} \frac{N(x)}{x^2(x-2k)^2(ax - x_{\text{th}})^2}, \\ N(x) &= 3(1+a)^4k^4 - 2a(6 + 19a + 16a^2 + 3a^3)k^3x \\ &\quad + 3a^2(9 + 10a + a^2)k^2x^2 - 2a^2(4+5a)kx^3 + 2a^2x^4. \end{aligned} \tag{7.72}$$

The potential (7.72) has no singularities at $r > 2k$ and is partly negative, which should in general lead to an instability. To prove it, one can use the method of time-domain integration [271] allowing for the time evolution of perturbations under prescribed initial and boundary conditions. The latter, as usual, must provide the absence of ingoing waves from the boundary, and in our case it is sufficient to simply require $\psi \to 0$ and $\chi \to 0$ as $x \to \pm\infty$ for all three branches of the anti-Fisher solution. The reason is that the effective potential V_{eff} as well as the regularized potentials vanish at large $|x|$, and therefore those modes of interest with $\omega^2 < 0$ should exponentially decay at large $|x|$.

Examples of plots for the potential (7.72) and the results of time-domain integration are shown in Fig. 7.1. By fitting of the profile we find that the perturbations grow approximately as $\psi \propto e^{0.25t/m}$.

In the limit $a \to 1$ the regularized potential still has a negative gap, but at $a = 1$ no growing mode is observed, and a stationary solution dominates at late times (Fig. 7.2), while at any nonzero $a - 1$ the perturbation does grow. This shows how the instability is "dying out" when approaching the Schwarzschild solution. One can recall that in the genuine Schwarzschild case there is no

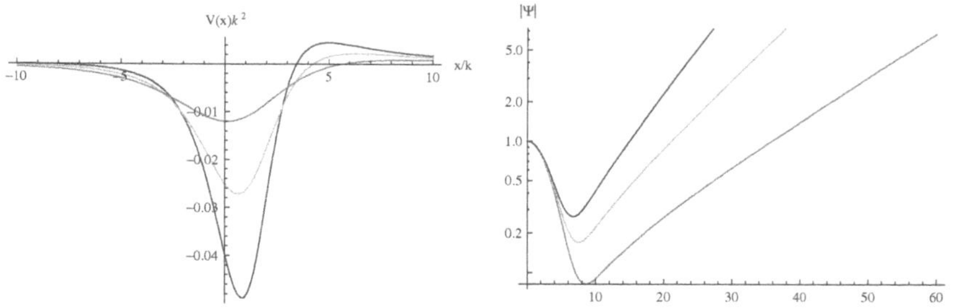

(a) (b)

Figure 7.1 The regularized effective potential W_A (a) and the time-domain profiles (b) for Branch A solutions with $a = 3/2, 2, 3$. Smaller values of a correspond to deeper potential wells and a more rapid growth of perturbations.

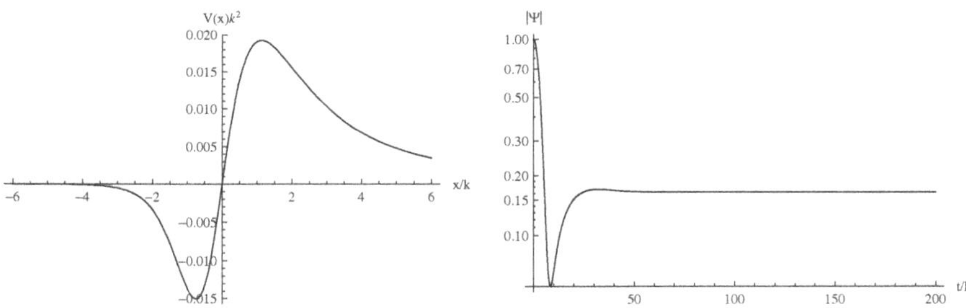

(a) (b)

Figure 7.2 The potential W_A (a) and the time-domain profile (b) for the Branch A solution with $a = 1$.

scalar field, and the modes we are considering here simply do not exist.

- **Branch B.** For the solution (7.66), the effective potential (7.68)

$$V_{\text{eff}}(x) = m \exp\left(-\frac{4m}{x}\right) \frac{2x^3 - 3mx^2 + 4m^2x - m^3}{(x-m)^2 x^4}, \tag{7.73}$$

is singular at $x_{\text{th}} = m$. We again find a static solution to (7.31):

$$\psi_{s0}(x) \propto \frac{x^2 \, e^{m/x}}{x - m}, \tag{7.74}$$

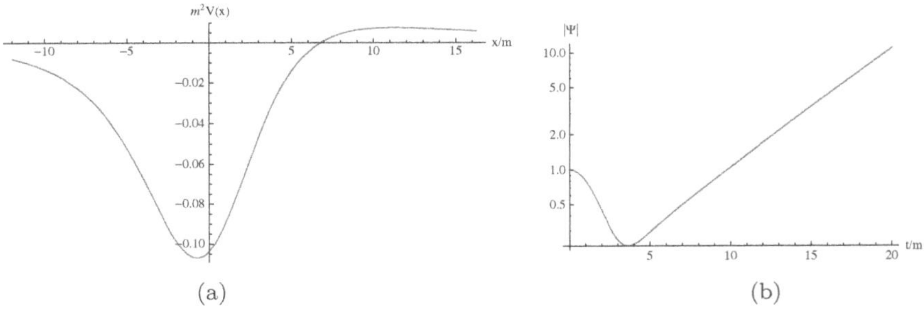

Figure 7.3 The potential W_B (a) and the time-domain profile (b) for the Branch B solution.

and perform the transformation described in Sec. 7.3.1. The regularized effective potential has the form of an inverse potential barrier:

$$W_B(x) = \exp\left(-\frac{4m}{x}\right)\frac{3m^2 - 10mx + 2x^2}{x^4}. \tag{7.75}$$

The perturbations grow approximately as $\psi \propto \mathrm{e}^{0.23t/m}$ (see an example in Fig. 7.3).

- **Branch C.** For the wormhole solution (7.67), the potential (7.68) has the form

$$V_{\text{eff}}(x) = \mathrm{e}^{-4mu} \times \frac{2k^4 + k^2(3m^2 - 2mx + 3x^2) - m(m^3 - 4m^2x + 3mx^2 - 2x^3)}{(m-x)^2(k^2+x^2)^2}. \tag{7.76}$$

It again exhibits a positive pole on the throat. To regularize it, we can take any of the two static solutions

$$\psi_{s-1}(x) \propto \frac{r(x)}{x-m}[k^2 + m^2 - mu(k^2 + mx)], \tag{7.77}$$

$$\psi_{s-2}(x) \propto \frac{r(x)}{x-m}(k^2 + mx). \tag{7.78}$$

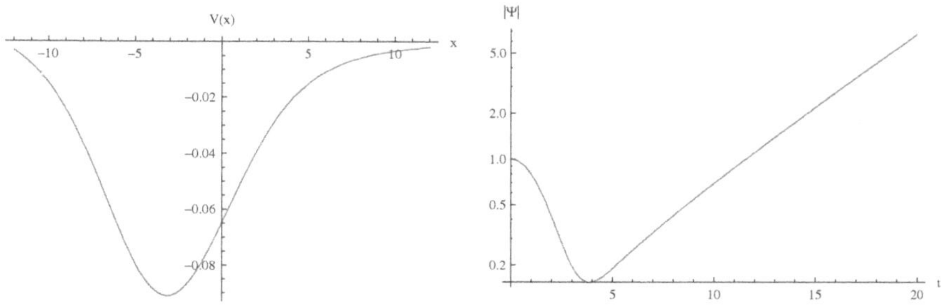

(a) (b)

Figure 7.4 The potential W_{C1} (a) and the time-domain profile (b) for Branch C ($k = m = 1$).

The first of them coincides with the solution found by González *et al.* in [260]. As a result, we reproduce their regular potential

$$W_{C1}(x) = e^{-4mu}\frac{N_1(x)}{(k^2+x^2)^2(k^2(mu-1)+m^2(ux-1))^2}, \tag{7.79}$$

where

$$\begin{aligned} N_1(x) = &-3k^6(mu-1)^2 + m^4(3m^2-10mx+2x^2)(ux-1)^2 \\ &+ k^2m^3(ux-1)(8x+2m^2u+m(5-15ux)) \\ &+ k^4m(mu-1)(2x+m^2u+m(9-12ux)). \end{aligned}$$

A further investigation revealing an unstable mode of perturbation is described in detail in [260] (Fig. 7.4).

The second static solution (7.78) can be used to demonstrate that, despite the different regularized potentials, the physical result, namely, the perturbation growth rate remains the same. Indeed, the regularized potential produced by the solution (7.78) reads

$$\begin{aligned} W_{C2}(x) &= e^{-4mu}\frac{N_2(x)}{(k^2+mx)^2(k^2+x^2)^2}, \\ N_2(x) &= -3k^6 + k^4m(m-12x) + k^2m^2(2m-15x)x \\ &\quad + m^2x^2(3m^2-10mx+2x^2), \end{aligned} \tag{7.80}$$

and leads to perturbations growing as shown in Fig. 7.5, at the same rate as in Fig. 7.4.

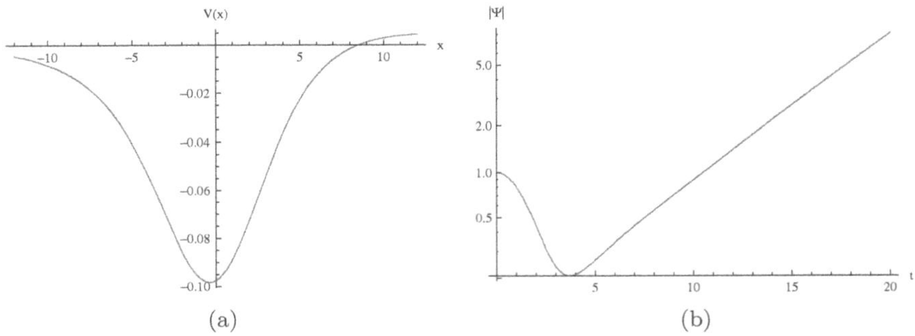

Figure 7.5 The potential W_{C2} (a) and the time-domain profile (b) for Branch C ($k = m = 1$).

We have demonstrated the instability of all anti-Fisher solutions to the Einstein–scalar equations under spherically symmetric perturbations, thus confirming and extending the conclusions of [260] made for Branch C solutions (wormholes). It turns out that in almost all cases the characteristic time of perturbation growth is of the order of the time needed for a light signal to cover a distance equal to the throat radius.

7.4.2 *Black universes*

A family of black-universe solutions to the scalar-Einstein equations with a nonzero self-interaction potential $V(\phi)$ has been considered in Sec. 5.1.2., Eqs. (5.13)–(5.15), see also (5.30)–(5.35). Black universes correspond to the metrics with the constants $m > 0$, $c < 0$. For numerical studies, we put there the length scale parameter $a = 1$, and the solution takes the form

$$A(x) = 1 + 3m + c(1 + x^2) + 3m(1 + x^2)\arctan x, \tag{7.81}$$

$$\phi = \pm\sqrt{2}\arctan x + \phi_0, \tag{7.82}$$

$$V = -c\frac{r^2 + 2x^2}{r^2} - 3m\left(\frac{3x}{r^2} + \frac{r^2 + 2x^2}{r^2}\arctan x\right), \tag{7.83}$$

where $r = \sqrt{1 + x^2}$.

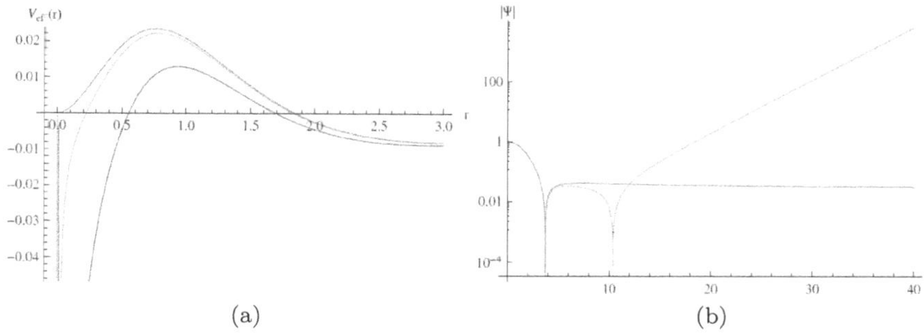

Figure 7.6 (a) Effective potentials $V_{\rm eff}(x)$ for radial perturbations of a black universe with $m = 1/3$, $c = -1, -1.001, -1.01$ (top to bottom). The potentials vanish at the horizon. (b) Time-domain perturbation evolution for $c = -1$ (stable) and $c = -1.001$ (unstable).

Since $A(0) = 1 + c$, black universes with $c \leq -1$ have no throat in the static region, their minimum of $r(x)$ being in a T-region; at $c = -1$ this minimum precisely coincides with the horizon. From Fig. 7.6 one can see that for $c = -1$ the effective potential for perturbations has a negative gap, however, time-domain integration proves that in this case the black universe is stable. For smaller values of c, there emerges an additional negative gap between the peak and the horizon, leading to an instability even for $c = -1.001$. At large negative c the potential peak disappears, and the potential becomes negative everywhere, which inevitably leads to an instability.

At $c > -1$ the throat $x = 0$ is in the static region. In this case, the numerically found regularized effective potentials have large negative gaps leading to instabilities (see Fig. 7.7). As c approaches zero, the growth rate of the time-domain profile decreases, still remaining positive because the effective potential remains negative in a wide region near the throat, making the perturbations grow. We can conclude that the only parameter for which a black universe can be stable is $c = -1$.

For axial perturbations with nonzero multipolarities ℓ it has been found that the effective potential for perturbations is positive-definite in all cases under consideration. Therefore, if we perturb a black hole outside the horizon in this manner, the perturbations are stable, as exemplified in Fig. 7.8.

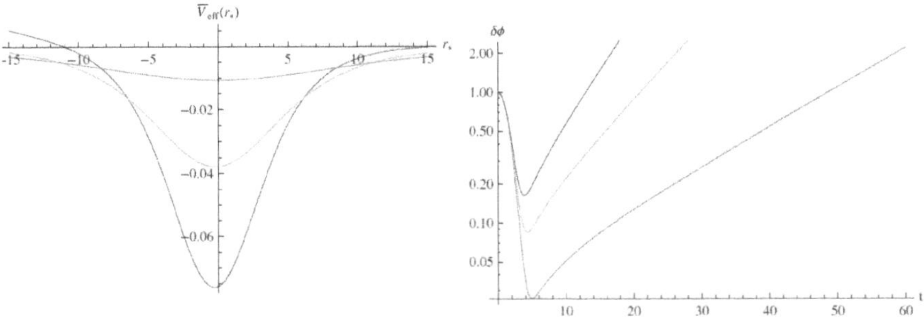

(a) (b)

Figure 7.7 Regularized effective potentials (a) upside down and time-domain profiles (b) bottom up for radial perturbations of a black universe with $m = 1/3$, $c = -0.1, -0.2, -0.3$.

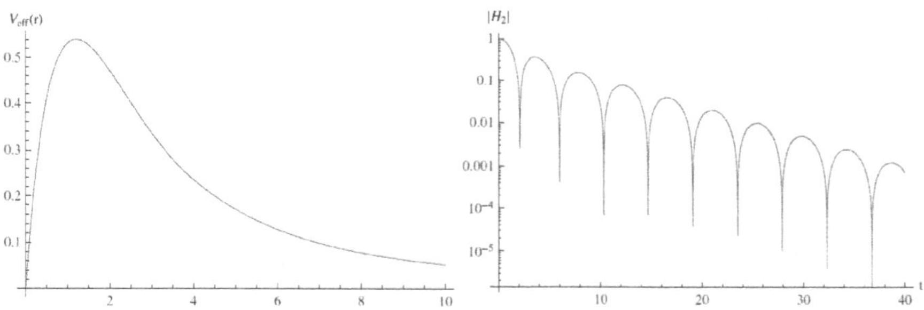

(a) (b)

Figure 7.8 The effective potential (a) and the time-domain profile (b) for axial gravitational perturbations ($\ell = 2$) of black universes with $m = 1/3$, $c = -1$. The potential vanishes at the horizon $r = 0$.

It has been shown that black universes with $c = -1$ ($a = 3\pi m/2$) are stable against polar monopole perturbations, and their response to external perturbations is dominated at late times by the quasinormal frequencies. The low-lying axial quasinormal frequencies have the smallest decay rates in the perturbation spectrum and thus dominate in a signal from the environment of such a BH at sufficiently late times. For the stable black universe under consideration, such frequencies have been calculated in [121] with the help of the WKB approach [306, 352, 491], widely used in similar calculations, see, e.g., [120, 340, 354, 448].

7.4.3 *Wormholes with a trapped ghost*

Consider, as an example, probably the simplest solution from the family (6.43), (6.51)–(6.55), that is, a symmetric, twice asymptotically flat wormhole obtained in the case $m = 0$. The effective potential (7.33) is written without fixing the coordinate in the background solution, while the present family uses the quasiglobal coordinate x, so it makes sense to rewrite (7.33) in its terms in the notations of the metric (6.43), (6.51):

$$V_{\text{eff}}(x) = A(x)\left[\frac{2hr^2\phi'^2}{r'^2}\Big(V - \frac{1}{r^2}\Big) + \frac{2rV'}{r'} - \frac{h_\phi V_\phi}{4h^2} + \frac{V_{\phi\phi}}{2h}\right] + A^2\frac{r''}{r} + AA'\Big(\frac{r'}{r} + \frac{h'}{h}\Big), \tag{7.84}$$

where the prime stands for d/dx, and we have replaced $P(\phi)$ with $V(\phi)$ since the solution does not contain an electromagnetic field. For calculations it is helpful to pass on to derivatives in x, e.g., $V_\phi = V'/\phi'$, etc. The result is

$$V_{\text{eff}}(x) = \frac{13 + 12x^2 + 3x^4}{9x^2(15 - x^2)^2(5 + x^2)^2(3 + 4x^2 + x^4)^4} \times \big(- 1316250 - 4168125x^2 - 3399750x^4 - 614610x^6 + 445878x^8 + 156460x^{10} - 53010x^{12} - 25998x^{14} - 1092x^{16} + 369x^{18}\big). \tag{7.85}$$

This function is plotted in Fig. 7.9: as expected, there is a positive pole at $x = 0$ (the throat) and negative poles at $x = \pm\sqrt{15}$, two transition surfaces $\mathbb{S}_{\text{trans}+}$ and $\mathbb{S}_{\text{trans}-}$, respectively.

The coordinates x and z are connected by the relation $dx/dz = A(x)$, where $A = Br^2$ and $B(x)$ is given by Eq. (6.52), hence, in our case ($m = 0$),

$$A(x) = \frac{26 + 24x^2 + 6x^4}{6(1 + x^2)(3 + x^2)}. \tag{7.86}$$

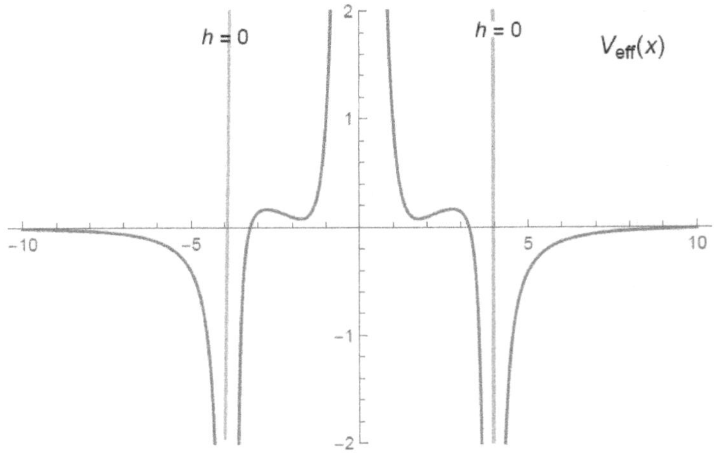

Figure 7.9 The effective potential $V_{\text{eff}}(x)$ for the background solution with a trapped ghost in the case $m = 0$ (a symmetric, twice asymptotically flat wormhole)

The integral $z = \int dx/A(x)$ leads to a transcedental equation for $x(z)$. Therefore, a transition to z in the perturbation equations can only be performed numerically. It looks more reasonable to pass on to the variable x in the master equation.

Furthermore, an important point is that a possible instability means that (i) there is an unstable mode in each of the three separate ranges of x: $\mathbb{R}_1 = (-\infty, -\sqrt{15})$, $\mathbb{R}_2 = (-\sqrt{15}, \sqrt{15})$, and $\mathbb{R}_3 = (\sqrt{15}, \infty)$, and (ii) the corresponding eigenvalue of the boundary-value problem is the same in all these ranges. Therefore, if there is no unstable mode in one of these ranges, we conclude that the background solution is stable under spherically symmetric perturbations.

In our case, due to symmetry of the problem, the equations in $\mathbb{R}_1$ and $\mathbb{R}_3$ coincide, and it is sufficient to consider only the ranges $\mathbb{R}_2$ and $\mathbb{R}_3$. In $\mathbb{R}_2$, which contains a throat, it is necessary to regularize the effective potential as described above. It is therefore reasonable to begin with $\mathbb{R}_3$, where such a regularization is not needed, and if no unstable mode is found there, we can already conclude that the background system is stable.

Thus a stability study should consist of the following steps:

1. Transform Eq. (7.35) with the potential (7.85) to the variable x.
2. Solve the boundary-value problem with the resulting equation in range $\mathbb{R}_3$ with the boundary conditions $|\delta\phi| < \infty$ at $x = \sqrt{15}$ and $\delta\phi \to 0$ as $x \to \infty$.
3. If item 2 does not reveal unstable modes, conclude that the background system is stable. If there are such modes, pass on to item 4.
4. Reformulate the S-transformation in terms of the variable x.
5. Find the function $S(x)$ realizing the S-transformation and the regularized effective potential W_{reg}.
6. Solve the boundary-value problem with the resulting equation in range $\mathbb{R}_2$ and the boundary conditions $|\delta\phi| < \infty$ at $x = \sqrt{15}$ and $x = -\sqrt{15}$.
7. Compare the eigenvalues ω^2 obtained in $\mathbb{R}_2$ and $\mathbb{R}_3$. If some of them coincide, the background system is unstable and decays with a rate corresponding to these ω^2. It is stable if there is no such coincidence.

Such a study for this and other trapped-ghost space–times is now in progress.

We can conclude [61, 84, 140, 363] that scalar fields may change their nature from canonical to ghost in a smooth way without creating any space–time singularities. This feature significantly widens the opportunities of scalar field dynamics and the possible choice of scalar–tensor theories of gravity, although here we have only considered minimally coupled scalars.

More specifically, trapped ghosts make possible spherically symmetric wormhole and regular black hole models where the ghost is present in some restricted strong-field region whereas in the weak-field region, where observers can live, the scalar has usual canonical properties. It can be speculated that if such ghosts do exist in Nature, they are all confined to strong-field regions ("all genies are sitting in their bottles"), but just one of them, having been released, has occupied the whole Universe and plays the part of dark energy (if dark energy is really phantom, which is more or less likely but not certain).

For static, spherically symmetric models in the Einstein–scalar field system under consideration, in addition to examples of exact solutions, some general properties have been found:

(i) Trapped-ghost solutions to the field equations are only possible with nonzero potentials $V(\phi)$.
(ii) If a solutions is twice asymptotically flat, and the Schwarzschild mass is positive at one of the infinities, it is negative at the other. Hence, mirror ($\mathbb{Z}_2$) symmetry with respect to a certain value of the radial coordinate is only possible if $m = 0$.
(iii) The transition surfaces from canonical to phantom behavior of the scalar field create a potential well for spherically symmetric perturbations, and the shape of this well (see Eq. (7.55)) does not depend on details of the model. Moreover, the natural requirement that perturbations must be finite leads to a constraint which, being added to the boundary conditions, substantially restricts the set of possible solutions and thus plays a stabilizing role.

The latter result is somewhat unexpected and, in our view, can raise an interest in trapped-ghost models. This opportunity can probably be used in attempts to solve various problems of gravitational physics and cosmology. Stability studies for specific trapped-ghost configurations like those described here make one of evident tasks for the near future.

7.5 Extensions and related problems

An important extension of the presently obtained results is connected with the fact that the equations due to (7.1) are not only the equations of GR with a scalar field but also the Einstein-frame equations of a wealth of extended theories (see Chapter 4): the general scalar-tensor (STT), $F(R)$ theories of gravity and hybrid metric-Palatini gravity (HMPG). The observable physics in such theories is usually described in the corresponding Jordan conformal frame, whose metric differs from that of the Einstein frame only by a conformal factor that varies from theory to theory. A transition from one frame to the other is, from the viewpoint of differential equations, simply a substitution, therefore the stability study can be safely performed in the Einstein frame. Then, if the conformal factor is everywhere regular and nonzero (including the end points of the coordinate range), the conformal mapping transfers a flat or AdS asymptotic to a similar asymptotic in another frame, and a horizon to a horizon. As a result, it also preserves the boundary conditions for regular perturbations, and the stability conclusions obtained in the framework of GR are readily extended to the corresponding solutions of these generalized theories. In particular, we can assert that all vacuum static, spherically symmetric solutions in STT connected with the (anti-)Fisher solutions by everywhere regular conformal factors, are unstable.

However, in many cases the conformal factors bear nontrivial features, i.e., they blow up or vanish somewhere, and this can affect the boundary conditions for the perturbation equations; in any such case a separate study is necessary. Consider, for instance, the counterpart of Fisher's solution (7.64) in the Brans–Dicke (BD) STT, in which the Jordan-frame metric and the scalar field are

$$g^{\rm J}_{\mu\nu} = (1/\Phi) g^{\rm E}_{\mu\nu}, \qquad \Phi = \exp(\phi/\sqrt{\omega + 3/2}), \tag{7.87}$$

where $\omega > -3/2$ is the BD coupling constant, $g^{\rm E}_{\mu\nu}$ and ϕ make the Einstein-frame solution (Fisher's) (7.64), and $|a| < 1$. What we then obtain is the so-called Brans class 1 solution [73]. In both frames, the value $u = 2k$ is a naked singularity (see more details on these solutions in [95, 137]). In the stability study, to formulate a boundary condition at this singularity in GR, we have used the minimal requirement (7.70) for Fisher's solution, $|\delta\phi/\phi| < \infty$, providing the validity of the perturbation scheme, and we then concluded that the background solution is unstable. In the BD theory, it is more reasonable to require that the perturbed conformal factor $1/\Phi$ should not behave worse than the unperturbed one, i.e., $|\delta\Phi/\Phi| < \infty$. However, since $\delta\Phi/\Phi \sim \delta\phi$, we arrive at the condition $|\delta\phi| < \infty$ which is more restrictive then (7.70) if $\phi \to \infty$, and this made us conclude in [95] that this BD solution is stable.

Another example of using the conformal mapping between Jordan and Einstein frames for stability studies can be found in [111, 112], where the instability was proved for electrically neutral and charged wormholes supported by nonminimally coupled scalar fields [30, 75, 112] (for the case of conformal coupling see Sec. 4.3). Then there is a drastic difference between the manifold structures in the two conformal frames: in the Einstein frame, without an electric charge, it is the Fisher solution with the metric $g_{\mu\nu}$ that has a singularity at $u = 2k$. In Jordan's, this singularity is removed due to the conformal factor, the solution is continued beyond this (now regular) sphere and has a flat asymptotic at the other end. Such wormholes proved to be unstable [111, 112], but their instability is of quite different nature than that described in this chapter and in [260]: it is related to a negative pole of the effective potential (7.69) at $u = 2k$ for Fisher's solution (7.64), $\varepsilon = +1$ and a similar singularity in its counterpart with an electric charge. Actually, it is the same singularity as that in $V_{\rm eff}$ for Fisher's solution, but for the latter it occurred at the boundary of the range of the solution whereas in continued solutions it is located somewhere in the middle.

The same transition takes place in the BH solution with a massless conformally coupled scalar field [39, 56] with the metric coinciding with the extreme Reissner–Nordström metric with a scalar source — see (4.57). This solution also involves a conformally continued space-time with respect to the Einstein frame, and there is again a singularity in $V_{\rm eff}$ on the transition surface, but the stability study involves other boundary conditions as compared to the wormhole case since one of the boundaries ($r = m$) is now a horizon. The occurrence of an infinitely deep potential well in $V_{\rm eff}$ for spherically symmetric perturbation of this object led to an instability conclusion [115], but a later, more careful study has shown that it is stable [404]. Many solutions of this kind are yet waiting for their stability studies.

In HMPG many spherically symmetric solutions are shown to be unstable on the basis of their Einstein-frame counterparts due to regularity of the corresponding conformal factor [88], but also for many of them, involving conformal continuation, additional studies are necessary.

A general state of affairs is that more and more solutions emerging in the literature, trying to describe real or hypothetic objects in the Universe, require stability investigations, and it is an often hard but quite necessary part of the theoretical work.

Part II

Cosmology

Chapter 8

Stages of the Universe's evolution

This chapter describes the standard theory of our Universe's evolution. The physical essence of the phenomena under study is emphasized. For understanding the material, it is sufficient to know the fundamentals of field theory, gravitation and quantum mechanics.

8.1 The cosmological principle and the Einstein equations

When discussing many problems concerning the properties of the Universe as a whole or its part accessible to observations, one can distract from specific properties of its separate regions and the matter that fills them. It is expressed in the so-called *cosmological principle,* according to which all points of the Universe and all directions in it are equivalent, or, in other words, the space is homogeneous and isotropic. Due to the Einstein equations, the matter distribution must also be homogeneous and isotropic. The cosmological principle agrees fairly well with observations: regions of the size $\geq$100 Mpc really differ very little from each other.

The assumption of homogeneity and isotropy of space and matter distributed in it drastically simplifies the calculations and the analysis of physical processes. The corresponding equations were very briefly discussed in Chapter 2. Let us dwell upon them in more detail.

The system contains dynamic variables related to matter (density, pressure, etc.) and gravity (the metric tensor components $g_{\mu\nu}$).

They obey the Einstein equations

$$G_\mu^\nu \equiv R_\mu^\nu - \frac{1}{2}\delta_\mu^\nu R = -8\pi G T_\mu^\nu - \Lambda\delta_\mu^\nu, \tag{8.1}$$

where R_μ^ν is the Ricci tensor and $R \equiv R_\alpha^\alpha$ is the scalar curvature; both of them are expressed in terms of the metric tensor and its first- and second-order derivatives with respect to the space–time coordinates. The stress-energy tensor (SET) T_μ^ν describes the properties of matter filling the space, see Chapter 2. The equations also take into account the so-called cosmological constant Λ, which is very probably observed in experiments. Equation (8.1) can be written in another equivalent form,

$$R_\mu^\nu = -8\pi G\left(T_\mu^\nu - \frac{1}{2}\delta_\mu^\nu T\right) + \delta_\mu^\nu\Lambda, \tag{8.2}$$

where $T \equiv T_\alpha^\alpha$. Due to the contracted Bianchi identities (2.46) and the Einstein equations, the SET satisfies the equalities (2.52)

$$\nabla_\nu T_\mu^\nu = 0, \tag{8.3}$$

which have the meaning of conservation laws (see Chapter 2 and the footnote to Eq. (2.52)), and the equations of motion for matter can be derived therefrom.

Owing to the assumption of spatial homogeneity and isotropy, the squared interval can be written in the Friedmann–Robertson–Walker (FRW) form

$$ds^2 = g_{\mu\nu}dx^\mu dx^\nu = dt^2 - a^2(t)\left[\frac{dr^2}{1-kr^2} + r^2 d\Omega^2\right], \tag{8.4}$$

where $d\Omega^2 = d\theta^2 + \sin^2\theta d\varphi^2$ is the length element on a unit sphere (as in the previous chapters). Here, a particular choice of the coordinates has been made: the synchronous (true cosmological) time t and the dimensionless spherical coordinates r, θ, φ. The behavior of the gravitational field is determined by the scale factor $a(t)$ (with the dimension of length) and the parameter k which can take three values depending on the sign of spatial curvature: $k = +1$ for spherical space, $k = 0$ for flat space, and $k = -1$ for hyperbolic space.

In the metric (8.4), the nonzero components of the Ricci and Einstein tensors are

$$R_0^0 = \frac{3\ddot{a}}{a}, \qquad R_{\underline{i}}^{\underline{i}} = \frac{1}{a^2}(a\ddot{a} + 2\dot{a}^2 + 2k),$$

$$G_0^0 = -\frac{3}{a^2}(\dot{a}^2 + k), \quad G_{\underline{i}}^{\underline{i}} = -\frac{1}{a^2}(2a\ddot{a} + \dot{a}^2 + k); \tag{8.5}$$

there is no summing over an underlined index.

Due to the symmetry properties of the tensor G_μ^ν and the Einstein equations, the tensor T_μ^ν automatically takes the form

$$T_\mu^\nu = \mathrm{diag}(\rho, -p, -p, -p), \tag{8.6}$$

where ρ and p are certain functions of the time coordinate. It is easy to verify that if the matter is a perfect fluid, then the coordinates chosen correspond to the comoving reference frame (RF) of the fluid, and the quantities ρ and p should be interpreted as its density and pressure.

Indeed, the SET of a perfect fluid is

$$T_\mu^\nu = (p + \rho)u_\mu u^\nu - p\delta_\mu^\nu, \tag{8.7}$$

where ρ and p are its density and pressure, respectively, and u^μ is the 4-velocity of its element. In the comoving RF (i.e., for an observer at rest with respect to the fluid), the 4-velocity has the form $u^\mu = (1, 0, 0, 0)$, and the SET takes the form (8.6).

From (8.5) it is evident that there are two different Einstein equations. It is convenient to choose for consideration the equation with the left-hand side G_0^0 because it contains only a first-order derivative of the scale factor a, and the trace $R = 8\pi GT + 4\Lambda$:

$$3\frac{\dot{a}^2}{a^2} + \frac{3k}{a^2} = 8\pi G\rho + \Lambda, \tag{8.8}$$

$$\frac{6}{a^2}(a\ddot{a} + \dot{a}^2 + k) = 8\pi G(\rho - 3p) + 4\Lambda. \tag{8.9}$$

One more first-order equation follows from the conservation law (8.3):

$$\frac{d\rho}{da} = -\frac{3}{a}(p + \rho),$$

or

$$\frac{d\rho}{dt} = -3H(p + \rho), \tag{8.10}$$

where the quantity $H := \dot{a}/a$, characterizing the expansion rate, is called the Hubble parameter.

It is easy to verify that Eq. (8.9) is a consequence of (8.8) and (8.10). Thus we have two independent equations (8.8) and (8.10) for three unknown functions $a(t)$, $\rho(t)$ and $p(t)$.

Evidently, a full solution of the equations depends on the properties of matter described by the equation of state that connects the energy density and the pressure, $p = p(\rho)$. Very frequently this relation can be chosen in the simplest linear form

$$p = w\rho, \tag{8.11}$$

where $w = \text{const}$ has different values for different kinds of matter:

- $w = -1$ corresponds to a cosmological constant,
- $w = 0$ to dustlike matter (or a gas of noninteracting particles),
- $w = 1/3$ to disordered radiation or an ultrarelativistic gas,
- $w = 1$ to matter of maximum rigidity compatible with causality, in which the velocities of sound is equal to that of light in vacuum.

Let us consider some geometric properties of FRW space–times with the metric (8.4) and an arbitrary scale factor $a(t)$.

Let one observer be located at the point $r = r_0 = 0$, and another one at a point with some $r \neq 0$. Then the last term in (8.4) makes it possible to calculate the area of the sphere $r = \text{const}$, at which the second observer is located at some time instant t: $S = 4\pi a^2(t) r^2$. Thus, even if the second observer is at rest, that is, $r = \text{const}$, the physical radius of the sphere is changing proportionally to the scale factor $a(t)$:

$$R(t) \equiv a(t)r. \tag{8.12}$$

Different clusters of galaxies are examples of such "second observers". The coordinate radius of such a cluster, r, is fixed, whereas the physical radius $R(t)$, with an observer at the centre of the sphere, is growing with time. An experimental confirmation of the physical radius growth is the frequency decrease of light emitted by the second observer and received by the first one (*the cosmological redshift*).

The distances between points of space can be measured in different ways. The results of measurements (even in Minkowski space) depend on the choice of the RF and, for a time-dependent metric, even on the choice of a spatial section (i.e., on the clock synchronization method) in the framework of the same RF. We will use the notion of instantaneous physical distance, i.e., the distance at a fixed time instant in the metric (8.4). This distance is measured by a set of observers in the RF which is at rest in the metric (8.4) with clocks synchronized by the cosmological time t. At measurements along the radius r, the physical distance between neighbours separated by the coordinate difference dr is $a(t)dr/\sqrt{1-kr^2}$. The total distance

$$R_{\text{inst}} = a(t)\int_0^r \frac{dr}{\sqrt{1-kr^2}}, \tag{8.13}$$

measured by all such observers at the same time instant, is equal to the instantaneous physical radius.

If we introduce, instead of r, the coordinate χ by the equality

$$\chi = \int_0^r \frac{dr}{\sqrt{1-kr^2}} = \begin{cases} \arcsin r, & k=1, \\ r, & k=0, \\ \sinh^{-1} r, & k=-1, \end{cases} \tag{8.14}$$

we obtain in terms of χ an alternative expression for the metric (8.4):

$$ds^2 = dt^2 - a^2(t)[d\chi + r^2(\chi)d\Omega^2],$$

$$r(\chi) = \begin{cases} \sin\chi, & k=1, \\ \chi, & k=0, \\ \sinh\chi, & k=-1, \end{cases} \tag{8.15}$$

and a very simple expression for the instantaneous distance (along the radius, at fixed values of θ and φ) between the observers $r = r(\chi_1)$

and $r = r(\chi_2)$:

$$R_{\rm inst} = a(t)(\chi_2 - \chi_1). \tag{8.16}$$

Consider the motion of a light signal emitted at $t = t_0$ on an arbitrary sphere, labelled by some r, and coming at $t = t_1$ to an observer located at the origin (centre) $r = 0$. In this case the motion occurs along the radius ($d\theta = d\varphi = 0$). From the condition $ds = 0$, according to (8.4) and (8.15), we have

$$\int_{t_0}^{t_1} \frac{dt}{a(t)} = \int_0^r \frac{dr}{\sqrt{1-kr^2}} = \chi. \tag{8.17}$$

Thus, the instantaneous physical distance $R_{\rm inst} = R_{\rm hor}(t_1, t_0)$ between the observer and the source (with the radial coordinate r_1 or χ_1), measured at t_1, is

$$R_{\rm hor}(t_1, t_0) = a(t_1)\chi_1 = a(t_1)\int_{t_0}^{t_1} \frac{dt}{a(t)}. \tag{8.18}$$

This quantity is called the *particle horizon radius* at the time t_1 for particles emitted at the time t_0. (Frequently, omitting the word "radius", it is called "the particle horizon" or even simply "the horizon" in cases where it cannot cause a confusion.)

A subtle point should be mentioned here: this distance measured at t_1 is always larger than the physical distance *covered* by the light signal as it travelled from the source to the observer. The latter is simply equal to $t_1 - t_0$ (or $c(t_1 - t_0)$ if one wishes to write the speed of light c explicitly), since the speed of light, measured locally at each point on the signal trajectory, is a universal constant, and the time t coincides with the physical time of the considered reference frame. The distance $R_{\rm hor}(t_1, t_0)$ is larger because the Universe has expanded while the signal travelled.[1] However, the distance $c(t_1 - t_0)$

[1]The distance *covered* by the signal and the distance to which it has *moved away* in an expanding background are quite different quantities. It can be understood from the following example. Imagine that you enter an escalator moving up, ascend along it at your usual pace and have made, say, 10 steps — but you are now as far away as 40 steps from the bottom at the expense of the escalator's own velocity!

is never used in cosmological calculations since other quantities like R_{inst} are much more convenient in both theoretical and observational relations.

It should be noted that the notion of a particle horizon has a few definitions at different levels of rigour. One of them is "the distance to which the light signal has moved away from its source in a certain time interval". For example, the particle horizon radius of our Universe is about 10^{28} cm for its lifetime since the beginning of expansion.

Let us also remind the reader that, apart from the notion of a particle horizon in cosmology, there are quite a number of horizons defined in the theory of gravity, which are above all used in black hole physics (see Chapters 3 and 4). Thus, the event horizon is sometimes defined as the locus of points that can be achieved by light signals, emitted from a fixed distant point, for infinite time. A typical example of such a horizon is the event horizon of the Schwarzschild BH.

8.1.1 *Some solutions for the scale factor*

Consider the time dependence of the scale factor $a(t)$ for some kinds of matter with the equation of state (8.11) in the homogeneous and isotropic space–time with the FRW metric (8.4). Recall that we can speak of spatial homogeneity and isotropy in the modern epoch only if we mean averaging of matter density at scales of the order of hundreds of megaparsecs.

The basic equations for our analysis are Eqs. (8.8) and (8.10). As will be seen from what follows, in many cases (including models of early stages of the evolution) the scale factor is much larger than unity, and the term k/a^2 in Eq. (8.8) may be neglected. Then Eq. (8.8) takes the form

$$\dot{a}^2 = \frac{8\pi G}{3}\rho a^2. \tag{8.19}$$

Equation (8.10) with (8.11) takes the form

$$\frac{d\rho}{da} = -3\gamma\frac{\rho}{a}, \quad \gamma := w + 1, \tag{8.20}$$

with the obvious solution

$$\rho = Ca^{-3\gamma} = Ca^{-3(w+1)}. \tag{8.21}$$

The constant C is found from the initial conditions: $C = \rho_{\text{in}} a_{\text{in}}^{3\gamma}$, where the subscript "in" refers to an initial instant at the evolutionary stage under consideration. Transitions between stages cannot be abrupt in reality, but this effect is often neglected.

Substituting the expression (8.21) to Eq. (8.19), it is easy to find the time dependence of the scale factor [327]:

$$a(t) = \left[a(t_{\text{in}})^{3\gamma/2} + \frac{2}{3\gamma}\sqrt{\frac{8\pi G}{3}} C(t - t_{\text{in}}) \right]^{2/(3\gamma)}.$$

Inserting the constant C, we obtain the following expression for the scale factor with different values of w or γ:

$$a(t) = a_{\text{in}} \left[1 + \frac{2}{3\gamma}\sqrt{\frac{8\pi G\rho_{\text{in}}}{3}} (t - t_{\text{in}}) \right]^{2/(3\gamma)}. \tag{8.22}$$

The influence of the properties of matter is reflected in the parameter γ.

It is conventional to divide the Universe evolution into five stages: inflation, reheating, the radiation dominated stage, the matter dominated stage, and the modern stage of accelerated expansion, also called secondary inflation. Every stage, depending on the dominating form of matter, is approximately characterized by its parameter γ.

Since the scale factor rapidly grows with time, the first term in the brackets can frequently be neglected. As a result, we obtain the compact expression

$$a(t) \simeq a_{\text{in}} \left[\frac{2}{3\gamma} H_{\text{in}} \cdot (t - t_{\text{in}}) \right]^{2/(3\gamma)}. \tag{8.23}$$

Here, $H_{\text{in}} \equiv H(t_{\text{in}}) = \sqrt{8\pi G\rho_{\text{in}}/3}$ is the value of the Hubble parameter at the beginning of this stage.

It is easy to obtain the time dependence of the Hubble parameter:

$$H(t) \equiv \frac{\dot{a}(t)}{a(t)} = \frac{2}{3\gamma t}. \tag{8.24}$$

(We note that the Hubble parameter was called the Hubble constant for some time after the expansion of the Universe was discovered, since only times close to the present epoch were only considered, and the value of H is almost invariable during this period.)

All stages, except for the inflationary ones, are well described by the above relations. The inflationary stages, approximately described by the de Sitter metric, have a number of important features. Before discussing them, it is useful to consider the basic properties of de Sitter space.

8.2 De Sitter space

At the beginning of last century, de Sitter studied the properties of a curved space which does not contain any matter. The only feature different from Minkowski space was the presence of a nonzero cosmological constant in Eq. (8.8). The peculiar properties of this space were only of academic interest for a long time. But now it is becoming evident that it is de Sitter space that describes fairly well both the initial, inflationary stage of our Universe and the modern accelerated stage. It is therefore useful and instructive to devote some time to learning its properties.

This kind of space was briefly discussed in Chapter 2, and now we will consider it in more detail. Since there is no matter, $p = \rho = 0$, Eq. (8.8) is greatly simplified,

$$\dot{a}^2 - H^2 a^2 = -k, \quad H^2 \equiv \frac{\Lambda}{3} \tag{8.25}$$

(assuming $\Lambda > 0$) and is easily integrated:

$$k = 0: \quad a(t) = a_0\, e^{\pm Ht}, \quad a_0 = \text{const}$$

$$\text{(a spatially flat Universe);} \tag{8.26}$$

$$k = 1: \quad a(t) = H^{-1} \cosh[H(t - t_0)], \quad t_0 = \text{const}$$

$$\text{(a closed Universe);} \tag{8.27}$$

$$k = -1: \quad a(t) = H^{-1} \sinh[H(t - t_0)], \quad t_0 = \text{const}$$

$$\text{(a hyperbolic Universe).} \tag{8.28}$$

The choice of constants (i.e., the choice of zero point for t) such that $a_0 = H^{-1}$, $t_0 = 0$, and the plus sign in (8.26) (it corresponds to the Universe expansion) lead to a convenient general expression for an asymptotic form of all three expressions for $a(t)$ at late times:

$$a(t) = H^{-1}\,\mathrm{e}^{Ht}, \quad t \gg H^{-1}. \tag{8.29}$$

Let us use this expression for an analysis and a deeper understanding of the general expressions obtained above. Let two observers be located at spatial points with fixed values of r: $r_1 = 0$ and $r_2 = r$, i.e., both are at rest in the RF in which the metric is written. Nevertheless, the physical distance which they measure, according to the expressions (8.12) and (8.29), grows exponentially with time.

It is also important to clearly understand the question of distances to which light signals escape in de Sitter space. Consider a flat Universe ($k = 0$) with the scale factor $a(t) = H^{-1}\,\mathrm{e}^{Ht}$ as the simplest case preserving all necessary features (by the way, all three models have the same asymptotic behavior). For flat models, the dimensionless coordinates r and χ that we have introduced coincide. Similarly to (8.17), we obtain the expression

$$\Delta r(t,t') = \int_t^{t'} d\tau/a(\tau) = \mathrm{e}^{-Ht} - \mathrm{e}^{-Ht'} \tag{8.30}$$

for the coordinate distance covered by a light signal which has started at the time t, by the time t'. Hence the maximum value of Δr accessible to a signal emitted at the time t, is

$$\Delta r_{\rm hor} \equiv \Delta r(t,\ t' \to \infty) = \mathrm{e}^{-Ht}. \tag{8.31}$$

Thus, the horizon is located at a finite coordinate distance, and even more than that, the later is the signal emitted, the smaller this distance is. In the coordinate sense, the horizon size is decreasing with time.

From the viewpoint of physical distances, the situation is different: for the time $(t' - t)$, the light signal moves away to the distance

$$R_{\rm phys}(t,t') = a(t')\Delta r(t,t') = H^{-1}[\mathrm{e}^{H(t'-t)} - 1].$$

With fixed initial time t, the distance grows exponentially, which is in drastic contrast with the usual situation in Minkowski space. Moreover, according to the equation

$$R_{\rm hor}(t) = R_{\rm phys}(t, t' \to \infty) = H^{-1}\, {\rm e}^{H(t'-t)}, \qquad (8.32)$$

the horizon size tends to infinity.

The notion of a horizon is of utmost importance in cosmology because it determines the dynamic processes in the early Universe. In the inflationary period, soon after the birth of the Universe, the space–time could be approximately viewed as the de Sitter one, and the horizon size was about 10^{-27} cm. Such was the size of causally connected regions. Recalling that the Compton length of the electron is about 10^{-11} cm, it is easy to see that the very existence of the horizon inserts certain peculiarity into the physical processes. At the modern stage, dark energy is dominating, and this most probably means that we are entering a de Sitter space. In this case, the modern horizon size is $\sim 10^{28}$ cm, and larger distances are unobservable.

8.2.1 *Is there real expansion in de Sitter space?*

Let us once again consider the de Sitter metric in the form (8.4) with flat 3-space ($k = 0$):

$$ds^2 = dt^2 - H^{-2}\, {\rm e}^{2Ht}(dr^2 + r^2 d\Omega^2), \qquad (8.33)$$

where $H = \sqrt{\Lambda/3}$. The scale factor is time-dependent, and therefore the distances between point objects at rest with respect to the cosmological RF also change.

Meanwhile, there is a coordinate transformation that brings the metric to a static form, see (3.18)

$$ds^2 = (1 - H^2 r^2) dt^2 - (1 - H^2 r^2)^{-1} dr^2 - r^2 d\Omega^2.$$

In this metric, the distances between bodies fixed in this static RF are invariable. There also exists an RF in which the distances decrease in time, e.g., this will be the case if we take the minus sign in Eq. (8.26).

Let us return to the metric (8.33) and consider test particle motion. The geodesic equation has the form

$$\frac{d^2x^i}{ds^2} + \Gamma^i_{\mu\nu}\frac{dx^\mu}{ds}\frac{dx^\nu}{ds} = 0. \tag{8.34}$$

Since the Christoffel symbol $\Gamma^i_{tt} = 0$, one of the solutions of this equation is $x^i(t) = \text{const}$. The coordinates of points (and their differences) are time-independent. On the other hand, physical measurements ($R_{\text{phys}}(t) = e^{Ht}R_{\text{phys}}(0)$) clearly indicate the growth of distances between particles at rest.

So, does the Universe really expand, while the de Sitter space is maximally symmetric and all its points-events are absolutely equivalent? This seems to mean that a transition from one time instant to another changes nothing and there is no expansion at all.

A simple explanation of this seeming paradox is that it is not the four-dimensional space–time that expands but the 3-space of that or other RF. Moreover, we have just seen that the time lines $x^i = \text{const}$ of the RF, in which the metric has the form (8.33), are geodesics — while the distances R_{phys} are growing — consequently, this bundle (congruence) of geodesics is expanding. Other geodesic congruences in the same space–time can contract; it depends on the initial data specified in Eq. (8.34).

One can present one more argument confirming the reality of spatial expansion in the metric (8.33). Consider a set of test particles homogeneously distributed in the whole space and immovable in the coordinates (8.33), i.e., those with the world lines $x^i = \text{const}$. It is important that they weakly interact with each other (e.g., gravitationally). If we discover that their interaction is diminishing, then it indicates unambiguously that the distances between them grow, i.e., the Universe is really expanding.

Recall that, in the same coordinates, the horizon size (8.31) is decreasing in time by the law e^{-Ht} while the coordinate distance between the particles is fixed. Then, for any two particles there is a time instant at which the horizon size will be smaller than the distance between them, which will indicate the absence of a causal connection between the test particles. In other words, the initially nonzero interaction between the particles must tend to zero in time, hence the Universe (more precisely, the 3-space of this particular RF) is expanding.

One could present a similar reasoning in terms of physical distances, and naturally the result will be the same.

8.3 Inflation

The idea of inflation, i.e., extremely rapid expansion of the Universe at its early stage, was advanced in the early 80s of the last century. The achievements of the inflationary paradigm in explaining the observable properties of the Universe made these ideas generally accepted. A great number of inflationary scenarios are known today, and it is a hard problem to single out among them the one that was realized in the past. The number of inflationary models is continually growing. In what follows, we discuss the properties of the inflationary models based on the idea of so-called chaotic inflation.

Decades have passed since the moment when it became clear that the well-known expansion of the Universe, beginning with the "hot" (radiation-dominated) stage, leaves many questions, and it became evident that there was a nontrivial development of the Universe before the hot stage. A full list of problems can be found in [377] and in most of cosmological textbooks. The basic ones are discussed in Sec. 8.6.

The way to a solution of this set of problems was outlined in the papers [258, 276, 513].

Some time later it became clear that inflation is able to explain the basic set of observational data. The advances of the inflationary scenario were so impressive that it was included into the standard cosmological model as its main component.

The simplest and the most widespread method of describing the inflationary period is as follows. Let us assume the existence of a scalar field (inflaton) which evolves together with the gravitational field created by it. Under certain conditions, to be discussed later on, there emerges a situation resembling the de Sitter one. Thus the size of space under the horizon is growing exponentially, and it is the main feature of the inflationary period.

The Lagrangian density of a system consisting of a scalar field and gravity is written in the form

$$L = \sqrt{-g}\left\{\frac{1}{16\pi G}R + \frac{1}{2}g^{\mu\nu}\partial_\mu\varphi\partial_\nu\varphi - V(\varphi)\right\}, \tag{8.35}$$

where $g = \det(g_{\mu\nu})$ and G is the gravitational constant.

We can say in advance that the inflationary process is effective with a large energy density of the scalar field. The latter can emerge due to quantum fluctuations. Let us estimate the size of a fluctuation using the uncertainty relation

$$\Delta E\Delta t \sim 1(= \hbar), \tag{8.36}$$

and that $\Delta E \simeq \epsilon(\Delta L)^3$. We are interested in regions where there is enough time to establish causal connections. Therefore the lifetime of such a fluctuation should not be too short. Let us suppose $\Delta t \sim \Delta L$, i.e., a light signal has time to fly through the whole region. Then the size of such a region and the energy density of the quantum fluctuation are simply related by

$$\Delta L \sim \epsilon^{-1/4}.$$

By modern views, for the beginning of inflation, the energy density must be of order of 10^{-12} in Planck units. The size of the region ΔL must be about 10^3 in Planck units, or $\sim 10^{-30}$ cm. We are coming to an understanding of the fact that quantum fluctuations of fields in the space that surrounds us permanently create regions with an increased energy density. From the viewpoint of an observer from outside, the lifetime of such a fluctuation is small. In the example under consideration it is of the order of 10^{-40} s. The spatial size of a region occupied by a fluctuation is $\sim 10^{-30}$ cm. Despite being so small, it is still larger than the Planck size, which allows for using the Einstein equations in their standard form in order to describe the processes inside such a region.

In general, a region of an increased energy density can be created artificially, for example, in heavy ion accelerators. Creation of such regions of arbitrary size, able to exist for a finite time, may become real in the future. Let such a region be formed with a size much larger than the horizon size. Let us ask ourselves: can external and internal observers make a coordinated decision on (non)expansion of this region? The answer is negative because light launched into such a region (or from it) can only cover a distance within the horizon size.

Further on, the viewpoint of an internal observer is considered.

The scalar field equation follows from (8.35):

$$\partial_\mu(\sqrt{-g}g^{\mu\nu}\partial_\nu\varphi) + \sqrt{-g}V'(\varphi) = 0. \tag{8.37}$$

The metric is assumed in the form (8.4). Due to spatial homogeneity, we also assume the homogeneity of the scalar field φ, i.e., $\varphi = \varphi(t)$. Equation (8.37) simplifies:

$$\ddot{\varphi} + 3H\dot{\varphi} + V'(\varphi) = 0, \quad H \equiv \dot{a}/a. \tag{8.38}$$

One more equation is obtained if one takes into account that the energy density of the scalar field is $\rho = \frac{1}{2}\dot{\varphi}^2 + V(\varphi)$. Then Eq. (8.8) in the form

$$H^2 = \frac{8\pi G}{3}\left(\frac{1}{2}\dot{\varphi}^2 + V(\varphi)\right), \tag{8.39}$$

is the second equation for the set of dynamic variables $\varphi(t)$, $a(t)$. In this case we can assume that the term $\Lambda/3$ in (8.8) is either zero or is already included in the definition of the potential V.

A key point for the inflationary process is a "slow" evolution of the scalar field φ (the inflaton). In this case, the behavior of the system (8.38), (8.39) takes place effectively in de Sitter space even if $\Lambda = 0$. Indeed, by analogy with ordinary material point mechanics, a slow motion takes place if the term responsible for friction, $3H\dot{\varphi}$, is large enough, i.e.,

$$3H|\dot{\varphi}| \gg |\ddot{\varphi}|. \tag{8.40}$$

This simplifies the set of equations even more. Indeed, using (8.40), we can write Eq. (8.38) as

$$3H\dot{\varphi} + V'(\varphi) \simeq 0. \tag{8.41}$$

Hence we arrive at the inequality

$$V'(\varphi) \sim 3H\dot{\varphi} \gg \ddot{\varphi}.$$

Multiplying the equation by $\dot{\varphi}$ and integrating, we obtain the desired inequality

$$\dot{\varphi}^2 \ll V(\varphi). \tag{8.42}$$

This inequality means that the kinetic energy is small as compared to the potential energy, and the latter therefore changes very little during the inflation $V \simeq \text{const}$, and moreover, due to Eq. (8.39), the Hubble parameter is also almost constant:

$$H \equiv \frac{\dot{a}}{a} \simeq \sqrt{\frac{8\pi G}{3}V(\varphi)}. \tag{8.43}$$

The solution to (8.43), $a(t) \propto \exp(Ht)$, means an (approximately) exponential growth of the scale factor and hence of physical distances, similar to de Sitter space. It is not surprising because an (approximately) constant potential of the inflaton φ can be interpreted as a cosmological constant.

The dynamics of a scalar field coupled to gravity is much more interesting than the dynamics of de Sitter space. In particular, the inequality (8.42) is eventually necessarily violated since the field φ, though slowly, moves towards a minimum of the potential. As soon as the potential becomes sufficiently small, the Hubble parameter H will be small as well. As a result, one can neglect the "friction" in Eq. (8.38), which means that the inflaton begins to move rapidly to the minimum of the potential, and the inflationary stage is over. Let us, however, now follow the processes at the inflationary stage. The dynamics of the inflaton φ is determined by Eq. (8.41) and that of the scale factor a by Eq. (8.43). The latter is solved in a general form:

$$a(t) = a(t_{\text{in}}) \exp\left[\int_{t_{\text{in}}}^{t} H(\varphi)dt\right]. \tag{8.44}$$

If the inflation began at $t = 0$ and has ended by $t = t_e$, then the initial spatial region has grown by a factor of $a(t_e)$. Conventionally, the Hubble parameter value $H_e \sim 10^{13}$ GeV, is used, which is in agreement with the observational data. For the overwhelming majority of estimates, it is sufficient to put $H(\varphi) \approx H_e = \text{const}$. In this case, the scale factor at the inflationary stage looks as follows:

$$a(t) \simeq a(t_{\text{in}}) \exp[H_{\text{end}}(t_{\text{end}} - t_{\text{in}})]. \tag{8.45}$$

Here, H_{end} is the Hubble parameter value at the end of the inflationary period, and t_{end} is the inflation ending time.

It is also useful to introduce the notion of e-folding number, N, defined as

$$N \equiv \ln\left[\frac{a(t)}{a(t_{\text{in}})}\right]. \tag{8.46}$$

In other words, if we know the number N, it means that by the present time the Universe has expanded by a factor of e^N.

If the inflation began when the scalar field had the value φ_{in}, then the number N is calculated as follows:

$$N \equiv \ln\left[\frac{a(t)}{a(t_{\text{in}})}\right] = \int_{t_{\text{in}}}^{t} H dt = \int_{\varphi_{\text{in}}}^{\varphi} H \frac{d\varphi}{\dot{\varphi}} = -\int_{\varphi_{\text{in}}}^{\varphi} \frac{3H(\varphi)^2 d\varphi}{V'(\varphi)}. \tag{8.47}$$

Let us discuss the slow-rolling condition (8.40) in more detail. To determine more explicitly under which conditions it is possible, let us do the following. From Eq. (8.41) written in the form

$$\dot{\varphi} = -\frac{V'(\varphi)}{3H(\varphi)},$$

we express the second-order derivative:

$$\ddot{\varphi} = \frac{V'(\varphi)}{3H^2(\varphi)} H'(\varphi)\dot{\varphi} - \frac{V''(\varphi)}{3H(\varphi)}\dot{\varphi}.$$

For the slow-rolling condition to hold, both terms in the right-hand side should be small as compared to $3H\dot{\varphi}$. Recalling the expression (8.43) for the Hubble parameter, we obtain the following conditions:

$$\varepsilon \equiv \frac{M_{\text{Pl}}^2}{16\pi} \frac{V'(\varphi)^2}{V(\varphi)^2} \ll 1; \quad \eta \equiv \frac{M_{\text{Pl}}^2}{8\pi} \left|\frac{V''(\varphi)}{V(\varphi)}\right| \ll 1. \tag{8.48}$$

The stress-energy tensor (SET) of the scalar field $\varphi(t)$ in the metric (8.4), as follows from the symmetry of the problem, coincides with the SET of a perfect fluid in its comoving RF (see Sec. 5.1), $T_\mu^\nu = \mathrm{diag}(\rho, -p, -p, -p)$, and for the scalar field the energy density ρ and the pressure p are expressed as

$$\rho = \frac{1}{2}\dot{\varphi}^2 + V(\varphi), \tag{8.49}$$

$$p = \frac{1}{2}\dot{\varphi}^2 - V(\varphi). \tag{8.50}$$

In the inflationary period, $\frac{1}{2}\dot{\varphi}^2 \ll V(\varphi)$, and this means an approximately de Sitter space with the corresponding behavior of the scale factor.

8.4 Post-inflationary stages

8.4.1 *Post-inflationary reheating of the Universe*

The stage right after inflation is probably the most difficult for analysis. Even its name (*reheating*) is not quite correct since there has not been any primary heating. It is this stage during which active high-energy particle production and their thermalization occurs, which, in the language of statistical physics, means heating of the plasma just born.

The inflationary period terminates when the "friction" becomes small, i.e., at $H_{\rm end} \lesssim m_{\rm infl}$ ($m_{\rm infl}$ is the inflaton mass). The inflaton field begins to rapidly oscillate near the minimum of the potential according to Eq. (8.38). It is this motion, alternating with time, that creates the particles. The oscillation energy gradually passes over to the energy of particles. Using the expressions for the pressure, $p = \frac{1}{2}\dot{\varphi}^2 - V(\varphi)$, and for the energy density, $\rho = \frac{1}{2}\dot{\varphi}^2 + V(\varphi)$, it is easy to transform this equation to the familiar form

$$\frac{d\rho}{dt} = -3H(t)(p + \rho). \tag{8.51}$$

The additional condition (the equation of state) $\rho + p = \gamma\rho$, holds only approximately, after averaging over the oscillation period. Meanwhile, the parameter γ depends on the model:

$$\gamma \simeq \frac{\int_0^T (\rho + p)dt}{\int_0^T \rho dt} = \frac{\int_0^T \dot{\varphi}^2 dt}{\int_0^T \left[\frac{1}{2}\dot{\varphi}^2 + V(\varphi)\right] dt} = \frac{\int_0^{\varphi_{\max}} \dot{\varphi} d\varphi}{\int_0^{\varphi_{\max}} d\varphi \left[\frac{1}{2}\dot{\varphi}^2 + V(\varphi)\right] / \dot{\varphi}}. \tag{8.52}$$

This expression can be further simplified if one assumes that, during a single oscillation period, the Universe expands very little, which is quite realistic. Then, according to the energy conservation law, we have $\dot{\varphi} = \sqrt{2[V(\varphi_{\max}) - V(\varphi)]}$. Near its minimum, the potential is approximated by the parabola $V(\varphi) \cong \lambda\varphi^2$, Eq. (8.52) is strongly simplified, and the quantity γ at the reheating stage turns out to be equal to unity,

$$\gamma \cong 1. \tag{8.53}$$

Let us determine the time dependence of the scale factor. Its initial value at the reheating stage coincides with that at the end of inflation, $a(t_{\text{in}}) = a_{\text{infl}}(t_{\text{end}}) = H(\varphi_U)^{-1} \, \mathrm{e}^{N_U}$. It is assumed that the visible part of the Universe has formed at N_U e-foldings before the end of inflation. The initial size of the Universe was equal to the inverse Hubble parameter, $H(\varphi_U)^{-1}$, while the inflaton value was φ_U. The energy density at the beginning of reheating was the same as at the end of inflation, $\rho(t_{\text{in}}) = \rho_{\text{infl}}(t_{\text{end}}) = V(\varphi_{\text{end}})$. Then Eq. (8.23) is written as

$$a_{\text{reh}}(t) = \left(\frac{2}{3} H_{\text{end}}\right)^{2/3} H(\varphi_U)^{-1} \, \mathrm{e}^{N_U} \, t^{2/3}. \tag{8.54}$$

Recall that it is usually assumed that $H_{\text{end}} \simeq H(\varphi_U) \simeq 10^{13}$ GeV.

During the rapid oscillations, the inflaton field produces particles, giving them a part of its energy. Suppose that the interaction energy density of the inflaton (φ) with fermions ψ and other scalar particles χ has the usual form,

$$L_{\text{int}} = h\varphi\bar{\psi}\psi + g\varphi\chi^2,$$

where g and h are coupling constants. Then the rate of change of inflaton energy density is

$$\dot{\rho} = \frac{d}{dt}\left[\frac{1}{2}\dot{\varphi}^2 + V(\varphi) + \Delta L\right] = \dot{\varphi}\ddot{\varphi} + \dot{\varphi}V'(\varphi) + \Delta\dot{L},$$

where ΔL is a term responsible for diminishing the inflaton energy density at its decay into other, secondary particles.

Suppose that the back reaction of the radiation on the inflaton dynamics is small [211]. Then, in the first approximation, according to the equations of motion, we obtain

$$\ddot{\varphi} \simeq -3H\dot{\varphi} - V'.$$

Substituting into the previous expression, we find the changing rate of the inflaton energy density

$$\dot{\rho} \simeq -3H(\rho + p) + \Delta\dot{L}.$$

Now the meaning of the last term is clear: it is the changing rate of inflaton energy density due to decay into other particles,

$$\Delta\dot{L} = -\Delta\dot{\rho},$$

where $\Delta\dot{\rho}$ is the secondary particles' energy density changing rate due to the inflaton decay. Expanding it in a Taylor series by the derivative of the field φ, we have

$$\Delta\dot{\rho}(\dot{\varphi}) = \Delta\dot{\rho}(0) + A\dot{\varphi} + \Gamma\dot{\varphi}^2 + \ldots \simeq \Gamma\dot{\varphi}^2.$$

The first term of the expansion is zero because a static field does not radiate, and the second one is zero because at oscillations of the inflaton the mean value of its derivative is (approximately) zero.

Let us also take into account that the pressure and energy density of the oscillating inflaton field satisfy the usual relation $p = w\rho$. Then,

$$\dot{\rho} \simeq -(3H + \Gamma)\dot{\varphi}^2 = -(3H + \Gamma)(p + \rho) \simeq -(3H + \Gamma)(w + 1)\rho.$$

Recalling that $H = \dot{a}/a$, we obtain the time dependence of the inflaton energy density as

$$\rho = \rho_{\text{end}} \left(\frac{a(t)}{a(t_{\text{end}})} \right)^{-3(w+1)} \exp\{-\Gamma(w + 1)(t - t_{\text{end}})\}, \qquad (8.55)$$

where the subscript "end" refers to the inflation ending time. This expression is quite understandable physically. The second factor means that the inflaton energy density is decreasing due to expansion of space, and the third one due to its decay into particles.

What is now happening to the energy density of the created particles of matter? The changing rate of the energy density of relativistic particles ρ_m is written as follows:

$$\begin{aligned} \dot{\rho}_m &= -3H(p_m + \rho_m) + \Gamma\dot{\varphi}^2 \\ &= -3H(w_m + 1)\rho_m + \Gamma(w + 1)\rho \\ &= 4H\rho_m + \Gamma(w + 1)\rho. \end{aligned} \qquad (8.56)$$

Here, it has been taken into account that $w_m = 1/3$ for relativistic particles. It is evident that two terms are competing: the density decrease due to expansion of space (the first term) and its increase due to inflaton decay.

The relativistic particles rapidly thermalize. When the oscillations begin, the energy inflow into the particle plasma is predominant, and the temperature is growing. As time passes, the oscillation intensity decreases, and the space expansion effect that cools the plasma becomes more important. Let us estimate the temperature of the plasma at which its density is the highest. Assuming $\rho_m \sim \rho$, we obtain from the previous equation $H \sim \Gamma$. Since the Hubble parameter and the cosmological time are related by $t \sim 1/H$ (see (8.24)), while the temperature is related to time as $T \sim \sqrt{M_{\text{Pl}}/t}$ (see Sec. 7.4.2 and Eq. (8.63) for more detailed consideration), we arrive at the following estimate:

$$T_{\text{reh}} \sim \sqrt{M_{\text{Pl}}/t} \sim \sqrt{M_{\text{Pl}} H} \sim \sqrt{M_{\text{Pl}} \Gamma}.$$

A large uncertainty in the calculation of the quantity Γ prevents us from determining the heating temperature unambiguously. It is usually supposed that it varies in a wide range between 10^4 and 10^{12} GeV.

8.4.2 *The radiation-dominated stage*

Up to now, we succeeded in expressing the energy density in terms of fields, and this allowed us to solve Eq. (8.41) and find the time dependence of the scale factor. But, by the end of the reheating stage, the inflaton oscillations have decayed, having created high-energy particles. The existence of the particle plasma indicate that it is desirable to introduce the notion of temperature, to use the standard tools of statistical physics and thermodynamics. The notion of temperature is only rigorous if the system is in equilibrium. It our case it is evidently not true because the Universe is expanding.

Nevertheless, the relaxation processes occur so rapidly that at each time instant the state of the Universe is close to equilibrium. To verify that, let us assume that it is the case and check if we come to a contradiction. We introduce the plasma temperature T at the time t. It will be shown below that the cosmological time and the temperature are related by $t \sim M_{\text{Pl}}/T^2$ (see Eq. (8.63)). An equilibrium is established due to particle collisions. The characteristic time of a collision between an electron and a photon, $t_{\gamma e}$, is $t_{\gamma e} \sim 1/(n\sigma v)$, where $n \sim T^3$ is the electron density, $\sigma \sim \alpha^2/T^2$ is the Compton cross-section, and $v \simeq 1$ is the electron velocity. We can use the notion of temperature only if $t_{\gamma e} \ll t$. Substituting the estimates presented in this paragraph, we obtain the applicability condition of the notion of temperature in the expanding Universe in the form $T \ll \alpha^2 M_{\text{Pl}} \sim 10^{17}$ GeV. It is known that the temperature at the end of inflation can hardly exceed 10^9 GeV. Thus an equilibrium in the medium is established very rapidly, at least due to Compton scattering, which justifies introducing the notion of temperature.

To find an explicit form of the scale factor $a_{\mathrm{RD}}(t)$ at the radiation-dominated stage, we use the general relation (8.22) with $\gamma = 4/3$. This value follows from the pressure-density relation $p = \rho/3$ for a medium consisting of relativistic particles and Eq. (8.20).

The initial conditions for this stage are $t_{\mathrm{in}} = t_{\mathrm{reh}}$, $\rho_{\mathrm{in}} = \rho(t_{\mathrm{reh}})$ and $a_{\mathrm{in}} = a(t_{\mathrm{reh}})$, where the subscript "reh" refers to the end of the previous stage, reheating . The final expression for the scale factor is

$$a(t) \equiv a_{\mathrm{RD}}(t) = a_{\mathrm{reh}}(t_{\mathrm{reh}}) \left[\frac{1}{2} H(t_{\mathrm{reh}})\right]^{1/2} (t - t_{\mathrm{reh}})^{1/2}. \tag{8.57}$$

Let us look how the temperature evolves with time. First of all, we recall that the energy density of the ultrarelativistic particles is related to the temperature T by

$$\rho = \frac{\pi^2}{30} g_* T^4, \tag{8.58}$$

where g_* is the number of types of particles, taking into account their statistical weight.

Let us further establish the important relationship between the temperature and the scale factor. On the one hand, at this stage $\gamma = 4/3$, and according to (8.21) we have $\rho \sim a^{-4}$. On the other hand, $\rho \sim T^4$. Therefore,

$$a(t) = \mathrm{const}/T(t). \tag{8.59}$$

The temperature decreases inversely proportionally to the scale factor.

Let us show that the entropy after the reheating stage (when the particles were created and hence the entropy grew) remains constant. Indeed, using the known relationship between the entropy density and the temperature

$$s = \frac{2\pi^2}{45} g_* T^3, \tag{8.60}$$

it is easy to obtain

$$S \sim s a^3(t) = \mathrm{const}. \tag{8.61}$$

Another useful relation, following from (8.59), is also important for making estimates in cosmological models:

$$T(t) = \frac{a(t_{\rm reh})}{a(t)} T_{\rm reh}. \tag{8.62}$$

Lastly, combining Eqs. (8.57)–(8.59), it is easy to obtain the explicit time dependence for the temperature of the medium filling the Universe:

$$T = \left(\frac{45}{32\pi^2 g_*}\right)^{1/4} \sqrt{\frac{M_{\rm Pl}}{t}}. \tag{8.63}$$

Let us stress that this relation is valid in the radiation-dominated stage only.

8.4.3 *The matter-dominated stage*

As the Universe expands, the medium temperature falls, massive particles become nonrelativistic, and the photon wavelengths grow. There comes a moment when the particles' rest energy becomes equal to their kinetic energy. Beginning with this moment, there begins the matter-dominated stage. From that time on, the pressure p becomes so small that it can be put equal to zero, and accordingly the parameter γ is equal to 1 in Eq. (8.20). By full analogy with the previous stages, the scale factor $a_{\rm MD}(t)$ can be found from Eq. (8.22) with $\gamma = 1$. The initial conditions for this stage are

$$t_{\rm in} = t_{\rm RD}, \quad \rho_{\rm in} = \rho(t_{\rm RD}) = \frac{\pi^2}{30} g_* T_{\rm RD}^4, \quad a_{\rm in} = a_{\rm RD}(t_{\rm RD}),$$

where $t_{\rm RD}$ corresponds to the end of the radiation-dominated stage.

It is of interest that Eq. (8.59) remains valid at this stage as well, with one reservation: now T is the photon temperature. The latter are called relic photons.

Before beginning to prove (8.59), let us discuss one more period of utmost importance in the life of the Universe, the so-called recombination period. The point is that at high temperatures electrons and protons cannot form hydrogen atoms because of their small binding energy. The Universe is filled with a plasma of charged particles which actively interact with photons. It means, by the way, that the emitted photons performed something like Brownian motion. The Universe was not transparent to electromagnetic radiation. But at a certain moment, when the temperature became low enough, hydrogen atoms began to form. Being neutral, they weakly interacted with photons allowing them to propagate to large distances. The time interval during which the atoms recombined and the medium was becoming transparent is called the recombination period. All the existing data say that the two periods, that of recombination and that of equality between the energy densities of matter and radiation, roughly coincided in time. In what follows, we will neglect their difference, which is small from the cosmological viewpoint.

Thus we suppose that the photons do not interact with matter after reaching the temperature $T_{\rm rec}$. Before this time, the photon distribution by energies was a Bose–Einstein distribution, which, by the recombination time, had the form

$$dN(t_{\rm rec}) = V_{\rm rec}\frac{E_{\rm rec}^2}{\pi^2}\frac{dE_{\rm rec}}{\exp(E_{\rm rec}/T_{\rm rec})-1}. \tag{8.64}$$

Here, dN is the number of photons with energies between $E_{\rm rec}$ and $E_{\rm rec}+dE_{\rm rec}$ inside the volume $V_{\rm rec}$. Since after recombination the photons' interaction with the ambient medium is regarded to be small, the photon energy decreases only due to the common expansion of space,

$$E(t) = p(t) = \frac{2\pi}{\lambda(t)} = \frac{2\pi}{\dfrac{a(t)}{a(t_{\rm rec})}\lambda(t_{\rm rec})} = \frac{a(t_{\rm rec})}{a(t)}E_{\rm rec}. \tag{8.65}$$

The number of photons is a conserved quantity, so that

$$dN(t) = dN(t_{\rm rec}),$$

while the volume V containing these particles grows by the law

$$V(t) = \left(\frac{a(t)}{a(t_{\rm rec})}\right)^3 V_{\rm rec}.$$

As a result, we obtain the photon distribution at arbitrary time t:

$$dN(t) = V(t)\frac{E(t)^2}{\pi^2}\frac{dE}{\exp\left[E(t)\dfrac{a(t)}{a(t_{\text{rec}})}\Big/T_{\text{rec}}\right]-1}. \tag{8.66}$$

The distribution preserves its Bose–Einstein form, and

$$T(t) = \frac{a(t_{\text{rec}})}{a(t)}T_{\text{rec}}. \tag{8.67}$$

Our statement has been proved: the law (8.59) really holds at both the radiation-dominated stage and the matter-dominated stage. The unknown constant in (8.59) can be determined from normalization for the present time: const $= a_0 T_0$, or for the recombination time, as has been done before.

8.4.4 *The modern stage of accelerated expansion (secondary inflation)*

More and more confidently, the observational data show that the density of dark energy, a substance uniformly distributed in space, contributes to the modern total energy density at about 70%. The most natural assumption is that this quantity is time-independent and is related to the cosmological constant Λ. This means in turn that in the present epoch the Universe is approximately described by a de Sitter model but with the energy density much smaller than in the inflationary period. More than that, with further expansion of space, the matter density decreases while the energy density related to the cosmological constant remains constant. Consequently, the Hubble parameter also tends to a constant value,[2]

$$H_\Lambda \to \sqrt{\frac{8\pi}{3}\frac{\rho_\Lambda}{M_{\text{Pl}}^2}}, \tag{8.68}$$

[2]The cosmological constant corresponds to a density to pressure ratio w equal to negative unity. The modern cosmological observations admit a certain interval of w values including -1. Let us present, as examples, two recent estimates: $w = -1.10 \pm 0.14(1\sigma)$ [343] (according to the 7-year WMAP data), and $w = -1.069^{+0.091}_{-0.092}$[520] (mainly from data on type Ia supernovae from the SNLS3 sample). Noteworthy, the central values of w in these estimates belong to the phantom range, $w < -1$.

and the scale factor to the function

$$a(t) = H_\Lambda^{-1}\, e^{H_\Lambda (t-t_0)}. \tag{8.69}$$

The Hubble parameter characterizes the rate of expansion, and to describe the change of this rate, i.e., the acceleration, one more dimensionless parameter is used, the deceleration parameter $q(t) := -a\ddot{a}/\dot{a}^2$. The parameter q was introduced when it was believed that the Universe is expanding with deceleration. Accelerated expansion corresponds to values $q < 0$.

The modern observational estimates of these parameters are different as given by different authors but are approximately in the following limits:

$$H_0 \approx 0.69\ \frac{\text{km}}{\text{c} \cdot \text{Mpc}}, \qquad q_0 \approx -1 \pm 0.4, \tag{8.70}$$

where the subscript "0" corresponds to the modern epoch. In the case (8.69) we obtain $q(t) = q_0 = -1$.

The inverse Hubble parameter characterizes the size of a causally connected region. An estimate on the basis of the observational data gives

$$H^{-1} \approx 10^{28}\,\text{cm} \approx 10^{62} M_{\text{Pl}}^{-1}.$$

It is this quantity that gives the size of the visible part of the Universe. If the Hubble parameter is really constant, then information on more remote parts of the Universe will never be available to us.

When using relations like (8.22) (such as Eqs. (8.77), (8.78), (8.57), (8.80) given below) for obtaining numerical estimates, there emerges a question connected with the particular value of the scale factor a_{in} at the beginning of each stage. Besides, the value of the coordinate distance r is not evident. Therefore a problem appears when calculating, e.g., the scale factor $a(t_0)$ at a given time instant t_0 and consequently the physical distance between two objects, equal to

$$R_0 = a(t_0) r. \tag{8.71}$$

It is this quantity that is measured by observers. The problem can be circumvented if one considers the expression (8.71) as a way to get rid of the comoving coordinate r by normalizing the scale factor to its present value $a_0 = a(t_0)$. Then the distance between two point objects at arbitrary time is expressed as

$$R(t) = \frac{a(t)}{a_0} R_0.$$

For example, the distance during the matter-dominated stage is

$$R(t) = \left(\frac{t - t_{\rm RD}}{t_0 - t_{\rm RD}} \right)^{2/3} R_0 \simeq \left(\frac{t}{t_0} \right)^{2/3} R_0.$$

The lifetime of the Universe t_0 and its size a_0 are already well defined.

The dimensionless scale factor is also frequently introduced:

$$\mathrm{a}(t) \equiv \frac{a(t)}{a_0}, \quad 0 < \mathrm{a}(t) \leq 1.$$

8.4.5 *Future of the Universe — is a Big Rip expected?*

The description of dark energy with the aid of the cosmological constant in the Einstein equations well agrees with observations in the modern epoch. However, the same observations leave quite a wide range of admissible values of the parameter w if one describes dark energy as a perfect fluid with the equation of state $p = w\rho$. For such a description, in general, the parameter w is time-dependent, while for $w = $ const the conservation law leads to Eq. (8.21), $\rho \sim a^{-3(w+1)}$.

It is easy to verify that an accelerated expansion, $q < 0$, requires a negative pressure, $w < -1/3$. Then, for a large scale factor a, the term with k in Eq. (8.82) is small relative to the other terms, so that *the late-time behavior of the scale factor does not depend on the spatial curvature*, and the dependence $a(t)$ can be described as follows:

(a) For $-1/3 > w > -1$, so that $\rho > |p|$ (the dominant energy condition holds), the so-called power-law inflation occurs:

$$a \sim t^{2/[3(w+1)]}, \quad q = -1 + \frac{3}{2}(w+1) > -1; \qquad (8.72)$$

(b) at $w = -1$, which corresponds to the cosmological constant $\Lambda > 0$, $\rho = \text{const} > 0$, exponential inflation is obtained:

$$a \sim e^{Ht}, \quad H = \text{const}, \quad q = -1; \tag{8.73}$$

(c) for $w < -1$ with the so-called phantom matter, hyperinflation takes place, ending with a singularity due to growth of the scale factor:

$$a \sim (t_* - t)^{-2/[3|w+1|]}, \quad q = -1 - \frac{3}{2}|w+1| < -1, \tag{8.74}$$

where t_* is the time of a singularity called the Big Rip.

As is clear from the estimates (8.70), from the observational viewpoint, all three variants are admissible. However, since the acceleration not only exists but increases, there is some preference for (b), and many papers discuss cosmological models where the dark energy "overcomes the phantom barrier", i.e., at some instant ($3{-}4{\cdot}10^9$ years ago) passes over from $w > -1$ to $w < -1$.

In case (c) matter behaves quite exotically: unlike usual matter, its density does not decrease as the volume grows, but even grows, and at the singular time both a and ρ blow up simultaneously.

The catastrophic growth of $a(t)$ makes all distances grow, beginning with extragalactic ones, then also interstellar, interplanetary, and finally intra-atomic ones, i.e., right before the singularity all kinds of matter should decay and even all composite particles.

Is such a sad fate of our Universe inevitable if a modern value $w < -1$ is after all confirmed? Fortunately, the answer is negative.

Indeed, if $w = p/\rho$ is time-dependent, one can assume that the dark energy is represented by a scalar field ϕ with the Lagrangian

$$L_s = \frac{1}{2}\varepsilon g^{\mu\nu}\phi_{,\mu}\phi_{,\nu} - V(\phi), \tag{8.75}$$

where $\varepsilon = \pm 1$ and $V(\phi)$ is the self-interaction potential of the field ϕ. These are the same normal and phantom scalar fields that we met in the previous chapters. At $\phi = \phi(t)$,

$$\rho = \frac{1}{2}\varepsilon\dot{\phi}^2 + V, \quad p = \frac{1}{2}\varepsilon\dot{\phi}^2 - V, \quad w = \frac{p}{\rho} = -1 + \frac{2\varepsilon\dot{\phi}^2}{2V + \varepsilon\dot{\phi}^2}. \tag{8.76}$$

Thus, a normal scalar field ($\varepsilon = +1$) with a positive potential V gives $w > -1$ while a phantom scalar field ($\varepsilon = -1$) with $V > 0$ leads to $w < -1$. However, if at large t the scalar field tends to an extremum of the potential $V_{\rm ext} > 0$ sufficiently rapidly, then $\dot{\phi} \to 0$ and $w \to -1$ as $t \to \infty$; $V_{\rm ext}$ behaves as an effective cosmological constant, and accordingly we obtain a de Sitter asymptotic (8.73) for the Universe evolution.

It is well known that a normal scalar field tends to a minimum of its potential ("rolls down" along the curve $V(\phi)$), whereas a phantom one, on the contrary, tends to a maximum of V ("climbs up" along the curve $V(\phi)$). If $V(\phi)$ has a maximum, then a phantom scalar field will tend to it in the course of the cosmological evolution. For this case, an exact result is known [236]: *if the potential $V(\phi)$ is bounded above, then its maximum represents a global attractor for cosmological solutions to the gravitational and scalar field equations.*

8.5 The scale factor in the general case

It is useful to put together the basic expressions for the scale factor at different stages. Since each stage is much longer than the previous ones, i.e., $t \gg t_{\rm in}$, we use the more compact but approximate equation (8.23).

The inflationary stage:

$$a(t) \equiv a_I(t) = H_U^{-1} \exp\left(\int H\,dt\right) \approx H_e^{-1}\,{\rm e}^{N_U t}. \tag{8.77}$$

The numerical values $N_U \approx 60$, $H_e \approx 10^{13}\,{\rm GeV}$ agree with observations for the majority of models.

The reheating stage:

$$a(t) \equiv a_{\rm reh}(t) = a_I(t_e)\left(\frac{2}{3}H_e\right)^{2/3}(t - t_e)^{2/3}. \tag{8.78}$$

The radiation-dominated stage:

$$a(t) \equiv a_{\mathrm{RD}}(t) = a_{\mathrm{reh}}(t_{\mathrm{reh}}) \left[\frac{1}{2} H(t_{\mathrm{reh}})\right]^{1/2} \quad (t - t_{\mathrm{reh}})^{1/2}. \tag{8.79}$$

The matter-dominated stage:

$$a(t) \equiv a_{\mathrm{MD}}(t) = a_{\mathrm{RD}}(t_{\mathrm{RD}}) \left[\frac{2}{3} H(t_{\mathrm{RD}})\right]^{2/3} \quad (t - t_{\mathrm{RD}})^{2/3}. \tag{8.80}$$

The modern accelerated expansion (secondary inflation), assuming that the dark energy is described by the cosmological constant Λ in the Einstein equations:

$$a(t) = H_\Lambda^{-1}\, \mathrm{e}^{H_\Lambda (t - t_0)}, \quad H_\Lambda = \sqrt{\Lambda/3}. \tag{8.81}$$

All these expressions are approximate because they take into account only one predominant source of the gravitational field at each stage.

The real content of the Universe is a mixture of radiation, matter and dark energy which is rather probably described by the cosmological constant. Let us obtain a general expression for the scale factor, using Eq. (8.8):

$$\frac{\dot{a}^2}{a^2} + \frac{k}{a^2} = \frac{8\pi G}{3}\rho + \frac{\Lambda}{3}. \tag{8.82}$$

The contribution from the dark energy density ρ_Λ is written separately as the Λ-term, $\Lambda/3 = (8\pi G/3)\rho_\Lambda$. The remaining part of the energy density consists of the matter density ρ_m and the radiation density ρ_r,

$$\rho = \rho_m + \rho_r. \tag{8.83}$$

Their dependence on the scale factor has already been discussed,

$$\rho_m(t) = \rho_m(t_0)/a^3(t), \quad \rho_r(t) = \rho_r(t_0)/a^4(t).$$

Substituting these quantities into (8.82), we obtain the equation for the scale factor

$$\frac{\dot{a}^2}{a^2} = \frac{8\pi G}{3}\left[\frac{\rho_m(t_0)}{a^3(t)} + \frac{\rho_r(t_0)}{a^4(t)} + \rho_\Lambda\right] - \frac{k}{a^2}. \tag{8.84}$$

Let us choose the additional condition in the form $a(t_0) = 1$. Numerical solution of this equation does not represent any problem. Let us transform it, introducing some definitions often used in practice.

First of all, we define the critical density at the modern epoch:

$$\rho_c \equiv \frac{3H_0^2}{8\pi G}, \tag{8.85}$$

where H_0 is the modern value of the Hubble parameter. At such a value of the energy density the Universe should be flat ($k = 0$) provided $\Lambda = 0$. In general, the mean density of the Universe $\rho \neq \rho_c$, therefore it is convenient to introduce the dimensionless density parameter

$$\Omega_{\text{tot}} = \rho/\rho_c. \tag{8.86}$$

Let us also introduce the partial density parameters Ω for the present time. Thus, for matter, radiation and the cosmological constant we have

$$\Omega_m = \frac{\rho_m(t_0)}{\rho_c}, \quad \Omega_r = \frac{\rho_r(t_0)}{\rho_c}, \quad \Omega_\Lambda = \frac{\rho_\Lambda(t_0)}{\rho_c}.$$

An expression for Ω_k, the density parameter related to the spatial curvature is constructed in a different way:

$$\Omega_k = -\frac{k}{H_0{}^2}.$$

Now Eq. (8.84) looks as follows:

$$\frac{\dot{a}^2(t)}{a^2(t)} \equiv H^2(t) = H_0^2[\Omega_k a^{-2}(t) + \Omega_m a^{-3}(t) + \Omega_r a^{-4}(t) + \Omega_\Lambda], \tag{8.87}$$

where $H_0 = H(t_0)$. Choosing $t = t_0$ in Eq. (8.87), we obtain the sum rule

$$\Omega_r + \Omega_m + \Omega_\Lambda + \Omega_k = 1.$$

Lastly, let us present the time dependence of the horizon size (in the present case, it means the distance to which a light signal can get for the time t):

$$R_{\rm hor}(t) = a(t) \int_0^t \frac{dt'}{a(t')}.$$

At the inflationary stage, $a(t) = \text{const} \cdot e^{Ht}$, where H is the Hubble parameter expressed in terms of energy density and assuming a constant value with good accuracy. The horizon size is $R_{\rm hor} = H^{-1}(1 - e^{-Ht})$, and at large times it tends to the inverse Hubble parameter, i.e., also a constant.

At other stages, we have the approximation $a(t) = \text{const} \cdot t^{\beta}$, where the constant $\beta < 1$. Then the Hubble parameter has the form $H(t) = \beta t^{-1}$, and the horizon size is $R_{\rm hor} = t/(1-\beta)$. Thus $H^{-1} = R_{\rm hor}(1-\beta)/\beta$, i.e., the inverse Hubble parameter coincides with the horizon size by order of magnitude.

8.6 Why do we need an inflationary period?

In scenarios of the Universe's birth studied before the inflationary idea was put forward, the first stage was that of a "hot" radiation-dominated Universe. All of them face some unsolvable problems. Let us enumerate the most significant ones.

1. The initial singularity. A study of the classical equations of GR applied to the Universe as a whole leads to the conclusion that its initial state was characterized by an infinite density. The inflationary idea does not completely solve this problem though indicates a way of its solution. Very probably, our Universe has originated from quantum fluctuations where the classical equations, and, the more so, the conclusions that rest on them, were unapplicable.
2. Our space is described with high accuracy by Euclidean geometry. Its causes are not evident.
3. Our Universe contains about 10^{22} stars. How could an object of such mass and volume emerge? Why does it exist for such a long time?
4. The cosmic microwave background (CMB) is highly isotropic. Why are its energy density deflections so small?

5. How did the energy density inhomogeneities that have led to galaxy formation appear?
6. Why does matter dominate in our Universe while the laws of Nature are almost symmetric with respect to changing particles to antiparticles?
7. At the early stage, at large energy densities, topologically stable states like magnetic monopoles should form. Their number and mass are large enough to be noticeable [577]. But such objects have not yet been discovered.

Let us discuss the main problems. The first of them is the Universe flatness problem.

8.6.1 *The flatness problem*

Let us recall the notion of the critical density defined as the density of matter such that the curvature parameter is $k = 0$ at $\Lambda = 0$. From Eq. (8.8) it directly follows

$$\rho_{\text{crit}}(t) \equiv \frac{3}{8\pi} M_{\text{Pl}}^2 H^2(t). \tag{8.88}$$

(Keep in mind that $G = 1/M_{\text{Pl}}^2$ and $\dot{a}/a = H$.) Since the constant Λ is small at least at the radiation-dominated stage, let us neglect it in Eq. (8.8), which results in

$$\left| \frac{\rho_{\text{crit}}(t) - \rho(t)}{\rho_{\text{crit}}(t)} \right| = \dot{a}^{-2}(t), \tag{8.89}$$

for $k = \pm 1$. The modern value of the energy density is very close to the critical value, [510], so that the inequality

$$\left| \frac{\rho_{\text{crit}}(t_0) - \rho(t_0)}{\rho_{\text{crit}}(t_0)} \right| = \dot{a}^{-2}(t_0) < 1, \tag{8.90}$$

is established confidently. From Eqs. (8.89) and (8.90) we obtain

$$\left| \frac{\rho_{\text{crit}}(t) - \rho(t)}{\rho_{\text{crit}}(t)} \right| = \left| \frac{\rho_{\text{crit}}(t_0) - \rho(t_0)}{\rho_{\text{crit}}(t_0)} \right| \frac{\dot{a}^2(t_0)}{\dot{a}^2(t)} < \frac{\dot{a}^2(t_0)}{\dot{a}^2(t)}.$$

In the absence of inflation, there are only two stages, the radiation- and matter dominated ones. The time dependence of the scale factor at both stages can be presented in the form

$$a(t) = \mathrm{const} \cdot t^n, \quad n = 1/2 \quad \text{or} \quad 2/3 < 1. \tag{8.91}$$

Consequently,

$$\left| \frac{\rho_{\mathrm{crit}}(t) - \rho(t)}{\rho_{\mathrm{crit}}(t)} \right| < \frac{\dot{a}^2(t_0)}{\dot{a}^2(t)} \sim \left(\frac{t}{t_0} \right)^{2(1-n)} \ll 1. \tag{8.92}$$

Then, if the dependence (8.91) is correct, at birth of the Universe, i.e., at $t \ll t_0$, the matter density must be equal to the critical one with inexplicable accuracy. The only way out from this situation is a violation of the law (8.91), at least at the initial stage. It is this way out that is provided by inflation.

Let the Universe pass an inflationary stage with the scale factor (8.44) $a(t) = H^{-1}\cosh(Ht)$. The Hubble parameter at inflation is almost constant, $H \simeq \mathrm{const}$. Then the time dependence of the ratio (8.89) is

$$\left| \frac{\rho_{\mathrm{crit}}(t) - \rho(t)}{\rho_{\mathrm{crit}}(t)} \right| = \sinh^{-2}(Ht).$$

This function exponentially decreases at the inflationary stage, and this explains its smallness in the post-inflationary period, see (8.92).

8.6.2 *The initial size of the Universe*

If now the Universe has the size $L_U \sim 10^{28}$ cm, at any t its size is

$$L(t) = \frac{a(t)}{a(t_0)} L_U = \frac{T_0}{T} L_U. \tag{8.93}$$

Let us estimate the smallest size of the Universe. From Eq. (8.93) it follows that it took place at the greatest temperature. The classical description is valid at energies, hence temperatures smaller than the Planckian energy, $T \leq M_{\mathrm{Pl}} \sim 10^{32}$ K. Substituting this value into (8.93) and taking into account the modern CMB temperature

$T_0 \sim 2.7\,\text{K}$, we obtain the minimum size of the Universe at birth:

$$L_{\text{min}} \sim 10^{-4}\,\text{cm}.$$

It remains an enigma why was the Universe born with such a large size as compared to the Planckian one.

The inflationary mechanism just allows the Universe to rapidly expand to the above size.

Moreover, if the Universe was born with the size $\sim 10^{-4}$ cm at the Planckian energy scale, then, at the same moment, the size of a causally connected region is about 10^{-33} cm. It means that the Universe consisted of $\sim 10^{87}$ causally disconnected regions. It is absolutely unclear why the energy density is the same in all these regions up to about 10^{-5}. It is the essence of the *horizon problem.*

The assumption of the existence of an inflationary period solves this and other problems, such as the *monopole problem* [577].

8.6.3 *Causal connections at inflation and after it*

The inflationary stage has a feature of great interest, distinguishing it from all other stages. Since an understanding of this feature is necessary for understanding the majority of subsequent processes, let us dwell upon it in more detail.

Consider two spatial points at rest in the cosmological RF. Suppose that at the initial moment t_{in} the distance between them is equal to $l(t_{\text{in}})$, and both are located in the same causally connected region, "under the horizon". Since the size of a causally connected region is determined by the inverse of the Hubble parameter H, at the first moment there holds the inequality

$$l(t_{\text{in}}) < H^{-1}(t_{\text{in}}).$$

Further on, the distance between the particles grow along with the scale factor $a(t)$

$$l(t) = a(t)r, \tag{8.94}$$

where the coordinate distance r does not change with time, and the Hubble parameter is by definition $H(t) \equiv \dot{a}(t)/a(t)$. Using these relations, it is easy to find the time dependence of the ratio $l(t)/H(t)^{-1}$.

This ratio of the distance between the particles and the horizon size as a function of time is the subject of our interest:

$$\frac{l(t)}{H^{-1}(t)} = \dot{a}(t)r. \tag{8.95}$$

In the inflationary period, the scale factor is $a(t) \propto e^{H_0 t}$, while the Hubble parameter is approximately constant, $H(t) \simeq H_0 = \text{const}$. Equation (8.95) acquires the form

$$\left(\frac{l(t)}{H(t)^{-1}}\right)_{\text{infl}} \propto e^{H_0 t},$$

so that the distance between two particles grows exponentially as compared to the horizon size. Even if these particle were born in a causally connected region, there will be an instant t_1 when they will turn out to be in causally disconnected regions, for which $l(t_1) > H^{-1}(t_1)$. It is said that at this moment the distance between the particles "has crossed the horizon".

Thus, the previously close particles turn out to be far from each other, in causally disconnected regions, at the expense of the properties of space during inflation.

What happens when the inflation is over? The scale factor behaves as

$$a(t) \propto t^{\beta},$$

where the parameter β is in the interval $0 < \beta < 1$ at any post-inflationary stage [see (8.57), (8.80)]. Let us calculate the ratio of interest using Eq. (8.95):

$$\left(\frac{l(t)}{H(t)^{-1}}\right)_{\text{power}} \propto t^{\beta-1}, \quad 0 < \beta < 1.$$

We see that the horizon size $H(t)^{-1}$ grows more rapidly than the distance between the two particles. At some time t_2 the horizon size becomes equal to the distance between these particles and later exceeds it. The particles are then again in a single causally connected region. All the above-said certainly refers not only to distances between particles but also to any other phenomena which do not strongly affect the metric, e.g., field fluctuations.

8.7 Basic properties of expanding space

8.7.1 *The redshift*

Here and henceforth the scale factor is normalized so that $a(t_0) = 1$.

Let a source emit a light signal with the wavelength $\lambda(t)$ at a time t, which is received on Earth with the wavelength λ. At emission, the scale factor is equal to $a(t)$, and at the moment of absorption it is naturally equal to unity.

The redshift is defined as

$$z = \frac{\Delta\lambda}{\lambda} = \frac{\lambda - \lambda(t)}{\lambda}. \tag{8.96}$$

Evidently,

$$z = \frac{1 - a(t)}{a(t)} \quad \Rightarrow \quad a(t) = \frac{1}{1+z}. \tag{8.97}$$

Let us suppose that the Universe is flat, $k = 0$. The coordinate distance $r(t)$ and the physical distance $R(t)$ to the source are

$$r(t) = \int_t^{t_0} \frac{dt'}{a(t')}, \tag{8.98}$$

$$R(t) = a(t)r(t) = \frac{1}{1+z} r(t). \tag{8.99}$$

We are interested in the physical distance (8.99). To find it, we will need an expression for $a(t)$ or the related $H(z)$. In more detail,

$$R(t) = a(t)\int_t^{t_0} \frac{dt'}{a(t')} = \frac{1}{1+z(t)}\int_t^{t_0} [1 + z(t')]dt'.$$

The substitution $t' \to z$ leads to the expression

$$dz = -\frac{\dot{a}}{a(t')^2}dt' = -\frac{H(t')}{a(t')}dt' = -H(z)[1 + z(t')]dt'.$$

The dependence $H(z)$ is known: according to (8.87),

$$H(z) = H_0\sqrt{\Omega_r(1+z)^4 + \Omega_m(1+z)^3 + \Omega_\Lambda}.$$

The quantity $\Omega_k = 0$ since we are considering a flat Universe with $k = 0$. Finally, the distance to the object that has emitted the light signal is

$$R(t) = \frac{1}{1+z}\int_0^z \frac{dz'}{H(z')}.$$

And the redshift z is determined from observations by the frequency shift between the emitted and absorbed signals according to (8.96).

Thus, in the expanding Universe, we can find the distance to the emitting object. We can also find the velocity of its separation from the observer due to cosmological expansion:

$$v(t) = \dot{R}(t) = \dot{a}(t)r(t) = H(z)R(t) = \frac{H(z)}{1+z}\int_0^z \frac{dz}{H(z)}.$$

The emission time t is also unambiguously related to the redshift z. This relationship can be easily found for any of the main periods using the expressions (8.78), (8.57), (8.80) and the relation (8.97).

Problem. Observers have detected radiation from a quasar with $z = 6.41$. Find: when and at what distance from the Earth was this light emitted? What is the speed of the quasar measured from the Earth?

Let us note that, apart from the cosmological redshift, another, gravitational redshift is also well known, and it is one of the classical observable effects of relativistic gravity; it is closely related to the time slowing-down near gravitating masses. It has no direct relationship with cosmology but must be taken into account for interpretation of astronomical observations.

8.7.2 *The luminosity distance*

An important role in observations is played by the notion of luminosity distance d_L. Let an object emit an energy E per unit time in the form of photons or gravitational waves. An observer on Earth detects the energy flux F. Since in the expanding Universe the wavelength is changing as the signal moves from the source to the observer, the seeming luminosity of the object is $L = E/a(t)$. It is simply the effect of photon energy decrease due to increased wavelength. It is natural to define the luminosity distance through the relation

$$F = \frac{L}{4\pi d_L^2}.$$

Then the relationship between the energy emitted by the source and that absorbed by the receiver on Earth at $t = t_0$ is

$$F dt_0 \cdot 4\pi r^2 = E\, dt = L\, a(t)\, dt. \tag{8.100}$$

Let us prove the relationship

$$\frac{dt}{dt_0} = a(t). \tag{8.101}$$

For a signal emitted at t and received at present, t_0, by an observer at the coordinate distance r, we have

$$\int_t^{t_0} \frac{dt'}{a(t')} = r.$$

The same after the time interval dt:

$$\int_{t+dt}^{t_0+dt_0} \frac{dt'}{a(t')} = r.$$

Comparing the two expressions for r, we obtain (8.101) and hence

$$F = \frac{La^2(t)}{4\pi r^2}.$$

Returning to (8.100) and (8.101), we obtain

$$d_L = \sqrt{\frac{L}{4\pi F}} = \sqrt{\frac{L}{4\pi La^2(t)/(4\pi r^2)}} = \frac{r}{a(t)} = \frac{R_{\text{phys}}}{a^2(t)},$$

and finally we obtain a relationship between the luminosity distance and the redshift in the form

$$d_L = \frac{R_{\text{phys}}}{(1+z)^2}. \tag{8.102}$$

8.7.3 *The velocity of particles in FRW space–time*

The properties of matter, its density and its equation of state, affect the global properties of the space itself and, in particular, its expansion dynamics, i.e., the time dependence of its scale factor. But the opposite is also true: it turns out that the expansion of space diminishes the velocities of particles. Let us prove it, for example, for massive nonrelativistic particles.

Let us first prove the Hubble law for arbitrary times and distances. Let there be two observers located at a coordinate distance r from each other, both at rest in the comoving RF, i.e., $r = \text{const}$. The physical distance varies in time, $R(t) = a(t)r$. Thus the physical velocity of one observer with respect to the other is also variable. Indeed,

$$v = \dot{R} = a(t)r = H(t)a(t)r = H(t)R. \tag{8.103}$$

Evidently, the Hubble parameter $H(t)$ can be considered to be approximately constant only at the present epoch.

Consider the motion of a particle from one observer to the other, located at the coordinate distance dr, under the same conditions, so that $dr = \text{const}$. Physically, according to the Hubble law, the second observer moves away from the first one with the velocity

$$dv_{\text{obs}} = H(t)dR,$$

where dR is the physical distance between them. A particle having the velocity $V(t)$ with respect to the first observer and passing near him at the time $t = t_0$ will reach the second observer in the time

$$dt = dR/V(t).$$

If the space were not expanding, the particle velocity $V(t)$ would be constant. In the expanding Universe it is not the case. In the time interval dt, the particle velocity is measured by the second observer who moves relative to the first one with the velocity dv_{obs}. He will find that the velocity of the particle passing by him is

$$V(t + dt) = V(t) - dv_{\text{obs}} = V(t) - H(t)dR.$$

This equality shows that

$$\frac{dV}{dt} = -H(t)\frac{dR}{dt} = -H(t)V = -\frac{V}{a}\frac{da}{dt}.$$

The last equality is simply substituting the definition of the Hubble parameter. Finally, we have the following trivial equation:

$$dV/V = -da/a,$$

whose solution is

$$V(t) = \text{const}/a(t). \tag{8.104}$$

The constant is, as usual, determined by the initial conditions.

Thus in the expanding Universe the velocity of a free particle decreases with time. The Newton laws hold only in Minkowski space in which $a(t) = 1$. Moreover, unlike both Newtonian mechanics and special relativity, there is a distinguished RF in nature: it is the one in which the CMB is isotropic.

Chapter 9

Field dynamics in the inflationary period

The inflationary paradigm has been developed for about 40 years, beginning with the papers [276, 513, 514]. It has allowed for successfully solving the basic cosmological problems, beginning with the earliest stages at which our Universe was formed and ending with the galaxy formation stage. The existence of an inflationary period in the history of our Universe seems inevitable since it explains a great number of observational facts [324, 377]. The first inflationary mechanisms [276, 514] were based on self-consistent equations of the scalar and gravitational fields. They solved in the most economic manner the key problems of the Big Bang theory: the horizon problem, the flatness problem, etc. The key point is that, under certain conditions, the interacting gravitational and scalar fields lead to an exponential growth of initially small spatial regions. The scalar "inflaton" field ultimately decays into fermions, thus heating the Universe [380].

Nevertheless, the simplest inflationary models usually require substantial improvement. For instance, their predictions concerning the CMB temperature fluctuations agree with reality only with unnaturally small parameters of the potential for the inflaton field which is responsible for the very opportunity of the inflationary process. On the other hand, its interaction with the matter fields must be sufficient for creating, in the post-inflationary period, the observed baryon and lepton numbers.

A further development of the theory has led to emergence of a large number of inflationary models containing additional fields, such as, for example, hybrid inflation [379] and inflation with pseudo-Nambu–Goldstone field [209]. Most of these models are based on interactions between the classical components of different fields. Meanwhile, their nature contains numerous kinds of fields whose interactions must lead to new phenomena during inflation. Thus, for instance, dissipation effects lower the probabilities of phase transitions. Some inflationary models include as one of their basic elements an interaction between the classical inflaton field with various sorts of particles it has created. This effect is a basis for the model of "warm inflation"; back reaction of the created particles on the dynamics of the inflaton field has been discussed in [211]. At present, the best agreement with the observations is provided by the Starobinsky model based on a nonlinear extension of GR.

9.1 Quadratic inflation

Consider inflation with a quadratic potential as one of the most promising models. The general equations have been presented in the previous chapter. It is well known that the inflaton field, as a number of other scalar fields, emerges in multidimensional theories in a natural way. The shape of the potential can be quite complicated and strongly depends on the theory chosen. However, our interest is in the shape of the potential near its minimum since our Universe was formed at the last stage of inflation, in the process of its completion. But near a minimum, any reasonable potential can be expanded in a Taylor series and approximated by its quadratic term. Thus we choose as an example the simplest, quadratic potential

$$V(\varphi) = \frac{1}{2}m^2\varphi^2. \tag{9.1}$$

The quadratic inflation is not confirmed by observations. However, it is worth being discussed for educational purposes. The general equations of Sec. 8.3 can be given in a specific form. In particular, the set of equations (8.38) and (8.39) is solved analytically. The solution has the form

$$\varphi(t) = \varphi_{\rm in} - \frac{mM_{\rm Pl}}{\sqrt{12\pi}}t. \tag{9.2}$$

Recall that this solution is valid as long as the inequality (8.40) holds. For the quadratic potential it reduces to the inequality

$$\varphi \gg \frac{M_{\rm Pl}}{2\sqrt{3\pi}}. \tag{9.3}$$

It is useful to know the field value at which the slow-rolling conditions are violated, i.e., when $\epsilon \simeq 1$, $\eta \simeq 1$, and therefore inflation terminates. It can be easily found that for the quadratic potential we are considering this happens at

$$\varphi_{\rm end} \approx \frac{1}{2\sqrt{\pi}}M_{\rm Pl}. \tag{9.4}$$

Thus, the φ field (the inflaton) should be at the Planck scale to make the conditions for an inflationary stage.

Let us estimate the growth of the spatial size of the filed fluctuation. To do that, one should find the number of e-folds N, since the size of the Universe grows by a factor of e^N. This number is easily obtained from Eq. (7.47) in our special case of a quadratic potential:

$$N = \frac{2\pi}{M_{\rm Pl}^2}(\varphi_{\rm in}^2 - \varphi_{\rm end}^2). \tag{9.5}$$

It is known from observations that $N \approx 60$. Now we can estimate the inflaton value at the moment when the modern horizon was created, i.e., at creation of the spatial region that we call the observable Universe:

$$\varphi_{\rm in} \approx M_{\rm Pl}\sqrt{\frac{N}{2\pi}} \sim 3M_{\rm Pl}. \tag{9.6}$$

It has been taken into account here that $\varphi_{\rm end} \ll \varphi_{\rm in}$ and that the number of e-folds known from observations is $N \approx 60$. It is now easy to estimate the value of the potential (9.1) at the creation of the modern horizon:

$$V(\varphi_{\rm in}) = \frac{1}{2}m^2\varphi_{\rm in}^2 \approx 4.5m^2M_{\rm Pl}^2 \approx 4.5 \cdot 10^{-12}M_{\rm Pl}^4.$$

The last equality has been obtained for the usually-supposed value of the inflaton mass, $m \approx 10^{-6} M_{\rm Pl}$. Thus the modern horizon of our Universe appeared at energies much smaller than the Planck one.

Let us see what is the size of the horizon that appeared at the Planck energy scale, at $V \sim M_{\rm Pl}^4$. In this case, the initial field value is $\varphi_{\rm Planck} \sim M_{\rm Pl}^2/m \sim 10^6 M_{\rm Pl}$. Substituting this value to the expression (9.5), we obtain

$$N \sim 2\pi \cdot 10^{12}.$$

So the linear size of the spatial region that emerged at the Planck energy scale has increased by a factor of $\exp(2\pi \cdot 10^{12})$. Its initial size was of the Planck order. Right after the end of inflation, the spatial size of such a region is

$$L \sim \exp(2\pi \cdot 10^{12}) \cdot 10^{-33}\,\text{cm} \sim 10^{2\cdot 10^{12}}\,\text{cm}.$$

9.2 Quantum fluctuations during inflation

9.2.1 *A brief analysis*

As is known, the classical evolution of systems is slightly changed by quantum fluctuations. In Minkowski space their role is insignificant in many cases because the quantum corrections are proportional to a small parameter, the Planck constant $\hbar$. In addition, according to Heisenberg's uncertainty principle, the larger a fluctuation, the smaller its existence time. The situation is radically different at the inflationary stage which is approximately described in terms of de Sitter space. The main property of inflation is that any spatial inhomogeneity is stretched along with the expansion of space. Field fluctuations are also inhomogeneities, and their behavior is quite unlike the behavior of fluctuations in Minkowski space. In de Sitter space, their size rapidly grows and exceeds the horizon size. This process resembles particle creation in a strong field. After creation, virtual particles fly apart instead of annihilating, as happens in Minkowski space. The additional energy is created by the work of the external field, the gravitational field in our case.

The hypothetic field that makes the inflationary stage possible is generally assumed to be scalar. One of the Einstein equations where this scalar field φ is a source has the form

$$\ddot{\varphi} + 3H\dot{\varphi} - H^{-2}\,\mathrm{e}^{-2Ht}\Delta\varphi + V'(\varphi) = 0. \tag{9.7}$$

Equation (9.7) follows from Eqs. (8.37) and (8.4) after simple calculations. Here, we have neglected the weak change of the Hubble parameter $H = \dot{a}/a$ at the inflationary stage, so that the scale factor is $a(t) = H^{-1}\,\mathrm{e}^{Ht}$.

Now we divide the field into "classical" (Φ) and "quantum" (q) parts:

$$\varphi(\mathbf{x}, t) = \Phi(\mathbf{x}, t) + q(\mathbf{x}, t). \tag{9.8}$$

The term $q(\mathbf{x}, t) \equiv \delta\phi(\mathbf{x}, t)$, representing small fluctuations, must be quantized, and to do so, we employ, as usual, the Fourier expansion

$$\delta\varphi(x, t) = \int \frac{d^3k}{(2\pi)^3}\left[a_k f_k(t)\,\mathrm{e}^{ikx} + \text{h.c.}\right],$$

where a_k is the annihilation operator for the mode k. Note that in this analysis we are using the comoving RF with a dimensionless coordinate x, and the wave vector k is also dimensionless. From the basic Eq. (9.7), putting $H \simeq \text{const}$, we obtain

$$\ddot{f}_k + 3H\dot{f}_k + \frac{k^2}{(H\,\mathrm{e}^{Ht})^2} f_k + V''(\Phi) f_k = 0. \tag{9.9}$$

During inflation the field is changing slowly, therefore we neglect the last term, which allows for finding a suitable solution to this equation

$$f_k(t) = \frac{H}{\sqrt{2k^3}}(i + k\,\mathrm{e}^{-Ht})\,\mathrm{e}^{ik\,\mathrm{e}^{-Ht}}. \tag{9.10}$$

The behavior of the solution is extremely curious. At large wave numbers it rapidly oscillates while at small wave numbers it tends to a constant. A critical wave number that separates the two regimes is $k_H = a(t)H(t) = \mathrm{e}^{Ht}$. This wave number is inversely proportional to the characteristic size of the region, which is in this case a fluctuation emerging at a certain instant at the inflationary stage.

The following picture is appearing. Quantum effects create a fluctuation of the φ field whose spatial size, l, is naturally smaller than the horizon size because a fluctuation can only appear in a causally connected region. The condition $l < 1$ means $k > 1$ (recall that we are using dimensionless coordinates and momenta).

Let us suppose that there initially emerged a small-scale field configuration with the size $l_0 \ll 1$ and the characteristic momentum $k_0 \gg 1$. The function f_k describing this configuration, (9.10), rapidly oscillates. The limiting momentum $k_H(t) = \mathrm{e}^{Ht}$ grows exponentially rapidly, and there must occur such an instant, t_H, that the characteristic momentum of the system will be comparable with the limiting momentum, $k_0 \sim k_H(t_H)$. Beginning with this time, the function f_k stops to oscillate and tends to a constant. It is clear that the values of this constant are randomly distributed. The distribution law is considered in the next section.

So far we have been dealing with comoving coordinates. Let us now look at these processes from the viewpoint of physical distances $R_{\text{phys}} = a(t)r = H^{-1}\,\mathrm{e}^{Ht}r$ and the corresponding physical momenta $P_{\text{phys}} = p/a(t) = H\mathrm{e}^{-Ht}k$. The limiting physical momentum at which the magnitude of a certain fluctuation is "frozen" is $P^H_{\text{phys}} = H\mathrm{e}^{-Ht}k_H = H$. The characteristic size is then $L_{\text{phys}} = 1/P^H_{\text{phys}} = 1/H$. Thus, a fluctuation emerging at the inflationary stage due to quantum effects develops in the following way: the field value fluctuates very rapidly while its spatial size exponentially grows until it becomes the inverse value of the Hubble parameter.

9.2.2 *A detailed consideration*

Let us return to the expansion (9.8). The classical part of Φ is associated with smooth and slow changes of the field. The following expansion on the basis of the Fourier transformation is often used:

$$\varphi(\mathbf{x},t) = \int \frac{d^3\mathbf{k}}{(2\pi)^{3/2}}[a_{\mathbf{k}}(t)\,\mathrm{e}^{-i(\mathbf{kx})} + a^\dagger_{\mathbf{k}}(t)\,\mathrm{e}^{i(\mathbf{kx})}],$$

$$\Phi(\mathbf{x},t) = \int_{|\mathbf{k}|<k^*} \frac{d^3\mathbf{k}}{(2\pi)^{3/2}}[a_{\mathbf{k}}(t)\,\mathrm{e}^{-i(\mathbf{kx})} + a^\dagger_{\mathbf{k}}(t)\,\mathrm{e}^{i(\mathbf{kx})}],$$

$$q(\mathbf{x},t) = \int_{|\mathbf{k}|>k^*} \frac{d^3\mathbf{k}}{(2\pi)^{3/2}}[a_{\mathbf{k}}(t)\,\mathrm{e}^{-i(\mathbf{kx})} + a^\dagger_{\mathbf{k}}(t)\,\mathrm{e}^{i(\mathbf{kx})}], \quad (9.11)$$

where the critical momentum k^* separates the "rapid" subsystem from the "slow", classical one. This choice is not unique and depends on the problem under consideration. In Minkowski space, the most convenient is the choice of basis functions of the form $a_{\mathbf{k}}(t) \sim e^{ik_0 t}$, which are solutions of the d'Alembert equation

$$\Box g(\mathbf{x}, t) = 0.$$

In de Sitter space, it is also convenient to choose the solutions of this equation as basis functions. After the Fourier transformation

$$\tilde{g}_{\mathbf{p}}(t) = \int d^3x\, e^{i(\mathbf{px})} g(\mathbf{x}, t),$$

the equation acquires the form

$$\frac{\partial^2 \tilde{g}_{\mathbf{p}}(t)}{\partial t^2} + 3H\frac{\partial \tilde{g}_{\mathbf{p}}(t)}{\partial t} + (H\mathbf{p})^2 e^{-2Ht}\tilde{g}_{\mathbf{p}}(t) = 0, \tag{9.12}$$

where H is the Hubble parameter. Let us note that the momentum $\mathbf{p}$ is dimensionless as well as the comoving coordinate $\mathbf{x}$. One often uses the momentum $\mathbf{P} \equiv H\mathbf{p}$, having the correct dimensionality. Then the set of solutions of Eq. (9.12) is expressed in terms of Hankel functions:

$$H^{(2)}_{3/2}(y) = [H^{(1)}_{3/2}(y)]^* = -\sqrt{\frac{2}{\pi y}}\, e^{-iy}\left(1 + \frac{1}{iy}\right).$$

The solution can be presented in the form

$$\tilde{g}_{\mathbf{p}}(t) = \frac{\sqrt{\pi}}{2} H\eta^{3/2}[c_1(p)H^{(1)}_{3/2}(\eta P) + c_2(p)H^{(2)}_{3/2}(\eta P)].$$

Here, the so-called conformal time has been introduced:

$$\eta \equiv -H^{-1}\, e^{-Ht}. \tag{9.13}$$

This quantity simplifies the equations when analyzing processes in de Sitter space. The two constants are determined from additional conditions. More precisely, it is natural to suppose that at distances much smaller than the horizon size, i.e., as $p \to \infty$, the solution must coincide with that in Minkowski space. It is possible if $c_1 = 0$, $c_2 = -1$. Thus the set of orthonormalized solutions has the form [377]

$$\tilde{g}_{\mathbf{p}}(t) = \frac{iH}{P^{3/2}\sqrt{2}}\left(1 + \frac{P}{iH}\, e^{-Ht}\right)\exp\left(\frac{iP}{H}\, e^{-Ht}\right). \tag{9.14}$$

An analysis of the behavior of $\tilde{g}_{\mathbf{p}}(t)$ indicates the existence of the threshold value of the momentum,

$$P^*(t) \equiv H\,\mathrm{e}^{Ht}. \tag{9.15}$$

At each time t, the solution oscillates at $P \gg P^*$ and tends to a constant at $P \ll P^*$:

$$\tilde{g}_{\mathbf{p}}(t) \simeq \frac{iH}{P^{3/2}\sqrt{2}}, \quad P \ll P^*. \tag{9.16}$$

To make clear the physical meaning of the result obtained, let us pass on to physical distances, $\mathbf{R}_{\rm phys} = a(t)\mathbf{r}$, and momenta. Evidently, the physical momentum $\mathbf{P}_{\rm phys}$ is connected with the coordinate one, $\mathbf{p}$, in the following way: $\mathbf{P}_{\rm phys} = \mathbf{p}/a(t) = \mathbf{P}/(a(t)H)$. At the inflationary stage, the scale factor is related to the Hubble parameter, $a(t) \simeq H^{-1}\,\mathrm{e}^{Ht}$. Consequently, the threshold value of the physical momentum $P^*_{\rm phys}$ is

$$P^*_{\rm phys} \equiv \frac{P^*}{a(t)H} = H.$$

The characteristic value of a fluctuation is $L_{\rm phys} \sim P^{-1}_{\rm phys}$. If this size is larger than the horizon size H^{-1}, the magnitude of the fluctuation tends to a constant. Note that the physical size $L_{\rm phys}$ of the fluctuation grows exponentially:

$$L_{\rm phys} \sim P^{-1}_{\rm phys} = \frac{a(t)H}{P} = \frac{1}{P}\,\mathrm{e}^{Ht}. \tag{9.17}$$

This behavior is just characteristic of de Sitter space. In Minkowski space, the lifetime of a quantum fluctuation is of the order $1/\Delta E$ (ΔE is the fluctuation energy). In de Sitter space, a quantum fluctuation, having been born, increases its size exponentially according to the expression (9.17). At the same time, its magnitude is determined by the expression (9.16).

Now we can return to the problem of splitting the scalar field φ into classical and quantum parts. The quantum part, see (9.11), can be written in the form

$$q(\mathbf{x},t) \equiv \int \frac{d^3\mathbf{p}}{(2\pi)^{3/2}} W(P,t)[\hat{a}_{\mathbf{p}}\tilde{g}_{\mathbf{p}}(t)\,\mathrm{e}^{-i(\mathbf{px})} + \hat{a}^{\dagger}_{\mathbf{p}}\tilde{g}^{*}_{\mathbf{p}}(t)\,\mathrm{e}^{i(\mathbf{px})}]. \tag{9.18}$$

Here, we have introduced the creation and annihilation operators $\hat{a}^{\dagger}_{\mathbf{p}}$ and $\hat{a}_{\mathbf{p}}$, according to the standard manner of field quantization. Instead of roughly cutting off the momentum by the condition $P > P^*$, we use the function $W(P,t)$ with the properties $W(P \to 0, t) \to 0$, $W(P \to \infty, t) \to 1$ [516]. This can be, e.g., the function

$$W(P,t) = \theta(P - \varepsilon H\,\mathrm{e}^{Ht}), \quad \varepsilon \ll 1. \tag{9.19}$$

As will be seen later, the final result does not depend on the small but otherwise arbitrary parameter ε. Substituting the expressions (9.8) and (9.18) to Eq. (9.7), we obtain

$$\frac{\partial\Phi}{\partial t} - \frac{1}{3H}\left[\mathrm{e}^{-2Ht}\Delta\Phi - \frac{\partial V(\Phi)}{\partial\Phi}\right] = y(\mathbf{x},t),$$

$$y(\mathbf{x},t) \equiv \left(-\frac{1}{3H}\frac{\partial^2}{\partial t^2} - \frac{\partial}{\partial t} + \frac{1}{3H}\,\mathrm{e}^{-2Ht}\Delta\right) q(\mathbf{x},t). \tag{9.20}$$

In this equation, we have made an approximation: taking into account the slow change of the field during inflation, we omit the second-order derivative and neglect the higher powers of the function $y(\mathbf{x},t)$. Equation (9.20) describes the dynamics of the Φ field subject to the random "force" y. The latter is assumed to be small, so that we can seek the solution in the form [465]

$$\Phi = \Phi_{\mathrm{det}} + \phi. \tag{9.21}$$

The main part of classical field Φ_{det} obeys the equation

$$\frac{\partial\Phi_{\mathrm{det}}}{\partial t} - \frac{1}{3H}\left[\mathrm{e}^{-2Ht}\Delta\Phi_{\mathrm{det}} - \frac{\partial V(\Phi_{\mathrm{det}})}{\partial\Phi_{\mathrm{det}}}\right] = 0, \tag{9.22}$$

whereas the random part ϕ entirely depends on quantum fluctuations according to the linear equation

$$\frac{\partial\phi}{\partial t} - \frac{1}{3H}[\mathrm{e}^{-2Ht}\Delta\phi - V''(\Phi_{\mathrm{det}})\phi] = y(\mathbf{x},t). \tag{9.23}$$

(We consider the limit $\Phi_{\text{det}} \gg \phi$, which is justified if the random "force" $y(\mathbf{x},t)$ is small.) Calculating the random "force" appearing due to quantum fluctuations, we obtain (see details in [465])

$$\begin{aligned} y(\mathbf{x},t) &\equiv \left(\frac{-1}{3H}\frac{\partial^2}{\partial t^2} - \frac{\partial}{\partial t} + \frac{1}{3H}\,\mathrm{e}^{-2Ht}\Delta\right) q(\mathbf{x},t) \\ &= \left(\frac{-1}{3H}\frac{\partial^2}{\partial t^2} - \frac{\partial}{\partial t} + \frac{1}{3H}\,\mathrm{e}^{-2Ht}\Delta\right) \int \frac{d^3\mathbf{p}}{(2\pi)^{3/2}}\theta\left(P - \varepsilon H\,\mathrm{e}^{Ht}\right) \\ &\quad \times \left[\hat{a}_{\mathbf{p}}\tilde{g}_{\mathbf{p}}(t)\,\mathrm{e}^{-i(\mathbf{px})} + \hat{a}^\dagger_{\mathbf{p}}\tilde{g}^*_{\mathbf{p}}(t)\,\mathrm{e}^{i(\mathbf{px})}\right] \\ &\simeq \frac{H^3\varepsilon}{\sqrt{2}}\,\mathrm{e}^{Ht}\int \frac{d^3p}{(2\pi p)^{3/2}}\delta(P - \varepsilon H\,\mathrm{e}^{Ht})[\hat{a}_{\mathbf{p}}\,\mathrm{e}^{-i(\mathbf{px})} - \hat{a}^\dagger_{\mathbf{p}}\,\mathrm{e}^{i(\mathbf{px})}]. \end{aligned} \tag{9.24}$$

Here, we have used Eq. (9.12), which substantially simplifies the calculations; omitted is the second-order derivative, proportional to the small parameter ε^2. We also have $P = \sqrt{\mathbf{P}^2} = \sqrt{(\mathbf{p}H)^2}$. In the last line we use the approximation (9.16). The validity of this assumption follows from smallness of the momentum that enters into the argument of the delta function.

Consider the correlator

$$D(\mathbf{x},t,\mathbf{x}',t') \equiv \langle 0\,|y(\mathbf{x},t), y(\mathbf{x}',t')|\,0\rangle\,.$$

Using the expression (9.24) obtained above and the properties of the creation and annihilation operators $a^\dagger$ and a, we can obtain an analytic expression for this quantity:

$$D(\mathbf{x},t,\mathbf{x}',t') = \frac{H^3}{4\pi^2}\delta(t-t')\frac{\sin\varepsilon|\mathbf{x}-\mathbf{x}'|\,\mathrm{e}^{Ht}}{\varepsilon|\mathbf{x}-\mathbf{x}'|\,\mathrm{e}^{Ht}}, \quad \varepsilon \ll 1. \tag{9.25}$$

9.2.3 *Homogeneous fluctuations*

According to the expression (9.25), the correlator $D(\mathbf{x},t,\mathbf{x}',t')$ turns out to be an abrupt function of the distance $|\mathbf{x}-\mathbf{x}'|$. Let us neglect the spatial derivatives in Eq. (9.23), which greatly simplifies the equations. So, there is a reason to consider a homogeneous distribution

$\Phi = \Phi(t)$ obeying the simple equation

$$\frac{\partial \Phi}{\partial t} + \frac{1}{3H}\frac{\partial V(\Phi)}{\partial \Phi} = 0, \tag{9.26}$$

$$\frac{\partial \phi}{\partial t} + \frac{m^2}{3H}\phi = y(t). \tag{9.27}$$

Here, we have introduced the notation

$$m^2 \equiv V''(\Phi_{\text{det}}).$$

For the quadratic potential

$$V(\phi) = V_0 + \frac{1}{2}m^2\phi^2,$$

this quantity is a constant, while in the general case of an arbitrary potential it is almost constant during inflation. The correlator (9.25) of the random function $y(t)$ in the limit $\varepsilon \ll 1$ can be approximated as follows:

$$\langle y(t_1)y(t_2)\rangle = D(\mathbf{x}, t, \mathbf{x}, t') = \frac{H}{4\pi^2}\delta(t_1 - t_2).$$

The emergence of a delta function in the right-hand side of the equation means the Gaussian distribution law of the random function $y(t)$ with the energy

$$W(y) = \text{const} \cdot \exp\left[-\frac{1}{2\sigma^2}\int y(t)^2 dt\right], \quad \sigma = \frac{H^{3/2}}{2\pi}.$$

Due to a linear relation between the functions ϕ and $y(t)$, see (9.27), their probability distributions are proportional to each other. It means that the probability of finding the value of the function $\phi(t)$ inside a certain small interval is equal to (see [241])

$$dP(\phi) = \text{const} \cdot \mathcal{D}\phi \exp\left[-\frac{1}{2\sigma^2}\int\left[\frac{\partial \phi}{\partial t} + \frac{m^2}{3H}\phi\right]^2 dt\right].$$

The measure has the form $\mathcal{D}\phi \equiv \prod_{i=1}^{N} d\phi(t_i)$, $N \to \infty$.

It is now easy to obtain the probability of finding the field value ϕ_2 at the moment t_2 under the condition that the quantity ϕ_1 at

t_1 is known. Evidently, it is necessary to integrate over all field values in the interval (t_1, t_2), except for the boundary points $\phi_1 \equiv \phi(t_1)$, $\phi_2 \equiv \phi(t_2)$, which leads to the expression

$$dP(\phi_2, t_2; \phi_1, t_1)$$
$$= \text{const} \cdot d\phi_2 \int_{\phi_1}^{\phi_2} \mathcal{D}\phi \exp\left[-\frac{1}{2\sigma^2}\int_{t_1}^{t_2}\left[\frac{\partial\phi}{\partial t} + \frac{m^2}{3H}\phi\right]^2 dt\right]. \tag{9.28}$$

The unknown constant is found from the normalization condition

$$\int_{-\infty}^{\infty} d\phi_2 P(\phi_2, t_2;\ \phi_1, t_1) = 1.$$

The functional integral (9.28) can be calculated exactly [241] using the saddle point method, which implies a search for an extremum of the action

$$\ddot{\phi} - \mu^2\phi = 0, \quad \mu \equiv \frac{m^2}{3H},$$

with the boundary condition

$$\phi(t_1) = \phi_1, \quad \phi(t_2) = \phi_2.$$

A solution of this equation reads

$$\phi(t) = A\,\mathrm{e}^{\mu t} + B\,\mathrm{e}^{-\mu t},$$
$$A = \frac{\phi_2 - \phi_1\,\mathrm{e}^{-\mu T}}{2\sinh(\mu T)}, \quad B = \frac{-\phi_2 + \phi_1\,\mathrm{e}^{\mu T}}{2\sinh(\mu T)}; \quad T = t_2 - t_1.$$

Substituting it into the integral in the exponential in the expression (9.28), we find the required probability as

$$dP(\phi_2, t_1 + T; \phi_1, t_1) = d\phi_2 \cdot \sqrt{\frac{r}{\pi}}\exp[-r(\phi_2 - \phi_1\,\mathrm{e}^{-\mu T})^2],$$
$$r \equiv \frac{\mu}{\sigma^2}\frac{1}{1 - \mathrm{e}^{-2\mu T}},$$
$$\mu = \frac{m^2}{3H} \simeq \text{const}, \quad \sigma = \frac{H^{3/2}}{2\pi} \simeq \text{const}. \tag{9.29}$$

In the massless field limit we arrive at a simpler expression

$$dP(\phi_2, t_1 + T; \phi_1, t_1) = d\phi_2 \sqrt{\frac{2\pi}{H^3 T}} \exp\left[-\frac{2\pi^2}{H^3 T}(\phi_2 - \phi_1)^2\right]. \tag{9.30}$$

This formula is widely used in different inflationary scenarios where the field evolution should be small and therefore, the second-order derivative of the potential, and hence the mass can be neglected.

The field evolution picture looks as follows. The field is assumed to consist of two components, the classical (deterministic) one and the quantum one; see the expression (9.21). The classical part Φ_{det} obeys the equations of motion (9.26) with a random "force" y. As shown above, the influence of this force is included in the random component ϕ and has the probability density (9.29). It is easy to calculate $\langle \phi^2 \rangle$ for estimating the time-depending deflections from the classical value Φ_{det}. Using the expression for the probability (9.29), we obtain

$$\begin{aligned} \langle \phi(t)^2 \rangle &= \int_{-\infty}^{\infty} \phi^2 \, dP(\phi, t_1 + t; \phi_1, t_1) \\ &= \frac{1}{2r} = \frac{\sigma^2}{2\mu}\left(1 - \mathrm{e}^{-2\mu t}\right). \end{aligned} \tag{9.31}$$

This expression can be substantially simplified in the case of a massless field. As already said, the field can be regarded as approximately massless during inflation, when the condition $m \ll H$ holds. Expanding the exponential in Eq. (9.31), we obtain

$$\sqrt{\langle \phi(t)^2 \rangle} = \sigma = \frac{H}{2\pi}\sqrt{Ht}. \tag{9.32}$$

In terms of e-folds, $N \equiv Ht$, we obtain a formula to be widely used in what follows:

$$\sqrt{\langle \phi(t)^2 \rangle} = \sigma = \frac{H}{2\pi}\sqrt{N}. \tag{9.33}$$

In particular, one can conclude that fluctuations with the magnitude $\sim H/2\pi$ are formed in the time interval $t \sim H^{-1}$ $(N = 1)$.

The expressions (9.15), (9.17) indicate that its spatial size becomes fixed at the scale $L_{\text{phys}} \sim H^{-1}$.

The spatial size of the fluctuation can also be estimated. To that end, let us note that the correlator (9.25) is not small at $t \sim H^{-1}$ if the comoving distance $|\mathbf{x} - \mathbf{x}'| \lesssim 1$. It means that those fluctuations with comoving size of the order of unity are significant. Their size in physical coordinates grows in the usual way,

$$L_{\text{fluct}} = a(t)|\mathbf{x} - \mathbf{x}'| \approx H^{-1}\,\mathrm{e}^{Ht}, \tag{9.34}$$

and is equal to H^{-1} at $t \sim H^{-1}$. This remark is very important for what follows. By the way, the results (9.33) and (9.34) can be easily reproduced even in Minkowski space. Indeed, choosing the Lagrangian for a massless scalar field, let us make order-of-magnitude estimates for the action

$$S = \int d^4x\sqrt{-g}\frac{1}{2}g^{\mu\nu}\partial_\mu\varphi\partial_\nu\varphi.$$

As a result, we obtain

$$S \sim \frac{(\Delta\varphi)^2}{H^{-2}}H^{-4}.$$

Here, $\Delta\varphi$ is the magnitude of a fluctuation emerging by the time $t \sim H^{-1}$. It has also been taken into account that the fluctuation size is of the order H^{-1}, which is true for a massless field propagating with the speed of light. The probability of such a fluctuation is of the order of unity under the condition that the action $S \sim 1$. Accordingly, we get the estimate

$$\Delta\varphi \sim H,$$

which coincides by order of magnitude with the result (9.33) obtained more rigorously.

There also exists another way of studying fluctuations, based on the fact that Eq. (9.27) is nothing else than the Langevin equation. Then one can at once make the statement that the probability density satisfies the Fokker–Planck equation

$$\frac{\partial P}{\partial t} = \frac{H^3}{8\pi^2}\frac{\partial^2 P}{\partial\phi^2} + \frac{m^2}{3H^2}\frac{\partial^2(\phi P)}{\partial\phi^2}. \tag{9.35}$$

The expression (9.29) for the probability obtained above is a solution of this equation ($dP = Pd\phi$ in our notations).

9.3 Influence of massive fields on the process of inflation

In this section, we discuss the influence of quantum fluctuations on the inflaton field's classical evolution. Our purpose will be to study the back reaction of the cloud of virtual particles created by the inflaton field on the evolution of the field itself. Let us stress that we are only considering interaction with virtual particles that do not relate to the ambient temperature. It is known that such a cloud, possessing inertial properties, can appreciably slow down the system evolution. This phenomenon has been well studied in solid state physics (the polaron effect [360]). The whole analysis is conducted in Minkowski space, i.e., without taking into account gravitational effects.

The inertial properties of the virtual cloud depend not so much on the shape of the interaction potential or on the kinds of particles as on the value of the coupling constant and on the mass of the virtual particles, because it is these parameters that determine the richness of space in particles of different kinds. It is also clear that the inclusion of new fields only strengthens the "braking" effect in the classical field evolution. Therefore in what follows, we consider the simplest form of interaction allowing for obtaining analytical results. Specifically, we consider, apart from the inflaton field φ, an additional scalar field χ and write the action in the form

$$S = \int d^4x\sqrt{-g}\left[\frac{1}{2}\varphi_{,\mu}\varphi^{,\mu} - V(\varphi) + \frac{1}{2}\chi_{,\mu}\chi^{,\mu} - \frac{1}{2}m_\chi^2\chi^2 - \varkappa\chi u(\varphi)\right]. \tag{9.36}$$

Here, $u(\varphi)$ is a polynomial with respect to φ with the power not higher than three for renormalizable theories. The linear term in the additional field χ is only needed in the interaction for finally obtaining observable analytical results, valid for an arbitrary coupling constant $\varkappa$. An interaction of this kind emerges in supersymmetric theories and is discussed in hybrid inflation models [391].

In the paper [211], such a form of interaction is used to study the back reaction of the created particles on the classical motion of the inflaton field. As indicated above, the inclusion of real opportunities, e.g., with fermions, will only enhance the braking effect. Since our interest is in the interaction with virtual χ-particles, consider the transition amplitude

$$A(\varphi_i, \chi_i = 0; \varphi_f, \chi_f = 0) = \int_{\varphi_i}^{\varphi_f} D\varphi \int_0^0 D\chi \exp[iS]. \tag{9.37}$$

We here assume that the χ field is sufficiently massive and, by the relevant time, is located near a minimum of its effective potential. Integrating in the field variable χ, we obtain an amplitude of the form (9.37) with the effective action

$$S_{\text{eff}} = \int d^4x\sqrt{-g}\left[\frac{1}{2}\varphi_{,\mu}\varphi^{,\mu} - V(\varphi)\right] + \frac{\varkappa^2}{2}\int d^4x \int d^4x'\sqrt{-g}u(\varphi(x))G(x,x')u(\varphi(x')), \tag{9.38}$$

where $G(x, x')$ is the Green fluctuation of the scalar field χ. This expression is exact but, due to the presence of a nonlocal term, which is physically interpreted as a contribution from a cloud of virtual χ-particles, exact analytical transformations are impossible. Therefore, as in the method of effective action [304], we expand the nonlocal term in (9.38) in powers of $x - x'$. Then the first several terms renormalize the initial parameters of the potential and the field variable φ (the wave function). Since the potential parameters are determined by the physical conditions at some energy scale, we will be interested in new terms that cannot be incorporated in renormalizations.

An explicit expansion of the expression (9.38) in a power series in $x - x'$ requires the integration of the Green function, which, if we take into account gravitational effects is a hard problem. Instead, let us use the Green function [51] written as

$$G(x, x') = \frac{1}{m_\chi^2}\delta(x - x') - \frac{1}{m_\chi^2}\frac{1}{\sqrt{-g}}\partial_\mu\sqrt{-g}\partial^\mu G(x, x'). \tag{9.39}$$

Substituting this expression into (9.38) and neglecting higher derivatives in φ, we obtain

$$S_{\text{eff}} = \int d^4x\sqrt{-g}\left[\frac{1}{2}\varphi_{,\mu}\varphi^{,\mu} - V_{\text{ren}}(\varphi) - \frac{1}{\sqrt{-g}}\frac{\varkappa^2}{2m_\chi^4}u(\varphi)\partial_\mu\sqrt{-g}\partial^\mu u(\varphi)\right], \tag{9.40}$$

with the corresponding equation of motion

$$\partial_\mu\sqrt{-g}\partial^\mu\varphi + \sqrt{-g}V'_{\text{ren}}(\varphi) + \frac{\alpha^2}{m_\chi^2}u'_\varphi(\varphi)\partial_\mu\sqrt{-g}\partial^\mu u(\varphi) = 0, \tag{9.41}$$

and energy density

$$E = T_{00} = \frac{1}{2}(\partial_0\varphi)^2 + V_{\text{ren}}(\varphi) + \frac{\alpha^2}{2m_\chi^2}(\partial_0 u)^2, \tag{9.42}$$

where we have introduced the dimensionless parameter $\alpha \equiv \varkappa/m_\chi$.

The renormalized effective potential $V_{\text{ren}}(\varphi)$ contains a sum of contributions from interactions with all possible fields. A contribution from the interaction with the χ field is easily found explicitly:

$$V_{\text{ren}}(\varphi) = V(\varphi) + \delta V_\varphi = V(\varphi) - (\alpha^2/2)u(\varphi)^2.$$

A shortcoming of the first model of chaotic inflation with the potential $\lambda_{\text{ren}}\varphi^4$ is the small value of the coupling constant, $\lambda_{\text{ren}}(\sim 10^{-13})$, at which there is an agreement with the observational data. It means that all terms in the expression for λ_{ren}, including δV_χ, should cancel with high accuracy. It is further proved that taking into account a renormalization of the kinetic term makes it possible, in particular, to weaken the substantially restrictions on the parameters of the theory due to observational data.

The terms similar to the last term of expression (9.40) emerge at the renormalization of any theory. They are usually interpreted as "quantum corrections to the parameters of the theory depending on the value of the field itself" [304]. In weak fields, a contribution from this term is negligibly small. But at the inflationary stage, at large field values, the last term in Eq. (9.40) can be quite significant.

The classical Eq. (9.41) can also be obtained in another way, using the classical equations that minimize the initial action (9.36),

$$\frac{1}{\sqrt{-g}}\partial_\mu\left(\sqrt{-g}\partial^\mu\chi\right)+m_\chi^2\chi+\varkappa\varphi^2=0,$$
$$\frac{1}{\sqrt{-g}}\partial_\mu\left(\sqrt{-g}\partial^\mu\varphi\right)+V'(\varphi)+2\varkappa\varphi\chi=0. \tag{9.43}$$

We here consider the case $u(\varphi) = \varphi^2$. The first Eq. (9.43) can be transformed to

$$\chi(x)=-\varkappa\int G(x,x')\varphi^2(x')dx'. \tag{9.44}$$

Substituting (9.39) and (9.44) to the second Eq. (9.43), we arrive, in the same slow motion approximation, at Eq. (9.41).

Let us note that the correction $\delta V = -(\alpha^2/2)\varphi^4$ to the potential has emerged from an analysis of the classical equation (9.43). On the other hand, quite the same answer can be obtained by calculating the first quantum correction to the potential of the φ field due to interaction with the χ field at zero 4-momenta of the external lines corresponding to quanta of the φ field. The last term in the expression (9.41) is usually interpreted as a "back reaction of radiation" [211].

During inflation, the field is regarded as homogeneous, $\varphi = \varphi(t)$, and Eq. (9.41) is greatly simplified. Let us write it for the specific case $u(\varphi) = \varphi^2$, also taking into account that the scale factor a is expressed in the usual way in terms of the Hubble parameter H, $a = \exp(\int H dt)$:

$$\frac{d^2\varphi}{dt^2}+3H\frac{d\varphi}{dt}+V'(\varphi)+\frac{4\alpha^2}{m_\chi^2}\left[3H\varphi^2\frac{d\varphi}{dt}+\varphi^2\frac{d^2\varphi}{dt^2}+\varphi\left(\frac{d\varphi}{dt}\right)^2\right]=0.$$

Due to a slow change of the φ field in time, we discard the terms proportional to $d^2\varphi/dt^2$ and $(d\varphi/dt)^2$, which is justified by the final result, and after that we arrive at the easily integrable equation

$$\left(3H+\frac{12H\alpha^2}{m_\chi^2}\varphi^2\right)\dot{\varphi}+V'(\varphi)=0. \tag{9.45}$$

Solving this equation for the potential $V(\varphi) = \lambda\varphi^n$, taking into account the usual relationship between the Hubble parameter and the potential,

$$H = \sqrt{8\pi V(\varphi)/3}/M_{\rm Pl},$$

we arrive at an expression for the implicit time dependence of the field variable φ:

$$t = \frac{\sqrt{24\pi}}{M_{\rm Pl}\lambda^{1/2}}\left[\frac{1}{n(2-n/2)}\left(\varphi_0^{2-n/2} - \varphi^{2-n/2}\right) + \frac{4}{n(4-n/2)}\frac{\alpha^2}{m_\chi^2}\left(\varphi_0^{4-n/2} - \varphi^{4-n/2}\right)\right]. \tag{9.46}$$

Here, the first term reproduces the result of the standard inflationary model while the second one takes into account the interaction with virtual χ-particles.

It is necessary to note that according to Eq. (9.45) there is an intermediate inflationary stage at which one already cannot neglect the second-order derivative in the φ field in the square brackets. Nevertheless, of greater interest is the first stage, and it is this stage that is studied below. The further consideration is performed under the assumption that the second term dominates in Eqs. (9.45) and (9.46), i.e., under the condition

$$2\alpha\varphi/m_\chi \gg 1. \tag{9.47}$$

For what follows, it is convenient to introduce the quantity

$$\varphi_c \equiv \frac{m_\chi}{2\alpha}, \tag{9.48}$$

which represents a boundary field value and separates the first stage of superslow motion of the φ field from the second stage of ordinary inflation. In both cases, the field dynamics is described by simple analytical expressions.

In the case $\varphi > \varphi_c$, the time dependence of the field is

$$\varphi(t) = \left[\varphi_0^{4-n/2} - t/f_n\right]^{1/(4-n/2)},$$

$$f_n = \frac{8\sqrt{6\pi}}{n(4-n/2)} \frac{\varkappa^2}{m_\chi^4 M_{\rm Pl}\lambda^{1/2}}. \qquad (9.49)$$

This expression is obtained under the "superslow-rolling" condition, which, according to Eq. (9.42), has a somewhat unusual form

$$\ddot{\varphi} \ll 12H\varphi^2\dot{\varphi}\frac{\alpha^2}{m_\chi^2}. \qquad (9.50)$$

It is important to note that, as expected, the field changing rate

$$\dot{\varphi} = \frac{m^2}{12\alpha^2\varphi^2} \cdot \frac{V'}{H},$$

obtained from (9.45) with (9.47), turns out to be much smaller than the usual value $\dot{\varphi} = V'/3H$. The first stage of inflation terminates when the condition (9.50) does not hold any more. Further on follows a period of ordinary inflation, which lasts as long as the condition $\ddot{\varphi} \ll 3H\dot{\varphi}$ holds. As estimates show, the second stage turns out to be very short.

Let us determine the values of quantum fluctuations emerging at the inflationary stage. The fluctuations of noninteracting fields have been studied well enough [377]. Meanwhile, for fields with self-interaction, only order-of-magnitude estimates are known [208].

Let us address the fluctuations emerging at the first stage of inflation for the potential $\lambda\varphi^4$. It can be done most simply if we introduce the auxiliary field $\widetilde{\varphi}$ and, by substituting $\widetilde{\varphi} = (\alpha/m_\chi)\varphi^2$, bring the action (9.40) to the form

$$S = \int dx\sqrt{-g}\left[\frac{1}{2}\partial_\mu\tilde{\varphi}\partial^\mu\tilde{\varphi} - \frac{1}{2}\tilde{m}^2\tilde{\varphi}^2\right], \qquad (9.51)$$

corresponding to a free field with the mass $\tilde{m} \equiv m_\chi\sqrt{2\lambda}/\alpha$. Such a substitution is admissible at this stage of inflation where the field values are manifestly greater than zero. For a massive field, the fluctuation value is known [143], $\Delta\widetilde{\varphi} = \sqrt{3/(8\pi^2)}H^2/\tilde{m}$. Also known is the restriction on the mass of the field $\widetilde{\varphi}$, obtained by comparison with the measurements of energy density fluctuations by COBE [43], $\delta\rho/\rho \approx 6\cdot 10^{-5}$ on the modern horizon scale: $\tilde{m} \sim 10^{-6}M_{\rm Pl}$. Expressing $\tilde{m}$ in terms of the initial parameters, we obtain a relationship between them:

$$\frac{m_\chi}{M_{\rm Pl}}\frac{\sqrt{\lambda}}{\alpha} \sim 10^{-6}. \tag{9.52}$$

Since it is natural to suppose $m_\chi \ll M_{\rm Pl}$, the inequality (9.52), obtained from the observational data, is not burdensome and holds in a wide range of parameters.

Let us determine the field value φ_U which gave rise to the causally connected region that created the visible part of the Universe. It is known that the number of e-folds necessary for explaining the observational data is $N_U \simeq 60$. Then, using, as always, the relationship $N_U = \int_{\varphi_U}^{\varphi_{\rm end}} H dt$, we get

$$\begin{aligned} N_U &= \int_{\varphi_U}^{\varphi_c} \frac{H(\varphi)}{\dot{\varphi}} d\varphi + \int_{\varphi_c}^{\varphi_{\rm end}} \frac{H(\varphi)}{\dot{\varphi}} d\varphi \\ &= \frac{2\pi\alpha^2}{M_{\rm Pl}^2 m_\chi^2}\left(\varphi_U^4 - \varphi_c^4\right) + \frac{\pi}{M_{\rm Pl}^2}(\varphi_c^2 - \varphi_{\rm end}^2). \end{aligned} \tag{9.53}$$

Taking into account the different time dependencies of the φ field at the first and the second stages of inflation, and also using the smallness of the field value $\varphi_{\rm end}$ right after the end of inflation as compared with its initial value φ_U, we obtain the desired expression

$$\varphi_U \simeq \left(\frac{N_U}{2\pi}\right)^{1/4}\sqrt{\frac{M_{\rm Pl} m_\chi}{\alpha}}. \tag{9.54}$$

We note that the visible part of the Universe can form at $\varphi < M_{\rm Pl}$, at lower energies as compared to quadratic inflation. It is explained by the "superslow" field motion at the first stage of inflation, allowing the Universe to expand to the necessary size.

All the above reasoning is correct if the field χ has a sufficiently large mass, so that, at inflation, it can be considered as being located at the minimum of its potential. It is known that a field begins to roll rapidly to a minimum if the Hubble parameter becomes smaller than its mass, $H < m$. The Hubble parameter is time-dependent, therefore let us give the necessary estimates for the initial moment of emergence of the visible Universe ($\varphi = \varphi_U$) and at termination of the first stage of inflation ($\varphi = \varphi_c$). Simple calculations lead to the following result:

$$m_\chi > H(\varphi_U) \mapsto \frac{\sqrt{\lambda}}{\alpha} \leq 0.1,$$
$$m_\chi > H(\varphi_c) \mapsto m_\chi \leq M_{\mathrm{Pl}} \frac{\sqrt{\lambda}}{\alpha^2}. \qquad (9.55)$$

These inequalities hold in a wide range of the parameter, at least near the termination time of the first stage of inflation. Quantum corrections to the mass m_χ of the χ field bound the mass from below, therefore we will further assume that $m_\chi \geq \varkappa$, or, which is the same, $\alpha \leq 1$.

Thus, the interaction between the inflaton field and the quantum fluctuations of a massive field of another sort created by the inflaton makes it possible to explain, in particular, the smallness of energy density fluctuations under sufficiently soft conditions on the parameters of the potential.

9.4 Suppression of vacuum decay by virtual particles

There have probably been a few phase transitions in the early Universe. To such transitions, one can attribute the one at the Grand Unification scale of $\sim 10^{17}$ GeV and the one at the electroweak interaction scale, ~ 100 GeV. So far, we have been considering the class of models based on chaotic inflation. However, there is also an inflationary scenario based on a first-order phase transition from which the Universe evolution begins [342]. In the field theory, first-order phase transitions happen by means of a decay of a metastable (false) vacuum into spherically symmetric domains filled by a true vacuum. These domains begin to expand and occupy more and more volume. A calculation of the probability of such a decay represents a nontrivial problem.

This problem was repeatedly discussed in the literature in different areas of physics, such as field theory, solid state physics, and the physics of phase transitions. Theoretical aspects of this phenomenon have been studied from a quantum-mechanical viewpoint for quite a long time [365]. Significant progress in these studies in quantum field theory began in the 70s, see [183], and these works laid the basis of the instanton approach to the calculations of vacuum decay. The same method is also applicable for calculating the probability of quantum-mechanical tunneling. The basic idea is that the integration contour is deformed in such a way that the essential contribution to the integral representing the transition amplitude be given by a single trajectory, the "instanton" $x_{\text{inst}}(t_E)$, where t_E is time in Euclidean space. The subbarrier transition amplitude obtained in this way contains the necessary suppressing factor $\exp(-S[x_{\text{inst}}])$.

Further progress in the subbarrier transition theory revealed the existence of numerous additional factors affecting the transition dynamics. These factors can appreciably change the physical picture of the transition and particular conclusions of one or other cosmological models.

One of the main factors is the temperature at which the subbarrier transition takes place. In cosmological phase transitions, the temperature corrections to the effective potential strongly distort its shape, so that at high temperatures such a transition can simply be absent [377]. As the temperature decreases, there emerge additional minima of the potential, and a phase transition becomes possible, though the parameters of the effective potential are different from those observed at present.

Another fundamental factor is the ambient medium. Interaction with it usually diminishes the tunneling probability. When studying phase transitions due to false vacuum decay, it is necessary to take into account energy dissipation during the phase transition itself, at the motion of the wall that separates the true vacuum from the false one. It is of particular importance in studying the electroweak transition at the early stage of the Universe evolution [382], when the density of the ambient medium is high.

A third factor is the interaction of a particle or a classical field with virtual particles of another sort, see [476]. The "heavy nucleon" model also confirms the conclusion that a cloud of scalar particles interacting with the nucleus affects the renormalizations of both its wave function and its mass.

A separate question is calculating quantum corrections to the vacuum decay probability in field theory. It turns out that the corrections, being small as compared with the main contribution, still lead to an appreciable change in the magnitude of the main effect, the subbarrier transition probability. The papers [350, 351] describe a method of summing the low levels which allow for a correct determination of the contribution of quantum corrections. In what follows, we consider the influence of a cloud of virtual particles on the classical evolution of a scalar field and the false vacuum decay probability.

We consider a first-order phase transition, where the scalar field making this transition is surrounded by a cloud of virtual particles of another sort. As shown below, such a cloud diminishes the probability of a subbarrier transition even at zero temperature, and the suppression can be strong enough to change the picture of the transition completely. It is sufficiently clear from a physical viewpoint because the cloud of virtual particles needs time for its state restructuring.

Consider a system with the action (9.36) in Euclidean space. Acting in full analogy with the previous sections, we obtain the effective Euclidean action for the main field φ:

$$S_{\text{eff}}^{E} = \int d^4x \left[-\frac{1}{2}\varphi_{,\mu}\varphi^{,\mu} - V(\varphi) \right] + \frac{\varkappa^2}{2} \int d^4x \int d^4x' u(\varphi(x)) G_E(x,x') u(\varphi(x')). \tag{9.56}$$

(Here and henceforth, the gravitational effects are ignored). For what follows it is convenient to transform this expression to

$$S_E = \int d^4x \left[\frac{1}{2}(\partial\varphi)^2 + V(\varphi) - \frac{\alpha^2}{2} u(\varphi)^2 \right] + \frac{\alpha^2}{2} \int d^4x \frac{\partial u(\varphi(x))}{\partial x_\mu} \frac{\partial u(\varphi(x))}{\partial x_\mu}. \tag{9.57}$$

The third term in this expression must be included in renormalization of the potential,

$$V_{\mathrm{ren}}(\varphi) = V(\varphi) - \frac{\alpha^2}{2} u(\varphi(x))^2, \quad \alpha := \varkappa/m_\chi,$$

while the last term can substantially affect the system behavior. It is this nonlocal term that reflects the influence of the cloud of virtual particles of the sort χ on the dynamics of the field ϕ. Here, we also take into account that for the Green function $G_E(x-x')$ the inequality

$$\int dx G_E(x-x') = 1/m_\chi^2 \tag{9.58}$$

is valid.

Let the potential V_{ren} have two minima and the initial field value φ_F correspond to a metastable state. In this case, the metastable state decays by formation and expansion of true vacuum bubbles with the field value φ_T. This process is described by the $O(4)$-invariant solution $\varphi_B(r)$ of the classical field equation with the boundary conditions $\varphi_B(0) = \varphi_T, \varphi_B(\infty) = \varphi_F$. The decay probability density was found for the first time in the paper [183]:

$$\Gamma/V = \left(\frac{S_E(\varphi_B)}{2\pi}\right)^2 \left|\frac{\mathrm{Det}'\,\hat{D}(\varphi_B)}{\mathrm{Det}\,\hat{D}(\varphi_F)}\right|^{-1/2} \mathrm{e}^{-S_E(\varphi_B)}, \tag{9.59}$$

where the core K of the operator $\hat{D}(\varphi)$ is

$$K(x,y) \equiv \frac{\delta^2 S_E(\varphi)}{\delta\varphi(x)\,\delta\varphi(y)}. \tag{9.60}$$

The calculation of the normalizing determinant in the denominator is usually not difficult. Since $\varphi_F = \mathrm{const}$, everything is reduced to a calculation of the well-known determinant like $\mathrm{Det}(-\partial^2 + \mathrm{const})$. However, in the case under consideration, the effective action is nonlocal, and the calculation of the normalizing determinant is a problem. If we introduce the definitions

$$\Omega^2 \equiv V''(\varphi_F) + \alpha^2 \left(\frac{du}{d\varphi_F}\right)^2,$$

$$M^4 \equiv \varkappa^2 \left(\frac{du}{d\varphi_F}\right)^2, \tag{9.61}$$

then the problem reduces to calculating the determinant of the operator $\hat{D}(\varphi_F)$, whose core has the form

$$K_F(x,y) = \{\delta(x-y)(-\partial_x^2 + \Omega^2) - M^4 G_E(x-y)\}. \tag{9.62}$$

This core is not diagonal, but, due to its particular features, can be diagonalized and after that is easily calculated (see [360]). Let us introduce the operator $\hat{G}_E^{-1} \equiv (-\partial_x^2 + m_\chi^2)$. Then,

$$\begin{aligned}\operatorname{Det} \hat{D}(\varphi_F) &= \operatorname{Det} G_E^{-1}\hat{D}(\varphi_F)/\operatorname{Det} G_E^{-1}\\ &= \operatorname{Det}[(-\partial_x^2 + \Omega^2)(-\partial_x^2 + m_\chi^2) - M^4]/\operatorname{Det} G_E^{-1}\\ &= \operatorname{Det}(-\partial_x^2 + \omega_1^2)\cdot \operatorname{Det}(-\partial_x^2 + \omega_2^2)/\operatorname{Det} G_E^{-1},\end{aligned} \tag{9.63}$$

where we have introduced the parameters

$$\omega_{1,2}^2 = \frac{1}{2}\left[\Omega^2 + m_\chi^2 - \sqrt{(\Omega^2 - m_\chi^2)^2 + 4M^4}\right]. \tag{9.64}$$

The problem of calculating the normalizing factor in the pre-exponential of the expression (9.59) has been reduced to calculations of known determinants of oscillator-type operators.

As a matter of fact, quantum corrections at high energies lead to serious alteration in the effective action and substantially affects the vacuum decay rate. Let us show that using as an example a potential with two minima, a local one and an absolute one,

$$V_{\text{ren}} = \frac{\lambda}{8}(\varphi^2 - a^2)^2 + \frac{\varepsilon}{2a}(\varphi - a). \tag{9.65}$$

The instantonic solution to the Euclidean equation of motion can be characterized as follows:

$$\varphi(x) = \varphi_{\text{inst}}(r) = A\tanh\left(\frac{M}{2}(r-R)\right) - B, \tag{9.66}$$

where $r^2 \equiv \sum_{\alpha=1}^{4} x_\alpha^2$, and the parameters R and M are arbitrary quantities determined by minimization of the action. The parameters A and B are found from the boundary conditions

$$\varphi_{\text{inst}}(r \to \infty) = \varphi_+, \quad \varphi_{\text{inst}}(r \to 0) = \varphi_-. \tag{9.67}$$

Here, φ_+ corresponds the right local minimum of the potential (9.65) (the false vacuum) and φ_- to the left, absolute minimum (the true vacuum).

The action (9.57) is written with neglected high-order derivatives in the field φ. This approximation is correct under the condition $\partial\varphi_B/m_\chi\varphi_B \ll 1$ (m_χ is the mass of χ particles that form the cloud). On the instanton trajectory, the derivative $\partial\varphi_B$ is proportional to the mass of a quantum of the main field, m_φ. Thus, a validity criterion for the approximation is

$$m_\varphi/m_\chi \ll 1. \tag{9.68}$$

In fact, the transition domain from false to true vacuum is strongly widened by retarded effects due the cloud of virtual χ-particles. Therefore, the above condition manifestly holds and is even too strict.

For making a numerical analysis, let us write down the classical equations whose solution is the O(4)-symmetric instantonic trajectory:

$$\begin{aligned} &\partial_r^2\varphi + \frac{3}{r}\partial_r\varphi - V'_{\text{ren}}(\varphi) \\ &\quad + \frac{\alpha^2}{m_\chi^2}u'(\varphi)\left[\frac{3}{r}u'(\varphi)\partial_r\varphi + u''(\varphi)(\partial_r\varphi)^2 + u'(\varphi)\partial_r^2\varphi\right] = 0. \end{aligned} \tag{9.69}$$

A solution to this equation is sought in the form (9.66) with the unknown parameters M and R. The parameters A and B are determined by the boundary conditions.

The results of the numerical calculations are shown in Fig. 9.1. The abscissa axis is chosen to coincide with the known result [183, 184], obtained without taking into account the cloud of virtual particles. It is seen from the plot that the action can increase by orders of magnitude and hence the phase transition probability

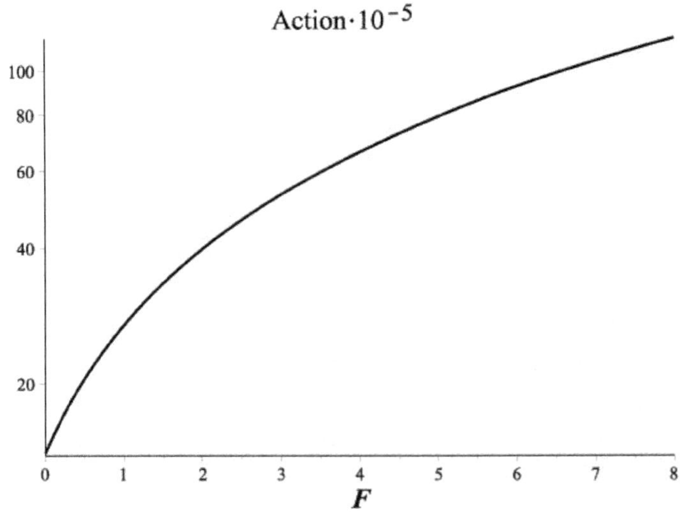

Figure 9.1 The dependence of the Euclidean effective action on the parameter $F = \varkappa^2/m_\chi^4$. The parameter values are $a = 1; \lambda = 0.1; \varepsilon = 0.01$ in units of $m_\chi = 1$.

turns out to be exponentially suppressed as compared with the usual predictions.

For a more specific estimate we assume that the action on the instantonic trajectory, without interaction with an additional field, is S_0, and that taking into account this interaction is ρS_0. Evidently, if this interaction is taken into account, the lifetime of the metastable state increases by a factor of $q \approx \exp[(\rho - 1)S_0]$. Using the plot of Fig. 9.1, we choose the intermediate value $\rho = 10$. The probability of a subbarrier transition is usually calculated in the quasiclassical approximation, such that $S_0 \gg 1$. Let us choose the minimally possible reasonable value of the action, $S_0 = 10$, which yields the estimate $q = \mathrm{e}^{90} \sim 10^{40}$. Such an abrupt increase in the lifetime of the metastable state is able to change the conclusions obtainable on the basis of a specific model. The action is usually much larger than the value chosen, and the suppression is still much stronger.

Evidently, the false vacuum decay probability is quite an uneasy problem, requiring the inclusion of first-order quantum corrections.

The latter can drastically change the estimated lifetime of a metastable state and, in particular, affect the rate at which our Universe was formed if that happened due to a first-order phase transition, or the rate of the electroweak transition.

9.5 Geometric approach to inflation

The results obtained above lead to an impression on the necessity of scalar fields or, more generally, matter field(s) for a successful inflation of space and particle production. Models describing the inflation have been elaborated, using other ideas like supersymmetry and attracting the inflationary idea for explanation of other cosmological problems like baryogenesis [209], dark matter [252] and primordial black holes [210, 326]. For reviews, see [14, 180] and references therein.

Nevertheless, it is tempting to reduce the number of species initially involved. The following discussion shows that we can restrict ourselves to geometry without inclusion of matter fields. The price is a more complicated form of the gravitational action.

9.5.1 *$f(R)$ models of inflation*

Gravity with higher derivatives is widely used in modern research despite the internal problems inherent to this approach [34]. Attempts to avoid the Ostrogradsky instabilities [568] are made [440], and extensions of the Einstein–Hilbert action attract much attention. The $f(R)$-gravity is one of the simplest extensions of the Einstein–Hilbert gravity, and it is free of Ostrogradsky instabilities. The reviews [156, 238] contain descriptions of $f(R)$ theories, including extension to the Gauss–Bonnet gravity.

In any case, quantization of gravity asks for a nonlinear geometric extension of the Einstein–Hilbert action. The first and most successful formulation of the inflationary model, the Starobinsky model [514], considers nonlinear geometric terms belonging to the $f(R)$ class of theories.

Reduction of the action to scalar-tensor gravity

Here, we discuss two methods to elaborate the model. The first method is based mostly on the conformal transformations that lead to the standard form of the gravitational action (see also Chapter 4). The price is the appearance of an additional dynamic variable in the form of a scalar field. The necessary formulas for D-dimensional space are presented in the subsection below. The second way consists in direct simulations on the basis of the dynamic equations. It is shortly considered in the next subsection.

Conformal transformations in D dimensions

Consider the action

$$S = \frac{M_{\rm Pl}^2}{2}\int d^D x\sqrt{|g|}f(R). \tag{9.70}$$

This action can be reduced to a scalar–tensor model in two steps. Firstly, consider the action depending on an auxiliary scalar field χ:

$$S_{ST} = \frac{m_D^{D-2}}{2}\int d^D Z\sqrt{|g_D|}[f'(\chi)R + f(\chi) - f'(\chi)\chi]. \tag{9.71}$$

It is assumed that $f' > 0$. One of the classical equations is

$$f''(\chi)(R - \chi) = 0. \tag{9.72}$$

So $\chi = R$, provided that $f''(\chi) \neq 0$. Substituting this into (9.71), we arrive at the initial action (9.70). These actions are equivalent **at the classical level.**

As the second step, perform the conformal transformation

$$g_{\mu\nu} = |f'|^{-2/(D-2)}\hat{g}_{\mu\nu}. \tag{9.73}$$

After some calculations, we obtain

$$S = \frac{1}{2}\int d^D x\sqrt{|\hat{g}_D|}[\hat{R} + (\partial\psi)^2 - 2V(\psi)] \tag{9.74}$$

(assuming $m_D = 1$), where

$$V(\psi) = e^{\frac{-D}{\sqrt{(D-1)(D-2)}}\psi} U(\chi), \quad U(\chi) \equiv \frac{1}{2}(f'(\chi)\chi - f(\chi)) \quad (9.75)$$

and

$$\psi \equiv \sqrt{\frac{D-1}{D-2}} \ln f'(\chi). \quad (9.76)$$

9.5.2 *The Starobinsky model*

The applications of this conformal transformation are wide and are considered in numerous publications. Here, we consider one of the applications concerning the Starobinsky model of inflation. This model at present provides the best fit to the observational data. Historically, this was the first model of inflation, but the models containing scalar field(s) were intensively discussed during a couple of decades. Such models are well elaborated and intuitively clear.

The model in question is described by the action (9.70) with $D = 4$ and

$$f(R) = R + R^2/6M^2. \quad (9.77)$$

As one can see from the previous subsection, the above conformal transformation leads to a scalar–tensor model of inflation, thus allowing us to apply the already known results. Indeed, the action (9.71) is the starting point for the inflationary models based on the scalar field dynamics. According to (9.75) and (9.76), the potential has the form

$$V(\psi) = \frac{3}{4} M_{\mathrm{Pl}}^2 M^2 (1 - e^{-\sqrt{2/3}\psi/M_{\mathrm{Pl}}})^2. \quad (9.78)$$

We have obtained the effective model containing the Einstein–Hilbert action and the scalar field contribution in the standard form. The scalar field potential possesses a profound minimum at $\psi = 0$ — a necessary element of the reheating just after the end of inflation. Also, the ground state energy should be zero with great accuracy due to the extreme smallness of the cosmological constant.

This action belongs to a set of inflationary models that differ in the form of the potential. Most of them are based on a small motion of fields and should satisfy some conditions which characterize the inflationary stage.

The slow-roll parameters are defined as

$$\epsilon = \frac{1}{2}\left(\frac{V'}{V}\right)^2, \tag{9.79}$$

$$\eta = \frac{V''}{V}, \tag{9.80}$$

$$\xi^2 = \frac{V'V'''}{V^2}. \tag{9.81}$$

The main parameters are the scalar spectral index n_s and the scalar/tensorial relative amplitude r, besides the running of the spectral index, which are given, in terms of the slow-roll parameters as

$$n_s = 1 - 6\epsilon + 2\eta, \tag{9.82}$$

$$r = 16\epsilon, \tag{9.83}$$

$$\frac{dn_s}{d\ln k} = -16\epsilon\eta + 24\epsilon^2 + 2\xi^2, \tag{9.84}$$

where k is the wave number.

According to [180], the observable values are

$$n_s \simeq 0.968, \quad r < 0.1, \quad \frac{dn_s}{d\ln k} = -0.0045 \pm 0.0067. \tag{9.85}$$

The potential (9.78) satisfies all of them for $M \simeq 10^{-5}$. Note that this potential has only one parameter, M, and nevertheless gives the best fit to the observations.

We see that the method of conformal transformations allows us to apply the well-known results elaborated for scalar fields, thus making the physical picture clearer.

The disadvantages of this approach are as follows: The relation of the Lagrangians before and after the conformal transformations is not clear at the quantum level. More general models with higher derivatives are not so easy to elaborate on. There are restrictions $f' \neq 0$ and $f'' \neq 0$.

The subsection below is devoted to the second way — numerical solution to the set of equations.

Direct solution of equations

As another approach, let us perform direct simulations in the framework of the Starobinsky model. Both approaches should give similar results. To this end, consider the four-dimensional theory described by the action

$$S[g_{\mu\nu}] = \frac{1}{2}m_{Pl}^2 \int d^4x\sqrt{|g_4|}\, f(R)\,. \tag{9.86}$$

The corresponding equations of motion in four dimensions are as follows:

$$\begin{aligned} &f_R'(R)R_{\mu\nu} - \frac{1}{2}\, f(R)g_{\mu\nu} + \Big[\nabla_\mu\nabla_\nu - g_{\mu\nu}\Box\Big]\, f_R'(R) = 0\,, \\ &\Box \equiv g^{\mu\nu}\nabla_\mu\nabla_\nu, \end{aligned} \tag{9.87}$$

with $\mu, \nu = 1, 2, 3, 4$. Taking into account the choice of the metric of the three-dimensional sphere

$$ds^2 = g_{\mu\nu}dx^\mu dx^\nu = dt^2 - \mathrm{e}^{2\alpha(t)}(dx^2 + \sin^2 x\, dy^2 + \sin^2 x\, \sin^2 y\, dz^2), \tag{9.88}$$

we obtain the set of equations

$$\begin{aligned} &6\dot\alpha(t)\dot R(t) f_R''(R) - 6(\ddot\alpha(t) + \dot\alpha^2(t)) f_R'(R) + f(R) = 0, \\ &2\dot R^2(t) f_R'''(R) + 2(\ddot R(t) + 2\dot\alpha(t)\dot R(t)) f_R''(R) \\ &\quad - (2\ddot\alpha(t) + 6\dot\alpha^2(t) + 4\,\mathrm{e}^{-2\alpha(t)}) f_R'(R) + f(R) = 0. \end{aligned} \tag{9.89}$$

The expression of the Ricci scalar for the metric (9.88) is

$$R(t) = 12\dot\alpha^2(t) + 6\ddot\alpha(t) + 6\mathrm{e}^{-2\alpha(t)}. \tag{9.90}$$

The last term is related to the sphere radius, $r(t) = \mathrm{e}^{\alpha(t)}$, which is supposed to be huge. Hence this term can be omitted, and the space is approximately flat.

Let us choose the form of the function as in the Starobinsky model,

$$f(R) = \frac{R^2}{6M^2} + R, \tag{9.91}$$

where the mass M is the model parameter,

$$M \sim 1.5 \cdot 10^{-5}\, M_{Pl} \left(\frac{50}{N_e}\right), \quad N_e = 55 \div 60\,. \tag{9.92}$$

A combination of the second equation of the system (9.89) and the expression of the Ricci scalar (9.90) gives a set of two equations for two variables R and α,

$$\begin{aligned} \ddot{R}(t) = &-2\dot{\alpha}(t)\dot{R}(t) - \frac{1}{12}R^2(t) + (\dot{\alpha}^2(t) + \mathrm{e}^{-2\alpha(t)} - M^2)R(t) \\ &+ 3M^2(\dot{\alpha}^2(t) + \mathrm{e}^{-2\alpha(t)}), \end{aligned} \tag{9.93}$$

$$\ddot{\alpha}(t) = -2\dot{\alpha}^2(t) - \mathrm{e}^{-2\alpha(t)} + \frac{1}{6}R(t). \tag{9.94}$$

Also, the first equation of the system (9.89) and Eq. (9.90) for the Ricci scalar give the first-order differential equation

$$\begin{aligned} R^2(t) &- 12\big(\dot{\alpha}^2(t) + \mathrm{e}^{-2\alpha(t)}\big)R(t) - 12\dot{\alpha}(t)\dot{R}(t) \\ &- 36M^2\big(\dot{\alpha}^2(t) + \mathrm{e}^{-2\alpha(t)}\big) = 0. \end{aligned} \tag{9.95}$$

This equation is an auxiliary condition for the system of the second-order equations for the unknown functions $\alpha(t)$ and $R(t)$. It can be used to connect the four initial conditions $\alpha(0), \dot{\alpha}(0), R(0), \dot{R}(0)$, so that only three independent conditions remain.

The result of numerical solution to the system (9.93) with the initial conditions

$$\begin{aligned} &\alpha(0) \equiv \alpha_0 = -\ln H_{infl} = \ln 10^6 \sim 13.8, \\ &\dot{\alpha}(0) \equiv \alpha_1 = H_{\text{infl}} \sim 10^{-6}, \quad \dot{R}(0) \equiv R_1 = 0. \end{aligned} \tag{9.96}$$

is presented in Fig. 9.2, see a discussion in [447]. Here, the Hubble parameter $H_{\text{infl}} = 10^{-6}$ in the Planck units. The initial value $R(0)$ follows from the auxiliary condition (9.95). The initial radius of the

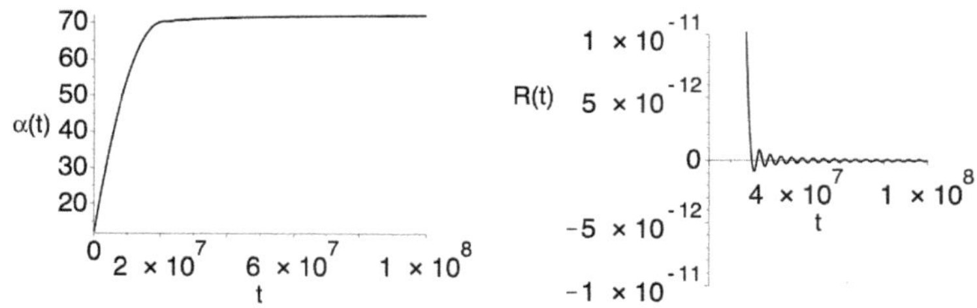

(a) (b)

Figure 9.2 Solution to the (9.93) system with the initial conditions (9.96).

whole space $\sim e^{\alpha(0)}$ is assumed to be very large due to its exponential growth starting from the Planckian scale.

The oscillations at the end of inflation can be explained if one substitutes (9.95) into the first equation of the system (9.93), with the result

$$\ddot{R}(t) + 3\dot{\alpha}(t)\dot{R}(t) + M^2 R(t) \simeq 0. \tag{9.97}$$

This equation describes free damped harmonic oscillator.

As a result, the direct numerical simulation and application of the conformal transformation give similar results and may be used if a Lagrangian is not very complicated.

Chapter 10

The large-scale structure

In the first two sections of this chapter, we briefly describe the current views of galaxy formation. The subsequent sections are devoted to exploring new opportunities provided to us by quantum fluctuations during inflation.

10.1 The cosmic microwave background

In 1964, Arno Penzias and Robert Wilson decided to carry out some research in radioastronomy and satellite communication. For testing the antenna, they chose the wavelength of 7.35 cm. Very soon it became clear that the antenna permanently detected an unexplained additional noise of unclear origin, and no way was found to get rid of it. The isotropy of the noise and its constancy in time meant that its source was located somewhere outside the Solar System. Even if its origin were related to our Galaxy, its intensity would vary due to the Earth's rotation and orbital motion around the Sun, which change the antenna's direction in space. So it was clear that the noise was of extragalactic origin. On measurement, the intensity of this radio signal corresponded to the radiation intensity of an absolutely black body with a temperature near 3 K. That is how the Cosmic Microwave Background (CMB), also called the relic radiation, was discovered.

What is the CMB? According the Big Bang theory, the Universe emerged approximately 14 billion years ago as a result of an enormous "bang" that created our space and time and the whole existing

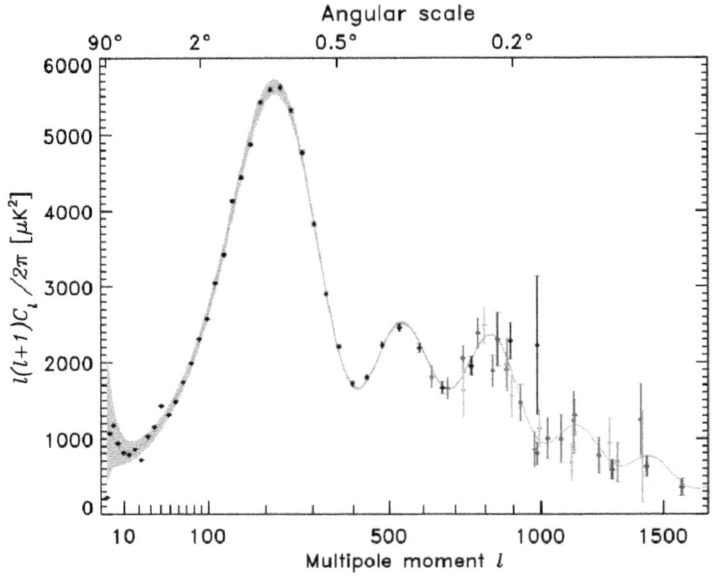

Figure 10.1 CMB temperature fluctuations.

matter and energy of our Universe. Up to an age of about 300,000 years, the young Universe consisted of a hot plasma of elementary particles and, certainly, photons. The common expansion of the Universe gradually cooled this medium, and as soon as the temperature fell down to a value of a few thousand degrees, the free protons and electrons formed neutral atoms. After that, the existing radiation in the form of photons propagated freely since they almost do not interact with neutral atoms. Due to further cosmic expansion, the wavelengths of this radiation continually increased. It is this radiation that was detected by Pensias and Wilson as an unknown noise.

The CMB has a spectrum corresponding to the temperature $T = 2.73\,\mathrm{K}$. CMB investigations at different angles have revealed surprisingly small values of its temperature fluctuations, about 10^{-5} in relative units. Nevertheless, these fluctuations bear valuable information on the early Universe (see Fig. 10.1).

Let us discuss at a qualitative level the effects leading to the temperature fluctuations $\Delta T(\theta, \varphi)$. For convenience, the perturbations are expanded in spherical harmonics

$$\frac{\Delta T(\theta,\varphi)}{T} = \sum_{l,m} a_{lm} Y_{lm}(\theta,\varphi), \tag{10.1}$$

and the plot of the quantity $l(l+1)C_l$, where $C_l = \langle\, |a_{lm}|^2 \,\rangle$ is studied.

This plot contains information of utmost importance. For example, we are now catching photons which were located near the horizon $l_{\rm rec}$ at recombination time, after which the Universe became transparent. (Recall that the horizon is the distance to which a light signal escapes in the expanding Universe from a certain time up to now.) While the Universe was opaque, the motion of massive particles was also hindered by friction due to interaction with photons.

On the other hand, in the same period of time there were fluctuations of the particle density and the gravitational potential. Subject to the gravitational field, the density fluctuations could have grown, but that did not happen because, as was said above, the motion of charged particles was strongly hindered. At recombination time, two events happened: (a) the friction disappeared, and particles began to move to the "centre" of the density fluctuation; in doing so, they are accelerated, which raises the local temperature; (b) photons from these heated regions move freely.

How does the size of the region of a density fluctuation affect further events?

1. Regions comparable with the horizon size at the recombination time, $L_{\rm rec} = 2H_{\rm rec}^{-1}$, are causally connected regions of the largest size at that time. These are regions having the largest gravitational well and hence the largest temperature contrast. Photons from such heated regions are now reaching us, and we must see a temperature peak in the plot. Let us estimate the size of these heated regions L_0 at present:

$$L_0 = \frac{1}{a(t)} L_{\rm rec} = \frac{T}{T_0} L_{\rm rec} \approx 300 \text{ Mpc}.$$

Since $H_{\rm rec}^{-1} \approx \frac{3}{2} t_{\rm rec}$, we have $T_{\rm rec} \approx 3000\,\text{K}$.

The angle at which we see such a region is

$$\theta = \frac{L_0}{L_U} \approx \frac{300}{9000} = 0.03 \text{ rad} \approx 2^\circ.$$

Consequently, a peak must be observed at about 2°. This acoustic, or Doppler peak, is clearly seen in the plot.

2. Regions much larger than the horizon size L_{rec} at recombination evolve insignificantly. Their mean temperature weakly depends on the size, and there is an evident plateau in the plot (the Sachs–Wolfe plateau) at large angles.

3. Regions much smaller than L_{rec}. After recombination, these regions had enough time not only to contract but also to expand due to internal pressure. Therefore, they are cooled a little, and we must see a minimum in the plot at smaller angles.

10.1.1 *Classical evolution of quantum fluctuations*

One of the most important results of the previous section is the following. At the inflationary stage, the density fluctuations created by quantum fluctuations rapidly increase their spatial size. At some instant, they become larger than the horizon size — the fluctuation crosses the horizon for the first time. By the end of inflation, the fluctuation scale can greatly exceed the horizon. The fluctuation magnitude is changing until the fluctuation scale crosses the horizon for the second time, now after the end of inflation. It turns out that in the period between two horizon crossings, there holds a simple relationship

$$\frac{\delta\rho}{p+\rho} \simeq \text{const}, \tag{10.2}$$

which we shall need in what follows. Let us first of all prove this relationship using Eq. (7.10) written in the form

$$\frac{d\rho}{p+\rho} = -\frac{3da}{a}.$$

Suppose there is a one-to-one correspondence between matter density and pressure, $\rho \leftrightarrow p$. Hence the pressure is a certain function of the density, $p(\rho)$, and the previous expression can be formally integrated:

$$\int_{\rho(t_0)}^{\rho(t)} \frac{d\rho}{p(\rho)+\rho} = -3\int_{a(t_0)}^{a(t)} \frac{da}{a}, \tag{10.3}$$

where t_0 is a time instant at the inflationary stage, and t is an arbitrary time satisfying the condition $t - t_0 \gg 1/H_e$, and H_e is the Hubble parameter at the end of inflation. If a quantum fluctuation creates a perturbation of the energy density $\delta\rho(t_0)$, then Eq. (10.3) should read

$$\int_{\rho(t_0)+\delta\rho(t_0)}^{\rho(t)+\delta\rho(t)} \frac{d\rho}{p(\rho)+\rho} = -3\int_{a(t_0)+\delta a(t_0)}^{a(t)+\delta a(t)} \frac{da}{a}, \tag{10.4}$$

which holds for the spatial region occupied by the fluctuation.

Let t_0 and t denote the times of the first and second horizon crossings by the fluctuation in question. Let us split both parts of Eq. (10.4) into a sum of integrals:

$$\int_{\rho(t_0)+\delta\rho(t_0)}^{\rho(t_0)} \frac{d\rho}{p(\rho)+\rho} + \int_{\rho(t)}^{\rho(t)+\delta\rho(t)} \frac{d\rho}{p(\rho)+\rho}$$
$$= -3\int_{a(t_0)+\delta a(t_0)}^{a(t_0)} \frac{da}{a} - 3\int_{a(t)}^{a(t)+\delta a(t)} \frac{da}{a}.$$

Here, Eq. (10.3), which is valid for a volume including the volume occupied by the fluctuation, is used in order to cancel the main terms in both parts of the equality. The remaining integrals can be easily estimated taking into account the smallness of the fluctuations:

$$\left(\frac{\delta\rho}{p+\rho} + 3\frac{da}{a}\right)_{t_0} \simeq \left(\frac{\delta\rho}{p+\rho} + 3\frac{da}{a}\right)_t.$$

The scale factor a grows exponentially since the beginning of the inflationary stage, so that we can neglect the second terms in both parts of the equation to obtain the required equality (10.2).

Equation (10.2), rewritten in the form

$$\left(\frac{\delta\rho}{p+\rho}\right)_{t_{\rm f}} \simeq \left(\frac{\delta\rho}{p+\rho}\right)_{t_{\rm in}}, \tag{10.5}$$

can be used for obtaining the spectrum of large-scale fluctuations. Here, $t_{\rm in} = t_0$ is the fluctuation emergence time at the inflationary stage, before the first horizon crossing, and $t_{\rm f}$ is the second horizon crossing time after the end of inflation at the matter-dominated stage. The left-hand side of this equality can be expressed in terms of the scalar field (inflaton) $\varphi = \varphi(t_{\rm in})$. The right-hand side of the equality is especially simple at the matter-dominated stage because the pressure is zero, $p = 0$. Let us write down the expression for $\delta\rho$ and $p + \rho$ at the inflationary stage:

$$\delta\rho = V'(\varphi)\delta\varphi, \quad p + \rho = \dot{\varphi}^2.$$

Now, the quantity $\delta\rho/\rho|_{t_{\rm f}}$ to be estimated is also expressed in terms of the inflaton field:

$$\left(\frac{\delta\rho}{\rho}\right)_{t_{\rm f}} \simeq \left(\frac{V'(\varphi)\delta\varphi}{\dot{\varphi}^2}\right)_{t_{\rm in}}.$$

In the slow-rolling approximation we have $\delta\varphi \approx H(\varphi)/2\pi$. Using the inflaton field equation, we obtain

$$\left(\frac{\delta\rho}{\rho}\right)_{t_{\rm f}} = \left(\frac{9H(\varphi)^3}{2\pi V'(\varphi)}\right)_{t_{\rm in}}. \tag{10.6}$$

At the inflationary stage, the Hubble parameter H is expressed in terms of the potential $V(\varphi)$, and we arrive at a formula for calculating the density fluctuation magnitude:

$$\left(\frac{\delta\rho}{\rho}\right)_{t_{\rm f}} = \frac{9}{5}\left(\frac{4}{3}\right)^{3/2}\frac{m\varphi^2}{M_{\rm Pl}^3}. \tag{10.7}$$

The fluctuations are characterized by values of the φ field at first horizon crossing. This formula includes, for completeness, the factor 2/5, which emerges when taking into account matter domination at the second horizon crossing. For what follows this factor is not very significant.

Another important parameter of the fluctuation is its characteristic size l. It depends on the times t_{in} and t_{f} of the first and the second horizon crossings, respectively. The fluctuation size at the time t such that $t_{\text{in}} < t < t_{\text{f}}$ can be estimated using a normalization to the size of the visible part of the Universe, L_U:

$$l = L_U \exp(N - N_U), \tag{10.8}$$

where N is the number of e-folds at which the scale l of this fluctuation is formed at the inflationary stage, and L_U denotes the size of our Universe, $L_U \approx 10^4\text{Mpc} \approx 10^{28}\,\text{cm}$.

On the other hand, the number of e-folds is expressed in terms of the inflaton value φ according to the expression (8.5):

$$N \approx \frac{2\pi}{M_{\text{Pl}}^2}\varphi^2,$$

where the potential is chosen in the form $V(\varphi) = (m/2)\varphi^2$, and the inequality $\varphi \equiv \varphi_{\text{in}} \gg \varphi_{\text{f}}$ is assumed. Substituting this expression and Eq. (10.8) into the expression (10.6), we arrive at a relationship between the fluctuation magnitude $(\delta\rho/\rho)_{t_{\text{f}}}$ and its size l [367]:

$$\begin{aligned}\left(\frac{\delta\rho}{\rho}\right)_{t_{\text{f}}} &= \left(\frac{\delta\rho}{\rho}\right)_{t_U}\left[1 + \frac{1}{N_U}\ln\left(\frac{l}{L_U}\right)\right] \\ &\simeq \left(\frac{\delta\rho}{\rho}\right)_{t_U}\left(1 + \ln\frac{l}{L_U}\right)^{1/N_U} \simeq \left(\frac{\delta\rho}{\rho}\right)_{t_U}\left(\frac{l}{L_U}\right)^{1/N_U}.\end{aligned} \tag{10.9}$$

Here, it is taken into consideration that the second term in the brackets is much smaller than unity at large scales, such that $l/L_U \geq 0.01$.

Consequently, the magnitudes of large-scale fluctuations very weakly depend on their spatial size. In this case, one speaks of an almost flat fluctuation spectrum. The above series expansion in $l \approx L_U$ leads to a power-law spectrum with the spectral index $n \equiv 1 - 2/N_U < 1$. Let us note that one can judge on the perspective of a specific model of inflation by the observed value of the spectral index.

10.2 The development of density fluctuations

10.2.1 *Density fluctuations in Minkowski space*

Let us follow the development of energy density fluctuations. We will suppose that the substance which is the carrier of energy can be considered as a perfect fluid with the stress-energy tensor

$$T_{\mu\nu} = pg_{\mu\nu} + (p+\rho)U_\mu U_\nu.$$

Here, U_μ are the velocity components of an elementary volume of the fluid. In the comoving reference frame,

$$U^t = 1, \quad U^i = 0.$$

The hydrodynamic equations of motion have the form

$$\begin{aligned}\frac{\partial\rho}{\partial t} + \nabla(\rho\bar{v}) &= 0,\\ \frac{\partial\bar{v}}{\partial t} + (\bar{v}\nabla)\bar{v} + \frac{1}{\rho}\nabla p + \nabla\varphi &= 0,\\ \Delta\varphi &= 4\pi G\rho,\end{aligned} \tag{10.10}$$

where φ is the Newtonian gravitational potential. The solution of this equation in the zero-order approximation is elementary:

$$\bar{v}_0 = 0, \quad \rho_0 = \text{const}, \quad p_0 = \text{const}.$$

In general, it is just an approximate solution, satisfactory when the influence of the gravitational field potential is small. Consider the behaviour of small fluctuations, marked with the subscript 1. Equations (10.10) are simplified:

$$\begin{aligned}\frac{\partial\rho_1}{\partial t} + \nabla(\rho_0\bar{v}_1) &= 0,\\ \frac{\partial\bar{v}_1}{\partial t} + \frac{v_s^2}{\rho_0}\nabla\rho_1 + \nabla\varphi_1 &= 0,\\ \Delta\varphi_1 &= 4\pi G\rho_1.\end{aligned} \tag{10.11}$$

The velocity of sound is $v_s^2 \equiv (\partial p/\partial\rho)_{\text{adiabat}} \simeq p_1/\rho_1$.

From Eqs. (10.11) it is easy to obtain the wave equation

$$\frac{\partial^2 \rho_1}{\partial t^2} - v_s^2 \nabla \rho_1 = 4\pi G \rho_0 \rho_1,$$

whose stability must be investigated. To this end we seek a solution in the form of plane waves

$$\rho_1(\bar{r}, t) = A\rho_0 \, \mathrm{e}^{-i\bar{k}\bar{r}+i\omega t}.$$

From the expression for the frequency

$$\omega = \pm\sqrt{k^2 v_s^2 - 4\pi G \rho_0}$$

it is evident that at small frequencies, such that

$$k < k_J \equiv \sqrt{4\pi G \rho_0 / v_s^2},$$

the solution is unstable.

The theory of gravitational instability was discussed for the first time by James Jeans in his paper "Stability of a spherical nebula" (1902), which considered quantitatively the gravitational instability of a gas cloud. The index "J" appearing in this section is related to Jeans's name.

The growing modes destroy the zero-order approximation in the form of a homogeneous energy density distribution. Under the impact of gravity, the structures become more dense. There occurs an exponential growth of fluctuations, ending with the formation of stars and stellar clusters. The size of an unstable region is, by order of magnitude,

$$l \equiv l_J > 2\pi / k_J.$$

One can also estimate the Jeans mass of matter M_J, above which the structure turns out to be unstable:

$$M_J \equiv \frac{4\pi}{3}\left(\frac{\pi}{k_J}\right)^3 \rho_0 = \frac{\pi^{5/2}}{6} \frac{v_s^2}{G^{3/2}\rho_0^{1/2}}.$$

10.2.2 *Density perturbations in the expanding Universe*

In the previous section, the cosmological expansion was not taken into account. Therefore the conclusions are correct only in a limited number of situations, for instance, for intragalactic gas condensation.

In a homogeneous Universe, the expansion is described by the time dependence of the scale factor $a(t)$. Then Eqs. (10.10) preserve their form, but their solution changes drastically. As a zero-order approximation, we will choose a solution in the form

$$\rho_0(t) = a^{-3}(t)\rho_0(t_0), \quad \vec{v}_0(t) = H(t)\vec{r}, \quad \nabla\varphi_0 = \frac{4\pi G}{3}\rho_0(t)\vec{r}.$$

We introduce small perturbations marking them by the subscript 1. Acting as above, we obtain

$$\ddot{\delta}_k + 2H\dot{\delta}_k + \left(\frac{v_s^2 k^2}{a^2(t)} - 4\pi G\rho_0(t)\right)\delta_k = 0,$$

where δ_k is the Fourier image of the relative density $\delta\rho/\rho \simeq \rho_1/\rho_0$. The velocity of sound v_s is proportional to the velocity of particles, therefore $v_s \sim 1/a(t)$. An unstable regime emerges at large wavelengths, i.e., at

$$k < k_J \equiv a(t)\sqrt{4\pi G\rho_0(t)/v_s^2(t)}.$$

Consider the growth of long-wave fluctuations, such that $k \ll k_J$. Recalling that at the matter-dominated stage we have the relationships

$$H(t) = \frac{\dot{a}}{a} = \frac{2}{3t}, \quad \rho_0(t) = \frac{1}{6\pi G t^2},$$

we obtain from the original equation that

$$\ddot{\delta}_k + \frac{4}{3t}\dot{\delta}_k - \frac{2}{3t}\delta_k = 0,$$

which leads to growing solutions:

$$\delta_k(t) = \delta_k(t_{\text{in}})\left(\frac{t}{t_{\text{in}}}\right)^{2/3}.$$

The long-wave fluctuations, having been formed, grow much slower than in a stationary Universe. Further on, these density fluctuations lead to the formation of galaxies and clusters of galaxies. Strong inhomogeneities in the energy density can be formed if the proposed Langrangian is not very simple. The subject is discussed below based on the primordial black holes as a topical example.

10.3 The baryonic asymmetry of the Universe

It is well known that, despite the symmetry between particles and antiparticles, baryons dominate in our Universe. It is one of the most important ingredients of the modern cosmological picture. At which moment did such an essential symmetry breakdown occur? What is the mechanism of baryosynthesis? A great number of reviews are devoted to this problem (see, e.g., [208, 324]), where quite various opportunities are discussed.

The whole previous presentation assumed homogeneity (on the average) of energy density and entropy fluctuations. In this chapter we will show that, under certain conditions, large-scale fluctuations of different kinds are also possible. This leads to new phenomena, such as macroscopically large antimatter regions and massive primordial black holes. The conditions for such phenomena originate in the course of inflation and develop in the post-inflationary period. The possible formation of antimatter regions has also been discussed earlier [171].

Since antimatter islands (regions) have not yet been found experimentally, the number of antibaryons in the whole Universe must be much smaller than the number of baryons. For the same reason, antimatter islands cannot be too large — otherwise we would observe annihilation photons at the edges of the corresponding regions.

On the other hand, if the size of antimatter regions is not sufficiently large, then annihilation at their boundaries should lead to their rapid evaporation, and it is just what happens in the majority of models of baryosynthesis. The processes of spontaneous CP symmetry breaking usually manifest themselves in the course of first- and second-order phase transitions that took place in the early Universe.

In this case, the antimatter regions turn out to be too small to be preserved until the present [208]. As has been shown in [325], a boundary separating an antimatter region from its baryonic environment moves due to annihilation not farther than by 0.5 pc by the termination of the radiation-dominated epoch. Thus, to be preserved up to the present time, the antimatter regions should be large enough. This in turn means that the future antimatter regions should have begun their formation at the inflationary stage, because the main property of this stage is the ability to effectively "stretch" the spatial lengths.

10.3.1 *Baryogenesis*

From the large number of models of baryosynthesis, we would like to choose the one [208, 209] where the inflationary period plays a substantial role.

This model postulates the existence of a complex scalar field χ, possessing a baryonic charge. Since at present the baryonic charge is concentrated in fermions (quarks), it is necessary to provide a transition of the baryonic charge from the χ field to the quarks. To this end, one introduces the interaction Lagrangian of the χ field with heavy quarks Q and leptons L, so that the full Lagrangian has the form

$$\begin{aligned} L = &-\partial_\mu\chi^*\partial^\mu\chi - V(\chi) + i\bar{Q}\gamma^\mu\partial_\mu Q + i\bar{L}\gamma^\mu\partial_\mu L \\ &- m_Q\bar{Q}Q - m_L\bar{L}L + (g\chi\bar{Q}L + \text{h.c.}). \end{aligned} \tag{10.12}$$

It is supposed that the heavy quarks Q and leptons L are connected with the ordinary quarks and leptons of matter fields. The fields χ and Q possess a baryon number but not a lepton number. On the contrary, the field L possesses the lepton number only. Consequently, the term in the Lagrangian (10.12), determining their interaction, leads to lepton number nonconservation [209]. The baryon number conservation follows from the $U(1)$ symmetry of the Lagrangian with respect to the transformations

$$\chi \to \exp(i\beta)\chi, \quad Q \to \exp(i\beta)Q, \quad L \to L. \tag{10.13}$$

Let us return to studying the dynamics of the scalar field χ that creates the baryon asymmetry. We choose the potential in the first-order approximation in the form

$$V(\chi) = -m_\chi^2 \chi^* \chi + \lambda_\chi (\chi^* \chi)^2 + V_0 = \lambda(|\chi|^2 - f^2/2)^2 + V_0, \tag{10.14}$$

where the constant V_0 is added for the minimum of the potential to be equal to zero. The "Mexican hat" potential (10.14) is the first, basic approximation. In what follows we will need quantum corrections to it.

The χ field can be presented in the form

$$\chi(\vartheta) = \frac{f}{\sqrt{2}} \exp\left(\frac{i\vartheta}{f}\right). \tag{10.15}$$

A violation of the $U(1)$ symmetry means that the radial component of the χ field takes the fixed value

$$f = m_\chi / \sqrt{\lambda_\chi},$$

and the variable ϑ in Eq. (10.15) now has the meaning of a massless scalar field since $V(\vartheta) = \text{const}$. The minimum of the potential is located on a circle of radius f. Further on we work with the dimensionless variable $\theta = \vartheta / f$.

The quantum corrections change the shape of the potential, so that its symmetry (10.13) is violated, and a small tilt of the "Mexican hat" is observed:

$$V(\theta) = \Lambda^4(1 - \cos\theta); \quad \Lambda \ll f. \tag{10.16}$$

This potential has a countable number of minima at $\theta = 2\pi N$, $N = 0, 1, 2, \ldots$. Note that the potential (10.16) is a convenient approximation of a more involved expression, see, e.g., [7]. The final form of our potential is

$$V_b(\chi) = \lambda(|\chi|^2 - f/\sqrt{2})^2 + \Lambda^4(1 - \cos\theta) + V_0. \tag{10.17}$$

For baryogenesis, of importance is the tilt of the potential (10.16) near the minimum which fixes the mass of the θ field:

$$m_\theta^2 = \frac{\Lambda^4}{f^2}. \tag{10.18}$$

Since the tilt is caused by quantum corrections, it is natural to assume a small value of the parameter Λ. It means that the contribution of the field θ to the energy density is small as compared with the full energy density at the inflationary stage. For this reason, the field θ is effectively massless, so that

$$m_\theta \ll H, \tag{10.19}$$

where H is the Hubble parameter at inflation. After the end of inflation, when the condition (10.19) is violated, the field θ oscillates near the minimum of the potential (10.16), and the energy $\rho_\theta \simeq \theta_i^2 m_\theta^2 f^2$ of the field θ passes on to the energy of created baryons and antibaryons [181, 209]. The sign of the baryonic charge in the final state depends on the initial value of phase at the time of symmetry breaking.

As a result of the symmetry breaking, the phase β of the χ field acquires the random value θ, and the effective Lagrangian has the form

$$\begin{aligned} L = &-\frac{f^2}{2}\partial_\mu\theta\partial^\mu\theta + i\bar{Q}\gamma^\mu\partial_\mu Q + i\bar{L}\gamma^\mu\partial_\mu L \\ &- m_Q\bar{Q}Q - m_L\bar{L}L + \left(\frac{g}{\sqrt{2}}f\bar{Q}L + \text{h.c.}\right) + \partial_\mu\theta\bar{Q}\gamma^\mu Q. \end{aligned} \tag{10.20}$$

For a spatially homogeneous field $\chi = (f/\sqrt{2})\,e^{i\theta}$, we have the following expression for the formed baryonic charge:

$$Q = i[\chi^* d\chi/dt - (d\chi^*/dt)\chi] = -f d\theta/dt. \tag{10.21}$$

It is clear that $Q > 0$ if $\dot{\theta} < 0$ in the process of classical motion to the zero of the phase θ. Consequently, a clockwise motion leads to a baryon excess, and a counterclockwise motion to an antibaryon excess.

Let us estimate the number of baryons and antibaryons created at oscillations of the phase θ with an arbitrary initial phase θ_i. An expression for the concentration of the baryons formed, $n_{B(\overline{B})}$, in the limit of a small phase θ_i has the form [209]

$$n(Q,L) = \frac{1}{V}\sum_{s_Q,s_L}\int \frac{d^3p}{(2\pi)^3 2p^0}\frac{d^3q}{(2\pi)^3 2q^0}|\langle Q(p,s_Q)L(q,s_L)|0\rangle|^2. \tag{10.22}$$

After standard calculations using the last term in (10.12) as an interaction Lagrangian, we obtain the density of created baryons

$$n_{B(\overline{B})} = \frac{g^2}{\pi^2}\int_{m_Q+m_L}^{\infty}\omega d\omega \left|\int_{-\infty}^{\infty} dt\chi(t)\,\mathrm{e}^{\pm 2i\omega t}\right|^2, \tag{10.23}$$

which is correct in the case $\chi(t\to-\infty)=\chi(t\to+\infty)=0$.

For the more general case $\chi(t\to-\infty)\neq 0$, $\chi(t\to+\infty)=0$, a result can be obtained by partial integration of the expression (10.23), which leads to

$$N_{B(\bar{B})} = \frac{g^2}{4\pi^2}\Omega_{\theta_i}\int d\omega\left|\int_{-\infty}^{\infty} d\tau\dot{\chi}(\tau)e^{\pm 2i\omega\tau}\right|^2, \tag{10.24}$$

where Ω_{θ_i} is a volume containing the initial phase θ_i. In obtaining Eq. (10.24), the surface term is equal to zero at $t=\infty$ due to Feynman conditions on radiation of the χ field at $t=-\infty$.

A numerical result can be approximately found by assuming the time dependence of the phase in the form

$$\theta(t) \approx \theta_i(1-m_\theta t), \tag{10.25}$$

during the first period of oscillations. Substituting (10.25) and (10.15) into (10.24), we obtain after simple calculations:

$$N_{B(\bar{B})} \approx \frac{g^2 f^2 m_\theta}{8\pi^2}\Omega_{\theta_i}k(\theta_i), \quad k(\theta_i) = \theta_i^2\int_{\mp\frac{\theta_i}{2}}^{\infty} d\omega\frac{\sin^2\omega}{\omega^2}, \tag{10.26}$$

where the sign at the lower limit of the integral corresponds to a baryonic or antibaryonic excess.

At the reheating time, the inflaton energy density is transformed to the energy density of relativistic particles. It is then assumed that the inflaton decays into light particles rather rapidly as compared with the oscillation period near a minimum (recall that due to the tilt of the potential the minimum is located at the point $\theta = 0$). Thus we have $\Gamma_{\text{tot}} \gg m_\theta$. Oscillations of the field θ begin when friction in the medium can be neglected, i.e., at $H \approx m_\theta$. The phase change with time leads to baryon and antibaryon creation as discussed above. The entropy density after thermalization is

$$s = \frac{2\pi^2}{45} g_* T^3, \tag{10.27}$$

where g_* is the total effective number of degrees of freedom. It is assumed that the temperature exceeds the electroweak symmetry violation scale.

At this temperature all degrees of freedom of the Standard Model are in equilibrium and $g_* \simeq 106.75$. The temperature is connected with the expansion rate as follows:

$$T = \sqrt{\frac{M_{\text{Pl}} H}{1.66 g_*^{1/2}}} \approx \frac{\sqrt{M_{\text{Pl}} m_\theta}}{g_*^{1/4}}. \tag{10.28}$$

In the last expression (10.28) it is assumed that relaxation processes begin when the condition $H \approx m_\theta$ holds. Using Eqs. (10.26)–(10.28), we obtain different expressions for baryon and antibaryon concentrations:

$$\frac{n_{B(\bar{B})}}{s} = \frac{45 g^2}{16\pi^4 g_*^{1/4}} \left(\frac{f}{M_{\text{Pl}}}\right)^{3/2} \frac{f}{\Lambda} k(\theta_i). \tag{10.29}$$

The function $k(\theta_i)$ includes the dependence of the magnitude and sign of the baryon asymmetry on the initial phase which is different in different spatial regions. The behaviour of this function can be easily analyzed numerically with the aid of the expression (10.26).

The expression (10.29) makes it possible to obtain the observed average baryonic asymmetry in the Universe, $n_B/s \approx 3 \cdot 10^{-10}$. In this model, it has been assumed that $f \geq H \simeq 10^{-6} M_{\text{Pl}}$. A natural value of the coupling constant is $g \leq 10^{-2}$, which leads to the observable baryon asymmetry under the reasonable requirement $f/\Lambda \geq 10^5$ [7].

10.3.2 *Large-scale fluctuations of the baryonic charge*

There appears an opportunity of interest if one takes into account the effect of phase fluctuations during inflation. To do so, let us consider in more detail the motion of the phase along the valley $|\chi| = f/\sqrt{2}$, having the shape of a circle. Let the phase $\theta = 0$ correspond to the north pole and $\theta = \pi$ to the south pole. The minimum of the potential is located at the north pole. Using Eq. (10.21), it is easy to show that preferred antibaryon production takes place if the field is moving to the north pole *counterclockwise*, while baryon production occurs if the field is moving *clockwise*.

Let us calculate the volume distribution of antimatter domains. The total volume of all antimatter domains formed by the end of the N_t-th e-folding before the end of inflation may be calculated using the recursive procedure: suppose the total volume of all the domains with the average phase $\bar{\theta}$ formed by that time is $V(\bar{\theta}, N_t)$. Then their total volume at the $(N_t - 1)$-th e-folding before the end of inflation is given by

$$V(\bar{\theta}, N_t - 1) = e^3\, V(\bar{\theta}, N_t) + [V_U(N_t) - e^3\, V(\bar{\theta}, N_t)] \cdot P(\bar{\theta}, N_t - 1) \cdot h, \tag{10.30}$$

where $V_U(N_t)$ is the volume of the Universe $V_U(N_t) \approx \mathrm{e}^{3\cdot(N_U - N_t)} H^{-3}$ at N_t e-folds before the end of inflation, $N_U \approx 60$ is the total number of e-folds during inflation, and $h = H/(2\pi f)$. $P(\bar{\theta}, N_t - 1)$ gives the phase distribution, which is Gaussian, see (8.33), (8.30):

$$P(\bar{\theta}, N_t) = \frac{1}{\sqrt{2\pi}\sigma_{N_t}} \cdot \exp\left(-\frac{(\theta_U - \bar{\theta})^2}{2\sigma_{N_t}^2}\right), \tag{10.31}$$

$$\sigma_{N_t} = \frac{H}{2\pi \cdot \overline{|\chi(N_t)|}} \cdot \sqrt{N_U - N_t}. \tag{10.32}$$

Here θ_U is the initial phase of the field.

Note that here N_t is the number of e-folds *before the end of inflation*, so it *decreases* with time; so if the moment $(t = 0)$ corresponds to the beginning of inflation, then $N_0 \equiv N_U \approx 60$, while at the end of inflation $N_\tau = 0$. Accordingly, the volume of antimatter domains at the beginning of inflation is $V(\bar{\theta}, N_U) = 0$.

The first term in Eq. (10.30), $e^3 V(\bar{\theta}, N_t)$, is the total volume of antimatter domains formed *before* the N_t-th e-folding. The second term

$$v(\bar{\theta}, N_t) = [V_U(N_t) - e^3\, V(\bar{\theta}, N_t)] \cdot P(\bar{\theta}, N_t - 1) \cdot h, \tag{10.33}$$

is the total volume of antimatter domains formed *during* the N_t-th e-folding. Since the initial volume of each domain is H^{-3}, the number of domains formed during a given e-folding is

$$n = \frac{v(\bar{\theta}, N_t)}{H^{-3}}. \tag{10.34}$$

Domains grow in size during inflation, so the earliest domains to form become the biggest at the end of inflation. Their linear sizes at present are determined by the equation

$$L(N_t) = 6 \cdot 10^3 e^{-(N_U - N_t)}. \tag{10.35}$$

Here $L(N_t)$ is the size (in Mpc) of the antimatter domain formed at the N_t-th e-folding.

Using the relations obtained, one can calculate the size distribution of matter and antimatter domains. However, to be realistic, our model should suppress in a consistent way the large-scale fluctuations of the baryonic charge. Otherwise such fluctuation would contribute to the CMB spectrum and would be observed. Indeed, according to the mechanism we are discussing, the CMB temperature fluctuations are proportional to the energy density of matter, $\delta\rho/\rho \simeq \delta\theta\Omega_B/\Omega_{\text{tot}}$. Here $\delta\theta = H/(2\pi f)$, see above, Ω_B is the relative baryonic energy density, and Ω_{tot} is the total relative energy density. Consequently, we have the restriction $\delta\theta \leq 10^{-3}$, at least at large scales. But in this case, for 60 e-folds during inflation, the phase will move by an angle of at most $\Delta\theta \sim 60 \cdot 10^{-3} \sim 0.1\,\text{rad}$ to zero, and a transition through the zero value is very unlikely.

Let us discuss how this difficulty could be overcome.

The first method is to take into account the interaction of χ with the inflaton φ, with a certain coupling constant g [327]. Then the potential (10.14) takes the form

$$V(\varphi, \chi) = \lambda(|\chi|^2 - f^2/2)^2 - g|\chi|^2(\varphi - cM_{\rm Pl})^2, \tag{10.36}$$

where λ, g and c are parameters of the potential. This potential also has the shape of a Mexican hat with a maximum at

$$|\chi| \equiv f_{\rm eff}(\varphi) = \sqrt{f^2 + \frac{g}{\lambda}(\varphi - cM_{\rm Pl})^2}. \tag{10.37}$$

Now, evidently, the position of the minimum is not constant. On the contrary, it strongly depends on the classical field φ, which changes in the course of inflation,

$$\varphi(t) = \varphi_U - \frac{m_\varphi M_{\rm Pl}}{2\sqrt{3\pi}} t, \tag{10.38}$$

for the quadratic potential $U(\varphi) = m_\varphi^2 \varphi^2/2$. Recall that the fluctuation magnitude of the phase θ of the field χ is inversely proportional to the quantity $f_{\rm eff}(\varphi)$. It is not hard to obtain the dependence of the inflaton value on the number of e-folds N:

$$\varphi = \varphi_N = \varphi_U - \frac{M_{\rm Pl}}{2\sqrt{3\pi}} N. \tag{10.39}$$

Here, it is taken into account that $m_\varphi \cong H$ and $N = Ht$. Thus, in N e-folds after Universe formation the effective scale $f_{\rm eff}$ (10.37) looks as follows:

$$f_{\rm eff}(N) = f\sqrt{1 + \frac{g}{\lambda}\frac{M_{\rm Pl}^2}{f^2}\left[\left(\frac{\varphi_U}{M_{pl}} - c\right) - \frac{N}{2\sqrt{3\pi}}\right]^2}. \tag{10.40}$$

Denoting $(\varphi_U/M_{pl} - c) \equiv N_f/(2\sqrt{3\pi})$, we arrive at the final expression of the form

$$f_{\rm eff}(N) = f\sqrt{1 + \frac{g}{12\pi\lambda}\frac{M_{\rm Pl}^2}{f^2}(N_f - N)^2}. \tag{10.41}$$

This expression has an important property: there is from the very beginning the large parameter $M_{\rm Pl}^2/f^2 \sim 10^{10}$, and the function $f_{\rm eff}(N)$ has a sharp minimum at $f_{\rm eff}(N_f) = f$ at a reasonable relationship between the parameters. Hence it immediately follows that the fluctuation magnitude of the phase θ

$$\langle\, \delta\theta \,\rangle = \frac{H}{2\pi f_{\rm eff}(N)}, \tag{10.42}$$

abruptly increases near the e-folding $N = N_f$.

Let us estimate the possible range of the parameters g and λ. The first inequality follows from the requirement that fluctuations should be sufficiently strongly suppressed, i.e., $f_{\rm eff} \gg f$, and from the expressions (10.37):

$$\langle\, \chi \,\rangle \sim \sqrt{g/\lambda} M_{\rm Pl} \gg f. \tag{10.43}$$

The second inequality is connected with the fact that the additional term inserted into the potential also renormalizes the inflaton mass,

$$m_{\rm ren}^2 = m^2 - g\langle\, \chi^2 \,\rangle.$$

The requirement that the renormalization should be small leads to the inequality $m^2 \gg g\langle\, \chi^2 \,\rangle \sim g^2 M_{\rm Pl}^2/\lambda$. Thus, the second inequality reads

$$m \gg g M_{\rm Pl}/\sqrt{\lambda}. \tag{10.44}$$

Choosing the numerical values of the parameters

$$f = 10^{-5} M_{\rm Pl}, \quad m = 10^{-6} M_{\rm Pl}, \tag{10.45}$$

and substituting them into the expressions (10.43), (10.44), we arrive at constraints on the parameters g and λ:

$$\sqrt{g/\lambda} \gg 10^{-5}, \quad g/\sqrt{\lambda} \ll 10^{-6}. \tag{10.46}$$

This set of inequalities is not very burdensome. Thus, if $\lambda \sim 1$, then $10^{-10} \ll g \ll 10^{-6}$. The smallness of the parameters is related to the choice of small parameters in (10.45), which is conventional for inflationary models. Under such restrictions on the parameters, the phase fluctuations have a sharp peak at some e-folding number, which we have denoted by N_f.

The second method of suppressing large-scale fluctuations is more cardinal and consists of the following. Recall that the Mexican hat potential has been postulated from the beginning. Its shape has no other foundation than convenience and effectiveness. But let us also recall the notion of a "landscape" that appears in string theory as well as in the mechanisms of cascade reduction and random potentials, to be discussed further in this book. As a whole, the landscape implies that the shape of the scalar field potential depends on a number of random circumstances such as the initial values of the field. Then the shape of the potential may be arbitrary. Let us employ this and slightly change the shape of the potential (10.14). Using the landscape ideas, we consider one of the possible forms of the potential

$$\begin{aligned} V(\chi) &= V_b(\chi) \cdot F(\chi), \\ F(\chi) &= \frac{C^2}{(|\chi|^a + C)^2}, \quad C, a = \text{const.} \end{aligned} \tag{10.47}$$

The function $F(\chi) \to [\chi \to 0]1$, so that $V(\chi) \to [\chi \to 0]V_b(\chi)$, and the modified potential is close to the original form V_b from (10.17) while χ is small.

Since the modified potential may be made less steep by varying the parameters a and C, the time required for the field to reach its minimum may be accordingly increased.

The χ field dynamics is governed by the equation of motion

$$\frac{\partial^2 \chi(t)}{\partial t^2} + 3H\frac{\partial \chi(t)}{\partial t} + \frac{\partial V(\chi)}{\partial \chi} = 0, \tag{10.48}$$

where H is the Hubble parameter. This differential equation can be solved numerically.

The number of e-folds since the beginning of inflation is $N = H \cdot t$. It can be seen that in the modified potential the field reaches its minimum ($|\chi|_{\min} = |f|/\sqrt{2}$) slower.

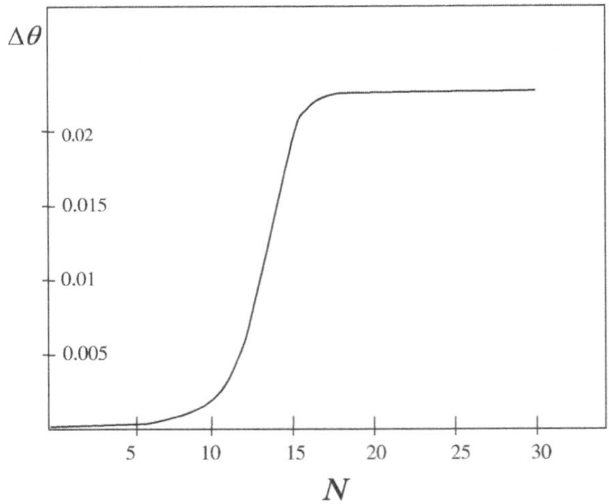

Figure 10.2 Computed values of $\Delta\theta$ as a function of N, the number of e-folds, $N = H \cdot t$. Parameter values: $H = 1, \lambda = 0.05, f = 10, a = 1.1, C = 10$.

The magnitude of the baryon fluctuation is proportional to the fluctuation $\Delta\theta$ of the phase θ which depends on time as

$$\Delta\theta(t) = \frac{H}{2\pi \cdot \overline{|\chi(t)|}} \tag{10.49}$$

each e-folding, where $\overline{|\chi(t)|}$ is the mean absolute value of the field χ during that e-folding. Figure 10.2 shows the evolution of $\Delta\theta$ in the modified potential. Due to the inverse relation between $\Delta\theta$ and χ, the modified potential allows the value of $\Delta\theta$ to be significantly smaller during the first e-foldings as compared to its value after $\sim$ 10th e-foldings. This is exactly what is needed to suppress large scale baryon fluctuations.

So, in this chapter we have considered an example of an inflationary model with inhomogeneous baryogenesis: islands of antimatter, sufficiently spatially extended, emerge here in a natural way. Their number and size and the antimatter density inside them are controlled by a few parameters of the model.

Besides the model of baryogenesis, there exist a number of other possibilities, see, e.g., [9, 208]. It would be of interest to apply to them the ideas outlined here.

10.4 Primordial black holes as the source of large-scale inhomogeneities

It is known that large galaxies contain central supermassive BHs (SMBHs) with masses in the range $10^6 M_\odot \div 10^9 M_\odot$. The most massive of them with masses $\sim 5 \cdot 10^9 M_\odot$ are observed in giant elliptical galaxies. Some SMHBs are observed in small galaxies, which looks strange. Many objects that can be interpreted as intermediate BHs (masses in the range $10^3 M_\odot \div 10^5 M_\odot$) have recently been discovered in globular clusters and even — possibly — in the central region of our Galaxy.

At the same time, it was conventionally believed that the PBH masses are bounded from above by a quantity of the order of a single solar mass. New mechanisms for forming massive and supermassive BHs long before the first stars are of interest.

A new wave of interest in PBHs has been induced by the recent discovery of gravitational waves from several BH mergers, with masses of the order of $\sim 30 M_\odot$. This discovery brings new puzzles — not only how BHs of such masses appear, but why the distance between them is small and why their spin is close to zero.

All these objects have a common feature: the standard explanations of their origin meet serious difficulties. The possibility of PBH formation was predicted in [576]. Then this scenario was grounded and developed further in [158, 170, 577] and in many other works. The paper [210] discusses the first model of supermassive BH formation. Wormholes can also be produced by the closed walls under some circumstances [195]. The number of events and papers on this subject grows quickly, so that we refer a reader to up-to-date reviews [161, 213].

The idea of PBH clusterization appears to be rather promising. Such structures could explain the early quasars observations [202], contribute to dark matter, produce gravitational wave bursts, can be a reason for reionization and pointlike cosmic gamma-ray sources. This idea gives a natural explanation of the origin of massive binary BHs. The possible existence of hundreds of BHs with a nearly solar mass within one parsec around the galactic center could also agree with the PBH cluster structure.

The cluster structure of the spatial distribution was discussed for the first time in [481] and is considered in [41, 205, 326]. It has been revealed that quantum fluctuations of a scalar field near a potential maximum or saddle point during inflation lead to the formation of closed domain walls. After inflation is over, the walls could collapse to BHs as the final state. The PBHs thus formed have a cluster structure, which is also discussed, e.g, in [16].

There is a substantial amount of inflationary models containing a potential of a shape appropriate for wall formation. The most known examples are natural inflation [209] and hybrid inflation [391, 478] (and its supergravity realization). The landscape string idea provides us with a wide class of potentials including saddle points, for example, see the review [222] and references therein. Nevertheless, the formation of closed walls is usually not discussed there.

10.4.1 *The basic idea*

In this section, we will rely on the models that lead to closed wall formation. To figure out the idea, let us assume that the potential of a real scalar field possesses at least two different vacuum states. In such a situation, there are two qualitatively different possibilities to distribute these states in the early Universe. The first possibility implies that the Universe contains approximately equal amounts of both vacuum states, a typical case at the usual thermal phase transition. It is supposed that after the phase transition, the two vacuum states are separated by a vacuum wall (kink). It leads to a wall-dominated Universe that contradicts the observational data. The alternative possibility corresponds to the case where the two vacuum states are populated with different probabilities. In this case, there emerge islands of a less probable vacuum state, surrounded by a sea of the other, more probable vacuum state. The scalar field's initial distribution, which breaks a symmetry between the vacuum states, is an essential condition for such an asymmetric final picture.

In the framework of our model, the formation of such islands starts at the inflationary stage. The fluctuations of the scalar field energy density are similar to those predicted by the standard inflation at its end. The only peculiarity is that the field in different areas tends to different vacuum states after inflation. That leads to the formation of islands of one vacuum embedded in the sea of another one. Nothing prevents them from being larger than the horizon size. The phase transition accompanied by the formation of the walls occurs only after the end of inflation, deeply in the FriedmanRobertsonWalker (FRW) epoch.

Therefore, the scalar field distribution which was formed during inflation allows the formation of closed walls. A wall becomes causally connected as a whole and begins to contract after horizon crossing. As a result, a closed wall's energy may be focused within a small volume inside the gravitational radius. That is a necessary condition for BH creation. The mass spectrum of BHs formed by this mechanism depends on the scalar field potential and the initial conditions for the scalar field at the beginning of inflation.

The next subsection is devoted to a more detailed discussion on the role of the quantum fluctuations in PBH formation.

10.4.2 *Field fluctuations near an extremum of the potential*

Fluctuations near a potential maximum

Some inflationary models suppose creation of our Universe either near a maximum of the inflaton field potential or near its saddle point(s) to supply slow rolling during a sufficient number of e-folds. As is discussed above, these models include the possibility of formation of macroscopically large closed walls from the scalar field. After the end of inflation, these closed walls collapse to BHs if these walls are large and heavy enough [327, 480]. This mechanism is realized in well-known models like Hybrid Inflation [379] and Natural Inflation [209]. A scalar field could be the inflaton itself or a supplementary field.

First of all, we consider a general mechanism of closed wall formation based on quantum fluctuations near a maximum of a real scalar field potential.

An evolving scalar field may be split into a classical part (mean field), governed by the classical equation of motion, and quantum fluctuations [516]. Let us choose a positive value for the initial field,

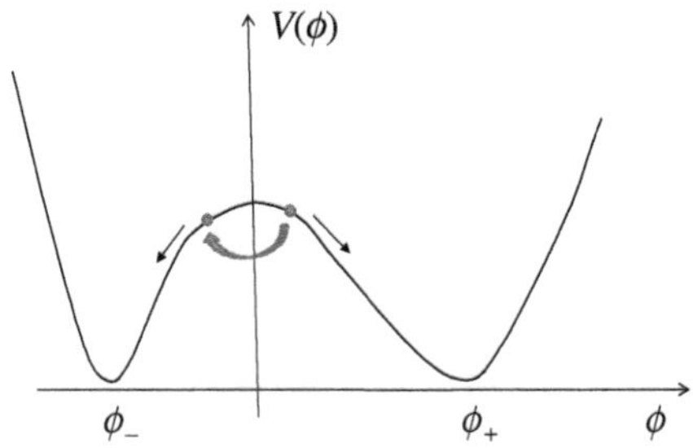

Figure 10.3 Quantum creation of walls during inflation. A right black point relates to the initial field value, $\phi_{\rm in}$.

$\phi_{\rm in} > 0$, as illustrated in Fig. 10.3. The mean field value will slowly grows with time, ultimately reaching a minimum of the potential at some value $\phi_+ > 0$ after the end of inflation. It means that a greater part of space will be finally filled with the field value $\phi = \phi_+$. Meanwhile, the field in some (small) spatial domain could jump over the maximum due to quantum fluctuations. Later on, the mean value of the field representing this fluctuation tends to another minimum of the potential, $\phi_- < 0$. As a result, the space at the final stage will be filled with the vacuum ϕ_+ while some spatial domains will be characterized by the field value $\phi = \phi_- < 0$. If one starts to move from inside the domain to the outside, the path will begin from a spatial point with ϕ_- and end at a point with ϕ_+. Hence, the spatial path must contain a point with the maximum value of the potential. It means that a wall is inevitably formed between such a spatial domain and the "outer" space with $\phi = \phi_+$.

So the "dangerous" values of fluctuations are those with $\phi \leq 0$. Such spatial domains will be surrounded by closed walls, and if their number is too large, it will strongly affect the dynamics of the early Universe. It is useful to calculate the probability of nucleation of these fluctuations. The latter depends on the initial field value $\phi_{\rm in}$ at the moment of creation of our Universe. The corresponding probability

$$P(\phi_{\rm in}, t) = \int_{\phi=-\infty}^{\phi=0} dP(\phi, t; \phi_{\rm in}, 0) \equiv \int_{\phi=-\infty}^{\phi=0} p(\phi, t; \phi_{\rm in}, 0) d\phi, \tag{10.50}$$

to find the field value $\phi \leq 0$ at some spatial point, for reasonable values of the parameters, is presented in Fig. 10.5. This probability determines the ratio of spatial volumes with different signs of the field. Notice that the probability of finding the field within a region Φ of arbitrary shape by the instant t is

$$P(\phi_{\text{in}}, t) = \int_{\Phi} dP(\phi, t; \phi_{\text{in}}, 0). \tag{10.51}$$

To facilitate the analysis, let us approximate the potential near its maximum by

$$V = V_0 - \frac{m^2}{2}\phi^2, \tag{10.52}$$

where the maximum is assumed at $\phi = 0$ without loss of generality. Then the probability to find a certain field value ϕ in the interval $d\phi$ has the form [327] (adapted to the case under consideration):

$$dP(\phi, t; \phi_{\text{in}}, 0) = d\phi \sqrt{\frac{a}{\pi(e^{2\mu t} - 1)}} \exp\left[-a\frac{(\phi - \phi_{\text{in}}e^{\mu t})^2}{e^{2\mu t} - 1}\right]. \tag{10.53}$$

Here, $a = \mu/\tilde{\sigma}^2$, $\mu \equiv m^2/(3H)$, and $\tilde{\sigma} = H^{3/2}/(2\pi)$, with the Hubble parameter $H \simeq \sqrt{8\pi V_0/(3M_{\text{Pl}})}$. From now on, we assume the relation $N \equiv Ht$ of the e-folds number N to the time t passed from the beginning of inflation.

The shape of the potential is often neglected when the inflaton fluctuations are studied. In this case, Eq. (10.53) is reduced to the well-known expression

$$p(\phi, t) = \frac{1}{\sqrt{2\pi}\sigma(t)} \exp\left(-\frac{(\phi - \phi_{\text{in}})^2}{2\sigma^2(t)}\right), \quad \sigma(t) = \frac{H}{2\pi}\sqrt{Ht}. \tag{10.54}$$

Fluctuations near a saddle point

Up to now, we did not pay attention to the energy density of both minima. We assume that the surface energy contained in the closed walls contributes to the PBH mass. Evidently, the energy density difference of the field in and out of the closed walls also contributes. The field fluctuations near saddle points provide us with a simpler

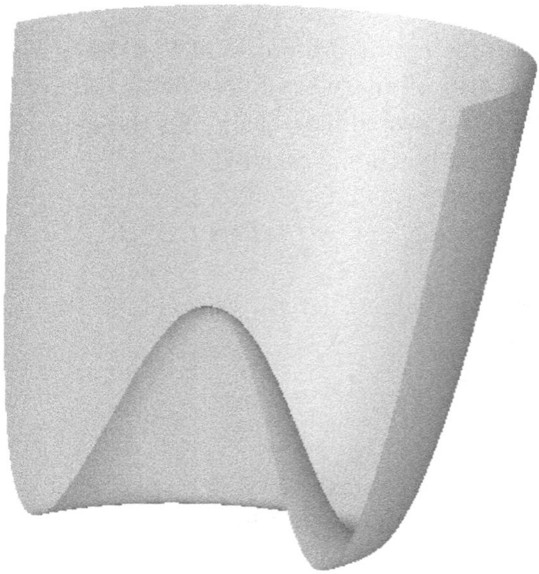

Figure 10.4 The tilted Mexican hat potential as a typical example of the potential with one saddle point and one minimum. The latter may be considered as the countable set of minima at $\theta = 2\pi n$, $n = 0, \pm 1, \pm 2, \ldots$, at the bottom of the potential, if the polar coordinates are used.

situation when the energies of the local minima are equal. The potential of such a sort is represented in Fig. 10.4.

The arguments leading to closed wall formation are mostly similar to those given above but more subtle. Let us suppose that the space under the modern horizon is filled with the field value near a saddle point of the potential. The space contains many causally disconnected regions, the number of which depends on a specific e-fold. Due to quantum fluctuations, the fields in one of these regions, $\mathcal{B}$, could overcome the saddle point, while in the other part of space, $\mathcal{U}$, the fields remain on the same side of the saddle point. Later on the fields in the region $\mathcal{B}$ roll down to the potential minimum. The region $\mathcal{B}$ is surrounded by the region $\mathcal{U}$, where the fields are on the other side of the saddle point. Let this fluctuation happen at the e-fold number N. Then, starting from the e-fold number $N + 1$, internal points of the domains $\mathcal{B}$ and $\mathcal{U}$ are causally disconnected because the distance between them becomes larger than the horizon H_{I}. This means that the fields inside these domains evolve independently.

The potential with a saddle point determines the steepest descent line l_{sd} passing through the saddle point (φ_s, χ_s). The space region $\mathcal{B}$ contains the fields on one side of the saddle point while the fields in another region $\mathcal{U}$ are placed on the other side of the saddle point. The space boundary $\partial\mathcal{B}$ of the domain $\mathcal{B}$ consists of points where $\varphi = \varphi_s$, $\chi = \chi_s$, and therefore the potential is maximum at those points of the line l_{sd}, which belong to the border $\partial\mathcal{B}$.

In the process of further evolution the fields in $\mathcal{B}$ and $\mathcal{U}$ will roll down into the potential minimum $(\varphi_{\min}, \chi_{\min})$, but their trajectories will lie on different sides of the saddle point. As a result, after a long period of independent evolution, the fields arrive at a configuration that encircles a local maximum of the potential. The values of the fields at spatial points x far from the boundary $\partial\mathcal{B}$ are close to $(\varphi_{\min}, \chi_{\min})$. The line l_{sd} contains the saddle point (φ_s, χ_s) with both its ends arriving at the same minimum $(\varphi_{\min}, \chi_{\min})$. Therefore, starting at one end of l_{sd} and moving to the other, the fields' values have to pass through (φ_s, χ_s) that defines the border $\partial\mathcal{B}$, which thus becomes a domain wall. The horizon grows quickly after the inflation is over. When the horizon crosses the scale of the domain wall, the latter starts shrinking, finally producing a PBH. This phenomenon was discussed in [205, 326].

The picture is similar to those for the potential with one maximum and two or more minima with equal energy density. Those fluctuations that overcome a saddle point of the potential will be surrounded by the walls which collapse in the following. All minima relate to one and the same physical minimum, so that their energy densities coincide, see Fig. 10.4.

10.4.3 *Size distribution of the closed walls*

Closed walls are formed at the post-inflationary stages. They collapse to BHs with masses proportional to the squares of the walls. The latter's sizes depend on an instant when the field crossed the potential maximum or the saddle point.

In this section, we discuss the recurrent formula necessary for the PBH mass spectrum calculation. The probability density (10.54) reads

$$p(\theta, N) = \frac{1}{\sqrt{2\pi}\sigma_N} \exp\left\{-\frac{(\phi_U - \phi^2)}{2\sigma_N^2}\right\}; \quad \sigma_N = \frac{H}{2\pi}\sqrt{N}. \quad (10.55)$$

Here time is expressed in terms of e-folds, $t = t_N = H^{-1}N$, $N = 1, 2, 3, \ldots, N_U \simeq 60$, and ϕ_U is the initial homogeneous distribution of the field within the horizon at the moment of the Universe formation.

Suppose that a volume $V(\bar{\phi})$ has been filled by the specific value $\bar{\phi}$ of the scalar field at an e-fold N_t. One e-fold later, such a volume is changed:

$$V(\bar{\phi}, N_t + 1) = e^3 V(\bar{\phi}, N_t) + \left[V_U(N_t) - e^3 V(\bar{\phi}, N_t)\right] p(\bar{\phi}, N_t + 1)\delta\phi. \quad (10.56)$$

The first term in the r.h.s. describes the space expansion. The second term gives the new volume nucleated during one e-fold. Here $V_U(N_t) \simeq e^{3N_t} H^{-3}$ is the Universe volume at the e-fold number N_t, and $\delta\phi = H/2\pi$ is the average field fluctuation at the end of one e-fold. The total number of causally separated volumes of the size H^{-1} can be estimated as

$$K_w(N_t + 1) = H^3\left[V(N_t + 1) - V(\bar{\phi}, N_t)\right]. \quad (10.57)$$

For our aims, we should choose $\bar{\phi}$ to be equal the critical value that delivers the maximum of potential. If the field fluctuation crosses the critical value, it will be surrounded by the wall after inflation. For example, $\bar{\phi} = \phi_{crit} = 0$ for the potential from Fig. 10.3.

The expression (10.57) gives the number of domains that appear during e-fold number $N_t + 1$. These domains will be surrounded by walls of approximately equal sizes that will be formed at the FRW stage. In the following, these walls could collapse into PBHs of equal masses.

10.4.4 *Wall formation and collapse*

Note that the kink connecting the two minima is formed after the end of inflation, when the field that fills the different space areas is settled in both minima $\phi_\pm$, see Fig. 10.3. During inflation, the fields move slowly, and the spatial derivatives of the field are small. The fluctuation that is shown in Fig. 10.3 (left black point) produces a spatiale domain. We will refer to such domains as Proto-Primordial Black Holes (PPBHs). They will be surrounded by a closed wall after the inflation is over and some time has passed. These walls expand to the moments of separation from the cosmological expansion, move with friction through the heated medium, and, finally, begin to contract to scales of the order of the wall width d. The BH appears if

$$d < r_g = 2GM, \tag{10.58}$$

where M is the wall mass.

Let the space domain overcome the potential maximum at an e-fold number $N = Ht$ during inflation. Its size R_w is about H^{-1}. Just after the end of inflation at $t = t_e$ the size has been increased exponentially

$$R_w(t_e) = H^{-1}e^{N_U - N}. \tag{10.59}$$

Denote N_0 as an e-fold of first flipping (leads to the most massive wall). The smallest sizes are formed at $N = Ht_e = N_U$. The space domain expands at $t > t_e$ up the moment of horizon crossing t_+,

$$R_w(t_+) = l_{hor} = 2t_+. \tag{10.60}$$

The rate of expansion is assumed like in RD stage, so that

$$R_w(t) = R_w(t_e)\sqrt{t/t_e}. \tag{10.61}$$

The question is how to fix t_e. The horizon which crosses the wall scale is nucleated during the last e-fold. Hence, to the instant of inflation ending, it lives the time $t_e = H^{-1}$. From (10.61),

$$R_w(t_+) = R_w(t_e)\sqrt{t_+ H}. \tag{10.62}$$

Comparing it with (10.60), $R_w(t_e)\sqrt{t_+H} = 2t_+$, we obtain

$$t_+ = \frac{e^{2(N_U - N)}}{4H}. \tag{10.63}$$

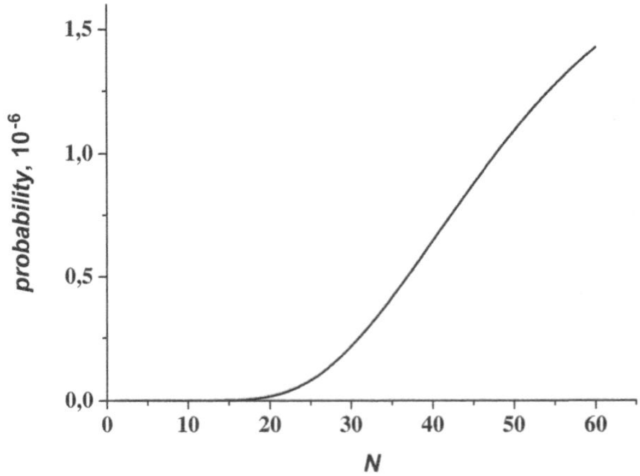

Figure 10.5 The part of space occupied by another vacuum state, depending on the time measured in e-folds. The initial field value is $\phi_{\text{in}} = \phi_{\text{max}} + 3H$, where ϕ_{max} is the field value at a maximum of the potential. The parameter m is chosen to be $m = 0.3H$, where H is the Hubble parameter at the top of the potential.

One can conclude that the wall size estimated at the moment of horizon crossing is

$$R_w(t_+) = 2t_+ = \frac{1}{2}H^{-1}e^{2(N_U - N)}.$$

The probability (10.53) is highly sensitive to the initial field value ϕ_{in}: the closer it is to the maximum of the potential, the greater is the part of the Universe that is covered by walls at a final stage. If the fraction of space within the walls is not very large, the resulting massive BHs formed from the walls could explain the early formation of quasars.

The mass and space distribution of the resulting black holes strongly depend on many factors: the choice of a specific model and the parameters of the corresponding Lagrangian; the initial conditions, the concentration, content and temperature of the media, which is responsible for the friction. Also, the shape of the closed walls is far from an ideal sphere. The same quantum fluctuations responsible for walls appearance lead to the fractal structure of the walls' shape. The closed wall plication was discussed in [326]. Here, we do not discuss this effect and concentrate on the main mechanism.

10.4.5 *PBH cluster structure — a specific example*

In this section, we show that the BH distribution, originated due to the field fluctuations near the extremum of its potential, reveals the cluster structure. This effect distinguishes the model in question from most of other models and leads to various observable consequences.

The mechanism of PBH formation discussed above can be responsible for the appearance of BHs in a wide range of masses, including the supermassive ones. Moreover, as was shown in [481] the cluster character of the PBH distribution is realized naturally. The physical explanation is quite clear. Indeed, according to the mechanism in question, the process of black hole formation begins when a field fluctuation overcomes the critical value (a maximum of the potential or its saddle point) at the inflationary stage.

The space area where the crossing took place is included in a bigger space area. The latter's field also has values close to the critical value because the field distribution during inflation is smooth. It means that the probability of such crossing is also not small for a bigger space area. Hence, the field in new areas will cross the critical value during the next e-fold with meaningful probability. That leads to the cluster structure of the BH distribution with BHs of different masses.

The present section contains a calculation of the BH spectrum around the central BH. The area we consider is much greater (by orders of magnitude) than the size of the first appeared PPBH (the space region that is the progenitor of the most massive BH). It contains a lot of smaller PPBHs. Thus is of interest to study the properties of the distribution of BHs with intermediate masses around the most massive ones.

To be more specific, let us choose the Lagrangian of the complex scalar field φ

$$L_{\varphi} = \frac{1}{2}|\partial\varphi|^2 - V(|\varphi|), \tag{10.64}$$

where $\varphi = r\,\mathrm{e}^{i\vartheta}$ with the potential in the form

$$V(|\varphi|) = \lambda(|\varphi|^2 - f^2/2)^2 + \Lambda^4(1 - \cos\theta), \quad \Lambda \ll f. \tag{10.65}$$

Here, λ and f are parameters of the Lagrangian. The potential has a countable set of saddle points at $\theta = (2n+1)\pi$ and is quite suitable for our goals.

We assume the mass of the radial field component, i.e., the value

$$m_r^2 \equiv d^2V/dr^2|_{r=f/\sqrt{2}},$$

to be sufficiently large, so that the complex magnitude of the field acquires a value somewhere in the circular valley $|\varphi| \cong f/\sqrt{2}$ before the end of inflation. Since the minimum of the potential (10.65) is almost degenerate, the field has the form

$$\varphi \simeq f/\sqrt{2} \cdot e^{i\theta(x)}. \tag{10.66}$$

The Lagrangian for the field $\vartheta(x) = f\theta(x)$,

$$L_\vartheta = \frac{1}{2}(\partial\vartheta)^2 - \Lambda^4(1 - \cos(\vartheta/f)), \tag{10.67}$$

is obtained by the substitution of the expression (10.66) into the Lagrangian (10.64). The kinetic term has the standard form, so that the average quantum fluctuation is $\delta\vartheta = H/2\pi$. Evidently, $\delta\theta = H/2\pi f$, which will be used later.

Consider a spatial region of the scale r at the e-fold number N_0 containing only one PPBH just appearing due to the quantum fluctuation. That means that the PPBH size is of the horizon scale $1/H$, usual for the size of the quantum fluctuations during the slow-roll inflation. By definitions, it contains the phase in the interval $\pi \div \pi + \delta\theta$. The scale r has been nucleated earlier at the e-fold $N_r < N_0$, so that

$$r = H^{-1}e^{N_0 - N_r} \tag{10.68}$$

contains $e^{3(N_0-N_r)}$ causally disconnected regions to the e-fold number N_0. Note that the formula

$$r = r_U e^{N_r - N_U} \tag{10.69}$$

is more adjusted for the post-inflationary stage. Here $r_U \sim 10^{28}$ cm is the size of the visible Universe nowadays and $N_U \sim 60$ is the duration of inflation measured in e-folds.

On the other hand, the well-known formula (10.55) written in the form

$$W = \frac{1}{\delta\theta\sqrt{2\pi(N_r - N)}} \exp\left[\frac{-(\theta_r - \theta)^2}{2\delta\theta^2(N_r - N)}\right] \tag{10.70}$$

represents the probability density to find the phase θ in the interval $\theta \div \theta + \delta\theta$ at an arbitrary e-fold number N within causally disconnected volumes with the size of the order of 1/H provided they have had the phase θ_r at the e-fold number N_r. Here, we have taken into account that

$$\delta\theta = \frac{H}{2\pi f}; \quad \sigma_N = \delta\theta\sqrt{N} \tag{10.71}$$

for the model (10.65). Our nearest aim is to find the value θ_r. To this end, notice that the number of causally independent domains acquiring the phase in the interval $\theta \div \theta + \delta\theta$ during the e-fold number $N > N_r$ is

$$K(N) = W(N, \pi)e^{3(N-N_r)}\delta\theta. \tag{10.72}$$

The second factor represents the number of causally independent domains of the size H^{-1} to the e-fold number $N > N_r$. By definition,

$$K(N_0) = 1 \tag{10.73}$$

(we have chosen an area with one and only one PPBH inside it to the e-fold number N_0). Equation (10.73) can be solved to obtain the phase θ_r at the e-fold number N_r,

$$\theta_r = \pi - \delta\theta\sqrt{\Delta N[6\Delta N - 2\ln(2\pi\Delta N)]}, \quad \Delta N = N_0 - N_r. \tag{10.74}$$

The number of PPBHs which have nucleated to a e-fold number N at the space scale r (see (10.68) or (10.69)) can be found with the use of Eqs. (10.70)–(10.72) and (10.74).

A few words about the surface energy density of the walls are necessary. This parameter strongly depends on the Lagrangian considered. Consider the cosine potential (10.67) for which the one-dimensional kink solution is well known [458]:

$$\theta(z) = -4\arctan\left[e^{\frac{\Lambda^2}{f}(z-z_0)}\right]. \tag{10.75}$$

It is assumed that the wall width d is much smaller than the wall size. Therefore, the wall is almost plane for an observer near the wall, and we can apply Eq. (10.75) for describing the distribution of phase across a plane border separating two areas with different vacuum states. Knowledge of the solution (10.75) allows one to find the wall's surface energy density

$$\mu = 4\Lambda^2 f. \tag{10.76}$$

The energy of this field configuration is concentrated in the surface of width $d = 2f/\Lambda^2$ forming a wall with the center at z_0.

The sections above contain a list of necessary and sufficient formulas to calculate the black holes' mass and space distribution. Examples of such spectra are represented in [41]. The resulting spectra have a peculiar feature: an exponential growth of BH number as their mass decreases. A physical reason is evident: closer to the end of the inflationary period, more closed walls are formed, and they are less "stretched" for the time remaining by the end of inflation. Recall that the mass of a wall is proportional to its area.

The wall's energy contributes to the energy density, meaning that such overdense regions should be separated from the cosmological expansion. It was shown in [205] that the cluster contains a central supermassive black hole that can serve as a seed for an early quasar.

10.4.6 *Processes inside the PBH clusters*

The PBH clusters could be a reason for various phenomena observed in cosmology and astrophysics, identified from different observations. These are the dark matter problem, the origin of supermassive BHs, the BH mergers detected by LIGO/Virgo, the unidentified cosmic gamma-ray sources, the reionization of the Universe, see [42].

In this section, we list the variety of effects to be studied in the nearest future.

On the intermediate-mass BHs

Special effort should be performed to suppress multiple production of small PBH. It has been done in [206]. The mass distribution obtained there is presented in Fig. 10.6. The PBH mass spectra are very sensitive to the initial values $\varphi_{\rm in}$ and $\theta_{\rm in}$. One can notice that a substantial amount of intermediate-mass BHs (IMBHs) is predicted for a specific choice of the parameter values. These IMBHs could be disposed in the galactic halos at significant distances from their centers. While "astrophysical" black holes are formed in the potential wells at centers of dark matter halos [203], the PBHs are, on the contrary, able to form an induced dark matter halo which looks like a dwarf galaxy [204, 205] with a sharp density growth towards the center.

It would be wrong to imagine that absolutely all IMBHs merge to form central SMBHs in modern galaxies. It is more likely that only some fraction will merge, while others should form an IMBH population in galactic halos.

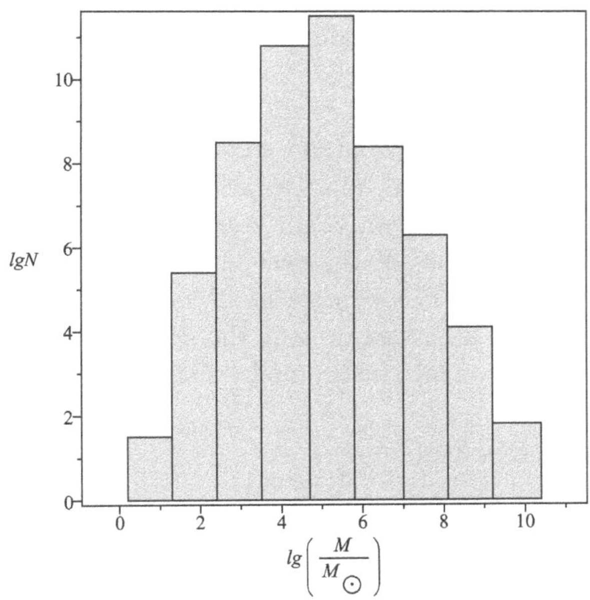

Figure 10.6 Primordial black hole mass distribution with the model parameters $\lambda = 0.005$, $\Lambda = 2.5H$, $f = 10H$, $\varphi_{\rm in} = 28H$, $\theta_{\rm in} = 2.5$.

The temperature inside the cluster

The mechanism of PBH cluster formation dicussed here can be tested by furure observations. Indeed, closed walls, being its necessary attribute, are nucleated having the shape far from the spherical one [31, 49, 329, 481]. The wall inhomogeneities are damped transferring the energy to the surrounding media and heating them. According to our estimate, the heating could be extremely effective, so that the temperature of matter around such a PBH cluster could be high. This allows one to separate our mechanism from others.

After separation from the cosmological expansion, the matter within the regions is heated or cooled due to variety of processes. One of them relates to the walls friction. Let us shortly discuss the others.

- The temperature of matter "freezes" after the cluster is separated from the cosmological expansion. It cools down due to a diffusive process which is much less effective as compared to the cooling due to the cosmological expansion.
- The walls of small masses cannot collapse into BHs. They disappear transferring their energy into the surrounding media.
- Those nuclear reactions that produce neutrino are responsible for cooling due to the energy outflow served by neutrino.
- Also, one should keep in mind the Hawking radiation of PBHs; nuclear reactions, radiation of the stars formed inside the region, gravitational dynamics of the system, shock waves, diffusion of matter, variation of the vacuum state while the region is born. Observation of local heated spots in the sky could be a substantial evidence in favor of the cluster model. Or not.

Supermassive black holes

The fact that each galaxy contains a supermassive BH is almost no longer in doubt. The standard mechanism of SMBH formation requires quite a long time to reach such a large BH mass and is hardly consistent with the SMBH existence at the redshift $z \gtrsim 6$. However, quasars are observed at higher redshifts ($z \sim 7$), and usually, they are associated with supermassive (as a rule $\sim 10^9\, M_\odot$) BHs. The problem of the early quasar origin is widely discussed nowadays. Figure 10.6 demonstrates that the mechanism under consideration can provide a sufficient amount of massive BHs. More detailed calculations may be found in [41, 202, 329].

PBH clusters and dark matter

As a dominant form of DM, PBHs were proposed a long time ago. By now, there have appeared many constraints on such candidates in the wide mass ranges. Nevertheless, all these constraints have been obtained basically for a monochromatic PBH mass distribution. They are not valid for PBH clusters.

There is another advantage of considering the (clustered) PBHs as dark matter. As was noticed in [178], their gravitational interaction allows circumventing the problems of the standard non-interacting CDM scenario like the cusp crisis, dwarf galaxies overproduction, etc.

Gravitational waves from PBH coalescence

By now, BH merger events have been detected by LIGO/Virgo [1]. It seems rather unnatural that pairs of heavy BHs ($\sim 30M_{\odot}$) with low initial spins were formed by the standard astrophysical mechanism. On the contrary, it is hard to avoid such events if the BHs are gathered within clusters.

The gravitational waves are a specific prospective tool for the cluster model validation. The PBH merges in pairs or in the clusters could provide the sources of GW bursts in addition to the binary BHs of stellar origin. One of the most interesting effects are the multiple events of GW bursts from the same PBH cluster. In contrast, binaries of stellar origin produce only isolated events from different galaxies. Therefore, future observations of expected multiple events may provide a crucial test of the PBH clustering discussed here.

PBH clusters as pointlike gamma-ray sources (GRS)

There is a large database on unidentified pointlike cosmic gamma-ray sources (PGRS) obtained by Fermi-LAT [5]. The majority of γ-ray sources are associated with active galactic nuclei, γ-ray pulsars and some other galactic sources. However, there is a significant part of GRS without counterparts found in other wavelengths and thus non-identified. They could have an unconventional origin, which attracts so far special interest. One of the possible explanations is based on the integral Hawking radiation of PBH clusters with a large number of small BHs (with masses $\lesssim 10^{15}$ g).

10.4.7 *Discussion*

The galaxies formation process includes quantum creation of energy density fluctuations, their spatial growth at the inflationary stage, growth of the density after the fluctuation comes under the horizon and the subsequent star formation. The stars are gradually accumulated near a center and merge in the standard picture, thus forming a massive black hole whose mass grows with time. However, the possibility of galaxy formation with supermassive BHs at the center is becoming more and more intriguing and unclear, especially in relation to the discovery of distant quasars with redshifts $z > 6$. Such an early formation of BHs with masses $\sim 10^9 M_\odot$ becomes a serious problem for the standard models of BH formation [201]. Thus, the scenarios with primordial massive BH formation [37, 159, 480, 576] attract attention as a possible alternative to the standard scenario. Such PBHs can serve as condensation centers for the baryons and dark matter [201, 202] with subsequent protogalaxy formation.

In this chapter, we have described a PBH formation mechanism as a result of collapse of closed field walls as well as a formation mechanism of these walls themselves. It has been shown that the shape of the BH mass spectra strongly depends on the parameters of the potential and the initial conditions.

The further evolution of a closed wall goes on as follows. The wall possessing an energy excess tends to the most energetically favorable state, thus heating the media. The fate of a small wall is of minor interest: they contract, different parts of walls collide with each other, and all their energy passes on to expanding waves of the field which had formed the wall. The situation with large walls is quite different. Their contraction delivers enormous energy to a small spatial volume. The gravitational field holds this energy, and a compact object is formed. A more detailed analysis shows that such objects, BHs, form clusters. No stars are required to form these massive primordial black holes! On the contrary, it is such a BH cluster that can create a gravitational potential well and collect there a sufficient number of baryons from which stars can later form.

A particular realization of this mechanism has been developed in the papers [202, 327, 328, 478, 480]. In [202], model parameters have been chosen such that large clusters of massive PBHs with central BH masses of $\sim 4 \cdot 10^7 M_\odot$ are formed from the very beginning. They grow by accretion up to values $\sim 10^9 M_\odot$ and thus describe the observed activity of the distant quasars.

In this scenario, there are several stages of BH and galaxy formation:

1. Formation of closed walls of a scalar field immediately after the end of inflation and their collapse to PBHs, which form a cluster [328, 480, 481]. The most massive PBH is formed at the center of the cluster after crossing the cosmological horizon.
2. Separation of the dense central region of the cluster from the cosmological expansion and its virialization. Many less massive PBHs surrounding the central one merge with it and increase its mass.
3. Separation of the internal regions of the cluster from the cosmological expansion and growth of a protogalaxy.
4. Gas cooling and star generation, merging of protogalaxies, and forming modern galaxies.

One cannot exclude the possibility that both galaxy formation scenarios, the standard one and the one with PBHs, are realized simultaneously. With all that, since at the initial stage the galaxies often merge, it is sufficient that a small fraction of protogalaxies possess massive BHs at their centers to explain the observed abundance of supermassive BHs at galactic centers.

Part III

Extra dimensions

Chapter 11

Multidimensional gravity with higher-order derivatives

The idea of extra dimensions, tracing back to the pioneering works of Thomas Kaluza [313] and Oscar Klein [337], became in the last decades a necessary component of almost all attempts to unify the four physical interactions. The most popular are models with dimensions $D = 10$ (superstrings) and 11 (M-theory), but there are quite a number of still higher-dimensional models. The bosonic sector of such theories usually contains, apart from the metric, different kinds of scalar and vector fields as well as antisymmetric form fields; meanwhile, it is well known that, in the extra (with respect to the usual four) dimensions, the D-dimensional metric components naturally create scalar and vector fields when reduced to the four-dimensional space–time (see, e.g., [543] and references therein).

It is the simplest case of such a reduction, leading to Einstein's gravity and Maxwell's electromagnetic field, that has been described in Kaluza's paper [313]. More complex unified multidimensional theories, providing the advent of boson fields that appear in the Standard Model of particle physics from multidimensional gravity, have been described in the book [543] (see there also references to other works in this area, including numerous papers by members of Yu.S. Vladimirov's group). A separate problem to be solved is geometrization of fermion fields.

Of special interest are multidimensional theories with a gravitational action containing terms nonlinear with respect to curvature.

Their emergence directly follows from quantum field theory in curved space–time [51, 263], and therefore it may not even be considered as an independent postulate.

According to [119, 134], multidimensional gravity with curvature-nonlinear Lagrangians leads to a diversity of low-energy theories owing to arbitrariness of the input parameters of the gravitational action and different variants of the initial data. The latter can include both the metric and the topology of extra dimensions that can emerge from quantum fluctuations at high energies. Thus even at fixed parameters of the initial effective Lagrangian, one can expect the emergence of many kinds of Universes with drastically different properties. Even under comparatively simple assumptions on the nature of multidimensional geometry, after reduction to the four observed dimensions there appear effective scalar fields with complicated potentials, whose properties to a large extent determine the physical properties of one or another universe.

Such a viewpoint is close to the concept of chaotic inflation, according to which there permanently emerge indefinitely many universes at fluctuations of a scalar field (inflaton). With a sufficiently sophisticated shape of the potential, the situation resembles the so-called landscape concept in string theory: the number of different vacua in heterotic string theory reaches 10^{1500} [372], the number of more realistic de Sitter vacua [311] is also huge though finite. We try to make a next step, explaining the origin of such potentials from the standpoint of multidimensional gravity.

Obtaining particular results in multidimensional gravity is substantially simplified by using the slow-change approximation [119, 134] that assumes small values of all derivatives (in a sense to be discussed below) and smallness of all energy densities as compared to Planck scales. Such assumptions turn out to hold even in early cosmology characterized by Grand Unification energies, to say nothing about processes at lower energies and in the realm of modern cosmology. This method allows for considering quite different kinds of gravitational actions and the wide ranges of parameters they contain.

In the Kaluza–Klein concept, the extra spatial dimensions remain invisible due to their extreme smallness. In the recent years, however, an alternative concept became a subject of active studies, namely, the brane-world concept, which considers our Universe as a distinguished three-dimensional (or four-dimensional if time is included) surface or layer, called the brane, in a multidimensional space–time where the extra dimensions have large or even infinite size. They remain invisible because the Standard Model fields (and hence the observers) are assumed to be concentrated on the brane while gravity (and as a rule only gravity) is allowed to propagate in the surrounding space, called the bulk. The history of such models is traced back to the early 80s, with papers by Akama [12], Rubakov and Shaposhnikov [475] and others. The outburst of interest in such models is basically related to some achievements of string theory and M-theory, in particular, with the well-known Hořava–Witten 11-dimensional model [295, 567], in which one of the extra dimensions has a much larger size than others. This approach was shown to be able to suggest a natural mechanism of solving the hierarchy problem in particle physics ([459] and others), while in weak fields the Newtonian law of gravity is preserved as required by modern experiments. Other unsolved basic problems of particle physics as well as cosmology have received significant progress in brane-world models.

A separate problem is the very number of extra dimensions which varies from zero to infinity in works by different authors. Introducing the notion of "extended superspace", we will arbitrarily vary this number. Originally, the concept of superspace meant the set of various geometries [560]; subsequently, the set of all possible topologies was included in it. Let us take one more step and extend the superspace to include spaces of various dimensionalities. To be more precise, let us define the extended superspace E as a direct product of superspaces $\mathbb{M}$ of various dimensionalities:

$$\mathrm{E} = \mathbb{M}_1 \times \mathbb{M}_2 \times \mathbb{M}_3 \times \cdots \times \mathbb{M}_D \ldots$$

Here, $\mathbb{M}_n$ is a superspace of dimension $n = 1, 2, \ldots, D$, which is the set of all possible geometries (up to diffeomorphisms) and topologies.

In this part, we discuss the approaches that facilitate the analysis of models on the classical level. Much attention is paid to the method of slow-change approximation. A variety of low-energy manifestations of the extra dimensions are considered.

11.1 Mathematical tools for multidimensional gravity with higher derivatives

Let us enumerate the methods which could be effectively applied to models based on multidimensional gravity. To our knowledge, there are four ways to move forward in the framework of gravity with higher derivatives in D dimensions:

1. **Reduction of the action to scalar-tensor gravity.** This way is effective for the action in the form

$$S = \frac{m_D^{D-2}}{2} \int d^D Z \sqrt{|g_D|} f(R). \tag{11.1}$$

and has been used in Chapter 9.

2. **Derivation of approximate equations provided that the system contains a small parameter.** This method is widely applied below in studying compact extra dimensions. In this case, the small parameter is the ratio of the average four-dimensional Ricci scalar and the internal Ricci scalar of extra dimensions, see Sec. 11.3.2.
3. **Direct solution of the equations of motion.** Discussed in Chapter 9, last section, where the size evolution of the Universe is studied.
4. **The method of trial functions.** This method is applied for complicated metrics, see Sec. 11.4.5 where a specific example is considered. The accuracy estimation is not an easy task.

11.2 Compact extra dimensions — a brief review

The idea that our world can have a dimensionality greater than four has appeared at the beginning of the last century. The idea has turned out to be both simple and profound. Simple because it is relatively simple to imagine, at least using two- and three-dimensional analogs. Thus, a sheet of paper is perceived (and used) as a two-dimensional object, but it is unambiguously three-dimensional for microbes. Evidently, due to a small size of, say, the fifth dimension, it can avoid our perception. The idea is deeply owing to its numerous and unexpected consequences. It is necessary to master the corresponding mathematical tools to obtain significant results. Therefore, in what follows we briefly present the necessary information.

Consider a D-dimensional differentiable manifold $\mathbb{M}_D$, which can be presented as a direct product,

$$\mathbb{M}_D = \mathbb{M}_4 \times \mathbb{M}_d. \tag{11.2}$$

The extra-dimensional space $\mathbb{M}_d$ is supposed to be compact and have a volume $V_d < \infty$. Let us introduce some coordinates in $\mathbb{M}_D$: $\{X\} = (\{x\}, \{y\})$, where the set x^μ, $\mu = 1, 2, 3, 4$ describes the four-dimensional space–time $\mathbb{M}_4$ and the set y_a, $a = 1, 2, \ldots, d$ the extra space $\mathbb{M}_d$. The interval is written in terms of the metric tensor as usual:

$$ds^2 = G_{MN} dX^M dX^N, \tag{11.3}$$

where the indices M, N run over the values $1, 2, \ldots, d + 4$. The volume of the extra space is expressed in terms of the positive-definite "internal" metric γ_{ab}:

$$V_d = V(\mathbb{M}_d) = \int d^d y \sqrt{\gamma} < \infty, \tag{11.4}$$

where $\gamma = \det(\gamma_{ab})$. The latter inequality is a necessary condition of compactness of this subspace.

11.2.1 *Scalar fields and extra spaces*

To demonstrate the opportunities, let us consider the behavior of a scalar field Φ defined in the full space $\mathbb{M}_D$, with the action

$$S = \int_{\mathbb{M}_D} d^D X \sqrt{|G|} \left[\frac{1}{2} (\partial_M \Phi \partial^M \Phi) - U(\Phi) \right]. \tag{11.5}$$

Varying the action in Φ with a fixed metric, we obtain the equation of motion

$$\Box_D \Phi + dU/d\Phi = 0.$$

Here, the D-dimensional d'Alembert operator is defined as follows:

$$\Box_D \Phi \equiv \nabla_M \nabla^M \Phi = \frac{1}{\sqrt{|G|}} \partial_M (\sqrt{|G|} \partial^M \Phi), \tag{11.6}$$

where $G = \det(G_{MN})$. If we restrict ourselves to small deflections from an equilibrium position (a minimum of the potential), we can approximate the potential with a quadratic expression $U(\Phi) \simeq M^2\Phi^2/2$. Then the equation of motion simplifies to take the form

$$(\Box_D + M^2)\Phi = (\Box_4 + \Box_d + M^2)\Phi = 0,$$

where $\Box_4$ and $\Box_d$ are d'Alembertian operators in the corresponding subspaces.

Let us introduce an orthonormal set of functions $Y_n(y)$, such that

$$\Box_d Y_n(y) = \lambda_n Y_n(y). \tag{11.7}$$

Since $\mathbb{M}_d$ is a compact space, the eigenvalues of the d'Alembert operator λ_n are discrete. The orthonormality condition is written as follows:

$$\int_{\mathbb{M}_d} d^d y \sqrt{|\gamma|} Y_n(y)^* Y_m(y) = \delta_{nm}. \tag{11.8}$$

Expanding Φ in a series in the eigenfunctions $Y_n(y)$,

$$\Phi(X) = \sum_n Y_n(y) \Phi_n(x). \tag{11.9}$$

and taking into account the expressions (11.7) and (11.8), it is easy to write the Φ field equation in the space $\mathbb{M}_4$:

$$(\Box_4 + \lambda_n + M^2)\Phi_n(x) = 0. \tag{11.10}$$

The fields $\Phi_n(x)$ are perceived by an observer as scalar fields with masses $M_n = M^2 + \lambda_n$. The nonobservability of scalar particles appearing from extra dimensions is usually explained by their large masses.

Let us consider a widespread example of extra space. Suppose that the extra space M_d has the geometry of a d-dimensional torus, $M_d = T^d \times \mathbb{S}^1 \times S^1 \times S^1 \times \cdots \times S^1$, $y^m \in [0,\ 2\pi R]$. Then the explicit form of the eigenvalues (11.7) is well known:

$$\lambda_n \equiv \lambda_{\{n_1,n_2,\ldots,n_d\}} = \sum_{j=1}^{d} \frac{n_j^2}{R^2}. \tag{11.11}$$

The eigenfunctions then have the form

$$Y_{\{n_1,n_2,\ldots,n_d\}}(y) = \frac{1}{\sqrt{V_T}} \exp\left[\frac{i\sum n_j y^j}{R}\right], \quad V_T = (2\pi R)^d. \tag{11.12}$$

If the radius R of the torus, characterizing the size of the extra space, is small, then all excitations with $j \geq 1$ correspond to supermassive particles.

11.2.2 *A Kaluza–Klein model with a single extra dimension*

A closed one-dimensional manifold (a closed curve) is, from the viewpoint of its internal geometry, a circle $\mathbb{S}^1$. Therefore, the structure (11.2) in the case $d = 1$ is

$$\mathbb{M}_5 = \mathbb{M}_4 \times \mathbb{S}^1. \tag{11.13}$$

It is convenient to represent the metric of the full space $\mathbb{M}_5$ as a matrix:

$$G_{AB} = \begin{pmatrix} G_{\mu\nu} & G_{\mu 5} \\ G_{5\mu} & G_{55} \end{pmatrix}, \tag{11.14}$$

where G_{55}, $G_{5\mu}$, $G_{\mu\nu}$ are the scalar, vector and tensor in the four-dimensional space $\mathbb{M}_4$.

How to extract gauge symmetry from extra dimensions

To understand the idea, let us make the restriction $G_{55} = \text{const} = -1$. Then transformations of the form

$$\begin{aligned} x'^{\mu} &= x^{\mu}, \\ y' &= y + \varepsilon(x), \quad \varepsilon(x) \ll 1 \end{aligned} \tag{11.15}$$

affect the metric in the following way:

$$G'_{\mu\nu} = \frac{\partial x^A}{\partial x'^{\mu}} \frac{\partial x^B}{\partial x'^{\nu}} G_{AB}, \quad G'_{\mu 5} = \frac{\partial x^A}{\partial x'^{\mu}} G_{A5},$$

while the coefficient G_{55} remains invariable. In our case of infinitesimal shifts,

$$\begin{aligned} G'_{\mu\nu} &= G_{\mu\nu} - G_{\nu 5}\partial_{\mu}\varepsilon - G_{\mu 5}\partial_{\nu}\varepsilon + O(\varepsilon^2), \\ G'_{\mu 5} &= G_{\mu 5} - G_{55}\partial_{\mu}\varepsilon = G_{\mu 5} + \partial_{\mu}\varepsilon. \end{aligned} \tag{11.16}$$

The very idea of extra dimensions has emerged from a desire to describe the gravitational field and the electromagnetic (gauge) field in a unified form. As components of the latter, one can take the off-diagonal elements of the metric tensor $G_{\mu 5}$, because, as is seen from (11.16), their transformations strongly resemble gauge ones. The situation is somewhat more complex with the choice of the four-dimensional metric. A choice of $G_{\mu\nu}$ as the metric tensor in the main subspace $\mathbb{M}_4$ does not seem good because it depends on the choice of the extra-space coordinate, (11.16). On the other hand, it is not hard to understand that

$$g_{\mu\nu} = G_{\mu\nu} + G_{\mu 5}G_{\nu 5},$$

is invariant under the infinitesimal transformations (11.15). It is this tensor that is convenient as a choice of the metric tensor in the main subspace $\mathbb{M}_4$. Denoting $B_{\mu} = -G_{\mu 5}$, we write the metric tensor in the form

$$G_{AB}(x) = \begin{pmatrix} g_{\mu\nu} - B_\mu B_\nu & -B_\mu \\ -B_\nu & -1 \end{pmatrix}. \tag{11.17}$$

The inverse matrix of G_{AB} is

$$G^{AB}(x) = \begin{pmatrix} g^{\mu\nu} & -B^\mu \\ -B^\nu & -1 + B_\lambda B^\lambda \end{pmatrix}. \tag{11.18}$$

The quantity B_μ, under the transformation (11.15), is transformed as

$$B'_\mu = B_\mu - \partial_\mu \varepsilon, \tag{11.19}$$

which is of particular interest: it is in this way that the vector potential of the electromagnetic field transforms under the action of the group $U(1)$ on the wave function of a charged particle. The greatest advantage of Kaluza–Klein theories is that gauge symmetries follow from the symmetry of a compact extra space instead of being additionally postulated.

The invariance of the action (11.5) under the transformations (11.15) causes the emergence of gauge invariance of this action under the transformations (11.19).

Let us look how the transformation (11.15) affects a real scalar field $\Phi(x, y)$. It can be represented by the Fourier series (11.9)

$$\Phi(x, y) = \sum_n \Phi_n(x) Y_n(y),$$

where $Y_n(y)$ is an orthonormalized set of eigenfunctions of the operator ∂_y^2 acting in the extra space $\mathbb{M}_1$. Since in our case it is a circle of a certain radius r, we have

$$Y_n(y) = \frac{1}{\sqrt{2\pi r}} \, \mathrm{e}^{-iny/r}.$$

The symmetry of the extra space (circle) manifests itself in the invariance of the functions $Y_n(y)$ under the shifts $y \to y + 2\pi r$. The scalar field is transformed by the substitution $y \to y + \varepsilon$ in the following way:

$$\Phi'(x, y) = \Phi(x, y + \varepsilon).$$

Comparing the Fourier transforms of both parts of the equality, we obtain

$$\Phi'_n(x) = \Phi_n(x)e^{i(n/r)\varepsilon}.$$

It is in this way that a charged scalar field transforms at gauge transformations. It means that the Fourier component $\Phi_n(x)$ can be perceived as a scalar field with the charge $q_n \equiv n/r$.

We see that $U(1)$ symmetry of the extra space implies the existence of the gauge 4-vector B_μ and charged particles in the space $\mathbb{M}_4$. It is clear that the Lagrangian in four dimensions that we are going to obtain will be invariant under gauge transformations since they are created by the transformations (11.15), under which the initial Lagrangian is manifestly invariant. Nevertheless, to obtain an explicit form of the low-energy action, some further reasoning is necessary.

Let the action of an arbitrary scalar field Φ have the form

$$S = \int d^5 z\sqrt{G}\left(-\frac{1}{2}\Phi\Box_G\Phi - \frac{1}{2}m^2\Phi^2\right).$$

The purely kinetic term in the action is written as

$$\frac{1}{2}\frac{\partial\Phi}{\partial x^A}\frac{\partial\Phi}{\partial x_A}.$$

The two expressions differ by a total derivative that does not affect the classical equations of motion for the field Φ. We choose the five-dimensional metric in the simplest form and approximate it as follows, see (11.17):

$$G_{AB}(x) = \begin{pmatrix} \eta_{\mu\nu} - B_\mu B_\nu & -B_\mu \\ -B_\nu & -1 \end{pmatrix}. \tag{11.20}$$

We here neglect the dependence of the gauge field $B_\mu(x)$ on the coordinate y of the extra space and the dynamics of the metric coefficient G_{55}. It means physically that, by assumption, all relaxation processes in the fifth dimension have finished. The d'Alembert operator is presented as

$$\Box_G = \eta^{\mu\nu}(\partial_\mu - B_\mu \partial_y)(\partial_\nu - B_\nu \partial_y) + \Box_y.$$

Substituting this expression into the action, we obtain

$$S = \frac{1}{2}\sum_n \int d^4x\sqrt{-g}\Phi_n^*(-\Box_n + m_n^2)\Phi_n,$$

where

$$\Box_n = g^{\mu\nu}(\partial_\mu - iq_n A_\mu)(\partial_\nu - iq_n A_\nu), \quad A_\mu \equiv \frac{-1}{\sqrt{2\varkappa}}B_\mu,$$

$$q_n = n(\sqrt{2\varkappa}/R), \quad m_n^2 = m^2 + n^2/R^2.$$

In the language of field theory, the resulting effective action describes a set of scalar fields with masses m_n^2 and charges q_n, interacting with the gauge field A_μ in Minkowski space.

Generalization

Up to now we have been considering the simplest version of a 5D theory where the metric coefficient G_{55} is coordinate-independent. Let us extend the boundaries of what was presented by assuming $G_{55} \neq \text{const}$.

It is convenient to introduce the field B_μ, generalizing the one introduced previously in the following way:

$$G_{\mu 5} = -\sqrt{|G_{55}|}B_\mu. \tag{11.21}$$

As will be shown a little later, the 5-metric can be written as

$$\begin{pmatrix} g_{\mu\nu} - e^{2\phi}B_\mu B_\nu & -\mathrm{e}^{\phi}B_\mu \\ -\mathrm{e}^{\phi}B_\mu & -e^{2\phi} \end{pmatrix}, \tag{11.22}$$

where we have introduced the scalar field ϕ:

$$G_{55} \equiv -e^{2\phi}. \tag{11.23}$$

This result makes it possible to single out the gauge symmetry in the five-dimensional Hilbert–Einstein action. Indeed, the Ricci scalar

R_5 of the 5D space is expressed in terms of the corresponding scalar R_4 of the 4D space as follows:

$$R_5 = R_4 - 2e^{-\phi}\Box e^{\phi} - \frac{1}{4}e^{2\phi}F_{\mu\nu}F^{\mu\nu}; \tag{11.24}$$

$$F_{\mu\nu} = \partial_\mu B_\nu - \partial_\nu B_\mu. \tag{11.25}$$

Then the 5D Einstein–Hilbert action, after some calculation, takes the form

$$\begin{aligned} S &= \frac{1}{16\pi G_5}\int_{\mathbb{M}_5} d^5X\sqrt{|G|}(R_5 - 2\Lambda) \\ &= \frac{1}{16\pi G_4}\int_{\mathbb{M}_4} d^4x\sqrt{|g|}\left(R_4 - 2\Lambda' - \frac{3}{2}\partial_\mu\phi\partial^\mu\phi - \frac{1}{4}\,\mathrm{e}^{3\phi}F_{\mu\nu}F^{\mu\nu}\right). \end{aligned} \tag{11.26}$$

The derivation of this formula is omitted due to its lengthiness. During the calculation, the following conformal transformation was done

$$g_{\mu\nu} \to \tilde{g}_{\mu\nu} = \mathrm{e}^{\phi}g_{\mu\nu}$$

and we have denoted $G_4 = G_5/(2\pi R)$, $\Lambda' = \frac{4}{3}\Lambda$. In (11.26) the metric $\tilde{g}$ is used.

Thus, having assumed the existence of a single extra dimension, we have obtained 4D gravity with a massless scalar field interacting with an Abelian gauge (electromagnetic) field. The theory with such an interaction is often called dilaton gravity, and the field ϕ is called the dilaton. If $\phi = \text{const}$, it is the usual Einstein–Maxwell theory.

The geometric meaning of $g_{\mu\nu}$

The extra space in the form of a circle has the symmetry (isometry group) $U(1)$, whose transformations are rotations by an arbitrary angle. The metric tensor $g_{\mu\nu}$ of the main (i.e., our) space is invariant under such transformations. It is this group that acts as a gauge group for the gauge field B_μ obtained from the geometry and interpreted as the electromagnetic field.

Owing to the simplicity of the extra space we were able to easily construct a suitable combination from the quantities $G_{\mu\nu}$, $G_{\mu 5}$ and G_{55}. In more involved cases this procedure is not so obvious. It is therefore necessary to come to a clearer understanding of the meaning of the quantities introduced.

Since the metric tensor G_{AB} has off-diagonal elements, the space $\mathbb{M}_4$ with the metric $G_{\mu\nu}$ is not orthogonal to $\mathbb{M}_1$. Let us find a space U_4 orthogonal to $\mathbb{M}_1$. We will seek a small displacement vector, orthogonal to $\mathbb{M}_1$, in the form

$$\vec{\delta} \equiv dx^\mu \vec{e}_\mu + dy\vec{e}_5, \tag{11.27}$$

where $\vec{e}_\mu$, $\vec{e}_5$ are a set of five unit vectors, and $\vec{e}_5$ is tangent to $\mathbb{M}_1$. For the orthogonality condition written as

$$(dx^\mu \vec{e}_\mu + dy\vec{e}_5) \cdot \vec{e}_5 = 0,$$

we find the displacement dy:

$$dy = -G_{55}^{-1} G_{5\mu} dx^\mu.$$

The absolute value of the displacement vector is

$$\vec{\delta}^2 = \delta^M G_{MN} \delta^N = (G_{\mu\nu} - G_{55}^{-1} G_{\mu 5} G_{\nu 5}) dx^\mu dx^\nu \equiv g_{\mu\nu} dx^\mu dx^\nu. \tag{11.28}$$

Here we have used the equalities $\delta^\mu = dx^\mu$, $\delta^5 = dy$, see (11.27). Now the meaning of the tensor $g_{\mu\nu}$ is clear: it determines the interval in a space orthogonal to $\mathbb{M}_1$.

Vielbeins

Let us introduce the vielbein formalism which is convenient for extension of the above results to compact extra spaces of d spatial dimensions.

It is well known that gravity can be described by a full set (basis) of linearly independent vielbein vectors e_A instead of the metric tensor. The index A is the number of the vector in $\mathbb{M}_5$, $A = 1, 2, \ldots, 5$. The metric tensor components g_{AB} are defined in terms of the vectors

e_A using their scalar product:

$$g_{AB} = e_A \cdot e_B.$$

The basis vectors can be chosen in such a way that the metric tensor defined with their aid have the canonical form $\eta_{AB} = \mathrm{diag}(1, -1, -1, \ldots, -1)$. The vielbein consisting of the vectors $e_{\bar{A}}$ such that $e_{\bar{A}} e_{\bar{B}} = \eta_{AB}$ is called the Lorentz basis. One basis can be expanded by another

$$e_A = b_A^{\bar{A}} e_{\bar{A}} = b_A^{\bar{\mu}} e_{\bar{\mu}} + b_A^{\bar{5}} e_{\bar{5}}. \tag{11.29}$$

We assume summing over repeated indices. The vectors e_5 and $e_{\bar{5}}$ are chosen to be collinear, $e_5 = b_5^{\bar{5}} e_{\bar{5}}$. Then Eq. (11.29) immediately implies that $b_5^{\bar{\mu}} = 0$. The expansion (11.29) can be presented in a matrix form, with the matrix

$$\begin{pmatrix} b_\mu^{\bar{\mu}} & 0 \\ b_\mu^{\bar{5}} & b_5^{\bar{5}} \end{pmatrix},$$

acting on the columns $(e_{\bar{\mu}}, e_{\bar{5}})$,$e_\mu, e_5$. The inverse matrix

$$\begin{pmatrix} h_{\bar{\mu}}^{\mu} & h_{\bar{\mu}}^{5} \\ h_{\bar{5}}^{\mu} & h_{\bar{5}}^{5} \end{pmatrix}$$

has an important property that

$$h_{\bar{5}}^{5} b_5^{\bar{5}} = 1.$$

The components of the metric tensor in $\mathbb{M}_4$ are expressed as follows:

$$\begin{aligned} G_{\mu\nu} &\equiv e_\mu \cdot e_\nu = b_\mu^{\bar{A}} b_\nu^{\bar{B}} e_{\bar{A}} e_{\bar{B}} = b_\mu^{\bar{\mu}} b_\nu^{\bar{\nu}} e_{\bar{\mu}} e_{\bar{\nu}} + b_\mu^{\bar{5}} b_\nu^{\bar{5}} e_{\bar{5}} e_{\bar{5}} \\ &= b_\mu^{\bar{\mu}} b_\nu^{\bar{\nu}} \eta_{\bar{\mu}\bar{\nu}} + b_\mu^{\bar{5}} b_\nu^{\bar{5}} \eta_{\bar{5}\bar{5}} = g_{\mu\nu} - b_\mu^{\bar{5}} b_\nu^{\bar{5}}, \end{aligned} \tag{11.30}$$

where we have introduced the tensor $g_{\mu\nu} \equiv b_\mu^{\bar{\mu}} b_\nu^{\bar{\nu}} e_{\bar{\mu}} e_{\bar{\nu}} = b_\mu^{\bar{\mu}} b_\nu^{\bar{\nu}} \eta_{\bar{\mu}\bar{\nu}}$, which is obviously invariant under coordinate transformations in the extra space. Besides,

$$G_{\mu 5} \equiv e_\mu \cdot e_5 = b_\mu^{\bar{A}} b_5^{\bar{B}} e_{\bar{A}} e_{\bar{B}} = b_\mu^{\bar{5}} b_5^{\bar{5}} e_{\bar{5}} e_{\bar{5}} = -b_\mu^{\bar{5}} b_5^{\bar{5}},$$
$$G_{55} \equiv e_5 \cdot e_5 = b_5^{\bar{5}} b_5^{\bar{5}} e_{\bar{5}} e_{\bar{5}} = -b_5^{\bar{5}} b_5^{\bar{5}}.$$

Defining the field B_μ by

$$B_\mu \equiv B_\mu^5 \equiv h_{\bar{5}}^5 b_\mu^{\bar{5}},$$

we arrive at the expressions obtained above (see (11.22), (11.23), where $b_5^{\bar{5}} = e^\phi$):

$$G_{\mu\nu} = g_{\mu\nu} - b_5^{\bar{5}} b_5^{\bar{5}} B_\mu B_\nu;$$
$$G_{\mu 5} = -b_5^{\bar{5}} B_\mu;$$
$$G_{55} = -b_5^{\bar{5}} b_5^{\bar{5}}.$$

11.2.3 *Kaluza–Klein models — the general case*

There is no reason for which it would be necessary to restrict oneself to a single extra dimension. As will be seen, an increased number of dimensions yields rich opportunities. Let us therefore consider a Riemannian space with an arbitrary number of dimensions, singling out, for evident reasons, the four-dimensional physical subspace.

In the previous section, we have considered splitting of the metric tensor using the vielbein method. This metric is readily extended to an arbitrary number of extra dimensions. In what follows we do things quite similar to the 5D variant.

We introduce the basis vectors: the coordinate basis e_A for splitting $\mathbb{M}_4 \times \mathbb{M}_d$, and $e_{\bar{A}}$, the Lorentz basis of the same space–time. One basis can be expanded by the other:

$$e_A = b_A^{\bar{A}} e_{\bar{A}} = b_A^{\bar{\mu}} e_{\bar{\mu}} + b_A^{\bar{a}} e_{\bar{a}}, \tag{11.31}$$

where again summing is assumed. The indices μ, ν refer to the space $\mathbb{M}_4$ while $a, b, \ldots$ to the space $\mathbb{M}_d$.

The expansion (11.31), by full analogy with the previous case, can be represented in a matrix form with the transition matrix

$$\begin{pmatrix} b^{\bar{\mu}}_{\mu} & b^{\bar{\mu}}_{a} \\ b^{\bar{a}}_{\mu} & b^{\bar{a}}_{a} \end{pmatrix}.$$

In the Lorentz basis, the scalar product $e_{\bar{\mu}} \cdot e_{\bar{a}} = 0$ by definition. Consequently, $b^{\bar{\mu}}_{a} = 0$, while the matrix

$$\begin{pmatrix} h^{\mu}_{\bar{\mu}} & h^{a}_{\bar{\mu}} \\ h^{\mu}_{\bar{a}} & h^{a}_{\bar{a}} \end{pmatrix},$$

is the most general form of the inverse matrix.

The components of the metric tensor G in $\mathbb{M}_4$ can be expressed in terms of the expansion coefficients of one basis with respect to the other:

$$\begin{aligned} G_{\mu\nu} \equiv e_{\mu} \cdot e_{\nu} &= b^{\bar{A}}_{\mu} b^{\bar{C}}_{\nu} e_{\bar{A}} e_{\bar{C}} = b^{\bar{\mu}}_{\mu} b^{\bar{\nu}}_{\nu} e_{\bar{\mu}} e_{\bar{\nu}} + b^{\bar{a}}_{\mu} b^{\bar{c}}_{\nu} e_{\bar{a}} e_{\bar{c}} \\ &= b^{\bar{\mu}}_{\mu} b^{\bar{\nu}}_{\nu} \eta_{\bar{\mu}\bar{\nu}} + b^{\bar{a}}_{\mu} b^{\bar{c}}_{\nu} \eta_{\bar{a}\bar{c}} = g_{\mu\nu} - b^{\bar{a}}_{\mu} b^{\bar{a}}_{\nu}. \end{aligned} \tag{11.32}$$

Besides,

$$G_{\mu a} \equiv e_{\mu} \cdot e_{a} = b^{\bar{A}}_{\mu} b^{\bar{C}}_{a} e_{\bar{A}} e_{\bar{C}} = b^{\bar{a}}_{\mu} b^{\bar{c}}_{a} e_{\bar{a}} e_{\bar{c}} = -b^{\bar{a}}_{\mu} b^{\bar{a}}_{a}, \tag{11.33}$$

and

$$G_{ac} \equiv e_{a} \cdot e_{c} = b^{\bar{a}}_{a} b^{\bar{c}}_{c} e_{\bar{a}} e_{\bar{c}} = -b^{\bar{a}}_{a} b^{\bar{a}}_{c}. \tag{11.34}$$

Here, we have used the basic property of the Lorentzian basis that $e_{\bar{A}} e_{\bar{C}} = \eta_{\bar{A}\bar{C}}$. Let us define the field B_{μ} as follows:

$$b^{\bar{a}}_{\mu} = b^{\bar{a}}_{a} B^{a}_{\mu},$$

or explicitly

$$B_{\mu} \equiv B^{a}_{\mu} \equiv h^{a}_{\bar{a}} b^{\bar{a}}_{\mu}.$$

It addition, it is useful to denote

$$G_{ac} \equiv \phi_{ac},$$

thus reminding the reader that this expression is a scalar with respect to transformations in M_4. Substituting this definition into (11.32), (11.33), (11.34), we obtain the desired formulae for splitting the metric tensor of multidimensional space:

$$G_{\mu\nu} = g_{\mu\nu} + \phi_{ac} B^a_\mu B^c_\nu, \tag{11.35}$$

$$G_{\mu a} = \phi_{ac} B^c_\mu. \tag{11.36}$$

Let us give an explicit form of coordinate transformations for these quantities. Thus, a transformation similar to the 5D case

$$x' = x'(x), \quad y' = y'(x, y),$$

leads to the following relations:

$$g'_{\mu\nu} = \frac{\partial x^\lambda}{\partial x'^\mu}\frac{\partial x^\rho}{\partial x'^\nu} g_{\mu\nu}, \quad G'_{ac} = \frac{\partial y^d}{\partial y'^a}\frac{\partial y^f}{\partial y'^c} G_{df}, \tag{11.37}$$

$$B'_\mu = \frac{\partial x^\nu}{\partial x'^\mu}\frac{\partial y'^a}{\partial y^c} B^c_\nu + \frac{\partial y'^a}{\partial y^c}\frac{\partial y^c}{\partial x^\nu}. \tag{11.38}$$

All these relations are valid without taking into account the symmetry of the extra space. Previously, when studying the 5D model, one could ignore the issue of form invariance of the extra space with respect to the transformation (11.15). Due to the simple geometry of the space $\mathbb{M}_1$, the form of its extra-space metric did not change under the transformations (11.15). If, however, the extra space is multidimensional, then arbitrary coordinate transformations, in general, affect the metric. For this reason, in the low-energy limit, the Lagrangian will not have the standard gauge-invariant form, although the off-diagonal components of the metric will be associated with vector gauge fields.

To obtain a gauge-invariant theory at low energies, it is necessary that the extra space possess the corresponding symmetry. Then the more narrow class of transformations applied to the metric of this space, the one that leaves the metric form-invariant, creates a gauge-invariant theory. We will see what are the consequences of this situation.

Killing vectors

As was previously shown, local transformations of the extra space $\mathbb{M}_d$ only,

$$y^a \to y'^a = y^a + \epsilon^a(x),$$

can be considered as transformations of a gauge field, at least in the 5D case. In general, this affects both the metric G_{AB} of the main space $\mathbb{M}_4$ and that of the extra space $\mathbb{M}_d$. While in the main space it is possible to single out the invariant part $g_{\mu\nu}$, transformations of the extra space metric is an undesirable feature. Therefore it makes sense to address coordinate transformations that leave the extra space metric invariable.

A coordinate transformation $y \to y'$ is called an isometry if it leaves the metric form-invariant:

$$G_{ab}(y) \to G'_{ab}(y') = G_{ab}(y'). \tag{11.39}$$

If such coordinate transformations form a group, it is called an isometry group $\mathcal{G}(\mathbb{M}_d)$ of the space $\mathbb{M}_d$. We will be interested in infinitesimal transformations.

Let there be a set of vectors $k_i^m(y),\ i = 1, 2, \ldots, r$ such that small shifts along them

$$y^m \to y'^m = y^m + k_i^m(y)\varepsilon^i(x), \quad \varepsilon \ll 1 \tag{11.40}$$

leave the metric of $\mathbb{M}_d$ invariable. The vectors $k_i^m(y)$ are called Killing vectors and are generators of the isometry of this space. It can be shown that these vectors satisfy the commutation relations

$$k_i^a\partial_a k_j^b - k_j^a\partial_a k_i^b = f_{ij}^l k_l^b. \tag{11.41}$$

The structure constants f_{ij}^l are determined by the specific isometry group. The maximum possible number of independent Killing vectors is related to the dimension of space in the following way:

$$\max A = \frac{d(d+1)}{2}. \tag{11.42}$$

The equations for determining the Killing vectors can be obtained from (11.39) and read

$$\nabla_a k_{ib} + \nabla_b k_{ia} = 0, \quad k_{ia} \equiv G_{ab} k_i^b(y). \tag{11.43}$$

They are called the Killing equations.

A non-Abelian gauge group as an isometry group of the extra space

Let us apply the above equations for studying a relationship between the symmetry of the extra space and gauge invariance of a low-energy theory in four dimensions. We choose the following coordinate dependence of the metric:

$$g_{\mu\nu} = g_{\mu\nu}(x), \quad G_{mn} = G_{mn}(y). \tag{11.44}$$

Furthermore, we expand the vector B_μ^m in a basis of Killing vectors k_i^n:

$$B_\mu^m = k_i^m(y) B_\mu^i(x). \tag{11.45}$$

This can be done because B_μ^m is a vector in the space $\mathbb{M}_d$, possessing the corresponding symmetry.

Now we can show that the isometry group in $\mathbb{M}_d$ is realized in $\mathbb{M}_4$ as a non-Abelian gauge group. To this end we consider the infinitesimal transformations

$$x'^\mu = x^\mu, \quad y'^\alpha = y^\alpha + k_i^\alpha(y)\varepsilon^i(x).$$

After some calculations, from the transformations (11.37) follows the transformation law for the object B_μ^i:

$$\delta B_\mu^i = -\partial_\mu \varepsilon^i + f_{jl}^i B_\mu^j \varepsilon^l.$$

It is precisely what is called a gauge transformation of the group G. The extra space metric $G_{ab}(y)$ is invariant under these transformations because k_α^i are Killing vectors of the extra space.

The standard gravitational action in D-dimensional space is

$$S_G = \frac{1}{2\varkappa_D}\int d^D X\sqrt{G}(R_D - 2\Lambda). \tag{11.46}$$

Lengthy calculations lead to the following expression for the Ricci scalar:

$$R_D = R_4 + R_d - \frac{1}{4}G_{ab}k_i^a k_j^b F^i_{\mu\nu}F^{j\mu\nu}, \tag{11.47}$$

$$F^i_{\mu\nu} = \partial_\mu B^i_\nu - \partial_\nu B^i_\mu - f^i_{jl}B^j_\mu B^l_\nu, \quad F^{j\mu\nu} = g^{\mu\lambda}g^{\nu\rho}F^i_{\lambda\rho}. \tag{11.48}$$

Substituting this expression into the action, we find

$$S_G = \frac{1}{2\varkappa}\int d^4x d^D y\sqrt{-g}\sqrt{G_d}\left(R_4 + R_d - \frac{1}{4}G_{ab}k_i^a k_j^b F^i_{\mu\nu}F^{j\mu\nu} - 2\Lambda\right).$$

Let us now use our assumption that the extra space metric $G_{ab}(y)$ does not depend on x. For that reason, one can normalize the Killing vectors as follows:

$$\frac{1}{2\varkappa}\int d^d y\sqrt{G}G_{ab}k_i^a(y)k_j^b(y) = \delta_{ij}.$$

Thus the purely gravitational action in D-dimensional space describes gravity and a gauge field in 4D space:

$$S = \int d^4x\sqrt{-g}\left[-\frac{1}{2\varkappa}R_4 - \frac{1}{4}F^i_{\mu\nu}(x)F^{i\mu\nu}(x) - 2\Lambda'\right]. \tag{11.49}$$

11.3 Multidimensional gravity as the effective theory

The Einstein–Hilbert action for a gravitational field, linear in curvature R, completely describes the physical phenomena at low energies where gravity is important. However, it is clear that quantum effects inevitably lead to nonlinear corrections in the expression for the action [215]. In this case, the action should contain terms with higher derivatives in the form of polynomials of various degrees in the Ricci scalar and other invariants. Thus, whatever gravitational action we take as the basis, after applying quantum corrections it takes a form like

$$S = \int d^N x(R + \varepsilon_1 R^2 + \varepsilon_2 R^3 + \varepsilon_3 R^4 + \cdots + \alpha_1 R_{AB} R^{AB} + \cdots),$$

with a set of unknown coefficients depending on the topology of space [148, 468, 518]. Of course, all field theories must be effective since we do not know all heavy states, for example at the Plank scale.

However, the problem is not that acute, since the nonlinear (in the Ricci scalar) theories can be reduced to a linear theory by a conformal transformation. In Sec. 11.1, we discuss several ways to elaborate a multidimensional model with higher derivatives.

In what follows we develop an approach, suggested by the authors, which allows for reducing multidimensional models with higher-order derivatives to the 4D Einstein–Hilbert action with scalar fields. In our papers [119, 134], we have suggested a more general method for reducing arbitrary Lagrangians to the standard Einstein–Hilbert form in the low-energy limit.

11.3.1 *f(R) theory*

Consider a ($D = d_0 + d_1$)-dimensional manifold $\mathbb{M}$, having the simplest geometric structure of a direct product, $\mathbb{M} = \mathbb{M}_0 \times \mathbb{M}_1$, with the metric

$$ds^2 = g_{\mu\nu} dx^\mu dx^\nu + \mathrm{e}^{2\beta(x)} b_{ab} dx^a dx^b, \tag{11.50}$$

where the extra-space metric components b_{ab} do not depend on x^μ, the observable space–time coordinates. For realistic models $d_0 = 4$, but in Eqs. (11.50)–(11.1) we leave an arbitrary d_0 for generality.

The D-dimensional Riemann tensor has the following nonzero components:

$$\begin{aligned}
&R^{\mu\nu}{}_{\rho\sigma} = \overline{R}^{\mu\nu}{}_{\rho\sigma},\\
&R^{\mu a}{}_{\nu b} = \delta^a_b\, B^\mu_\nu, \qquad B^\mu_\nu := \mathrm{e}^{-\beta} \nabla_\nu (\mathrm{e}^\beta \beta^\mu),\\
&R^{ab}{}_{cd} = \mathrm{e}^{-2\beta} \overline{R}^{ab}{}_{cd} + \delta^{ab}{}_{cd} \beta_\mu \beta^\mu,
\end{aligned} \tag{11.51}$$

where capital Latin indices run over all D values, an overbar marks quantities obtained separately from $g_{\mu\nu}$ and b_{ab}, $\beta_\mu \equiv \partial_\mu \beta$, and

$\delta^{ab}{}_{cd} \equiv \delta^a_c \delta^b_d - \delta^a_d \delta^b_c$. The nonzero components of the Ricci tensor and the scalar curvature read

$$
\begin{aligned}
R^\nu_\mu &= \overline{R}^\nu_\mu + d_1\, B^\nu_\mu,\\
R^b_a &= \mathrm{e}^{-2\beta}\overline{R}^b_a + \delta^b_a[\Box_{d_0}\,\beta + d_1(\partial\beta)^2],\\
R &= \overline{R}[g] + \mathrm{e}^{-2\beta}\overline{R}[b] + 2d_1\,\Box_{d_0}\,\beta + d_1(d_1+1)(\partial\beta)^2, \qquad (11.52)
\end{aligned}
$$

where $(\partial\beta)^2 \equiv \beta_\mu\beta^\mu$, $\Box_{d_0} = g^{\mu\nu}\nabla_\mu\nabla_\nu$ is the d_0-dimensional d'Alembert operator, and $\overline{R}[g]$ and $\overline{R}[b]$ are the Ricci scalars obtained from $g_{\mu\nu}$ and b_{ab}, respectively.

Suppose that b_{ab} describes a compact d_1-dimensional space of nonzero constant curvature, i.e., a sphere ($k = 1$) or a compact d_1-dimensional hyperbolic space[1] ($k = -1$) with a fixed curvature radius r_0 normalized to the D-dimensional analogue m_D of the Planck mass, i.e., $r_0 = 1/m_D$ (here, as everywhere, we use the natural units, with the speed of light c and Planck's constant $\hbar$ equal to unity). We have

$$
\begin{aligned}
\overline{R}^{ab}{}_{cd} &= k\, m_D^2\, \delta^{ab}{}_{cd},\\
\overline{R}^b_a &= k\, m_D^2\,(d_1 - 1)\delta^b_a,\\
\overline{R}[b] &= k\, m_D^2\, d_1(d_1-1) = R_b. \qquad (11.53)
\end{aligned}
$$

The scale factor $b(x) \equiv \mathrm{e}^\beta$ in (11.50) is thus kept dimensionless; R_b has the meaning of a characteristic curvature scale of the extra dimensions.

Consider a sufficiently general curvature-nonlinear theory of gravity with the action

$$
S = \frac{1}{2} m_D^{D-2} \int \sqrt{{}^D g}\, d^D x\, [f(R) + L_m], \qquad (11.54)
$$

[1]Compact hyperbolic spaces of constant curvature on the basis of the conventional Lobachevsky space $\mathbb{H}^d$ are isometric to $\mathbb{H}^d/\Gamma$, where Γ is a nontrivial discrete subgroup of the isometry group of $\mathbb{H}^d$, see, e.g., [218]. On possible applications of such (3D) spaces in cosmology see, e.g., [420] and references therein.

where $f(R)$ is an arbitrary smooth function, L_m is a matter Lagrangian and ${}^D g = |\det(g_{MN})|$. The extra coordinates are easily integrated out, and the action is reduced to $d_0 = 4$ dimensions:

$$S = \frac{1}{2}\mathcal{V}[d_1]\, m_D^{D-2} \int \sqrt{{}^4 g}\, d^4x\; e^{d_1\beta}\,[f(R) + L_m], \tag{11.55}$$

where ${}^4 g = |\det(g_{\mu\nu})|$ and $\mathcal{V}[d_1]$ is the volume of a compact d_1-dimensional space of unit curvature. Here metric (11.50), is kept in mind.

11.3.2 *Slow-change approximation — the Einstein frame*

Equation (11.55) describes a 4D theory which is nonlinear in curvature and contains a nonminimal coupling between the effective scalar field β and the curvature. Significant progress can be achieved by using an expansion in the smallness parameter $L_d/L_D \ll 1$, where L_d and L_D are the characteristic sizes of the extra and basic spaces. We will also consider only such geometries whose Ricci scalars satisfy the condition

$$R_d \gg R_D. \tag{11.56}$$

These two conditions do not contradict each other for such widely used geometries as, e.g., a d-dimensional sphere.

Let us simplify the theory in the following manner:

(a) Express everything in terms of four-dimensional variables and $\beta(x)$; note that now

$$\begin{aligned} &R = R_4 + \phi + f_1, \\ &R_4 = \overline{R}[g], \quad f_1 = 2d_1\,\Box_D\,\beta + d_1(d_1+1)(\partial\beta)^2, \end{aligned} \tag{11.57}$$

and we have introduced the effective scalar field

$$\phi(x) = R_b\, e^{-2\beta(x)} = k d_1(d_1 - 1) m_D^2\, e^{-2\beta(x)}. \tag{11.58}$$

Recall that we have $k = \pm 1$ for positive and negative curvature in d_1 extra dimensions, respectively, so that ϕ has different signs in these cases by definition.

(b) Suppose that all quantities are slowly varying, i.e., consider each derivative ∂_μ (including those in the definition of $\overline{R}$) as an expression containing a small parameter ε; neglect all quantities of orders higher than $O(\varepsilon^4)$ (see [215]). More specifically,

$$f_1 \ll R_n \tag{11.59}$$

and

$$|\partial_\mu g_{AB}| \sim \epsilon |\partial_a g_{AB}|. \tag{11.60}$$

(c) Perform a conformal mapping leading to the Einstein conformal frame, where the 4-curvature appears to be minimally coupled to the scalar ϕ.

In the decomposition (11.57), both terms f_1 and R_4 are regarded small in our approach, which actually means that all quantities, including the 4D curvature, are small as compared to the D-dimensional Planck scale. So the only term which is not small is ϕ, and we can use a Taylor decomposition of the function $f(R) = f(\phi + R_4 + f_1)$:

$$f(R) = f(\phi + R_4 + f_1) \simeq f(\phi) + f'(\phi) \cdot (R_4 + f_1) + \cdots , \tag{11.61}$$

with $f'(\phi) \equiv df/d\phi$. Substituting it into Eq. (11.55), we obtain up to $O(\varepsilon^2)$

$$S = \frac{1}{2}\mathcal{V}[d_1]\, m_D^{D-2} \int \sqrt{{}^4 g}\, d^4x\, e^{d_1\beta}[f'(\phi)R_4 + f(\phi) + f'(\phi)f_1 + L_m], \tag{11.62}$$

where β is related to ϕ according to (11.58). The expression (11.62) is typical of a scalar–tensor theory (STT) of gravity in a Jordan frame. We consider this as the "physical" frame which gives us the Planck mass, in particular

$$M_P^2 = \mathcal{V}[d_1] e^{d_1\beta_m} f'(\phi_m); \quad \mathcal{V}[d_1] = \frac{2\pi^{\frac{n+1}{2}}}{\Gamma(\frac{n+1}{2})}. \tag{11.63}$$

Here, ϕ_m is the stationary state at the low energies.

The choice of the Jordan frame as the physical one is controversial, and the discussion about this important point is still an object of intense debate. Arguments are evoked in the literature, for example, in favor of the Einstein frame as the physical frame, one of them is that the energy of the nonminimally coupled scalar field may allow negative energy configurations, see, for example, [237] for a discussion on this topic; see also a discussion of this issue in Chapter 4 of our book.

However, when matter is present, emerge new aspects of the problem since a conformal transformation implies that a minimal coupling between matter and the scalar field in the Jordan frame leads to a nonminimal coupling between them in the Einstein frame. It is argued that this fact does not change the physics but only the interpretation [214]. Here, we consider a pure gravitational model without matter and scalar fields. The Ricci scalar of the extra space is interpreted by an observer as the effective scalar field. According to [214]. the results obtained by the use of the gravitational section only do not depend on the frame.

For an analysis of the scalar field dynamics and, in particular, for finding stationary points, it is helpful to pass on to the Einstein frame where ϕ is minimally coupled to gravity. After the conformal mapping

$$g_{\mu\nu} \mapsto \widetilde{g}_{\mu\nu} = |\Phi(\phi)| g_{\mu\nu}, \quad \Phi(\phi) = \mathrm{e}^{d_1\beta} f'(\phi), \tag{11.64}$$

with the corresponding transformation of the scalar curvature

$$R_4 = |\Phi| \left(\widetilde{R}_4 + \frac{3}{\Phi}\widetilde{\Box}\Phi - \frac{3}{2\Phi^2}(\widetilde{\partial}\Phi)^2 \right), \tag{11.65}$$

(the tilde marks quantities obtained from or with $\widetilde{g}_{\mu\nu}$) the action (11.62) acquires the form

$$S = \frac{1}{2}\mathcal{V}[d_1]\, m_D^{D-2} \int \sqrt{\widetilde{g}}\, d^4x \left\{ [\operatorname{sign} f'(\phi)]\ [\widetilde{R}_4 + K(\phi)] \right.$$
$$\left. - V(\phi) + \frac{\mathrm{e}^{-d_1\beta}}{f'(\phi)^2} L_m \right\}, \tag{11.66}$$

with the kinetic (K) and potential (V) terms

$$K(\phi) = \frac{3}{2}\frac{(\partial\Phi)^2}{\Phi^2} - 2d_1\frac{\Phi^{,\mu}\beta_\mu}{\Phi} + (\partial\beta)^2,$$
$$V(\phi) = -\,\mathrm{e}^{-d_1\beta}\, f(\phi)/f'(\phi)^2. \tag{11.67}$$

It remains to express everything in terms of a single scalar variable, say, ϕ. We can write the action (11.66) in the form

$$S = \frac{\mathcal{V}[d_1]}{2} m_D^{D-2} \int d^4x\sqrt{\widetilde{g}}\,(\mathrm{sign}\, f')L,$$
$$L = \widetilde{R}_4 + \frac{1}{2}K_{\mathrm{Ein}}(\phi)(\partial\phi)^2 - V_{\mathrm{Ein}}(\phi) + \widetilde{L}_m, \tag{11.68}$$

$$\widetilde{L}_m = (\mathrm{sign}\, f')\frac{\mathrm{e}^{-d_1\beta}}{f'(\phi)^2}L_m; \tag{11.69}$$

$$K_{\mathrm{Ein}}(\phi) = \frac{1}{2\phi^2}\left[6\phi^2\left(\frac{f''}{f'}\right)^2 - 2d_1\phi\frac{f''}{f'} + \frac{1}{2}d_1(d_1{+}2)\right], \tag{11.70}$$

$$V_{\mathrm{Ein}}(\phi) = -(\mathrm{sign}\, f')\left[\frac{|\phi| m_D^{-2}}{d_1(d_1-1)}\right]^{d_1/2} \frac{f(\phi)}{f'(\phi)^2}. \tag{11.71}$$

In (11.65)–(11.68), the indices are raised and lowered with the metric $\widetilde{g}_{\mu\nu}$; everywhere $f = f(\phi)$ and $f' = df/d\phi$.

An important remark is necessary. The action (11.62) describes the field evolution at high energies in the Jordan frame. When the scalar field is settled in its minimum, we should express the 4D Planck mass according to (11.63). At the same time, the scalar field is evolving during inflation, and it is worth using the Einstein frame to facilitate the analysis. In this case, we should consider

$$m_4 \equiv \sqrt{V_n} \tag{11.72}$$

as the effective Planck mass during inflation.

Equations (11.68)–(11.71) are valid for both positive ($\phi > 0$) and negative ($\phi < 0$) curvature of the extra dimensions. Minima of the potential V_{Ein}, defined in the Einstein frame, determine stationary points of the scalar field, and they remain stationary points after a transition to any other conformal frame, including the original Jordan frame. It is the Einstein frame that determines the scalar field behavior since its dynamics near an extremum is governed directly by the potential $V(\phi)$ and only implicitly by the metric.

If ϕ resides at a minimum of the potential V_{Ein}, this potential turns into an effective cosmological constant. Being applied to the present cosmological epoch, it can determine the observable dark energy density that drives the accelerated expansion of the Universe. Alternatively, in principle, it may drive inflation in the early Universe.

The potential (11.71) valid for any function $f(R)$ looks quite complex to have some nontrivial extrema. In addition, we have obtained rather a complex form of the kinetic term (11.70). Its properties are known to be as important for the field dynamics as the shape of the potential. For example, as shown in [363], zeros and singular points of the kinetic term may describe stable states of a scalar field.

In our expressions, both the effective potential (11.71) and the kinetic term (11.70) are singular at the values of ϕ where $f'(R)$ [the factor before R_4 in the Jordan-frame action (11.62)] is zero. Unlike many papers restricted to $f'(R) > 0$, we also include models with $f'(R) < 0$. As will be seen below, this opens new promising possibilities such as new minima of the effective potential at which the extra dimensions may be stabilized.

In (11.68)–(11.71), we have changed the sign of the Lagrangian in the case $f' < 0$; to preserve the attractive nature of gravity for ordinary matter, the matter Lagrangian density should appear with an unusual sign from the beginning. As a result, the sign of the whole action of gravity and matter will be unusual. However, there is no effect on the matter equations of motion, and the conventional form of the effective Einstein equations at a stationary value of ϕ will also be preserved. Only the action of the ϕ field itself can be unusual, according to Eqs. (11.70) and (11.71). It should also be noted that the common sign of the total action does not affect quantum transitions as well. Indeed, the transition amplitude is expressed in the path integral technique as $\int \exp(iS[q])Dq$ where $q(t)$ is some dynamic variable. The transition probability

$$\iint \exp\left(iS[q_1] - iS[q_2]\right) Dq_1 Dq_2,$$

is invariant under the substitution $i \to -i$ (with interchanging the integration variables q_1 and q_2).

The following remark is in order here. There is a well-known conformal transformation [34, 35, 220, 231, 320, 395, 397, 562] that leads the action (11.1) to a form characteristic of the Einstein theory with a minimally coupled scalar field χ (instead of an additional degree of freedom connected with R), whose potential is expressed in terms of the function $f(R)$. This is a D-dimensional Einstein picture. The theory thus obtained can be further reduced by one more conformal transformation, now depending on $\beta(x)$, to the 4D Einstein picture. This has been done, in particular, in the papers [274, 275] for quadratic gravity. Our method is technically simpler and is more universal since (see below) it is able to work with other invariants as well, not only $f(R)$. In cases where our approximation works, the results of the two approaches coincide, as shown in [134].

11.3.3 *The first generalization — a more general form of the Lagrangian*

We have used the simplest extension of the Einstein–Gilbert action, quadratic gravity, so that the function $f(R)$ has the form

$$f(R) = aR^2 + bR + c. \tag{11.73}$$

The parameter b is $b = 1$ without loss of generality.

The method described above allows for considering wider classes of Lagrangians. Let us demonstrate that by adding terms proportional to the Ricci tensor squared $R_{AB}R^{AB}$ and the Kretschmann scalar $\mathcal{K} = R_{ABCD}R^{ABCD}$. By common views, these and other high-order curvature terms appear due to quantum corrections, and it seems natural to include them on equal footing with cR^2. Now the action has the form

$$S = \frac{\mathcal{V}[d_1]}{2\varkappa^2} \int d^4x \sqrt{{}^4g}\, e^{d_1\beta}(R + aR^2 + c_1 R_{AB}R^{AB} + c_2\mathcal{K} + c),$$

where the internal variables have been integrated out in full analogy with Eq. (11.1). For the metric (11.50), it is easy to obtain expressions for $R_{AB}R^{AB}$ and $\mathcal{K}$:

$$R_{AB}R^{AB} = \overline{R}_{\mu\nu}\overline{R}^{\mu\nu} + 2d_1\overline{R}_{\mu\nu}B^{\mu\nu} + \mathrm{e}^{-4\beta}\overline{R}_{ab}\overline{R}^{ab} + 2\,\mathrm{e}^{-2\beta}\overline{R}[h] \times [\Box\,\beta + d_1(\partial\beta)^2] + d_1[\Box\,\beta + d_1(\partial\beta)^2]^2, \tag{11.74}$$

$$\mathcal{K} = \overline{\mathcal{K}}[g] + 4d_1 B_{\mu\nu}B^{\mu\nu} + \mathrm{e}^{-4\beta}\overline{\mathcal{K}}[h] + 4\,\mathrm{e}^{-2\beta}\overline{R}[h](\partial\beta)^2 + 2d_1(d_1-1)[(\partial\beta)^2]^2. \tag{11.75}$$

In the slow-change approximation, in the same manner as before, we obtain the 4D effective Lagrangian

$$\sqrt{{}^4g}L = \sqrt{{}^4g}\,\mathrm{e}^{d_1\beta}[R_4(1+2a\phi) + \phi + c_{\mathrm{tot}}\phi^2 + c + (1+2a\phi)f_1 + 2c_1\phi\,\Box\,\beta + 2(c_1d_1 + 2c_2)(\partial\beta)^2], \tag{11.76}$$

where

$$c_{\mathrm{tot}} = a + \frac{c_1}{d_1} + \frac{2c_2}{d_1(d_1-1)}.$$

The conformal mapping (11.64) leads to the Einstein-frame Lagrangian (11.68) with the kinetic and potential terms

$$K_{\mathrm{Ein}}(\phi) = K^{(2)}_{\mathrm{Ein}}(\phi) + \frac{c_1+c_2}{2\phi(1+2a\phi)}, \tag{11.77}$$

$$V_{\mathrm{Ein}}(\phi) = -\,\mathrm{sign}(1+2a\phi) \times \left[\frac{|\phi|}{d_1(d_1-1)}\right]^{d_1/2}\frac{c_{\mathrm{tot}}\phi^2 + \phi + c}{(1+2a\phi)^2}, \tag{11.78}$$

where the term $K^{(2)}_{\mathrm{Ein}}(\phi)$ is taken from (11.70).

The presence of the parameters c_1 and c_2 adds freedom in choosing the shape of the potential. The kinetic term also acquires a more complex form which can significantly affect the scalar field dynamics [363]: zeros of the kinetic term can represent stationary values of ϕ. The ϕ field can be captured in the vicinity of such points in addition to minima of the potential.

In particular, as shown in [118], cosmological models with $\phi \to 0$ as $t \to \infty$, like those discussed in Sec. 11.4.1, become viable.

As a result, we obtain a class of spatially flat cosmologies where both the observable scale factor $a(\tau)$ and the extra-dimensional one, $b(\tau)$ grow exponentially at late times, but $b(\tau)$ grows sufficiently slowly to leave variations of the effective gravitational constant G in the limits compatible with observations. Such models predict substantial changes in the physical laws in the remote future due to further growth of the extra dimensions.

To satisfy the tight experimental constraints on the temporal variation of the effective gravitational constant G, the parameters of the theory (11.74) should satisfy the condition [118]

$$12\left[\frac{d_1+2}{d_1}+\frac{8(c_1+c_2)}{d_1^2}\right] \gtrsim 100. \tag{11.79}$$

Thus, if we leave only $f(R)$ in the initial theory, it is impossible to satisfy the constraint (11.79): to do so, it is necessary to invoke the Ricci tensor squared or/and the Kretschmann scalar with the initial constants c_1 and c_2 such that

$$c_1 + c_2 \gtrsim d_1^2. \tag{11.80}$$

Under the condition (11.80), the model is potentially viable since it combines a de Sitter expansion of the observable Universe with sufficiently slow variations of the gravitational constant.

11.3.4 *The second generalization — several extra factor spaces*

Let us consider nonlinear multidimensional gravity in more general spaces, admitting a reduction to smaller dimensions, namely, in a D-dimensional space–time $\mathbb{M}$ with the structure

$$\mathbb{M} = \mathbb{M}_0 \times \mathbb{M}_1 \times \cdots \times \mathbb{M}_n, \tag{11.81}$$

where $\dim \mathbb{M}_i = d_i$, and the metric

$$ds_D^2 = g_{ab}(x)dx^a dx^b + \sum_{i=1}^{n} \mathrm{e}^{2\beta_i(x)} g^{(i)}, \tag{11.82}$$

where (x) denotes the dependence on the first d_0 coordinates x^a; $g_{ab} = g_{ab}(x)$ is the metric in $\mathbb{M}_0$, and $g^{(i)}$ are x-independent d_i-dimensional metrics of the factor spaces $\mathbb{M}_i$, $i = \overline{1,n}$.

The initial action is taken in the form

$$S = \frac{1}{2} m_D^{D-2} \int \sqrt{{}^D g}[f(R) + c_1 R_{AB}R^{AB} + c_2 \mathcal{K} + L_m], \quad (11.83)$$

where $f(R)$ is an arbitrary function of the scalar curvature R of the space $\mathbb{M}$; c_1, c_2 are constants; R_{AB} and $\mathcal{K} = R_{ABCD}R^{ABCD}$ are the Ricci tensor and the Kretschmann scalar of the space $\mathbb{M}$, respectively; and L_m is the matter Lagrangian. Capital Latin indices encompass all D coordinates, small Latin ones $(a, b, \ldots)$ the coordinates of the factor space $\mathbb{M}_0$, and a_i, $b_i, \ldots$ the coordinates of the spaces $\mathbb{M}_i$.

We will suppose that the "Kaluza–Klein" (KK) factor spaces $\mathbb{M}_i$ are compact and small in size, to exclude their observability by modern experimental tools.

The D-dimensional Riemann tensor has the nonzero components

$$\begin{aligned}
R^{ab}{}_{cd} &= \overline{R}^{ab}{}_{cd}, \\
R^{aa_i}{}_{bb_i} &= \delta^{a_i}_{b_i}\, B^a_{b\,(i)}, \quad B^a_{b\,(i)} := \mathrm{e}^{-\beta_i} \nabla_b(\mathrm{e}^{\beta_i} \beta_i^{;a}), \\
R^{a_ib_i}{}_{c_id_i} &= \mathrm{e}^{-2\beta_i} \overline{R}^{a_ib_i}{}_{c_id_i} + \delta^{a_ib_i}{}_{c_id_i} \beta_{i,a}\beta_i^{;a}, \\
R^{a_ib_k}_{c_id_k} &= \delta^{a_i}_{c_i}\delta^{b_k}_{d_k}\beta_{i,a}\beta_k^{;a}, \quad i \neq k. \qquad (11.84)
\end{aligned}$$

The overbar marks quantities obtained from the factor space metrics g_{ab} and $g^{(i)}$ taken separately, $\beta_{,a} \equiv \partial_a \beta$; it is denoted, as before, $\delta^{ab}{}_{cd} \equiv \delta^a_c \delta^b_d - \delta^a_d \delta^b_c$ and similarly for other kinds of indices.

The nonzero components of the Ricci tensor and the scalar curvature are

$$\begin{aligned}
R^b_a &= \overline{R}^b_a + \sum_i d_i\, B^b_{a(i)}, \\
R^{b_i}_{a_i} &= \mathrm{e}^{-2\beta_i} \overline{R}^{b_i}_{a_i} + \delta^{b_i}_{a_i} [\Box \beta_i + \beta_{i,a}\sigma^{,a}], \quad \sigma := \sum_i d_i \beta_i, \\
R &= \overline{R}[g] + \sum_i \mathrm{e}^{-2\beta_i} \overline{R}_i + 2\Box\sigma + (\partial\sigma)^2 + \sum_i d_i (\partial\beta_i)^2, \qquad (11.85)
\end{aligned}$$

where $\sum_i$ means $\sum_{i=1}^n$; $\sigma := \sum_i d_i\beta_i$; $(\partial\sigma)^2 \equiv \sigma_{,a}\sigma^{,a}$ and similarly for other functions; $\Box = g^{ab}\nabla_a\nabla_b$ is the d_0-dimensional d'Alembert operator; $\overline{R}[g]$ and $\overline{R}_i$ are the Ricci scalars corresponding to g_{ab} and $g^{(i)}$, respectively.

Let us assume that the factor spaces $\mathbb{M}_i$ are d_i-dimensional compact spaces of constant nonzero curvature $K_i = \pm 1$, i.e., spheres ($K_i = 1$) or compact d_i-dimensional hyperbolic spaces ($K_i = -1$) with a fixed curvature radius r_0, normalized to the D-dimensional analogue m_D of the Planck mass, i.e., $r_0 = 1/m_D$ (recall that we use the natural units $c = \hbar = 1$). Then we have

$$\overline{R}^{a_ib_i}{}_{c_id_i} = K_i\, m_D^2\, \delta^{a_ib_i}{}_{c_id_i},$$
$$\overline{R}^{b_i}_{a_i} = K_i\, m_D^2\, (d_i - 1)\delta^{b_i}_{a_i}, \quad \overline{R}_i = K_i\, m_D^2\, d_i(d_i - 1). \tag{11.86}$$

The scale factors $r_i(x) \equiv \mathrm{e}^{\beta_i}$ in (11.82) are dimensionless.

11.3.5 *Slow-change approximation — reduction to d_0 dimensions*

Suppose that all quantities are slowly varying, i.e., consider each derivative ∂_a (including those in the definition of $\overline{R}$) as an expression containing a small parameter ε, and neglect all quantities of orders higher than $O(\varepsilon^2)$.

Then we have the following decompositions:

$$R = \phi + \overline{R}[g] + f_1 \quad f_1 := 2\Box\sigma + (\partial\sigma)^2 + \sum_i d_i(\partial\beta_i)^2;$$

$$f(R) = f(\phi) + f'(\phi)(\overline{R}[g] + f_1) + O(\varepsilon^4);$$

$$R_{AB}R^{AB} = \sum_i d_i\phi_i^2 + 2\sum_i d_i\phi_i[\Box\beta_i + (\partial\beta_i, \partial\sigma)] + O(\varepsilon^4);$$

$$\mathcal{K} = 2\sum_i \frac{d_i\phi_i^2}{(d_i - 1)} + 4\sum_i d_i\phi_i(\partial\beta_i)^2 + O(\varepsilon^4), \tag{11.87}$$

where

$$\phi_i := K_i m_D^2 (d_i - 1)\, \mathrm{e}^{-2\beta_i}, \quad \phi := \sum_i d_i \phi_i. \tag{11.88}$$

The symbol $(\partial\alpha, \partial\beta)$ means $g^{ab}\alpha_{,a}\beta_{,b}$, and $f'(\phi) = df/d\phi$.

As a result, neglecting $O(\varepsilon^4)$ and integrating out all $\mathbb{M}_i$, we obtain the action reduced to d_0 dimensions:

$$\begin{aligned} S &= \frac{1}{2}\mathcal{V}\, m_D^{D-2} \int \sqrt{g_0}\, d^{d_0}x\, \{\mathrm{e}^\sigma f'(\phi)\overline{R}_0 + K - 2V(\phi_i) + \mathrm{e}^\sigma L_m\}, \\ K &= f'(\phi)\, \mathrm{e}^\sigma f_1 + 2\,\mathrm{e}^\sigma \sum_i d_i\phi_i [c_1 \Box\beta_i + c_1(\partial\beta_i, \partial\sigma) + 2c_2(\partial\beta_i)^2], \\ -2V(\phi_i) &= \mathrm{e}^\sigma \left[f(\phi) + \sum_i d_i \phi_i^2 \left(c_1 + \frac{2c_2}{d_i - 1} \right) \right], \end{aligned} \tag{11.89}$$

where $g_0 = |\det(g_{\mu\nu})|$ and $\mathcal{V}$ is a product of volumes of n compact d_i-dimensional spaces $\mathbb{M}_i$ of unit curvature. The expression (11.89) is typical of a (multi)scalar–tensor theory (STT) of gravity in a Jordan frame.

Subtracting a full divergence, we get rid of second-order derivatives in (11.89), and the resulting kinetic term takes the form

$$\begin{aligned} K = f'\,\mathrm{e}^\sigma &\left[-(\partial\sigma)^2 + \sum_i d_i(\partial\beta_i)^2 \right] - 2f''\,\mathrm{e}^\sigma(\partial\phi, \partial\sigma) \\ &+ 4\,\mathrm{e}^\sigma(c_1 + c_2)\sum_i d_i\phi_i(\partial\beta_i)^2, \end{aligned} \tag{11.90}$$

where $f'' = d^2 f/d\phi^2$.

Transition to the Einstein frame

For further analysis, it is helpful to pass on to the Einstein frame using the conformal mapping (11.64). The expression with the scalar curvature in (11.89) transforms as follows:

$$\sqrt{g_0}\, e^{\sigma} f'(\phi)\overline{R}_0 = \sqrt{g_0}\Phi\overline{R}_0 = (\text{sign}\,\Phi)\sqrt{\widetilde{g}}\left[\widetilde{R} + \frac{d_0 - 1}{d_0 - 2}\frac{(\widetilde{\partial}\Phi)^2}{\Phi^2}\right] + \text{div}, \tag{11.91}$$

where the tilde marks quantities obtained from or with $\widetilde{g}_{\mu\nu}$ and div denotes a full divergence which does not contribute to the field equations. The action (11.89) acquires the form

$$S = \frac{1}{2}\mathcal{V} m_D^{D-2} \int \sqrt{\widetilde{g}}\, d^{d_0}x\{[\text{sign}\, f'(\phi)][\widetilde{R} + K_E] - 2V_E(\phi_i) + L_{m(E)}\}, \tag{11.92}$$

with the kinetic and potential terms

$$\begin{aligned} K_E = {} & \frac{1}{d_0 - 2}\left(\partial\sigma + \frac{f''}{f'}\partial\phi\right)^2 + \left(\frac{f''}{f'}\right)^2 (\partial\phi)^2 \\ & + \sum_i \left[d_i + \frac{4}{f'}(c_1 + c_2)\phi_i\right](\partial\beta_i)^2, \end{aligned} \tag{11.93}$$

$$\begin{aligned} -2V_E(\phi_i) = {} & e^{-2\sigma/(d_0-2)}|f'|^{-d_0/(d_0-2)} \\ & \times \left[f(\phi) + \sum_i d_i\phi_i^2\left(c_1 + \frac{2c_2}{d_i - 1}\right)\right], \end{aligned} \tag{11.94}$$

where the tildes are omitted though the metric $\widetilde{g}_{\mu\nu}$ is used, and the indices are raised and lowered with $\widetilde{g}_{\mu\nu}$. The matter Lagrangian is

$$L_{m(E)} = e^{-2\sigma/(d_0-2)}|f'|^{-d_0/(d_0-2)} L_m. \tag{11.95}$$

The quantities β_i and σ are expressed in terms of the n fields ϕ_i, whose number coincides with the number of factor spaces.

11.4 Possible manifestations of extra dimensions in the modern world

The introduction of extra dimensions is justified by the fact that their existence explains a great number of observed effects in the most economic way. In what follows we will present different applications of the general results of mathematical nature described above. In all our constructions we do not use any fields other than gravity. It should be noted that, although the idea of extra dimensions is often connected and even identified with string theory, in our consideration the idea of strings is not employed. We should also warn the reader that we are not building any final and complete theory here. Our goal is to demonstrate the wealth of opportunities that follow from the idea of extra dimensions. Therefore, we do not fix a particular form of the initial Lagrangian and the number of extra dimensions.

It is assumed that a D-dimensional space–time is born as a result of quantum fluctuations from space–time foam. Thus there can appear spaces with different topologies, geometry and dimensionality. A calculation of probabilities of such processes does not seem productive at the present stage, and we simply use the assumption that these probabilities are not zero.

11.4.1 *Self-stabilization of an extra space*

One of the shortcomings of the idea of extra dimensions is the difficulty in explaining why they are stable. The reasons why the size of extra dimensions remains small and finite instead of collapsing to the Planck value or expanding to visible values remain vague. We will show how this problem can be solved in the framework of our approach.

In what follows we consider pure gravity ($L_m = 0$) and use the units $m_D = 1$, thus dealing with dimensionless quantities.

As the first example we consider the action (11.54) with the function $f(\phi) = \phi + a\phi^2 + c$, $\quad a,\ c =$ const. Then (11.70) and (11.71) give the effective potential (11.78)

$$V_{\rm Ein}^{(2)}(\phi) = -\frac{\operatorname{sign}(1+2a\phi)}{[d_1(d_1-1)]^{d_1/2}}|\phi|^{d_1/2}\frac{a\phi^2+\phi+c}{(1+2c\phi)^2}, \tag{11.96}$$

and the coefficient $K_{\rm Ein}$ of the kinetic term

$$K_{\rm Ein}^{(2)}(\phi) = \frac{1}{(1+2a\phi)^2\phi^2} + \left[a^2\phi^2(d_1^2 - 2d_1 + 12) + d_1^2 a\phi + \frac{1}{4}d_1(d_1+2)\right]. \tag{11.97}$$

The potential is plotted in Fig. 11.3.

The existence of minima of the potential indicates possible stabilization of the size of extra dimensions. Indeed, if a field is in a minimum of the effective potential, there is a solution of the form $\phi(t) = \text{const}$. And since the size of the extra space is $L = b \sim 1/\sqrt{\phi}$, it means that this size is fixed.

There is one more minimum of the potential V_{Ein} with $f' > 0$, and located at the point $\phi = 0$. The asymptotic $\phi \to 0$ corresponds to growing rather than stabilized extra dimensions: $b = e^{\beta} \sim 1/\sqrt{|\phi|} \to \infty$. A model with such an asymptotic growth at late times may still be of interest if the growth is sufficiently slow and the size b does not reach detectable values by now.

Thus the behavior of the system is drastically different in different ranges of ϕ and depends on the numerical values of the initial parameters. We also confirm that gravity alone can stabilize the size of extra dimensions, without the need for introducing other fields with specific forms of potentials. Positive values of Λ_{eff} (vacuum energy density) were found in the range where $f' < 0$. As will be seen below, it is a feature of quadratic gravity only.

It is known that the existence of local minima of the potential corresponds to a possible false vacuum decay. As an example, Fig. 11.3 shows the effective potential (11.78), emerging under the condition that the extra space has the topology of a sphere. It is evident that the field located at a metastable minimum will eventually pass on to the basic state with $\phi = 0$. It means that after the phase transition the size of the extra space will grow infinitely. Nevertheless, the Universe existing in a metastable state can survive for quite a long time if the probability of tunneling to the lower minimum is small.

Since we have been working in the Einstein conformal frame, the problem of varying physical constants (above all, the effective Newtonian constant of gravity G_{eff}) did not emerge. One should note that even remaining in the Einstein frame, we could assume $m_D \neq m_4$, which would affect the estimated boundary between the classical and quantum worlds. In a more general framework, interpreting another conformal frame (possibly but not necessarily the original Jordan frame) as the observational one, we obtain a dependence of the constant G_{eff} (hence the current Planck mass $m_4 = G_{\text{eff}}^{-1/2}$) on the size of extra dimensions, which in general cannot only be time-dependent but also vary from point to point in space. In the cosmological context, models with variable G_{eff} should not only satisfy the observational bounds on the variation rate $\dot{G}_{\text{eff}}/G_{\text{eff}}$ ($\lesssim 10^{-13}$) but also take into account the effect of $G(t)$ on stellar evolution and processes in the early Universe. (Therefore, models with self-stabilizing extra dimensions like those discussed in [119, 134] can be more attractive.) In still more general models of this sort even the Planck constant $\hbar$ can be variable. A discussion of these problems can be found, e.g., in [219, 482, 547].

11.4.2 *Relation of the number of extra dimensions and the cosmological constant*

We have discussed some low-energy theories corresponding to special cases of the metric (11.82) at different choices of the initial action. It is not surprising that some values of the parameters are suitable for describing some properties of our Universe. More opportunities emerge if we vary the structure of extra dimensions, including the number of extra factor spaces, their dimension and curvature. In such cases it is possible to obtain a set of substantially different low-energy theories *even with a fixed initial Lagrangian.*

Let us recall once again the well-known shortcoming of chaotic inflation in its simplest, quadratic form. According to the observable temperature fluctuations of the CMB, the inflaton mass should be of the order $10^{-6}\,M_{\text{Pl}}$. Its smallness requires an explanation which so far does not exist. Consider the effective potential (11.94), created by the initial action (11.68). Its shape is presented in Fig. 11.1 for some values of the number of extra dimensions d. It is evident that the values of d can be chosen in such a way as to obtain the required inflaton mass. Numerical estimates lead to the following results: the second-order derivative of the potential at its minimum (which determines the inflaton mass) is $\sim 0.2m_D$ for $d = 3$, $\sim 1.5 \cdot 10^{-3} m_D$ for $d = 5$ and $\sim 0.8 \cdot 0^{-5} m_D$ for $d = 7$. The initial Lagrangian does not contain small parameters. Nevertheless the required value of the inflaton mass emerges at a classical level at a suitable choice of the parameter d.

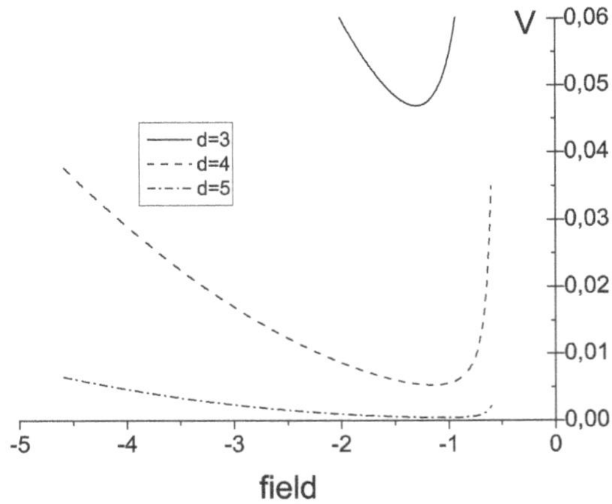

Figure 11.1 Effective potential $V(\phi)$ for the theory (11.68) with $f(R)$ of the form (11.73) for different numbers d of extra dimensions. The parameter values: $c = 1$, $\Lambda = -0.4$.

The strong influence of the number d on the form of the low-energy Lagrangian is not surprising because the potential (11.94) contains the factor $\sim d^{-d}$. Thus, if the stationary state is $\phi = \phi_0$, and the dimensionless parameters $|\phi_0| m_D^{-2}$ and $F'(\phi_0)$ are of the order of unity, then the effective cosmological constant $\Lambda_{\text{eff}} = V_{\text{Ein}}(\phi_0)$ is connected with the function $F(\phi_0)$ (which can be close to m_D^2) as follows:

$$\Lambda_{\text{eff}}/F(\phi_0) \sim [d(d-1)]^{-d/2}. \tag{11.98}$$

It is of interest that $d^{-d} \approx 10^{-123}$ for $d = 67$. It means that space-time fluctuations creating a (67+4)-dimensional space can cause the creation of space with the observable vacuum energy density $10^{-123}\, m_4$. The extreme smallness of Λ_{eff} is thus related to the number of extra dimensions of a space randomly created by quantum fluctuations at the Planck level.

Figure 11.2 shows another example of d-dependence on the effective potential shape. It is evident that the very existence of a minimum of the potential, indicating a possible stable configuration, is d-dependent. Thus, if the Universe was born with a single extra factor space of negative curvature, then $\phi < 0$, see Fig. 11.2. It is clear that with the initial field value $\phi < -1$ its mean value in such a

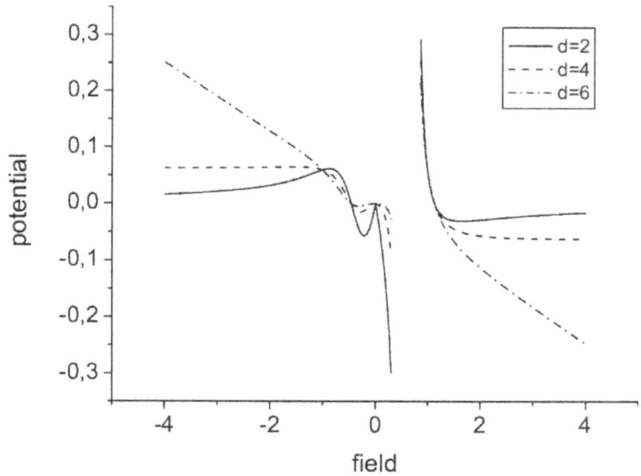

Figure 11.2 Potentials at different numbers d of extra dimensions, obtained from the Lagrangian (11.68) with $F(R) = R + cR^2 + w_1R^3 + w_2R^4 - 2\Lambda$. The parameters: $c = 0,\ w_1 = 0,\ w_2 = -1,\ \Lambda = -0.25$. The curves are adjusted to a single scale.

Universe tends to infinity if $d = 2$, to a constant if $d = 4$ and to zero if $d = 6$. If the Universe was born with $-1 < \phi < 0$, then the mean value of the field is localized at a minimum of the potential, and the size of the extra space remains fixed.

11.4.3 *Rapid particle creation in the post-inflationary period*

Let us address one more problem that can be successfully solved in the class of theories under consideration.

According to [207], rapid oscillations of the inflaton field immediately after inflation create the particles which form the matter content of the modern Universe. It is known [341, 501] that the particle and entropy production mechanism as well as heating of the Universe can be ineffective if the coupling constant of the inflaton and matter fields is too small. On the other hand, a strong coupling leads to significant quantum corrections of the inflaton potential parameters, which casts doubt on their smallness that follows from the observational data on CMB temperature fluctuations. This problem is solved, for instance, by hybrid inflation which includes one more hypothetical scalar field (see Chapter 8). During inflation, the energy

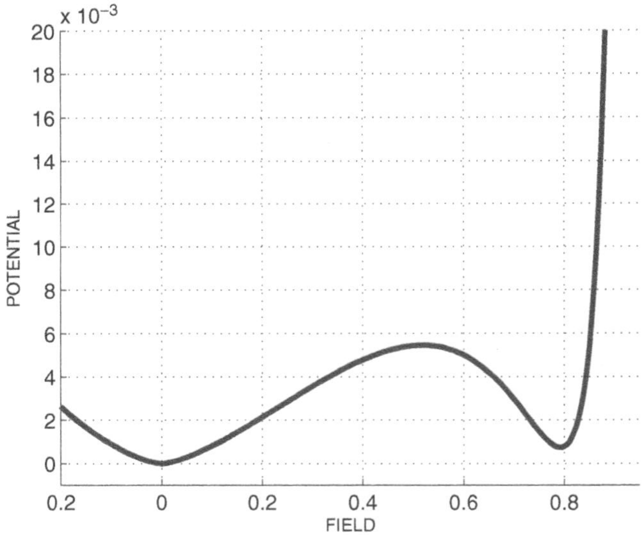

Figure 11.3 The effective potential (11.105). The parameter values: $c = -0.5$, $\Lambda = 0.2$, $c' = -0.626$, and $d = 2$.

density changes slowly until the classical field value reaches a bifurcation point, after which the field begins to rapidly oscillate near a minimum of the potential, producing the required particle number. Unfortunately, this model also leads to black hole over-production, which is a serious shortcoming [478]. Another promising mechanism of effective particle production (parametric resonance) is described in [341]. One more mechanism, connected with quantum production of supermassive (GUT scale) particles, sufficient for explaining the observed wealth of matter, has been considered by Grib and Pavlov [264–267].

Nonlinear multidimensional gravity provides one more opportunity.

Consider the potential and kinetic terms of the effective Lagrangian (11.68) with the function $f(R)$ of the form (11.73) These functions are plotted in Fig. 11.4. The nontrivial ϕ dependence of the kinetic term strongly affects the scalar field dynamics. The effect is especially evident if a minimum of the kinetic term coefficient coincides with a minimum of the potential, as in Fig. 11.4.

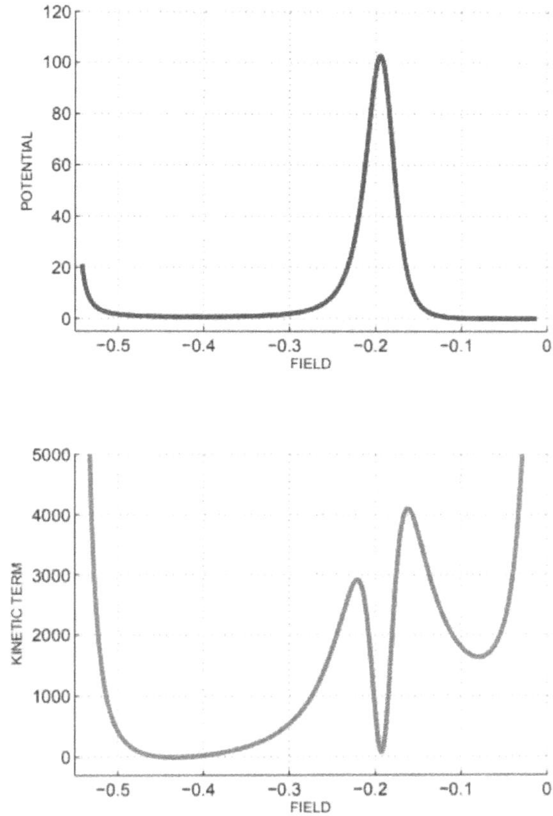

Figure 11.4 The potential and kinetic terms of the effective 4D theory (11.1) with the function $f(R)$ taken from the capture for Fig. 11.2 with the parameters $d = 3, c = 6.0, w_1 = w_2 = 0, \Lambda = 2$.

For illustration, consider a simplified scalar field model with the potential and the coefficient in the kinetic terms of the form

$$V(\phi) = \frac{1}{2}m^2\phi^2$$

$$K(\phi) = K_1 \cdot (\phi - \phi_{\min})^2 + K_{\min}, \quad K_1, \ K_{\min} > 0. \quad (11.99)$$

(The function $K(\phi)$ is positive and can be reduced to unity by a field redefinition $\phi \to \tilde{\phi}$, but this equivalent formulation is more convenient in the present case.)

A numerical solution of the corresponding classical equations for a spatially flat universe,

$$H^2 = \frac{\varkappa^2}{3}\left[\frac{1}{2}K(\phi)\dot{\phi}^2 + V(\phi)\right],$$

$$K(\phi)(\ddot{\phi} + 3H\dot{\phi}) + \frac{1}{2}K_\phi(\phi)\dot{\phi}^2 + V_\phi(\phi) = 0, \tag{11.100}$$

(the subscript ϕ denotes $d/d\phi$, and H is the Hubble parameter) is presented in Fig. 11.5 for the quadratic potential (11.99) and two different kinetic terms, with $K = 1/2$ and the one given in (11.99). The parameter values are given in the figure caption. Each curve describes the inflaton behaviour before and after the end of inflation. In the nonstandard kinetic term, $K(\phi) = 1/2$ at the end of the inflationary stage (the first point of curves' intersection in Fig. 11.5). It is clear that in the model with a variable kinetic term inflation begins at smaller energies as compared to the conventional case.

Let us also note the difference in the oscillation frequency of the inflaton field. The number of created particles is proportional to this

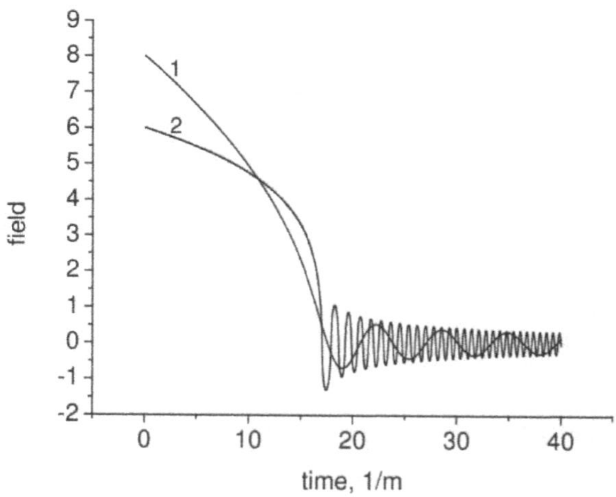

Figure 11.5 The inflaton behavior at the end of inflation for two models: (1) The standard kinetic term, $K = 1/2$; (2) The kinetic term of the form (11.99). The parameter values $K_0 = 0.1, \phi_{\min} = 0.1, K_{\min} = 0.005$ are given in Planck units, and time is measured in inverse inflaton field masses.

frequency, see the review [209] and an additional discussion in [327]. This conclusion can also be grounded as follows.

After the end of inflation, the inflaton oscillation amplitude is small compared with the Planck scale, and the coefficient $K_{\text{eff}} \sim K_{\text{min}}$ in the effective kinetic term is small due to the choice of the parameters. The effective Lagrangian containing an interaction of the inflaton with a certain scalar field χ

$$L_{\text{eff}} \simeq \frac{1}{2} K_{\text{min}} \dot{\phi}^2 + g\phi\chi\chi, \tag{11.101}$$

can be transformed to the standard form by redefining the inflaton field:

$$L_{\text{eff}} \simeq \frac{1}{2}\dot{\phi}^2 + \frac{g}{\sqrt{K_{\text{min}}}}\phi\chi\chi. \tag{11.102}$$

A small value of K_{min} increases the coupling constant (by orders of magnitude for the chosen parameter values) and consequently leads to intense particle creation. In this way one overcomes the difficulty discussed above: a weak interaction constant at the inflationary stage does not contradict a rapid particle production immediately after the end of inflation.

The field behavior near $K = 0$

The function $K(\phi)$, the coefficient of the kinetic term in the scalar field Lagrangian, can in general possess both zeros and poles, and their neighborhood is of special interest. Let us discuss the possible state of the system near a zero value of $K(\phi)$ [363]. If it is a simple zero, then the kinetic term changes its sign when crossing it. The case $K(\phi) = K_{\text{Ein}} > 0$ corresponds to a normal scalar field while the case $K(\phi) < 0$ corresponds to a phantom field with negative kinetic energy.

In general, $f(R)$ theories lead only to positive values of $K(\phi)$. Indeed, it is easy to see that the expression in square brackets in Eq. (11.70) is always positive, with a minimum equal to $d(d+3)/3$.

In more general theories containing other curvature invariants, this is not the case. Let us consider as an example the simplest nonlinear theory with the function

$$f(R) = cR^2 + R - 2\Lambda \tag{11.103}$$

and other second-order nonlinear terms: the Ricci tensor squared $R_{AB}R^{AB}$ and the Kretschmann scalar $\mathcal{K} = R_{ABCD}R^{ABCD}$, so that

$$S = \int d^D x \sqrt{^D g}[f(R) + c_1 R_{AB}R^{AB} + c_2 \mathcal{K}]. \tag{11.104}$$

Calculations in the model (11.104) with the function (11.103) and the metric (11.50), similar to those made above, lead to an effective scalar-tensor theory (11.68) with kinetic (11.77) and potential (11.78) terms

$$V_{\text{Ein}}(\phi) = -\operatorname{sign}(1+2c\phi)[d(d-1)]^{-d/2} \cdot |\phi|^{d/2} \frac{c'\phi^2 + \phi - 2\Lambda}{(1+2c\phi)^2},$$

$$c' = c + \frac{c_1}{d} + \frac{2c_2}{d(d-1)}, \tag{11.105}$$

$$K_{\text{Ein}}(\phi) = \frac{1}{\phi^2(1+2c\phi)^2}\left[c^2\phi^2(d^2 - 2d + 12) + d^2 c\phi + \frac{1}{4}d(d+2)\right]$$
$$+ \frac{c_1 + c_2}{2\phi(1+2c\phi)}. \tag{11.106}$$

The dashed lines A and B in Fig. 11.6 mark zeros of the kinetic term. If a field fluctuation has emerged to the left of point A, the field tends to a minimum of potential at the point $\phi = 0$, which corresponds to a growing size of the extra space. Some fluctuations are created in the range between points A and B, where the kinetic term is negative. Then the field ϕ, being phantom, moves towards a growing potential, to point B. Evidently, point B is an attractor without being a minimum of the potential. A universe for which this situation takes place has a nontrivial basic state, which should be discussed in more detail. To this end, consider a simplified situation:

$$S = \int d^4 x \left[\frac{1}{2}K(\phi)(\partial\phi)^2 - V(\phi)\right], \tag{11.107}$$

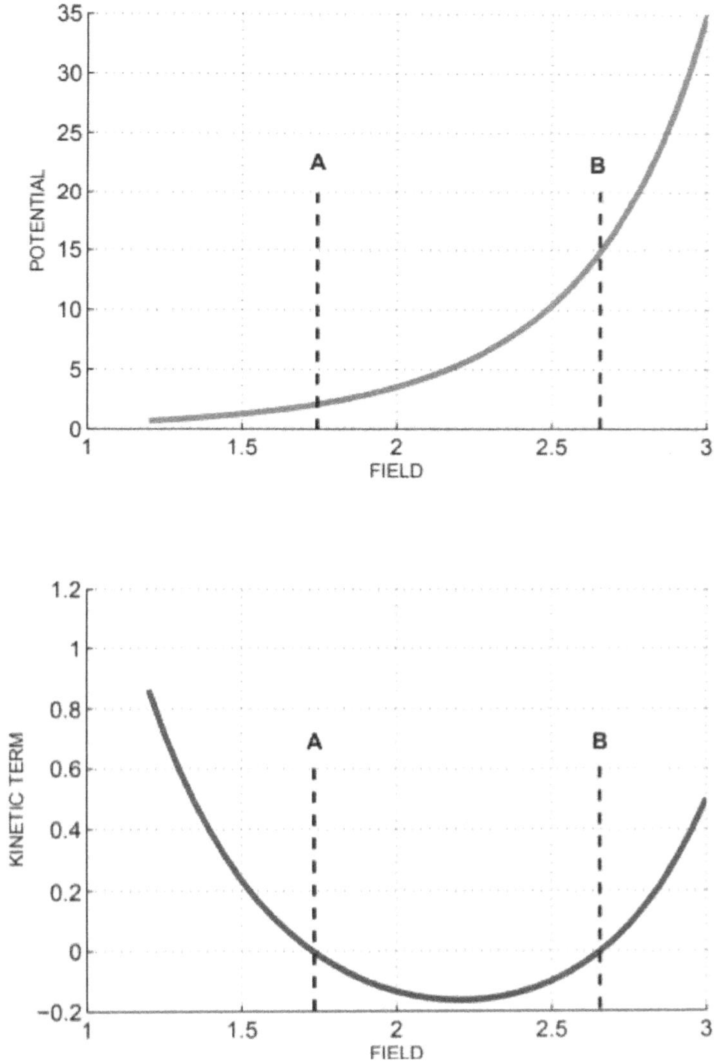

Figure 11.6 The potential (11.105) and the coefficient in the kinetic term (11.106) for the theory (11.104) and $f(R)$ from (11.103). The parameter values are: $d = 3$, $c = -0.12$, $\Lambda = 2$, $c_1 + c_2 = -5.5$, $c' = -0.626$.

with $K(\phi_{\rm crit}) = 0$ and, without loss of generality, put $\phi_{\rm crit} = 0$. Then the behavior of $K(\phi)$ and $V(\phi)$ at small ϕ is $K(\phi) = k\phi + o(\phi)$, and $V(\phi) = V(0) + h\phi + o(\phi)$, where it is assumed that $k > 0$, $h > 0$, referring to point B in Fig. 11.6. Evidently, near the critical point the field tends to it irrespective of where its initial value is chosen, to the left or to the right of it. Nevertheless, a classical motion near the critical point $\phi = 0$ is absent. Indeed, the classical equation for ϕ in curved space–time has the form

$$K(\phi)\,\Box\,\phi + \frac{1}{2}K_\phi(\partial\phi)^2 = -V_\phi, \tag{11.108}$$

or, according to the above expressions for K and V at small ϕ,

$$k\phi\,\Box\,\phi = -\left(h + \frac{1}{2}k(\partial\phi)^2\right) + o(\phi). \tag{11.109}$$

Let $(\partial\phi)^2 > 0$ (as in the cosmological case where $\phi = \phi(t)$). Then the right-hand side of Eq. (11.109) is smaller than $-h < 0$ and does not tend to zero as $\phi \to 0$. Hence the second-order derivative in $\Box\,\phi$ tends to infinity because $\phi \to \phi_{\rm crit} = 0$. Introduction of additional terms with higher-order derivatives into the Lagrangian does not improve the situation. Indeed, in a stationary state, where all derivatives are equal to zero, the classical equation (11.109) with $h > 0$ has no solution. It means that the kinetic energy of a homogeneous basic state cannot be equal to zero.

11.4.4 *Multiple factor spaces and a spatially varying size of the extra dimensions*

Recall that *a priori* we do not assume a fixed number of extra dimensions, their geometry, and topology. All this emerges at the Planck scale due to quantum fluctuations. The more complex the emerging structure of the extra space, the richer the possibilities.

Consider a space with the metric (11.82) and two extra factor spaces: $\mathbb{M} = \mathbb{M}_4 \times \mathbb{M}_{d_1} \times \mathbb{M}_{d_2}$. Using the same action (11.104), we must now introduce two scalar fields in order to describe the low-energy limit of the theory. The effective potential in the Einstein picture has the form

$$V_{\text{Ein}}(\phi_1,\phi_2) = -\frac{1}{2}\,\text{sign}(f'(d_1\phi_1 + d_2\phi_2))\frac{|\phi_1|^{d_1/2}}{[(d_1-1)]^{d_1/2}}\frac{|\phi_2|^{d_2/2}}{[(d_2-1)]^{d_2/2}}$$
$$\times\frac{f(d_1\phi_1+d_2\phi_2) + d_1\phi_1^2\left[c_1+\dfrac{2c_2}{d_1-1}\right]+d_2\phi_2^2\left[c_1+\dfrac{2c_2}{d_2-1}\right]}{[f'(d_1\phi_1+d_2\phi_2)]^2}. \tag{11.110}$$

Figure 11.7 presents a potential with two valleys along the mutually perpendicular directions $\phi_1 = 0$ and $\phi_2 = 0$, each of them corresponding to an infinitely large size of one of the factor spaces $\mathbb{M}_{d_1}$ or $\mathbb{M}_{d_2}$. Of greater interest is the local minimum, where both factor spaces have a finite size. The Universe can be located for a sufficiently long time in this metastable state, as in the case with one extra space with a simpler structure discussed above.

An opportunity of interest emerges if we suppose that the Universe was born at point B in Fig. 11.7. During inflation, the field moves from point B along the arrow, and quantum fluctuations create different field values in causally disconnected domains. The fate of

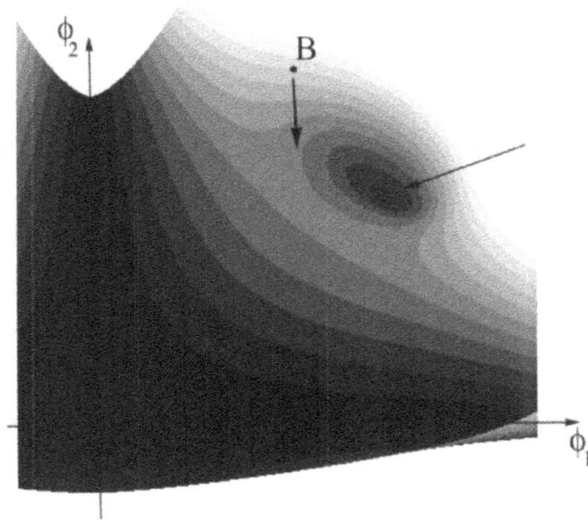

Figure 11.7 The effective potential for a model with the extra space $\mathbb{M}_{d_1}\times\mathbb{M}_{d_2}$, $d_1 = d_2 = 3$, with the parameters $c = -0.5$, $\Lambda = 0.2$, $c_1 = c_2 = -0.38$. View from above, lower levels look darker. A local minimum is marked by a long arrow.

a domain depends on the field value inside it. Even if the majority of domains get into the metastable minimum at the end of inflation, some part of them rushes to one of the valleys. In this case our Universe must contain a certain number of domains with a macroscopically large extra space. Their number and sizes strongly depend on the initial conditions.

If in our Universe there are domains with a large extra space, then the laws of low-energy physics in such domains are different from those we know. Thus, if a star flies into such a domain, the balance of forces inside it will be violated, and the star will either collapse or explode.

Thus we have shown that the same basic theory can lead to drastically different low-energy theories, depending on the structure of extra dimensions and the initial data.

11.4.5 *Funnel solutions — the method of trial functions*

This method is applied if a system in question is too complicated, see, e.g., [59]. The mathematically correct way to deal with gravitational systems is as follows. One should take the equations of motion for a metric of the most general form and solve it with the desired initial and boundary conditions. The freedom in the coordinate choice facilitates the analysis, but the system often remains too complicated to be solved. The problem is usually solved by choosing a limited set of metrics substituted into an action. The classical equations are obtained by variation of the metric belonging to the chosen set. A well-known example is the minisuperspace model of Universe creation. The method of trial functions is one of such a sort. Its essence is to choose an appropriate set of metrics as some functions depending on parameters. The action appears to be a function of the set of unknown parameters. The classical set of equations is the reduced to a set of algebraic equations for these parameters.

Let us illustrate it by a specific solution that nontrivially connects two subspaces. Consider a manifold M with the topology $T \times M_1 \times M_2 \times M_3$, where M_1 is one-dimensional infinite space, and M_2, M_3

are two-dimensional spheres. We study the pure gravitational field action in the form on page 446 with the metric

$$ds^2 = A(u)dt^2 - A(u)^{-1}du^2 - e^{2\beta_1(u)}d\Omega_1^2 - e^{2\beta_2(u)}d\Omega_2^2. \quad (11.111)$$

Here, $A(u)$, $\beta_1(u)$, and $\beta_2(u)$ are members of the limited set of metrics depending on the radial coordinate u, $-\infty < u < \infty$. The action contains invariants depending on these functions. For example, the Ricci scalar has the form (see [390, 485] for details)

$$R = 2e^{-2\beta_1} + 2e^{-2\beta_2} - A'' - 4A'\left(\beta_1' + \beta_2'\right)$$
$$-2A\left(3\beta_1'^2 + 3\beta_2'^2 + 4\beta_1'\beta_2' + 2\beta_1'' + 2\beta_2''\right), \quad (11.112)$$

with the prime denoting d/du.

Let us study the metric that represents a transition between the domain with (large M_2/small M_1) subspaces and the domain with subspaces (large M_1/small M_2) with the Minkowski metric at the asymptotes:

$$A(u \to \pm\infty) \to 1, \quad (11.113)$$

and

$$\beta_1(u \to +\infty) \to \ln u, \quad \beta_1(u \to -\infty) \to \ln r_0,$$
$$\beta_2(u \to -\infty) \to \ln u, \quad \beta_2(u \to +\infty) \to \ln r_0. \quad (11.114)$$

Here, $r_0 = e^{\beta_c}$ is the radius of extra space as $u \to \pm\infty$. The value of $r_0 = 1/\sqrt{-c}$ is connected with the physical parameter c according to Eqs. (11.58)–(11.129).

For numerical simulations, we use the Ritz method which means that the functions $A(u), \beta_1(u), \beta_2(u)$ in (11.111) are approximated by trial functions chosen in the form

$$A(u) = 1 - \frac{\xi_2}{\sqrt{\xi_1^2 + u^2}},$$
$$e^{\beta_1(u)} = r_0 + \frac{1}{2}\left(u + \sqrt[4]{\xi_3^4 + u^4}\right),$$
$$e^{\beta_2(u)} = r_0 - \frac{1}{2}\left(u - \sqrt[4]{\xi_3^4 + u^4}\right), \quad (11.115)$$

keeping in mind the conditions (11.113), (11.114). According to [390], the parameters ξ_1, ξ_2, ξ_3 should satisfy the equation

$$\left(\frac{2\xi_2}{c} + \xi_1^2\xi_2 + 3\sqrt{-c}\xi_3^4\right)(1-4c^2) + 3c_1(-c)^{3/2}\xi_3^4 = 0. \quad (11.116)$$

In what follows, we consider ξ_1, ξ_3 as independent quantitiess, and ξ_2 as their function. The parameters ξ_1, ξ_2, ξ_3 are to be determined by minimizing the action.

Now, the action $S(\xi_1, \xi_3)$ appears to be a function of two variables after the substitution of the trial functions (11.115) into action (11.68). Classical solutions should satisfy the equations

$$\frac{\delta S}{\delta g_{ab}} = 0 \quad (11.117)$$

or, according to the Ritz method [251],

$$\frac{\partial S}{\delta \xi_1} = 0, \quad \frac{\partial S}{\delta \xi_3} = 0. \quad (11.118)$$

One of the ways to solve this system is to find a minimum of the auxiliary function

$$\Omega(\xi_1, \xi_2(\xi_1, \xi_3), \xi_3) = \left(\frac{\partial S}{\partial \xi_1}\right)^2 + \left(\frac{\partial S}{\partial \xi_3}\right)^2. \quad (11.119)$$

The minimization procedure for $\Omega(\xi_1, \xi_2(\xi_1, \xi_3), \xi_3)$ gives the minimum at $\xi_1^* = 9.96, \xi_2^* = 9.57, \xi_3^* = 1.85$, see Fig. 11.8. The minimum is very profound so that the trial functions, see Fig. 11.9, are chosen properly.

For an external observer, the funnel looks like a microscopic object with a mass of Planck order.

The method of trial functions (the Ritz method) is a useful tool for studying complicated systems provided that the boundary conditions are known.

11.5 Extra dimensions and inflation

In this section, we consider an inflationary model starting from a multidimensional Lagrangian and gravity with second-order curvature

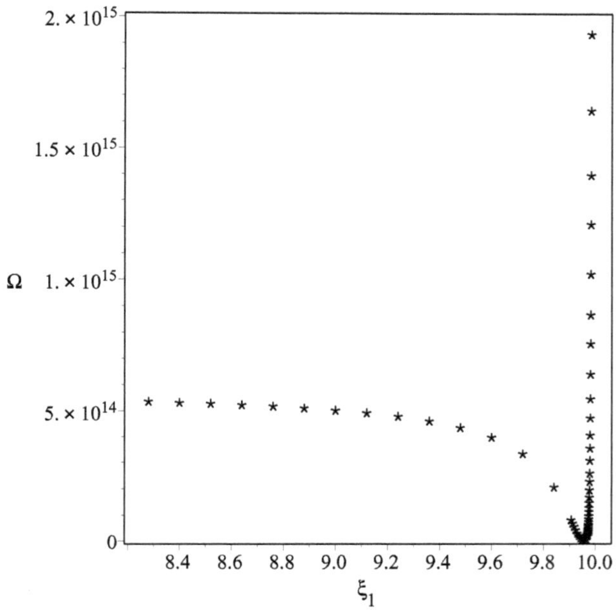

Figure 11.8 The auxiliary function Ω vs. the minimization parameter ξ_1 for $\xi_2^* = 9.57, \xi_3^* = 1.85$ and $f(R)$-parameters $a = -2, b = 1, c = -0.02$. The minimum of Ω corresponds to $\xi_1^* = 9.96$. $\Omega(\xi_1^* = 9.96, \xi_3^* = 1.85) = 5.3 \times 10^{11}$.

terms. The observable values of the spectral index and the tensor-to-scalar ratio are obtained for specific values of the model parameters.

The inflation gives information which should be reproduced by the scenario. More definitely, it is assumed that the effective potential should provide us with the reheating stage and appropriate values of the scalar spectral index n_s and the scalar/tensorial relative amplitude r. Here, we describe a successful inflationary model based on a non-linear theory of gravity in higher dimensions without the matter field.

11.5.1 *The model*

The following discussion is based on the results presented in Sec. 11.3.3 with the notations $d_0 = 4$, $d_1 = n$, $D = n + 4$. Let us consider a D-dimensional manifold $\mathbb{M}$, having the simplest geometric structure of a direct product, $\mathbb{M} = \mathbb{M}_4 \times \mathbb{M}_n$, with the

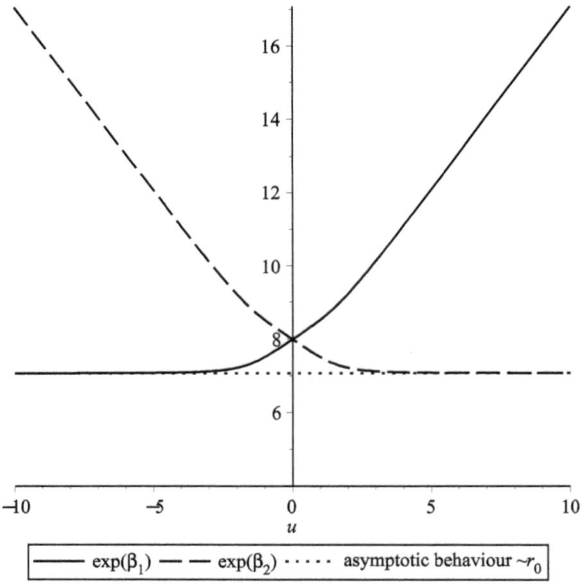

Figure 11.9 Radii of two-dimensional subspaces vs. the radial coordinate u for the parameter values $a = -2, b = 1, c = -0.02$, $\xi_3 = 1.85$.

metric

$$ds^2 = dt^2 - a(t)^2\big(dr^2 + r^2(d\theta^2 + \sin^2(\theta)d\varphi^2\big) - e^{2\beta(t)}d\Omega_n^2. \tag{11.120}$$

Here, we assume that extra space is n-dimensional and maximally symmetric, with positive curvature.

First of all, we should modify our action (11.74) using the results of the slow-change approximation and developed in the previous sections as well as the definition (11.73). The final form of the action is

$$S = \frac{m_4^2}{2}\int d^4x\sqrt{-g_4}\mathrm{sign}(2a\phi + b)\{R + K(\phi)\phi_{;\rho}\phi^{;\rho} - 2V(\phi)\}, \tag{11.121}$$

where

$$m_4 = \sqrt{V_n} = \sqrt{\frac{2\pi^{\frac{n+1}{2}}}{\Gamma(\frac{n+1}{2})}}, \quad m_D = 1 \tag{11.122}$$

with the definitions

$$K(\phi) = \frac{1}{\phi^2(2a\phi + b)^2}\left\{\left[\left(6 - n + \frac{n^2}{2}\right)a^2 + 2(c_1 + c_2)a\right]\phi^2 + \left[\frac{n^2}{2}ab + (c_1 + c_2)b\right]\phi + \frac{n(n+2)}{8}b^2\right\}, \tag{11.123}$$

$$V(\phi) = -\frac{\mathrm{sign}(2a\phi + b)}{2(2a\phi + b)^2}\left[\frac{\phi}{n(n-1)}\right]^{n/2}\left\{\left(a + \frac{c_V}{n}\right)\phi^2 + b\phi + c\right\}, \tag{11.124}$$

and

$$c_V = c_1 + 2\frac{c_2}{n-1}. \tag{11.125}$$

11.5.2 *Limits on the model parameters*

It is worth foreseeing how this model is related to the present epoch, which implies several conditions. At this stage, the space–time metric is (almost) Minkowskian due to the smallness of the cosmological constant. Necessary conditions are as follows:

$$V(\phi_m) = 0; \quad V'(\phi_m) = 0. \tag{11.126}$$

The inequalities

$$\phi_m > 0, \quad f'(\phi_m) > 0, \quad K(\phi_m) > 0, \quad V''(\phi_m) = m^2 > 0, \tag{11.127}$$

are necessary to consider the scalar field ϕ as a field with the standard properties. The function $f(R)$ has been chosen in the form (11.73) The parameters m_D, a, b, c, c_1, c_2 are not arbitrary. We may normalize the parameter m_D to equate it or one of other parameters to unity. It is assumed in Sec. 11.5.4 that $b = 1$.

Keeping in mind the quadratic form (11.73) of the function f, one can easily solve the algebraic equations (11.126) with the result

$$\phi_m = -\frac{b}{2(a + c_V/n)} \tag{11.128}$$

and with the following relation between the Lagrangian parameters

$$c = \frac{b^2}{4(a + c_V/n)}. \tag{11.129}$$

Our space–time with FLRW metric is surely not Minkowskian, so that the relation (11.129) holds only approximately.

Consider the inequalities (11.127) in more detail. The first inequality in (11.127) gives

$$-\frac{b}{2\,(a + c_V/n)} > 0. \tag{11.130}$$

The second inequality in (11.127) leads to

$$f'(\phi_m) > 0 \to \frac{bc_V/n}{(a + c_V/n)} > 0, \quad \text{or} \quad c_V < 0. \tag{11.131}$$

The third inequality in (11.127) leads to

$$\left[\left(12a^2 + 4ac_V + c_V^2\right) n + 2c_V^2 - 4c_V(c_1 + c_2)\right] > 0, \tag{11.132}$$

and the fourth equality in (11.127) gives

$$\begin{aligned} V''(\phi_m) &= -\frac{bn^2}{c_V^2}\left(\frac{a + c_V/n}{b}\right)^3 \left[-\frac{b}{2(a + c_V/n)n(n-1)}\right]^{n/2} \\ &\equiv m^2 > 0 \quad \text{i.e., } b > 0. \end{aligned} \tag{11.133}$$

As the result, the parameters of action must satisfy the conditions

$$b > 0, \quad c_V < 0, \quad a + c_V/n < 0. \tag{11.134}$$

Now we are ready to consider the inflationary stage in the framework of our model.

11.5.3 *Slow rolling and all that*

There are two main inflationary parameters in the model

$$\bar{\mathcal{L}} = \sqrt{-g}\{R + \psi_{;\rho}\psi^{;\rho} - 2V(\psi)\}. \tag{11.135}$$

with the standard form of the kinetic term. There is the scalar spectral index n_s and the scalar/tensor relative amplitude r, besides the running of the spectral index, which are given in terms of the slow-roll parameters as

$$n_s = 1 - 6\epsilon + 2\eta, \tag{11.136}$$

$$r = 16\epsilon, \tag{11.137}$$

$$\frac{dn_s}{d\ln k} = -16\epsilon\eta + 24\epsilon^2 + 2\xi^2, \tag{11.138}$$

k being the wavenumber.

The slow-roll parameters are defined as

$$\epsilon = \frac{1}{2}\left(\frac{V'}{V}\right)^2, \tag{11.139}$$

$$\eta = \frac{V''}{V}, \tag{11.140}$$

$$\xi^2 = \frac{V'V'''}{V^2}. \tag{11.141}$$

The primes are derivatives with respect to ψ. The parameter ξ is connected with the derivatives of the more usual parameters ϵ and η as will be seen later. These formulas are derived keeping in mind the Lagrangian (11.135) and should be adapted for our action (11.121) with $K(\phi) \neq 1$. It can be done by a change of variables according to the equation

$$\frac{d\phi}{d\psi} = \frac{1}{\sqrt{K}} \tag{11.142}$$

which connects the two scalar field ϕ and ψ. The slow-roll parameters (11.139), (11.140), (11.141) acquire the form

$$\epsilon = \frac{1}{2}\left(\frac{V_{,\psi}}{V}\right)^2 = \frac{1}{2K(\phi)}\left(\frac{V_{,\phi}}{V}\right)^2,$$
$$\eta = \frac{V_{,\psi\psi}}{V(\phi)} = \frac{1}{V(\phi)\sqrt{K}}\frac{\partial}{\partial\phi}\left(\frac{V_{,\phi}}{\sqrt{K}}\right),$$
$$\xi^2 = \frac{V_{,\psi}V_{,\psi\psi\psi}}{V^2} = \frac{V_{,\phi}}{V(\phi)^2K(\phi)}\frac{\partial}{\partial\phi}\left\{\frac{1}{K(\phi)}\left(V_{,\phi\phi} - \frac{1}{2}\frac{K_\phi}{K}V_{,\phi}\right)\right\}, \tag{11.143}$$

for the Lagrangian (11.121).

During inflation, the two slow-roll parameters must be small:

$$|\varepsilon| \ll 1, \quad |\eta| \ll 1. \tag{11.144}$$

The amount of inflation is given by the number of e-folds of the scale factor,

$$N \simeq -\int_{\phi_{in}}^{\phi_f} \frac{KV}{V_{,\phi}}d\phi, \tag{11.145}$$

which is valid for the slow-roll regime.

Now we have all necessary formulas needed to adjust the parameters of our model.

11.5.4 *Comparison with observations*

In this section, we apply the results obtained above to build the inflationary model (11.121) which does not contradict the observable values of the spectral index and the tensor-to-scalar ratio (11.136).

The appropriate parameters

$$n = 2, \quad b = 1, \quad a = -2, \quad c_V = -8, \quad c_K = 15000,$$
$$\left(c_V = c_1 + 2\frac{c_2}{n-1}, \quad c_K = c_1 + c_2\right) \tag{11.146}$$

were found as solutions to the system (11.136)–(11.138) provided that the observational parameters (11.148) are known. The parameter b

is equal to unity according to the remark made under the expression (11.68). The parameter c can be found from the expression (11.129). The combinations c_V and c_K are included in the functions V and K separately, which strongly facilitates the selection of the parameters. The restrictions (11.127)–(11.133) are also taken into account. The potential and the kinetic term of the model are presented in Fig. 11.10.

The present horizon could be nucleated at any slope of the potential. We suppose that it happens on the left of the potential minimum $\phi_m \simeq 0.083$. The calculations indicate that the initial field value $\phi_{in} \simeq 0.0588$, where the slow-roll parameters are small, is an appropriate starting point for the inflation. The final field value $\phi_f \simeq 0.0808$ for which the inflation ends, $\varepsilon(\phi_f \simeq 0.0808) \lesssim 1$, $\eta(\phi_f \simeq 0.0808) \lesssim 1$, are related to the initial field value by Eq. (11.145),

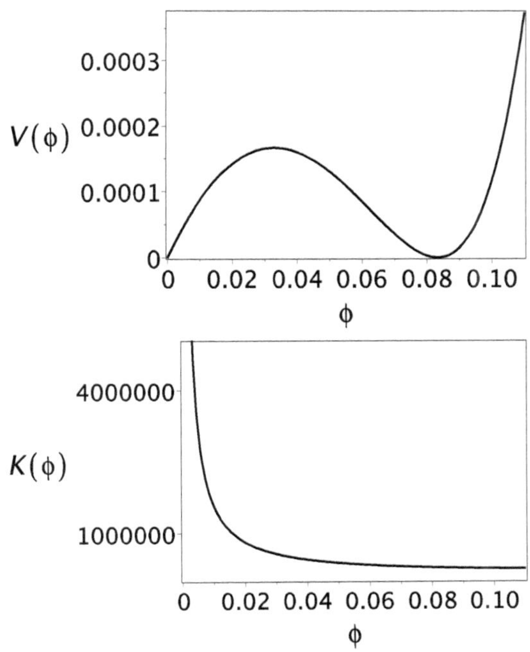

Figure 11.10 The form of the potential (left) and the kinetic term (right) for the parameters $n = 2, b = 1, a = -2, c_V = -8, c_K = 15000$. The potential minimum is at the point $\phi_m \simeq 0.083$, $m_D = 1$.

where the e-folds number is assumed to be $N \simeq 60$. Inserting the value $\phi_{in} \simeq 0.0588$ into the system (11.143), we come to our prediction for the spectral index (11.136), the tensor-to-scalar ratio (11.137) and the running spectral index (11.138)

$$n_s \simeq 0.968, \quad r \simeq 0.07, \quad \frac{dn_s}{d\ln k} \simeq 0.00051 \tag{11.147}$$

for large-scale fluctuations. According to [180], the observed values are

$$n_s \simeq 0.968 \pm 0.006, \quad r < 0.1, \quad \frac{dn_s}{d\ln k} = -0.0045 \pm 0.0067, \tag{11.148}$$

so that the inflationary model discussed here is valid. The time dependence of n_s and r can be seen from Fig. 11.11. The field evolves starting from the initial value $\phi_{in} \simeq 0.0588$ to the final one $\phi_f \simeq 0.0808$.

The plot of the potential presented in Fig. 11.10 can be used to check the inequality (11.56) to assure the self-consistency of our model. Indeed, according to the plot, $V \simeq 5 \cdot 10^{-5}$, $R_2 = \phi \simeq 0.07$ in the middle of inflation. At the same time, at the inflationary stage, $R_4 \simeq 10H^2 \simeq 10^2 V \simeq 5 \cdot 10^{-3}$ and $R_2 = \phi \simeq 7 \cdot 10^{-2}$ according to Fig. 11.11. The system (11.143) and the relation (11.129) were used to determine the numerical values of the Lagrangian parameters a, c, c_1, c_2. The accuracy of the obtained parameter values is about $R_4/R_2 \sim 0.1$ judging by our approximation (11.129) that

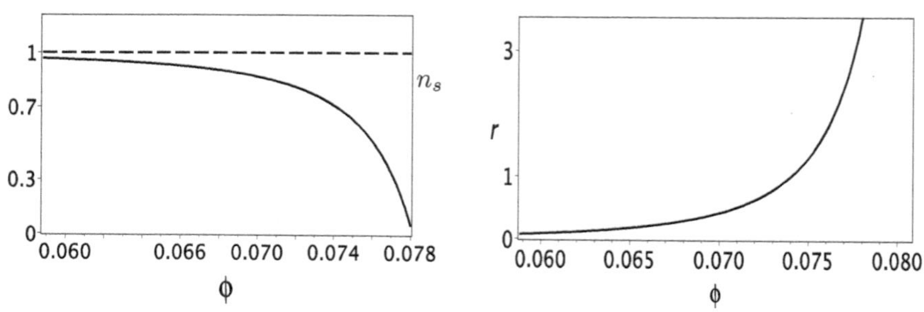

Figure 11.11 $n = 2, b = 1, a = -2, c_V = -8, c_K = 15000, \phi_m \simeq 0.083$. Smaller field values are related to a larger cosmological scale. The observable values (11.148) are found for $\phi \sim 0.06$.

looks satisfactory. In fact, a more detailed analysis indicates that the accuracy is even better. Indeed, the quadratic term which has been omitted in the Taylor expansion in the expression (11.76) is $\delta c = \frac{1}{2} f''(R_n) R_4^2 \sim \frac{a}{2}(10H^2)^2 \sim 10^{-5}$. Being constant during the inflation, it can be considered as additive to the parameter c presented in the expression for the function $f(R)$, see (11.73). The value of the parameter c can be found from (11.129). It appears to be equal to $c = 1/(4(a + c_V/2)) = -1/24$, which means that it is determined with a good accuracy, $\delta c/c \sim 10^{-3}$.

In this chapter, we express all numbers in the units $m_D = 1$, with the effective Planck mass m_4 determined by (11.72). After the inflation is over and the scalar field has settled in the minimum of its potential, one should restore more physical units valid for the Jordan frame, i.e., the Planck units using the relation (11.63). For the chosen parameter values (11.146), the relation

$$M_{\rm Pl} = \sqrt{V_n e^{n\beta_m} f'(\phi_m) m_D} \tag{11.149}$$

can be used to obtain the value of the D-dimensional Planck mass. For 2D extra space ($n = 2$, $V_2 = 4\pi$),

$$M_{\rm Pl} = \sqrt{V_2 e^{2\beta_m} f'(\phi_m) m_D} = \sqrt{8\pi \left(2a + \frac{1}{\phi_m}\right) m_D} \sim 10 m_D, \tag{11.150}$$

which means that the D-dimensional Planck mass m_D is by an order of magnitude smaller than the Planck mass in the Jordan frame.

The inflationary model was studied in the framework of nonlinear multidimensional gravity without matter fields. The Ricci scalar of the extra space plays the role of the inflaton. The parameters of the model are adjusted to satisfy the observed values of the spectral index and the tensor-to-scalar ratio. The model has a large parameter ($c_K \sim 10^5$), a common feature of inflationary models. The model leads to a complex form for the effective potential and kinetic factor that give promising results. We have given an example where the model studied here agrees with the observational constraint. A complete analysis of the space parameter is, of course, necessary.

11.6 Cosmological limits on the parameters of extra dimensions

The compact extra space is characterized by two main parameters, the D-dimensional Planck mass m_D and the scale l_d of compact extra dimensions. Both of them vary in a range from the Planck scale to the scale $\sim$1 TeV for m_D and to $\sim 10^{-18}$ cm for l_d. depending on a specific model. These constraints are set on the LHC collider. At the same time, there exists another collider created by Nature itself — the early Universe, or more definitely its first stage of evolution, the inflation.

Here we show that the case $l_d \gg H^{-1}$ contradicts the observations provided that the slow-rolling inflation is valid. Therefore, if the compact space does exist, its size should be smaller than $\simeq H_I^{-1} \simeq 10^{-27}$ cm. This limit is much stronger than those obtained from the LHC.

11.6.1 *Limit to the size of extra dimensions*

In this section, we discuss the influence of compact extra dimensions on the properties of the inflationary stage. A particular model of inflation does not matter. The facts which only will be used is that the 4-metric is almost de Sitter, and the Hubble parameter is almost constant during inflation. It will be shown that the size of the extra space must be smaller than the inverse Hubble scale during inflation in order not to contradict the inflationary features mentioned above.

Consider the D-dimensional action

$$S = \frac{m_D^{D-2}}{2} \int d^4x d^d y \sqrt{|g|} \left[R_D + \frac{1}{2} \partial_A \chi g^{AB} \partial_B \chi - \frac{1}{2} \mu^2 \chi^2 - 2\Lambda_D \right] \tag{11.151}$$

for a scalar field $\chi(x, y)$ which plays the role of the inflaton, m_D is the D-dimensional Planck mass. The metric is chosen in the form

$$ds^2 = dt^2 - a(t)^3 dl_3^2 - l_d^2 d\Omega_d^2, \quad l_d = \text{const.} \tag{11.152}$$

where l_d is the radius of the maximally symmetric d-dimensional extra space. The choice of such a kind of metric makes much easier the analysis and is widely used in the literature.

To proceed, let us reduce (11.151) to the 4D effective action. The standard way is to expand the function $\chi(x,y)$ in the eigenfunctions $Y_q(y)$,

$$\chi(t,x,y) = \sum_q \chi_q(x) Y_q(y), \tag{11.153}$$

where

$$\Box_d Y_q(y) = M_q^2 Y_q(y), \quad \frac{1}{v_d}\int d^d x\sqrt{|g_d|}Y_q(y)Y_p^*(y) = \delta_{qp}. \tag{11.154}$$

Integrating out the extra coordinates y, we obtain the action in the form

$$S = \int d^4x\sqrt{-g_4}\left[\frac{M_P^2}{2}R_4 + \sum_q \left(|\partial_\mu \chi_q|^2 - (M_q^2+\mu^2)|\chi_q|^2\right)\right] \tag{11.155}$$

with the notations

$$M_P^2 = v_d m_D^{d+2}; \quad \Lambda_D = \frac{1}{2}m_D^2 d(d-1)l_d^{-2};$$
$$M_q = \frac{q}{l_d}; \quad q = 0,1,2,\dots, \tag{11.156}$$

where q enumerates the excited states. We will neglect in the following the bare mass μ as compared to the Kaluza–Klein mass tower M_q for simplicity. The more general case leads to the same conclusion. The effective potential looks as follows:

$$V = \sum_{q=0}^{\infty}\frac{1}{2}M_q^2\chi_q^2(t). \tag{11.157}$$

There is a great variety of models describing the first stage of our Universe formation. One of the most well-known branches is based on the chaotic inflation with slowly moving scalar field(s). The latter is a reason for rapid, almost exponential space expansion while the horizon size remains unchanged if measured in the physical coordinates. This means that the space is divided into a quickly growing number of disconnected areas. Each spatial volume is bounded by its own horizon $l_h = H^{-1}$. The value $l_h \simeq 10^{-27}$ cm and consequently $H \simeq 10^{13}\text{GeV} \simeq \text{const}$ are in agreement with observations.

It will be shown in this section that the process of inflation described above cannot be realized if the size of the compact extra space l_d is larger than the horizon size. More definitely, we show that if

$$H \simeq \text{const} \tag{11.158}$$

and

$$n = \frac{l_d}{l_h} = l_d H \gg 1 \tag{11.159}$$

we come to a contradiction.

According to chaotic inflation [378] for the action (11.155), the fields evolve as

$$\chi_q(t) = \chi_{q,in} e^{-(M_q^2)/(3H)\ t}. \tag{11.160}$$

The KK tower contains light modes ($M_q < H$) and heavy modes ($M_q > H$). The latter go to the potential minimum before the inflation starts. The light modes of the scalar field are active during inflation. The number of modes is limited below according to the inequality

$$M_q < H \quad \Rightarrow \quad q < l_d H \equiv n. \tag{11.161}$$

The upper limit on the extra space size that has been reached at LHC is $l_d \sim 10^{-18}$ cm. In this case, the number of active modes is $n \sim l_d H \sim 10^9$. Therefore, the sum (11.157) contains 10^9 terms and may be approximated by an integral. For further estimation, suppose that the initial field values do not depend on q:

$$\chi_{q,in} \simeq \chi_{in}. \tag{11.162}$$

It gives the following integral estimate of the potential in (11.157):

$$\begin{aligned} V &= \sum_{q=0}^{\infty} \frac{1}{2} M_q^2 \chi_q^2(t) \simeq M_1^2 \chi_{in}^2 \int_0^\infty \mathrm{e}^{-\frac{2M_1^2 q^2}{3H} t} q^2 dq \\ &= \frac{1}{4} \chi_{in}^2 M_1^2 \left(\frac{3}{2M_1^2}\right)^{3/2} \left(\frac{H}{t}\right)^{3/2}, \end{aligned}$$

and we arrive at the equation for the Hubble parameter

$$H = \sqrt{\frac{8\pi}{3M_{Pl}^2}V} \simeq \sqrt{\frac{\pi}{M_{Pl}^2}\chi_{in}^2\left(\frac{3}{2M_1^2}\right)^{1/2}\left(\frac{H}{t}\right)^{3/2}}. \tag{11.163}$$

Solving this algebraic equation for H we obtain

$$H \sim t^{-3}. \tag{11.164}$$

Such a time dependence of the Hubble parameter is unacceptable for the inflationary period. Therefore, if the inequality (11.159) is true, the slow-roll inflation is absent. We come to the conclusion that the size of the extra space is bounded above,

$$l_d \leq 1/H \simeq 10^{-27}\,\text{cm}. \tag{11.165}$$

It is a strong limit for the size of extra space. Nevertheless, it is not a firm no-go theorem. There are some ways to escape the contradiction. One of them is to change the equality (11.162) which is a natural but quite uncertain estimate. For example, we may choose the initial conditions as $\chi_{1,in} \neq 0$, $\chi_{q,in} = 0$, which are realized with small probability. In this case, the standard inflationary scenario may occur. Some inflationary models need not slow-rolling, but they are not very common.

The inequality (11.165) also allows us to impose restrictions on the D-dimensional Planck mass. Indeed, according to (11.156),

$$M_P^2 = v_d m_D^{d+2} \simeq l_d^d m_D^{d+2} < H^{-d} m_D^{d+2}. \tag{11.166}$$

This leads to a restriction in the form

$$m_D > \left(\frac{H}{M_P}\right)^{d/(d+2)} M_{\text{Pl}}. \tag{11.167}$$

11.6.2 *Limit on the D-dimensional Planck mass*

In this subsection, we argue for a strong limit on the D-dimensional Planck mass imposed by the inflationary period of the early Universe. The arguments are quite general, so that a specific inflation model and the extra space structure do not matter. The only thing that we need is the stationarity of the extra dimensions. A way to stabilize the extra dimensions has been discussed in this chapter.

Consider the action

$$S_D = \frac{m_D^{D-2}}{2}\int d^D x(R_D - 2\Lambda_D + \cdots) \tag{11.168}$$

in D dimensions, $D = 4 + d$.

At high energies, the D-metric g_D in V_D is quite uncertain. On the other hand, the modern physical laws are assumed to be described by a specific stationary metric. As an example, it could be a direct product $V_4 \times V_d$ of 4D FRW space V_4 and a d-dimensional compact space V_d. Therefore, the transition $V_D \to V_4 \times V_d$ should take place. It happens at the D-dimensional Planck scale $l_q \equiv 1/m_D$, when quantum corrections still dominate. There is another parameter that plays a significant role, the Hubble parameter, H. The latter decreases with time from a value H_I at the beginning of inflation to almost zero nowadays.

The size $l_h = 1/H$ of a causally connected area at the inflationary stage is the classical value by definition. The classical description of the inflating Universe gives promising results. On the other hand, if the quantum scale l_q is larger than the classical value l_h, the latter has no physical meaning. It means that the inequality

$$l_q < l_h \quad \Rightarrow \quad m_D > H \tag{11.169}$$

should take place. Knowledge of the Hubble parameter gives a lower bound for the D-Planck mass,

$$m_D > 10^{13}\,\text{GeV}. \tag{11.170}$$

The conclusion is that any calculations of H under the condition $l_q > l_h$ (or $m_D < H$) cause a serious question.

Nevertheless, let us suppose that $m_D \simeq H$ holds somewhere in the middle of the inflationary stage at the e-fold number N, and we formally calculate the horizon scale l_h. It gives another interesting possibility. At this moment, we have e^{3N} independent causally connected areas where the transition $V_D \to V_4 \times V_d$ has happened. As was discussed in [485], the choice of those d dimensions which should be compactified is random. Therefore, each causally connected 4D volume is endowed with its own coordinate directions that are compact. It leads to the production of funnels [390] that are observed as pointlike objects with mass of the order of the Planck mass. Let us estimate their number. It is known that e^{3N} ($N \simeq 60$) causally disconnected areas are formed by the end of inflation. It means that the total mass of funnels in our Universe is huge, which contradicts the observations and confirms the inequality (11.169).

There are two lower limits — (11.167) and (11.169). One can conclude that the D-dimensional Planck mass m_D is strongly bounded below if the standard inflationary scenario is valid.

11.7 Conclusion

On reduction to four dimensions, nonlinear multidimensional gravity leads to a sophisticated scalar field dynamics, with stable and metastable minima of the potential and a nontrivial form of the kinetic term. Some such minima look promising in the sense of describing the modern Universe evolution, with an accelerated expansion of the external space and stable extra dimensions. The latter leads to constant values of the fundamental constants or to their very slow drift, as is the case if the system is approaching its stable state. In other viable models the extra dimensions are variable, but the variations are small and do not leave the bounds of current experimental constraints.

Among the unsolved problems are the choice of the physical conformal frame and the necessity of fine tuning of the parameters for obtaining the observed small value of the cosmological constant.

Chapter 12

The emergence of physical laws

A widespread opinion is that physics is formed at high energies; the idea of the inflationary processes in the early Universe is a well-known example. The absence of quantum gravity forces us to work at sub-Planckian energies as the highest scale. Hence, the future research aims at discovering or guessing the structure of the theory at the sub-Planckian scale. An evident difficulty is that we live at low energies, where the observable physical parameter values could strongly deviate from their initial values. Moreover, the structure of the Lagrangian at high energies could be different from that at low energies, for example, due to quantum corrections. One of the aims of fundamental physics is to postulate a Lagrangian depending on the primary parameters and to fix them using their connection with observational results. It means that we have to calculate the parameter values at low energies, starting from those at high energies.

At the same time, it is important to take into account the fine tuning of physical parameters as the comprehensive observational fact.

12.1 Fine tuning of the physical parameters

One of the impressing and seemingly underestimated observational facts is the total fine tuning of the Universe parameters. This means that all microscopic parameters, the particle masses and coupling constants, not only have the values at which an intelligent life can appear in the Universe, but the admissible values of these parameters are confined to very narrow ranges. Tiny deflections from their real values would lead to catastrophic consequences for the existence of complicated structures. Before proceeding with the discussion, let us consider a few examples.

12.1.1 *The electron and the properties of the Universe*

The characteristic masses of fermions are of the order of 1 GeV. Only the electron has a mass three orders of magnitude smaller. Imagine that its mass became a few times larger. It could seem that nothing significant should happen, all the same it is still a very light particle. However, an electron mass increase by a factor of three would lead to catastrophic consequences for the Universe. Indeed, the mass difference between the neutron and the proton is $\Delta m = m_n - m_p = 1.28\,\text{MeV}$, whereas the mass of a "new electron" amounts to $m_E = 3m_e = 1.53\,\text{MeV}$. It means that the neutron becomes stable because its decay via weak interaction

$$n \to p + e + \bar{\nu}_e$$

is energetically forbidden. Moreover, the allowed reaction

$$p + e \to n + \nu_e,$$

will destroy the stability of hydrogen atoms. Consequently, immediately after the recombination period, practically all protons and electrons are converted to neutral particles. Charged particles simply disappear long before the formation of the first stars. In this case stars will not be formed at all since the gravitational contraction of clouds is efficient only if there is an energy dissipation process.

Thus, an insignificant growth of the electron mass leads to the absence of stars in the Universe and consequently the absence of any, to say nothing of intelligent, life.

12.1.2 *The carbon level*

Fine tuning works so unconditionally that with its aid one can even estimate unknown parameters. The most well-known example is Fred Hoyle's prediction.

Carbon in stars is formed in two stages. At first two alpha particles merge, forming the unstable isotope beryllium-8. Then a third alpha particle joins this beryllium, forming a carbon nucleus. But beryllium-8 rapidly decays. Therefore, in forming the Universe, it would be reasonable to provide a resonant interaction between beryllium-8 and an alpha particle. Nature did precisely that: the energy level equal to 7.65 MeV is remarkable in that the energy of an excited state of the carbon nucleus is only 0.3 MeV higher than than the summed mass of the alpha particle and the beryllium nucleus. This 0.3 MeV is compensated by the kinetic energy of colliding particles, resonantly increasing the efficiency of the reaction. It is these considerations that led Fred Hoyle to his theoretical prediction of the energy of this level in 1953. When our Universe was only forming, Nature must have already "known" about the future necessity of this level for the existence of carbon life.

12.1.3 *Slow reactions in stars*

Let us mention one more interesting example of fine tuning of the Universe parameters. Consider the question: what should the Fermi constant G_F be for the emergence of our type of intelligent life? Experience tells us that it requires about five billion years for that. It is the time — not less than that — for which nuclear reactions should last in a star like the Sun. Thus the time of the first reaction in the proton–proton cycle,

$$p + p \to {}^2H + e^+ + \nu_e \tag{12.1}$$

should be $t = 1/(\sigma v n) \sim 5$–10 billion years, where $v \sim \sqrt{2T/m_p}$ is the mean proton velocity in a star, while the mean concentration of protons inside a star is $n \sim 1.5\,\text{g/cm}^3$. Nuclear reactions begin at temperatures $T \approx 10^7$. The cross-section of the first reaction of the proton–proton cycle, occurring due to the weak interaction, is $\sigma \sim G_F^2 E^2 \sim G_F^2 T^2$.

Collecting all that, we obtain an expression for the Fermi constant

$$G_F \sim \frac{m^{1/4}}{(nt)^{1/2} T^{5/4}}, \tag{12.2}$$

connecting it with such parameters as the mean stellar concentration of protons and temperature, the proton mass and the intelligent life emergence time, or, more precisely, the stellar lifetime. Substituting the numerical values, we obtain $G_F \sim 10^{-5}\,\text{GeV}^{-2}$, which coincides by order of magnitude with the experimental value.

Let us note that nuclear reactions usually occur rapidly. The small value of the Fermi constant makes it possible to slow down the stellar nuclear processes, beginning with the reaction (12.1). Let us look how efficient this slowing-down is. The solar luminosity is known: the energy per second released by the Sun is $L_\odot = 3 \cdot 10^{26}\,\text{J/s}$. Consequently, 1 kg of solar matter emits

$$\varepsilon_\odot \sim L_\odot / M_\odot \sim 10^{-7}\,\text{J/(s} \cdot \text{kg)}.$$

Compare this with the specific energy release of a human being of 80 kg:

$$\varepsilon_{\text{homo}} \sim \frac{2000 \cdot 10^3 \cdot 0.01 \cdot 4.18}{24 \cdot 3600 \cdot 80} \sim 0.02\,\text{J/(s} \cdot \text{kg)}.$$

Here, for the estimation, it is assumed that only 1% of the energy absorbed by man, 2000 kcal, passes over to heat. Owing to the smallness of the Fermi constant, stellar nuclear reactions turn out to be by many orders of magnitude less efficient as compared with chemical reactions.

Question. *If chemical reactions inside a human being occur much more rapidly than nuclear reactions in stars, why is the temperature of a human body so much lower than the solar temperature?*

The weak interaction should be really very weak in order that the neutron live long enough. In addition, during supernova explosions, neutrinos, owing to the weak interaction, are able to carry away the main part of energy from the central part of the star, allowing its inner layers to rapidly contract. But on the other hand, the weak interaction constant cannot be too small. Otherwise, first, the number of neutrinos created in an exploding star would be too small, and second, the outer layers of such a star would not receive sufficient energy from the neutrinos to fly away in space.

12.1.4 *Stellar lifetime*

Heavy elements are formed in the stars but they must be eventually ejected to space, in order that future stellar generations with planetary systems could form. Therefore the parameters of the theory should be selected in such a way that at least a certain part of the first stellar generation have a short lifetime ending with an explosion. But the next stellar generations should live for a long time (10 billion years) to make possible the emergence of intelligent life. Nature has been able to satisfy these contradictory requirements.

The long life of the next stellar generation rests on the slow proton-proton cycle (here is where the weak interaction is necessary). The short lifetime of the first stars takes place due to large masses plus primordial helium burning. But how to create the primordial helium?

12.1.5 *Passage through a needle's eye*

To create helium, neutrons are necessary. The latter, created when the Universe was hot, should eventually decay. How to preserve them? Can we preserve neutrons inside deuterium?

$$p + n \to D + \gamma.$$

But the inverse reaction prevents:

$$D + \gamma \to p + n,$$

it is efficient at high temperatures. The binding energy of deuterium is not large, 2.234 MeV, and at temperatures higher than 10^9K deuterium is rapidly destroyed. The threshold temperature 10^9K of the Universe is achieved in 300 s after the Beginning. If the neutrons had the "usual" lifetime like 10^{-6} s, their concentration by that time would be vanishingly small.

A way out is a long-lived neutron (900 s)! After this time, the temperature decreases enough, so that the energy of photons becomes insufficient to destroy the deuterium nuclei. The remaining 24% of neutrons turn out to be sufficient.

12.1.6 *It is good time to ask a question — what should an Ultimate Theory look like?*

It seems that there are two possible answers to this question:

(A) The laws of Nature are strictly derived from an initial postulate and simply cannot be different. The whole history of science can be viewed as an approach to this postulate.
(B) There are many universes with different laws, and we live in one of them. This approach is gaining more and more adherents, but is it constructive? Suppose that somebody wishes to create a universe like ours, at least purely theoretically. The above examples are designed to convince the reader that it is not a simple task. Indeed, the number of problems to be encountered by the brave creator is immense. Starting from the beginning she (he) must choose properties of all particles and their interactions keeping in mind the future evolution. The sketch looks as follows.

12.1.7 *Choosing particles and interactions*

The basic components of complex structures are protons and electrons, which already give rise to the simplest of the atoms, the hydrogen atom. More massive nuclei are also necessary, for instance, carbon, the base of organic compounds. But massive nuclei consisting of only protons will immediately decay because of the protons' electromagnetic repulsion. One will have to introduce a new interaction providing attraction. We call it the strong interaction. But then too large nuclei will form, which is also bad, as the matter density formed will be too large. Therefore we make the strong interaction short-range.

The situation has become much better, but the nuclei, though slowly, still decay. If, due to fluctuations, one of the protons flies a little farther from the others, the strong interaction cannot withstand the electric repulsion, the proton escapes, and the nucleus decays. It is evident that the goal is close, only a little is needed to make nuclei heavy enough and stable. The last stroke: we introduce the neutron, having no charge but taking part in the strong interaction.

Now the nuclei consist of protons and neutrons and are stable up to the nucleus of iron. All that should have been provided by Nature from the very beginning.

To create complex structures, one needs a large volume, and the expansion solves this problem. But the expansion rate cannot be arbitrary, for too many effects depend on it. The most basic is apparently star formation.

The expansion of space leads to such cooling of the medium at which no intelligence could live. The contradiction can be avoided by creating a locally warmed medium. This function is performed by the stars. Besides, the stars are factories of heavy elements. The long-term existence of the stars is one more problem that the nature had to solve.

It seems evident that the idea of a "multiverse" is a good basis for solving the fine-tuning problem, be it "science" or not.

More examples of fine tuning of the Universe — briefly

We further enumerate the parameters that need to be in a narrow range to make complex structure creation in the Universe possible.

1. The strong interaction constant.
   Larger: no hydrogen and light nuclei, e.g., carbon.
   Smaller: nuclei heavier than hydrogen are unstable.
2. The weak interaction constant.
   Larger: (a) too much primordial helium in the first stars $\Rightarrow$ too large amounts of heavy elements are created in the stars; (b) supernova explosions are absent. The nuclei remain in stellar remnants.
   Smaller: (a) too little primordial helium in the first stars; (b) supernova explosions are absent. The nuclei remain in stellar remnants.
3. The gravitational constant.
   Larger: the stars are too hot and therefore burn too rapidly (while five billion years are needed to create intelligence).

Smaller: cold stars. Consequently, no nuclear reactions and their product, heavy elements.

4. The electromagnetic interaction constant.
   Larger: the atoms are smaller (all structures become more dense). The nuclei are unstable.
   Smaller: too weak chemical bonds.
5. The ratio of the electromagnetic interaction constant to the gravitational constant.
   Larger: stellar masses are high and therefore their lifetimes are small.
   Smaller: no massive stars and therefore no heavy elements.
6. m_e/m_p.
   Larger: weak chemical bonds.
   Smaller: weak chemical bonds.
7. The Universe expansion rate.
   Larger: galaxies have no time to form.
   Smaller: the Universe collapses before the stars can form.
8. The mean baryonic energy density.
   Larger: too much primordial helium, the stars burn too rapidly.
   Smaller: too little primordial helium, hence too little amounts of heavy elements.
9. Primordial density fluctuations.
   Larger: the Universe would consist of black holes.
   Smaller: the stellar formation process is suppressed.
10. Mean interstellar distance.
    Larger: the heavy element density is insufficient for making solid planets.
    Smaller: unstable planetary orbits.
11. Beryllium-8 decay rate.
    Larger: absence of elements heavier than beryllium.
    Smaller: the nuclear reaction rate increases, leading to small stellar lifetimes.
12. The initial excess of nucleons vs. antinucleons.
    Larger: too large baryon density. Mostly black holes form.
    Smaller: baryonic matter insufficient for stellar formation.

13. Frequency of supernova explosions.
    Larger: life on the planets is destroyed.
    Smaller: heavy elements insufficient for making solid planets.
14. Dark matter density.
    Larger: an early collapse of the Universe.
    Smaller: galaxy formation suppressed.
15. The neutron–proton mass difference.
    Larger: smaller neutron lifetime, hence too little primordial helium.
    Smaller: rapid neutron decay. Heavy elements are absent.
16. The number of spatial dimensions (3).
    Larger: planets are not kept near stars, no stationary orbits.
    Smaller: hard to provide the viability of complex structures.

All the above considerations were aimed at convincing the reader that choosing the parameters for forming complex structures in the Universe is a task of formidable difficulty. Admissible ranges of their variation are extremely narrow. Whatever the future theory, the latter statement will refer to it and its parameter set. All that means that the probability of a random implementation of such a parameter set is close to zero. From this viewpoint, the idea of multiple universes with different properties is more attractive. The question is "only" the mechanism of their formation. Multidimensional gravity gives such an opportunity, and that is what is discussed in what follows.

12.2 The origin of symmetries and fundamental constants

Despite the advances made in fundamental physics, the existence of a large number of theories and approaches suggests that there are problems in describing the observational and experimental data. The idea of compact extra dimensions allows one to effectively explain a considerable number of phenomena and indicates a direction of further development in the theory in spite of the absence of a direct experimental confirmation of their existence.

Here we will study the problems associated with emergence of symmetries and fundamental constants at the early stage of the evolution of the Universe using the approach proposed in our previous works [134, 482]. This approach is based on the assumption that there are extra dimensions forming a compact space with the properties determining the observed low-energy physics. A solution to the first problem within the Kaluza–Klein approach is well known: gauge symmetries arise as a consequence of the corresponding symmetries of the extra space (see, e.g, [54]). Therefore, the problem is reduced to justification of the choice of symmetric spaces among a set of extra spaces with an arbitrary geometry. More precisely, we assume that, at some instant of time on the Planck scale, the four-dimensional Riemannian space M_4 [283, 536, 566] arises as a result of quantum fluctuations. Simultaneously, there appears a compact extra d-dimensional space $\mathbb{M}_d$. The set of possible geometries of the extra space $\mathbb{M}_d$ is at least a continuum. However, a geometry with a high degree of symmetry is observed because the existence of gauge symmetries is beyond question. The probability of its emergence is negligible, and hence there should exist a selection mechanism that separates the appropriate geometries. One of possible variants based on statistical considerations will be analyzed below.

The second problem is related to the fundamentality of the parameters $\hbar$ and G [219] and consists in the following. The geometric approach to the theory implies the presence of only one scale of the dimension of length ℓ. The parameters of the initial Lagrangian constructed only from the metric tensor are proportional to ℓ^n, where n is an integer. The question arises on the instant of time at which the Planck and gravitational constants appear as independent fundamental parameters and on their relationship with the initial parameters of the theory. An answer to this question for the gravitational constant is well known and, as a rule, is presented in the form

$$M_P^2 = m_D^{D-2} V_d. \tag{12.3}$$

Here, M_P is the Planck mass, m_D is a parameter with the dimension of mass or, more precisely, inverse length, and $V_d < \infty$ is the volume of the extra space $\mathbb{M}_d$. The similar problem of fundamentality of the

Planck constant $\hbar$ is less well understood, even though some interesting works have appeared in this direction. In particular, Volovik [547] discussed the place of the Planck constant in modern theory, analyzed different methods of including this constant in the field equations. However, a specific implementation of these hypothetical variants was not given.

Below, we discuss a mechanism of emergence of the Planck constant together with the gravitational constant without a detailed discussion of the possibilities this entails. As a result, we can reveal the relationships of the constants $\hbar$ and G determined from low-energy experiments with the parameters of the initial Lagrangian describing multidimensional gravity with higher derivatives.

Continuing the ideology developed in Chapter 11, we assume that spaces with an appropriate geometry arise in space–time foam with some (maybe small) probability. We are interested in spaces with the geometry of the direct product

$$\mathbb{M}_D = \mathbb{M}_{D_1} \times \mathbb{M}_{d_1}, \quad D_1 \geq 4, \;\; d_1 \geq 2 \tag{12.4}$$

and the volumes $V(D_1) \gg V(d_1)$. Transitions with changes in the geometry are conveniently described in terms of the path integral technique [283, 536]. For this purpose, the superspace $\mathcal{M}_D = (\mathbb{M}_D; g_{ij})$ is defined as a set of metrics g_{ij} in the space $\mathbb{M}_D$ up to diffeomorphisms. On a spacelike section Σ we introduce the metric h_{ij} (for details, see [566]) and define the space of all Riemannian $(D-1)$-metrics in the form

$$\mathrm{Riem}\,(\Sigma) = \{h_{ij}(x)\big| x \in \Sigma\}.$$

The transition amplitude from the section $\Sigma_{\rm in}$ to the section Σ_f is an integral over all geometries allowable by the boundary conditions:

$$A_{f,\rm in} = \langle\, h_f, \Sigma_f | h_{\rm in}, \Sigma_{\rm in} \,\rangle = \int_{h_{\rm in}}^{h_f} Dg \exp[iS(g)]. \tag{12.5}$$

The absence of the Planck constant $\hbar$ in the exponential function is a result of choosing the appropriate units of measurement. Nonetheless, it will be shown below that the Planck constant naturally emerges after the definition of the dimensional units simultaneously with the gravitational constant G.

12.2.1 *Fundamental constants and the properties of an extra space*

In this section, we follow the ideology developed in Chapter 11 and choose the action in the form (see, e.g., [426, 482, 507, 517]),

$$S = N_0 \int d^D y \sqrt{-G} f(R; a_n). \tag{12.6}$$

The parameters a_n, N_0 take on specific values after the choice of the length unit. The dimensionality of the arbitrary function $f(R)$ is conveniently chosen to coincide with the dimension of R, i.e., it is l^{-2}. Then, the dimension of N_0 is l^{2-D}. Note that the normalization constant N_0 equals $1/(16\pi G)$ only in the low-energy limit for four-dimensional space.

The metric of the space $\mathbb{M}_D$ is written in the form [134, 160]

$$\begin{aligned} ds^2 = G_{AB} dX^A dX^B &= g_{ab}(x) dx^a dx^b - e^{2\beta(x)} \gamma_{ij}(y) dy^i dy^j \\ &= \mathrm{N} dt^2 - g_{\mu\nu}(x) dx^\mu dx^\nu - e^{2\beta(x)} \gamma_{ij}(y) dy^i dy^j. \end{aligned} \tag{12.7}$$

Here, g_{ab} is the metric of the subspace $\mathbb{M}_4 = \mathbb{R} \times \mathbb{M}_3$ with the signature $(+---)$, $e^{2\beta(x)}$ is the curvature radius of the compact subspace $\mathbb{M}_d$, and $\gamma_{ij}(y)$ is its positive-definite metric. For a specified foliation of space by spacelike surfaces, we can always choose the normal Gaussian coordinates which are used in the last equality in the expression (12.7). We assume that the coordinates x, y have the dimension ℓ, time is measured in seconds, and the lapse function N is an unknown parameter with the dimension of $(\ell/s)^2$. The metrics g_{ab} and γ_{ij} are dimensionless.

As shown in the next section, the extra space with an arbitrary geometry evolves to a space with a maximum number of Killing vectors possible for the topology under consideration. In this respect, in the set of subspaces $\mathbb{M}_d$ we choose maximally symmetric spaces of constant curvature R_d, related to the curvature parameter k in a typical manner,

$$R_d(\gamma_{ij}) = \mathrm{e}^{-2\beta(x)} kd(d-1)\ell^{-2}. \tag{12.8}$$

Now, let us use the result obtained in Chapter 11, for transforming the action (12.6) to that of the scalar-tensor form of gravity (11.66) Assuming the existence of a minimum of the potential (11.71) at $\phi = \phi_m$ and using the slow-change approximation, the action (12.6) can be presented near this minimum in the form

$$S \simeq N_0 \mathcal{V}_d c_\ell \int dt d^3 x_\ell \left[R_\ell + \frac{1}{2} K(\phi_m)(\partial_\ell \phi)^2 - U(\phi_m) \right.$$
$$\left. - \frac{1}{2} U''(\phi_m)(\phi - \phi_m)^2 \right], \tag{12.9}$$

where

$$(\partial_\ell \phi)^2 = (\partial \phi / c_\ell \partial t)^2 - (\partial \phi / \partial x_\ell)^2.$$

Here, we have taken into account that $\sqrt{g} = c_\ell$ in four-dimensional Minkowski space. The index l designates the used unit of length, and c_ℓ is the speed of light in the chosen units.

The yet unknown unit l is

$$l = \alpha \cdot \mathrm{cm}, \tag{12.10}$$

where α is a parameter. In Eq. (12.9), we pass over to the standard units of length:

$$S = N_0 \mathcal{V}_d c \alpha^2 \int dt d^3 x \left[R_4 + \frac{1}{2} K(\phi_m)(\partial \phi)^2 - \alpha^2 U(\phi_m) \right.$$
$$\left. - \alpha^2 \frac{1}{2} U''(\phi_m)(\phi - \phi_m)^2 \right]. \tag{12.11}$$

Here, $x_\ell = \alpha x$, $c_\ell = \alpha c$, $R_\ell = R_4/\alpha^2$, and $(\partial_\ell \phi)^2 = (\partial \phi)^2/\alpha^2$.

Since the expression (12.6) is a low-energy limit of the action (12.11), it should adequately describe purely gravitational phenomena. Moreover, the expression (12.11) makes it possible to explain the origin of the inflaton potential and the cosmological constant. The effective action (12.11) contains the initial parameters of the theory without using fundamental parameters, such as the Planck constant $\hbar$ and the gravitational constant G. However, from practical and historical viewpoints, it is more convenient to explicitly introduce these parameters. In terms of the developed approach, and this can be done by imposing the constraints

$$N_0 \mathcal{V}_d c\alpha^2 = \frac{c^4}{16\pi G\hbar}, \tag{12.12}$$

$$\alpha^2 U(\phi_m) = \frac{16\pi G}{c^4}\Lambda. \tag{12.13}$$

Here, Λ is the observable vacuum energy (dark energy) density. It is important that such quantities as the volume of the extra space $\mathcal{V}_d$ and the minimum value of the potential $U(\phi_m)$ acquire specific values only at $\phi = \text{const}$. If the field ϕ is identified with the inflaton, a nontrivial situation arises. The inflationary stage is completed before the scalar field (inflaton) reaches a potential minimum. Consequently, at the inflationary stage, when the Universe was forming and the field ϕ varied with time, the quantities G and $\hbar$ should also have been time-dependent. In this case, the theory of gravity is an effective theory valid at low energies (see also [546]).

Let us introduce the definition

$$m_\Phi \equiv \alpha\sqrt{\frac{U''(\phi_m)}{K(\phi_m)}}, \tag{12.14}$$

and, under the assumption $K(\phi_m) > 0$, change the variables:

$$\sqrt{\frac{c^4}{16\pi G}K(\phi_m)}(\phi - \phi_m) = \Phi.$$

As a result, we arrive at the conventional form of the action for a scalar field Φ interacting with gravity, that is,

$$S = \frac{1}{\hbar}\int dt\, d^3x \left(\frac{c^4}{16\pi G}R_4 + \frac{1}{2}(\partial\Phi)^2 - \Lambda - \frac{1}{2}m_\Phi^2\Phi^2\right). \tag{12.15}$$

Now, it becomes clear that the parameter m_Φ has the meaning of the scalar field mass.

Equations (12.13) and (12.12) allow us to solve the problem of separate emergence of the gravitational and Planck constants, which appear to be functions of the parameters of the theory N_0, $\{a_n\}$. The set of parameters $\{a_n\}$ is also implicitly included in the expressions for the quantities $\mathcal{V}_d, U(\phi_m)$. We do not consider here the origin of the lapse function and hence the speed of light. Eliminating the insignificant parameter α from Eqs. (10.12), (12.13) and (12.14), we obtain

$$\frac{U(\phi_m)K(\phi_m)}{U''(\phi_m)} = \frac{16\pi G}{c^4}\frac{\Lambda}{m_\Phi^2}, \tag{12.16}$$

$$N_0\mathcal{V}_d(\phi_m)\frac{K(\phi_m)}{U''(\phi_m)} = \frac{c^3}{16\pi G\hbar m_\Phi^2}. \tag{12.17}$$

Here, the right-hand sides contain measurable quantities, whereas the left-hand sides involve quantities depending on the initial parameters of the theory. At energies on the Planck scale, when the field ϕ did not reach the potential minimum, the use of the modern values of the fundamental constants G and $\hbar$ requires some care. Moreover, assuming that the vacuum energies Λ are different in different regions of the Universe [212, 254, 482], the fundamental constants G and $\hbar$ are also different according to the expressions (12.16), (12.17). A discussion of this problem and references can be found in [547].

At the inflationary stage, when ϕ varied, the quantities G and $\hbar$ also varied. The dependence of these quantities on the field can be determined from (12.16) and (12.17) through the change $\phi_m \to \phi$, which are true for small deviations of the field from the equilibrium position. As a result, we have

$$\begin{aligned} G = G(\phi) &\simeq \frac{c^4 m_\Phi^2}{16\pi\Lambda}\frac{U(\phi)K(\phi)}{U''(\phi)}, \\ \hbar = \hbar(\phi) &\simeq \frac{c^3}{16\pi G(\phi)\mathcal{V}_d(\phi)N_0}. \end{aligned} \tag{12.18}$$

The time dependence of the inflaton field ϕ in different inflationary models is well understood [392], hence the dynamics of the fundamental parameters can be described by choosing a specific inflationary model.

12.2.2 *Why is the extra space symmetric?*

The transition amplitude (12.5) describes the birth of a space of type (12.4), where $D_1 = 4$ and there is no preferred geometry for the subspace $\mathbb{M}_{d_1}$. The set of various metrics has the cardinality of continuum, so that the measure of the symmetrical geometries is zero.

In this section, we propose a symmetrization mechanism for compact spaces. It is shown that the space with a metric that slightly differs from the metric

$$ds^2 = \sigma^2[N(t)^2 dt^2 - a(t)^2 d\Omega_3^2], \quad \sigma^2 = \frac{1}{12\pi^2 M_{\rm Pl}^2}. \qquad (12.19)$$

asymptotically tends to (12.19) in the course of cosmological expansion.

The transition amplitude (12.5) generating a four-dimensional space is, as a rule, calculated under the assumption of a special form of the metric h_f on the hypersurface Σ_f with an interval of the type [537, 566].

Let us accept the known postulate that the evolution of any closed system is accompanied by an increase in its entropy. It should be noted that the entropy of a subsystem can decrease. In particular, the Hawking evaporation of a black hole decreases its entropy but increases the entropy of the Universe as a whole. Below, we will consider a space with the geometry (12.4) under the assumption that the subspace volume is considerably larger than the volume of the compact subspace $\mathbb{M}_{d_1}$.

Entropy of a compact space

First and foremost, we will demonstrate that the entropy of a compact space reaches a minimum on a class of maximally symmetric spaces. The initial definition of the entropy in our work coincides with the known definition by Boltzmann, who related the entropy to the number Ω of states of a system: $s = k_B \ln \Omega$. A discussion of other possibilities can be found, for example, in [142, 315]. The notion of the number of states is correctly defined at a quantum level, where the set of energy levels and the degree of their degeneracy are known. However, quantization of the space $\mathbb{M}_d$ means quantization of a gravitational field, which by itself remains an unsolved problem. In this respect, we will restrict ourselves to calculating the number of states from a classical viewpoint. It is sufficient because we are interested in relative quantities (see also [512]).

The entropy of an arbitrary compact space $\mathbb{M}_d$, in which matter fields are absent, is a functional $s[G]$ of the metric tensor G. The number of states is determined by an observer outside the system. It is intuitively clear that the higher the symmetry of an object (in our case, a compact space), the smaller the statistical weight of the object. Let us prove this statement nevertheless. To calculate the statistical weight Ω, the space $\mathbb{M}_d$ is embedded in the space $\mathbb{R}^N$, which can be done if $\mathbb{M}_d$ is sufficiently smooth and N is sufficiently large [221]. It is assumed that the external observer is located in the space $\mathbb{R}^N$. Each point of the space $\mathbb{M}_d$ is described by the internal coordinate $y \in \mathbb{M}_d$ and the external coordinate $x \in \mathbb{R}^N$. We choose a specific point $P \in \mathbb{M}_d$ and fix its coordinate $x_P \in \mathbb{R}^N$. Then, we choose the set of basis vectors e_k, $k = 1, 2, \ldots, N$ in $\mathbb{R}^N$. We require that all d tangent linearly independent vectors at the point P, $e_a^{(P)}$, $a = 1, 2, \ldots, d$, which form the coordinate basis in $\mathbb{M}_d$, should be included in this set. Using this basis, we calculate the components of the metric tensor $G_{ab}^{(P)}(x_P)$ of the space $\mathbb{M}_d$. We fix the tangent space $T(P)$ spanned by the vectors $e_a^{(P)}$.

Then we choose another point $Q \in \mathbb{M}_d$, with the coordinates $x_Q \in \mathbb{R}^N$ and the tangent space $\mathbb{T}(Q)$. By moving along the curve $l_{PQ} \subset \mathbb{M}_d$, the point Q is displaced to the point with the coordinates x_P, so that the tangent space $\mathbb{T}(Q)$ coincides with the tangent space $\mathbb{T}(P)$.

The new components of the metric tensor $G_{ab}^{(Q)}(x_P)$ of the space $\mathbb{M}_d$ are calculated at the same point $x_P \in \mathbb{R}^N$. If the metric tensors do not coincide, $G_{ab}^{(Q)}(x_P) \neq G_{ab}^{(P)}(x_P)$, the observer in the space $\mathbb{R}^N$ will fix a new "microstate" and increase the statistical weight of the space $\mathbb{M}_d$ by unity. If the equality

$$G_{ab}^{(Q)}(x_P) = G_{ab}^{(P)}(x_P), \tag{12.20}$$

is valid, the number of microstates remains unchanged. It should be noted that the condition (12.20) corresponds to the existence condition of a Killing vector along the curve PQ. Therefore, the presence of Killing vectors decreases the statistical weight of a compact space. Maximally symmetric spaces have a minimum entropy.

Decay of excitations of a compact space

The compact subspace $\mathbb{M}_{d_1}$ can be considered as a subsystem of the space $\mathbb{M}_D$. Now, we show that, if the subspace volumes satisfy the inequality $V_{D_1} \gg V_{d_1}$, there is an entropy flow from the subspace $\mathbb{M}_{d_1}$ to the subspace $\mathbb{M}_{D_1}$, the entropy of $\mathbb{M}_{d_1}$ tends to a minimum, and its geometry tends to a maximally symmetric one. Furthermore, the entropy of the entire system, i.e., the space $\mathbb{M}_D$ increases.

As a rule, the entropy of a closed system changes along with conservation of the system energy. Since gravitational energy depends on the space topology, we will restrict our consideration to transitions without its changes. In [134, 275], it has been demonstrated that, in the framework of gravity nonlinear in the Ricci scalar, at least local minima of the energy density exist. In this case, there is a set of energy levels, and each geometry of the extra space can be represented in the form of an expansion in eigenfunctions of the d'Alembert operator of the background metric. In terms of the Kaluza–Klein theory, the eigenvalues of this operator contribute to the mass of excitations, which are interpreted as particles in the space $\mathbb{M}_4$. For a compact space with the geometry of a circle of radius r, the mass of the lightest particle is $m_1 = 1/r$, and it is experimentally bounded below by a value of several TeV. It is evident that its decay into light particles propagating in our space should be accompanied by an entropy increase. In the process, the geometry of the extra space relaxes to a state that has a minimum entropy and is characterized by the absence of excitations.

As an illustration, we consider a space of the type (12.4), where $\mathbb{M}_{D_1} = \mathbb{M}_4 \times \mathbb{M}_{d_2}$, $\mathbb{M}_{d_2} = \mathbb{S}_1$, $\mathbb{M}_{d_1} = \mathbb{S}_1$ with the metric g_{MN}, which differs from the diagonal metric $\eta_{MN} = \mathrm{diag}(1, -1, -1, -1, -r_1, -r_2)$ only slightly, so that $g_{MN} = \eta_{MN} + h_{MN}(x, y_1, y_2)$. The size of one of the extra spaces is considerably larger than the size of the other extra space: $r_2 = \beta r_1$, $\beta \gg 1$. The dynamic equation for the field h_{MN} can be written in the form [54]

$$\eta^{AB}\partial_A\partial_B h_{MN}(x, y_1, y_2) = 0.$$

By substituting the field h in the form of the expansion in eigenfunctions of the d'Alembert operators $Y_n(y_1)$, $Y_n(y_2)$ of both subspaces (in our case, circles),

$$h_{MN}(x, y_1, y_2) = \sum_{n_1,n_2} h_{MN}^{(n_1,n_2)}(x)Y_{n_1}(y_1), Y_{n_2}(y_2),$$

we obtain the equation for the components

$$\left(\Box_x + \frac{n_1^2}{r_1^2} + \frac{n_2^2}{r_2^2}\right) h_{MN}^{(n_1,n_2)}(x) = 0. \tag{12.21}$$

The microstates of the subspaces $\mathbb{M}_{d_1}$ and $\mathbb{M}_{d_2}$ are characterized by the integers n_1 and n_2. The first excited state of the subspace $\mathbb{M}_{d_1}$ has a mass $m_1 = 1/r_1$. A decay of this state into two excited states of the subspace $\mathbb{M}_{d_2}$ with masses $m_2 = n_2/r_2$ and $m_2' = n_2'/r_2$, $n_2, n_2' = 0, 1, 2, \ldots, [\beta]$ and zero momenta occurs with energy conservation,

$$m_1 = m_2 + m_2' \tag{12.22}$$

and an increase in the entropy of the entire system. This increase is associated with the fact that the number of different microstates of the space $\mathbb{M}_{d_2}$ that satisfy the condition (12.22) is of the order of $\Omega_2 = \beta/2 \sim r_2/r_1$. In this case, the geometry of the subspace $\mathbb{M}_{d_1}$ tends to a maximally symmetric geometry because the number of excitations in the subspace tends to zero and the entropy tends to a minimum.

It is obvious that the number of microstates increases with an increase in the volume of the subspace $\mathbb{M}_{d_2}$. Inclusion of states with different four-momenta in the Minkowski space $\mathbb{M}_4$ substantially enhances the effect.

Let us increase the dimension of the compact space with a large volume and consider a manifold of the type (12.4) where $\mathbb{M}_{D_1} = \mathbb{M}_4 \times \mathbb{M}_{d_2}$, $\mathbb{M}_{d_2} = \mathbb{S}_2$, $\mathbb{M}_{d_1} = \mathbb{S}_1$. It is known that the excitation mass in $\mathbb{S}_2$ is $m(l) = \sqrt{l(l+1)}/r_2$, and the degeneracy factor is $2l+1$.

The decay of an excitation of the space $\mathbb{S}_1$ into two excited states of the space $\mathbb{S}_2$ also proceeds with energy conservation: $m_1 = m(l) + m(l')$. It is easy to see that the number of allowed energy states with zero momentum is also proportional to the volume of the large space: $\Omega_2 \simeq \beta^2 = r_2^2/r_1^2$.

Therefore, if quantum fluctuations generate a space of the type $\mathbb{M}_a \times \mathbb{M}_b$*, the entropy flow is directed toward the subspace of larger size. The geometry of the subspace of smaller size tends to a maximally symmetric geometry compatible with its topology. The existence of gauge symmetries in the main space appears to be a purely statistical effect.*

The transition rate to a symmetric state

The entropy of the entire system consisting of two subspaces (12.4) increases as a result of transformations of particles (excitations) of the compact subspace $\mathbb{M}_{d_1}$ to particles propagating in the subspace M_{D_1}. The compact extra space that, at the initial time instant, had an arbitrary geometry, acquires a maximally symmetric shape during the entropy transfer to the main space. In this case, the entropy of the entire system (including the extra and main spaces) increases while the entropy of a subsystem (the compact extra space) tends to a minimum. The situation resembles the third law of thermodynamics, according to which the entropy of a body tends to zero with a decrease in the thermostat temperature.

The subspace $\mathbb{M}_{D_1}$ in (12.4) does not necessarily have a direct-product structure. The main requirement to this subspace is the presence of a large number of energy levels that contribute to the statistical weight of the system at fixed temperature. The Minkowski space M_4 itself exhibits this property to a full extent.

Let us estimate the "symmetrization" rate of the extra space. Weak deviations of the geometry from an equilibrium configuration can be interpreted as excited states of mass m_1 (see, e.g., [17]). Since it is the only scale, it should be expected that the probability of decay will satisfy the relation $\Gamma \sim m_1 \sim 1/L_d$, where L_d is the characteristic size of the extra space. Setting $L_d \leq 10^{-17}$ cm, we find that the lifetime of the excited state is $t_1 \sim L_d \leq 10^{-27}$ s. Therefore, the extra space transformed into the most symmetric state long before the onset of the primordial nucleosynthesis but maybe after completion of the inflationary stage. However, states that correspond to the first excited level of the Kaluza–Klein tower with a lifetime of the order of 10^5 s or more are also considered [468]. Under these conditions, the theory acquires gauge invariance well after the nucleosynthesis stage, which creates new possibilities and problems. In order that a Kaluza–Klein particle be stable, it is necessary to make additional assumptions that complicate the structure of the extra space [464].

12.3 Cascade birth of Universes in multidimensional spaces

In this section, the formation mechanism of universes with distinctly different properties is considered in the framework of pure gravity in a space of $D > 4$ dimensions.

The dynamics of our Universe is well described by a modern theory containing 30 to 40 parameters. This number, whose value is determined experimentally, is too large for a theory to be considered as an ultimate one. In addition, it is well known that the range of admissible parameter values is extremely narrow (fine tuning of the parameters) for the birth and existence of such complex structures as our Universe, which is hard to explain. Extensive literature is devoted to a discussion of this problem, see, e.g., [10]. One way of solving it is based on the assumption of multiple universes with different properties [311, 372, 477]. Rich opportunities of justifying this assumption are contained in the idea of multidimensionality of our space itself. The number of extra dimensions has long been a subject of debate. Thus, the Kaluza–Klein model originally contained one extra dimension. At present, for example, infinite-dimensional spaces [166] and even variable-dimensional spaces [55] are being discussed. The concept of superspace is extended to a set of superspaces with different, unbounded from above numbers of dimensions [482]. Based on the introduced extended superspace, we suggest a formation mechanism of Universes with distinctly different properties and the emergence mechanism for the Planck scale. We discuss the probability of quantum transitions that produce lower-dimensional subspaces.

Let us define according to [482] the superspace $\mathcal{M}_D = (\mathbb{M}_D, g_{ij})$ as a set of metrics g_{ij} in the space $\mathbb{M}_D$ up to diffeomorphisms. On a spacelike section Σ, let us introduce a metric h_{ij} (see, e.g., [566] for details) and define the space of all Riemannian $(D-1)$-metrics:

$$\mathrm{Riem}(\Sigma) = \{h_{ij}(x)\,|\,x \in \Sigma\}. \tag{12.23}$$

The topologies of the sections $\Sigma_{\rm in}$ and Σ_f can be different. We will be concerned with quantum transitions in which the topology of the hypersurface Σ_f is a direct product of subspaces, $\mathbb{M}_{D-1-d} \times \mathbb{M}_d$. The space $\mathbb{M}_d$ is assumed to be compact.

The entire analysis is performed in the framework of nonlinear gravity in a space of $D > 4$ dimensions without including any matter fields. We discuss the emergence of the Planck scale and its relationship with the inflaton mass. The reduction to a lower-dimensional space is made in several steps to produce a cascade. Different cascades give rise to four-dimensional spaces with different effective theories and different numbers of extra dimensions.

The parameters of the low-energy theory turn out to depend on the topology of the extra spaces and vary in a wide range (see also [185, 245]), though the parameters of the original theory are fixed. This also applies to such fundamental concepts as, for example, the Planck mass and the topology of the extra space. The absence of matter fields postulated here at the outset is a fundamental point. It is suggested that the metric tensor components of the extra (super)space at low energies will be interpreted as matter fields in the spirit of Kaluza–Klein theories.

12.3.1 *Simultaneous formation of space–time and the parameters of the theory*

In this section we investigate the problems of emergence of the fundamental constants as well as other parameters at an early stage of the Universe's evolution on the basis of the idea of multiple universes. This idea generally implies the existence of an initial Lagrangian with specific parameters and a potential density with numerous local minima. Each of these minima corresponds to a certain low-energy effective theory with its own unique set of parameters. Which of the minima will be ours depends on the initial conditions. Therefore, the observed low-energy physics depends not only on the initial parameters of the Lagrangian but also on the initial metric tensor of the created space–time.

This idea is usually developed in the framework of string theory [523], which incorporates, among other assumptions, the existence of extra dimensions. Here we develop a purely geometric approach [134], postulating only the existence of extra dimensions. We use the cascade reduction mechanism, introduced in [482], to explain the fine tuning of the parameters in our Universe without invoking any string theory assumptions other than the existence of extra dimensions.

The essence of our idea is as follows. Consider a certain multidimensional space. Due to quantum fluctuations, in some of its regions, the geometry of a direct product of two subspaces can arise. Suppose the curvature of one of these subspaces significantly exceeds that of the other; let us refer to the former as the *extra* subspace and to the latter as the *main* one. Quantum fluctuations in some region of a newly formed main subspace similarly divide it into a direct product of two subspaces. Considering further divisions, we arrive at a "chain" of space partitions. We will call every such partition a *reduction* of space. Multiple consecutive reductions will be called a *reduction cascade*.

Thus a *reduction cascade* consists of several steps reducing the effective dimensionality of space. It will be shown that every step of a cascade changes the parameters of the Lagrangian. Therefore, by choosing different cascades we can obtain different final Lagrangians starting from fixed initial parameters. Each Lagrangian corresponds to universes with distinct properties. It is this possibility that is usually associated with the concepts of a landscape, i.e., numerous low-energy vacua. Our goal is, in a sense, the opposite: we try to assess the set of all parameters of the initial Lagrangian leading to the observed fundamental constants.

We will show that there are numerous initial Lagrangians leading to the observable physics, and the low-energy physics depends not only on the initial parameters, but also on the properties of compact spaces in a particular cascade. Therefore a variation of the initial parameters may be compensated by an appropriate variation of the properties of the cascade, leaving the low-energy physics almost unchanged. This diminishes the importance of a search for the "unique" Lagrangian of the Theory of Everything.

We also discuss the relationships between the parameters of the initial Lagrangian and the fundamental constants $\hbar$ and G determined in low-energy experiments.

12.3.2 *Reduction cascades*

Let us discuss the main idea in detail. Consider a D-dimensional space $\mathbb{M}_D$. Due to its metric fluctuations, new spatial regions with various geometries are continually born within it. We will be interested in regions with a direct product geometry of the form

$$\mathbb{U}_1 = T \times \mathbb{M}_{d_1} \times \mathbb{M}_{D_1}, \quad \mathbb{U}_1 \subset \mathbb{M}_D. \tag{12.24}$$

Here, T is the timelike direction, $\mathbb{M}_{d_1}$ is a compact space of d_1 dimensions, and $\mathbb{M}_{D_1}$ is the main space whose metric fluctuations are studied at the next step of the cascade. We emphasize that the geometry changes not in the whole space $\mathbb{M}_D$ but only in its small region $\mathbb{U}_1$. All physical processes are considered from the viewpoint of an observer located inside the subspace $\mathbb{M}_{D_1}$.

Let us restrict our consideration to quantum fluctuations satisfying the following relation for Ricci scalars in the subspaces $\mathbb{M}_{d_1}$ and $\mathbb{M}_{D_1}$:

$$R_{d_1} \gg R_{D_1}. \tag{12.25}$$

For the simplest geometries, the larger the curvature of space, the smaller its volume. This is the type of geometries we will be studying. But the smaller the system volume, the faster the relaxation processes in it (see a discussion in [482]). Therefore, due to the condition (12.25), the processes in $\mathbb{M}_{d_1}$ advance much faster than those in $\mathbb{M}_{D_1}$. We will discuss the conditions needed to stabilize the volume of $\mathbb{M}_{d_1}$ and its geometry. The theory then becomes effectively D_1-dimensional, with d_1 compact extra dimensions. The parameters of the initial D-dimensional Lagrangian are renormalized, with their new values depending on the properties of the compact extra space $\mathbb{M}_{d_1}$.

What is described above is the first step of the cascade. The following steps are similar: due to the metric fluctuations in some volume of the newly formed space M_{D_1}, there arises a new geometry

$$\mathbb{U}_2 = T \times \mathbb{M}_{d_2} \times \mathbb{M}_{D_2}, \quad \mathbb{U}_2 \subset T \times \mathbb{M}_{D_1}, \tag{12.26}$$

resulting in segregation of another compact space $\mathbb{M}_{d_2}$. We choose those spaces among many for which the relation

$$R_{d_2} \gg R_{D_2}, \tag{12.27}$$

is satisfied similar to (12.25). The parameters of the reduced Lagrangian are renormalized once more; their values become dependent on the properties of the compact extra space $\mathbb{M}_{d_2}$ as well.

The succeeding steps form a cascade:

$$\begin{aligned} &\mathbb{M}_{D_1} \to \mathbb{M}_{D_2} \times \mathbb{M}_{d_2}; \\ &\mathbb{M}_{D_2} \to \mathbb{M}_{D_3} \times \mathbb{M}_{d_3} \to \cdots \to \mathbb{M}_3 \times \mathbb{M}_{d_{\text{final}}}. \end{aligned} \tag{12.28}$$

Cascades differ from each other by the properties of their compact subspaces: their volume, topology, and geometry.

Each reduction consists of two stages: the first is the quantum formation of space of the form (12.24); the second is the classical evolution of this space, which results in stabilization of the compact extra space $\mathbb{M}_{d_i}$.

12.3.3 *A step of the cascade in detail*

The action in the space $\mathbb{M}_D$ is chosen in the form,

$$\begin{aligned} &S_D = N_0 \int d^D X \sqrt{|G^{(D)}|} f(R; a_n), \\ &f(R; a_n) = \sum_n a_n R^n, \quad a_1 = 1, \end{aligned} \tag{12.29}$$

where $G^{(D)} \equiv \det(G_{AB})$, R is the Ricci scalar, N_0 and $\{a_i\}$ are constants. Let us refer to S_D as the first generation action. The standard form is $f(R; a_n) = R - 2\Lambda$, i.e., $a_1 = 1$, $a_0 = -2\Lambda$.

The first step of a cascade begins in a region of small volume by forming a subspace (12.24) with a metric of the form

$$ds^2 = G_{AB}dX^A dX^B = \mathrm{N}dt^2 - g_{ab}(x)dx^a dx^b - \mathrm{e}^{2\beta(x)}\gamma_{ij}(y)dy^i dy^j, \tag{12.30}$$

where $g_{ab}(x)$ is the spatial part of the metric in $\mathbb{M}_{D_1}$, $\gamma_{ij}(y)$ is a positive-definite metric of the extra space $\mathbb{M}_{d_1}$ and $\mathrm{e}^{2\beta(x)}$ is a scaling factor (see [119, 160]).

Let us find the probability of producing such a space, approximating it by a linear dependence on R. Recall that owing to the form of the chosen metric (12.30), the relations (12.25) are valid.

As shown in [483], an extra space with an arbitrary geometry evolves into a space with a maximum number of Killing vectors for a given topology. So we can choose a maximally symmetric space with a constant Ricci scalar R_{d_1} and a curvature $k = \pm 1$.

The volume V_{d_1} of an internal space of unit curvature depends on its geometry since it is expressed in terms of its internal metric:

$$V_{d_1} = \int d^{d_1}y\sqrt{|G^{(d_1)}|}. \tag{12.31}$$

Let us emphasize that it sufficies to demand that only a small fraction of the volume of the initial space $\mathbb{M}_D$ has the geometry (12.30). The following discussion is concerned with the point of view of an internal observer in $\mathbb{M}_{D_1}$, who cannot get any information from outside.

We will use the slow-change approximation suggested in [134], see Sec. 11.3.2:

$$|R_{d_1}| \gg |R_{D_1}|, \quad |R_{d_1}| \gg |f_{\mathrm{der}}|. \tag{12.32}$$

For expansion in a Taylor series,

$$\begin{aligned} f(R_D) &= f(R_{D_1} + R_{d_1} + f_{\mathrm{der}}) \\ &\simeq f(R_{d_1}) + f'(R_{d_1})R_{D_1} + f'(R_{d_1})f_{\mathrm{der}}. \end{aligned} \tag{12.33}$$

Also, the determinants satisfy the relation

$$|G^{(D)}| = e^{2\beta(x)} \cdot |G^{(D_1)}| \cdot |G^{(d_1)}|. \tag{12.34}$$

Substituting these expressions into the action (12.29) and carrying out certain computations given explicitly in [134], we obtain a Lagrangian of a scalar-tensor theory of gravity in D_1 dimensions:

$$S_{D_1} = N_0' \int d^{D_1}x \sqrt{|G^{(D_1)}|}\,(\text{sign}\, F') \left[R_{D_1} + \frac{K}{2}(\partial\phi)^2 - U\right], \tag{12.35}$$

$$K(\phi) = 3\left(\frac{f(\phi;a_n)''}{f(\phi;a_n)'}\right)^2 - \frac{d_1}{\phi}\frac{f(\phi;a_n)''}{f(\phi;a_n)'} + \frac{d_1}{4\phi^2}(d_1+2), \tag{12.36}$$

$$U(\phi) = -(\text{sign}\, f(\phi;a_n)')\left[\frac{|\phi|}{d_1(d_1-1)}\right]^{d_1/2}\frac{f(\phi;a_n)}{F'(\phi;a_n)^2}. \tag{12.37}$$

In accordance with (12.32), we have kept only terms linear in the Ricci scalar.

The classical evolution stage

Let us confine ourselves to those initial parameters which lead to potentials having minima. Our numerical calculations indicate that such parameters do exist.

After nucleation, classical dynamics of these subspaces implies that the compact extra subspace evolves so that the field ϕ approaches the value ϕ_m, corresponding to a minimum of the potential $U(\phi)$.

As follows from the definition of the field ϕ (10.150) and the expression for the interval (12.30), the characteristic size of the space $\mathbb{M}_{d_1}$ is proportional to $e^{\beta} \propto 1/\sqrt{\phi}$. Therefore, the size of a compact *extra* space quickly stabilizes when the field reaches the value $\phi = \phi_m$. This corresponds to $R_{d_1} \to \text{const}, f_{\text{der}} \to 0$. Thus in the following discussion we can use the conditions

$$R_{d_1} = \text{const}, \quad f_{\text{der}} = 0, \tag{12.38}$$

which greatly simplify the calculations.

The equation for the scaling factor of the main subspace during a de Sitter stage has the form [160]

$$\frac{D_1(D_1-1)}{2}\left(\frac{\dot{a}}{a}\right)^2 = \Lambda - \frac{D_1(D_1-1)}{2}k. \qquad (12.39)$$

We assume that the size of an extra space $\mathbb{M}_{d_1}$ has stabilized and $U(\phi) \simeq U(\phi_m) \equiv \Lambda$. Consequently, the scaling factor depends on time as

$$a(t) \propto \mathrm{e}^{Ht}, \quad H = \frac{2\Lambda}{D_1(D_1-1)},$$

at large t. The size of the main subspace rapidly increases.

Thus, we have a D_1-dimensional quickly expanding space and a d_1-dimensional compact extra space. The curvature-linear approximation is sufficient to obtain this result, but to advance further we will need a more accurate expression for the reduced action. The latter could be derived using the conditions (12.38). Expanding the relation $f(R;a_n) = f(R_{d_1} + R_{D_1}; a_n)$ into a Taylor series and integrating over the extra-dimensional coordinates, we obtain

$$S_{D_1} = N_0' \int d^{D_1}X\sqrt{|G^{(D_1)}|}f(R_{D_1};\tilde{a}_n). \qquad (12.40)$$

Here, the new parameters $\tilde{a}_n$ are functions of R_{d_1}, $n > 0$.

We arrive at the second generation of the action (12.40), which is similar to the first generation of the action (12.29) with changed numerical values of the parameters $\tilde{a}_n$. The dimensionality of the main space has been reduced, $D_1 < D$, and a new compact extra space of $d_1 = D - D_1$ dimensions has been formed.

The size of the space $\mathbb{M}_{D_1}$ is much larger than the size of $\mathbb{M}_{d_1}$. In this discussion we are concerned only with quantum fluctuations creating spaces satisfying such a relation. The subsequent dynamics further increases their disparity.

The second step of a cascade is analogous to the first one with substitution $\mathbb{S}_D \to \mathbb{S}_{D_1}$: in a small region of the space $\mathbb{M}_{D_1}$ there occurs a quantum fluctuation which creates a subspace with the topology $\mathbb{M}_{D_2} \times \mathbb{M}_{d_2}; D_2 + d_2 = D_1$. As a result of classical dynamics, the size of the space $\mathbb{M}_{d_2}$ is stabilized while the space $\mathbb{M}_{D_2}$ expands.

If we do not want to be concerned with excitations of the compact space $\mathbb{M}_{d_1}$, we should only consider such quantum fluctuations in $\mathbb{M}_{D_1}$ that satisfy $R_{d_2} \ll R_{d_1}$.

12.3.4 *Quadratic gravity as an explicit example*

In this section we will be guided by the formulas derived in Chapter 10. The initial gravitational field action includes all powers of the Ricci scalar and other invariants. A vast majority of the works on the subject uses some finite polynomial in the Ricci scalar. The choice of a particular polynomial may be justified as follows. Consider a quantum fluctuation that produces a geometry with a characteristic value R_0 of the scalar curvature. Then the initial Lagrangian (12.29) may be approximated by a finite polynomial:

$$f(R, a_n) \simeq \sum_{k=-K}^{K'} b(R_0)_k (R - R_0)^k. \tag{12.41}$$

The specific values of K and K' are chosen according to the author's purposes. The coefficients $b_k(R_0)$ depend on the location of the expansion (12.41) and vary in a wide range.

Consider quadratic gravity in the space $\mathbb{M}_D$ with the action of the form

$$S_D = \frac{N_0}{2} \int d^D X \sqrt{|G^{(D)}|} \left[R_D(G_{AB}) + C R_D^2(G_{AB}) \right.$$
$$\left. + C_1 R_{AB} R^{AB} + C_2 \mathcal{K} - 2\Lambda \right] + \int_{\partial M_D} K d^{D-1}\Sigma, \tag{12.42}$$

where we have also included the Ricci tensor squared and the Kretschmann scalar $\mathcal{K} = R_{ABCD} R^{ABCD}$. The boundary term ($\partial \mathbb{M}_D$) introduced by Hawking and Gibbons does not affect the classical dynamics and can be ignored in what follows [157].

Let us try to find the values of the parameters that allow the formation of a universe similar to ours. In this case a set of parameters of the Lagrangian $\{a_n\}$ (see (12.29)) is $\{C, C_1, C_2, \Lambda\}$.

Following the steps outlined above, we will find the form of the action (12.42) reduced to the space $\mathbb{M}_{D_1}$. An action pertaining to only $\mathbb{M}_{D_1}$ can be recovered by integrating the action in $\mathbb{M}_D$ (12.42) over $\mathbb{M}_{d_1}$.

The Ricci scalar can be expressed using Eq. (12.33):

$$R_D(G_{AB}) = R_{D_1}(g_{ab}) + \phi(x), \tag{12.43}$$

where we have set $f_{\rm der} = 0$, see (12.38). The decompositions of $R_{AB}R^{AB}$ and $\mathcal{K}$ are given by (see [134])

$$\begin{aligned} R_{AB}R^{AB} &= R_{ab}R^{ab} + \mathrm{e}^{-4\beta(x)} \cdot R_{\mu\nu}R^{\mu\nu}, \\ \mathcal{K} &= \mathcal{K}(g_{ab}) + \mathrm{e}^{-4\beta(x)} \cdot \mathcal{K}(\gamma_{ij}), \end{aligned} \tag{12.44}$$

where variables with the indices A, B correspond to the metric G_{AB}, those with the indices a, b to g_{ab}, and with μ, ν to the metric γ_{ij}, (i.e., they correspond to the spaces $\mathbb{M}_D$, $\mathbb{M}_{D_1}$ and $\mathbb{M}_{d_1}$, respectively).

To advance further, recall that we are considering the d_1-dimensional metric γ_{ij} of a constant curvature k, so that we can express the Riemann tensor, the Ricci tensor, and the Ricci scalar in $\mathbb{M}_{d_1}$ space in terms of its curvature:

$$R^{\mu\nu}{}_{\rho\eta} = k\,\delta^{\mu\nu}{}_{\rho\eta}, \quad R^{\nu}_{\mu} = k\,(d_1 - 1)\delta^{\nu}_{\mu}, \quad R'_{d_1} \equiv k\,d_1(d_1 - 1), \tag{12.45}$$

where $\delta^{\mu\nu}{}_{\rho\eta} = \delta^{\mu}_{\rho}\delta^{\nu}_{\eta} - \delta^{\mu}_{\eta}\delta^{\nu}_{\rho}$. R'_{d_1} represents the characteristic curvature scale of the extra dimensions. The expressions for the squared Riemann and Ricci tensors are derived from (12.45); the former is by definition the Kretschmann scalar:

$$\begin{aligned} R_{\mu\nu}R^{\mu\nu} &= d_1[k(d_1 - 1)]^2, \\ \mathcal{K}(\gamma_{ij}) \equiv R_{\mu\nu\rho\eta}R^{\mu\nu\rho\eta} &= 2d_1(d_1 - 1)k^2. \end{aligned} \tag{12.46}$$

Now we can rewrite (12.44) in the form:

$$
\begin{aligned}
R_{AB}R^{AB} &= R_{ab}R^{ab} + \frac{1}{d_1}\phi(x),\\
\mathcal{K} &= \mathcal{K}(g_{ab}) + \frac{2}{d_1(d_1-1)}\phi(x).
\end{aligned}
\tag{12.47}
$$

After plugging (12.43) and (12.47) into the action (12.42) and grouping the terms we obtain

$$
\begin{aligned}
S_D &= \frac{N_0}{2}\int d^D x\sqrt{|G^{(D)}|}\Big\{R_{D_1}(g_{ab})(1+2C\phi) + CR^2_{D_1}(g_{ab}) - 2\Lambda\\
&\quad + C_1 R_{ab}R^{ab} + C_2\mathcal{K}(g_{ab}) + \phi + \left(C + \frac{C_1}{d_1} + \frac{2C_2}{d_1(d_1-1)}\right)\phi^2\Big\}\\
&= \frac{N_0}{2}\int d^D x\sqrt{|G^{(D)}|}\cdot\mathcal{L}(g_{ab}).
\end{aligned}
\tag{12.48}
$$

The expression in brackets has been denoted $\mathcal{L}(g_{ab})$ for convenience; it does not depend on the coordinates of the extra space $\mathbb{M}_{d_1}$.

To find the action in $\mathbb{M}_{D_1}$ we have to integrate (12.48) over the space $\mathbb{M}_{d_1}$ using the appropriate volume definition (12.31). Substitution of Eq. (12.34) for $\sqrt{|G^{(D)}|}$, Eq. (11.58) for $e^{2\beta(x)}$ yields:

$$
\begin{aligned}
S_D &= \frac{N_0}{2}\int d^D x\sqrt{|G^{(D)}|}\cdot\mathcal{L}(g_{ab})\\
&= \frac{N_0}{2}\int d^{D_1}x\, e^{2\beta(x)d_1}\sqrt{|G^{(D_1)}|}\cdot\mathcal{L}(g_{ab})\int d^{d_1}x\sqrt{|G^{(d_1)}|}\\
&= \frac{N_0 V_{d_1}}{2}\int d^{D_1}x\sqrt{|G^{(D_1)}|}\left(\frac{R'_{d_1}(G^{(d_1)})}{\phi(x)}\right)^{d_1/2}\cdot\mathcal{L}(g_{ab}).
\end{aligned}
\tag{12.49}
$$

Suppose that there exists a minimum of the potential $U(\phi)$. The field $\phi(x)$ rapidly relaxes to it and stays fixed during the low-energy processes (see [483] for a discussion). This case is the most natural since the relaxation time is proportional to the scale of the extra space $\mathbb{M}_{d_1}$, which is small as compared to the scale of the space $\mathbb{M}_{D_1}$.

Assuming that these conditions are satisfied, let us make a conformal transformation of the form (see, e.g., [98])

$$\begin{aligned} g_{ab} &= |f(\phi_m)|^{-2/(D_1-2)}\,\widetilde{g}_{ab}, \\ f(\phi) &\equiv \phi^{-d_1/2}(x)[1+2C\phi(x)], \\ R_{D_1} &= |f(\phi_m)|^{2/(D_1-2)}\widetilde{R}_{D_1}, \\ R_{ab}R^{ab} &= |f(\phi_m)|^{4/(D_1-2)}\widetilde{R}_{ab}\tilde{R}^{ab}, \\ \mathcal{K} &= |f(\phi_m)|^{8/(D_1-2)}\widetilde{\mathcal{K}}, \\ \sqrt{|G^{(D_1)}|} &= |f(\phi_m)|^{-D_1/(D_1-2)}\sqrt{|\widetilde{G}^{(D_1)}|}, \end{aligned} \tag{12.50}$$

which, being applied to Eq. (12.49), brings us to the initial form of the action (compare with (12.42)):

$$\begin{aligned} S_{D_1} = \frac{N_0^1}{2}\int d^{D_1}x\sqrt{|G^{(D_1)}|}\{&R_{D_1}(g_{ab}) + C^{(D_1)}R_{D_1}(g_{ab})^2 \\ &+ C_1^{(D_1)}R_{ab}R^{ab}(g_{ab}) + C_2^{(D_1)}\mathcal{K}(g_{ab}) - 2\Lambda^{(D_1)}\}, \end{aligned} \tag{12.51}$$

where $N_0^1 = N_0 V_{d_1}$ and the tildes were omitted for short. The new parameters are expressed in terms of the old ones as follows:

$$C^{(D_1)} = \mathrm{sign}(f(\phi_m))\,|f(\phi_m)|^{(4-D_1)/(D_1-2)}\phi_m^{-d_1/2}C, \tag{12.52}$$

$$\begin{aligned} \Lambda^{(D_1)} &= \mathrm{sign}(f(\phi_m))|f(\phi_m)|^{-D_1/(D_1-2)}\phi_m^{-d_1/2} \\ &\quad\times\left[\Lambda - 1/2\left(\phi_m + \left(C + \frac{C_1}{d_1} + \frac{2C_2}{d_1(d_1-1)}\right)\phi_m^2\right)\right], \end{aligned} \tag{12.53}$$

$$C_1^{(D_1)} = \mathrm{sign}(f(\phi_m))\,|f(\phi_m)|^{(4-D_1)/(D_1-2)}\phi_m^{-d_1/2}C_1, \tag{12.54}$$

$$C_2^{(D_1)} = \mathrm{sign}(f(\phi_m))\,|f(\phi_m)|^{(8-D_1)/(D_1-2)}\phi_m^{-d_1/2}C_2. \tag{12.55}$$

where $f(\phi)$ is defined in (12.50). Recall that we are considering the case where the field ϕ is already at its minimum, $\phi = \phi_m$, so the kinetic terms are neglected. Equations (12.52)–(12.55) connect the old and new parameters after a single reduction.

Thus a single step of reduction to a space of smaller dimension only changes the numerical values of the initial parameters. Meanwhile, the form of the action remains invariable.

The next reduction leads to similar relations for the parameters $C^{(D_2)}$, $C_1^{(D_2)}$, $C_2^{(D_2)}$, $\Lambda^{(D_2)}$ with the substitutions $C \to C^{(D_1)}$, $C_1 \to C_1^{(D_1)}$, $C_2 \to C_2^{(D_1)}$, $\Lambda \to \Lambda^{(D_1)}$. Thus we have obtained recurrence formulas for the parameters.

Again, the action (12.51) for a subspace $\mathbb{M}_{D_1}$ coincides in its form with the initial action (12.42) for $\mathbb{M}_D$, but with renormalized parameters $C^{(D_1)}$, $C_1^{(D_1)}$, $C_2^{(D_1)}$, $\Lambda^{(D_1)}$.

12.3.5 *Numerical computations*

We generate the random initial parameter set $\{C, C_1, C_2, \Lambda\}$ and try to find the cascade that leads to the 4-dimensional space-time. The cascade is completely specified by the dimensions of the spaces that undergo reductions. For instance, we take a 14-dimensional space as our starting point and consider the reduction to an 11-dimensional one, which stabilizes when the field ϕ reaches its minimum. This new space is in turn reduced to an 8-dimensional one, and finally to the 4-dimensional one. Thus, we have a cascade that could be designated as $\langle 15 \to 11 \to 8 \to 4\rangle$.

We have carried out numerical simulations to assess the density of the initial parameters $\{C, C_1, C_2, \Lambda\}$ that can lead to the observable 4-dimensional space-time. The greater this density, the less precisely one has to specify the set of initial parameters of the theory. In the event that every set of parameters $\{C, C_1, C_2, \Lambda\}$ leads to the observable 4D space-time, there is no need to specify the initial parameters at all, because all of them can lead to our space-time. If this could be done for the initial space-time of any dimensionality, then dimensionality also does not have to be specified.

These reduced space–times have their own sets of parameters $\{C, C_1, C_2, \Lambda\}$. After each reduction the parameters are transformed according to (12.52)–(12.55). Using these recurrence formulas we can obtain the last set of parameters before the final reduction. Plugged into (12.37) and (12.36) they give the potential and kinetic energies at the last step of reductions. Finally, if this parameter set satisfies the conditions (12.52)–(12.55), we can conclude that they correspond to the observable four-dimensional space–time.

A set of initial parameters $\{C, C_1, C_2, \Lambda\}$ is represented by a point in the parametric hyperspace (C, C_1, C_2, Λ). We shall refer to those parameter sets that lead to observable constants (i.e., satisfy (12.52)–(12.55)) as the *solutions*. We are interested in the density of solutions in the parametric hyperspace. To assess the density visually we have to use its plane projections — for instance, projections onto planes (C_1, C_2) or (C, Λ). Figure 12.1 presents the results of numerical computation — projections of a four-dimensional solution image onto planes (C, C_1) and (C, Λ).

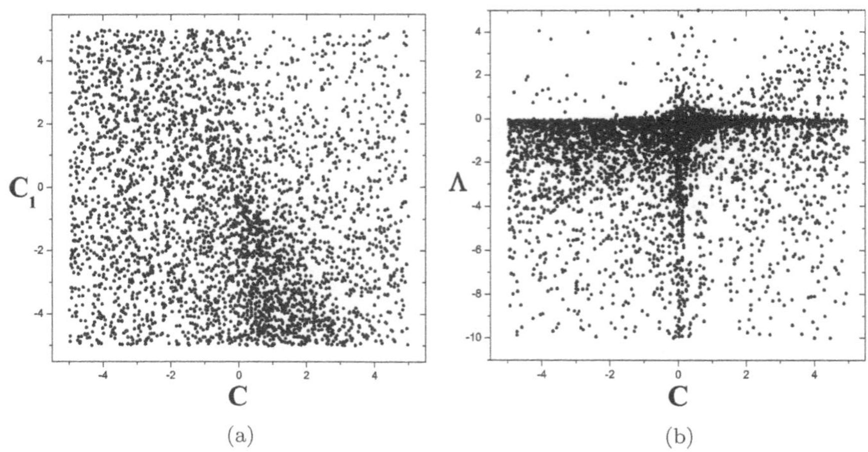

Figure 12.1 The points indicate those parameter values that lead to the observable Universe. All points lead to the same final state. Each reduction path connects the final state and unique initial set of the parameters. The results of numerical computations: plane projections of the points in the hyperspace (C, C_1, C_2, Λ) corresponding to solutions: (a) a projection onto the plane (C, C_1), (b) a projection onto the plane (C, Λ).

It can be seen that there are numerous solutions. This means that the observed low-energy physics is reproduced by a wide range of initial parameters of our theory.

Recall that we took into account only the simplest geometries of the hyperspaces undergoing reductions — only the maximally symmetrical ones. Including other geometries into consideration would yield additional solutions. If these solutions form a continuum in the parametric space, any point in that space (i.e., any set of initial parameters) would lead to the observable physics.

12.4 Discussion

The prospective theory, called a "Theory of Everything" (TOE), should solve the problem of fine tuning in the Universe. How are the initial parameters of TOE chosen and why do they allow very complex structures to arise? A way to approach this problem was found by proposing the existence of numerous vacua with different properties — a "landscape" in the modern terminology. The landscape derived in the scope of string theory is an important advance, but the questions to be answered still remain. Is the concept of "strings" indispensable or will the assumption of multiple dimensions alone suffice? How crucial are the values of the initial parameters? What is the probability of getting to a particular vacuum? Which additional values besides the metric do we need to include?

In the framework of the theory described in this chapter, these questions can be answered as follows:

- For a low-energy "landscape" to exist, we only need to assume the existence of multiple dimensions. The rest of the string theory tools as well as incorporation of additional fields are unnecessary. The cascade reduction mechanism effectively produces the low-energy physics with various parameters.
- Changes in the initial parameters do not substantially affect the probability of universe formation.

It has been shown that there are numerous values of initial parameters of the theory which can be "connected by a cascade" with observed fundamental constants. Particular numerical values of the initial parameters are therefore not as important as was previously thought.

The "landscape" concept implies that numerous low-energy theories with various properties are originated from a theory with unique initial parameters. Here we are concerned with the opposite scenario that the observed physics is derived from numerous initial theories. This "inverse landscape" model puts to doubt the significance of a search for the unique Lagrangian of a TOE.

A set of all such values is rather large although it does not constitute a continuum in the parametric hyperspace. This may be attributed to the limited subset of all possible cascades that we have studied — only those consisting of absolutely symmetric compact spaces. Evidently, an extension of this subset to include all possible geometries surely increases the number of acceptable parameters.

Conjecture:

For any given set of initial parameters, there exists a cascade of reductions leading from a multidimensional space to a universe of our type.

The credibility of this hypothesis should be fully examined in future research. The inhomogeneous extra dimensions revealed recently [141, 422, 484] could serve the prominent tool for this activity.

We have already studied the conditions making it possible to create universes with different properties. Extra dimensions and quantum fluctuations are necessary attributes.

The suggested approach also makes it possible to formulate some quantitative predictions. The main point is that the measure of the set of universes with an exact value of any of the parameters is zero. Therefore, for instance, an exact equality of the cosmological constant to zero is excluded. For the same reason there should not exist real scalar fields having potentials with the degenerate minima. The latter statement is true for any other sorts of fields in the absence of symmetries prescribed at the outset.

12.5 Final remark

The assumption of the extra dimensions existence turns out to be amazingly fruitful. On its basis, the many-worlds idea finds quite a natural explanation, as well as the fact of utmost importance, the fine tuning of the parameters of our Universe. Extra dimensions are a source of physical fields. It seems probable that a source of all phenomena and theories presented in the book is in the extra dimensions, and even such a vague fact as the elegance of correct physical theories becomes if not explained then at least more understandable.

It seems that the ideas on which modern physics is based can be combined in a direction related to extra dimensions. In this case, effort should be directed at studying the structure of an extra space, determining the properties of the Universe.

Appendix

A.1 A controversy between adherents of multiple universes (M) and an ultimate unified theory (U)

Where: a gravitational conference

When: coffee break

Who: two scientists

U: The whole development of science tends to unify the postulates. A success is obvious (the Standard Model, for instance). Do you want to entirely abandon all that?

M: But why do you think that this trend will always be successful? A wish to create a "theory of everything" — isn't it like a wish to make a theory of mass of the planet Earth? Well, by the way, the Standard Model has been created tens of years ago, and there is not much progress since then.

U: Well, the development of the inflationary idea is quite a significant progress. At least, we are moving in a positive direction by unifying the theories and testing their predictions experimentally. But how can you verify that there are many worlds? If, as you admit yourself, it is impossible to get to another universe, there cannot be any proof of your truth. So it does not sound better than creating a theory of the Earth's mass.

M: It should be recognized that it seems a weak point of the idea of multiple universes. But in this case, for instance, the photoelectric effect theory also has no right to exist: indeed, you cannot get into a solid body and make sure that the photon is really absorbed by the electron. In both cases one should accept indirect experimental proofs. On the other hand, your way of creating a unified theory will never lead to a final result. The next question will be: why does your Theory of Everything contain these parameter and not others? For example, if M-theory explains all the existing phenomena, the next question will be: why there are eleven or ten dimensions?

U: But who told you that the cognition process is finite? An asymptotic approximation to the Unified theory is also not the worst perspective. At least as compared with what you suggest. If all theories are possible, what is the reason to study anything?

M: Not quite so. The wealth of stars does not prevent us from studying the Sun. There are a lot of things to be understood in our own Universe. An asymptotic tendency to a unified theory is quite a worthy undertaking. But it does not promise an answer to an experimental fact: the finest selection of the unified theory parameters as though specially to provide conditions for emergence of the intelligence.

U: The unified theory must answer this question as well, although it looks trivial in the many-world concept. But still, to end with: where are all these universes situated?

M: It's really a difficult, almost philosophical question. We'll be able to answer it next to understanding where is our Universe.

A.2 Why do correct theories look elegant?

The idea of multiple universes with different properties also enables us to come to an understanding of such questions, quite far from physics, as the title of this section. Certainly, there cannot be a rigorous proof, not least because the notion of "elegance" is not strictly defined. Therefore let us begin with filling this gap. A criterion of elegance is the degree of economy of the means needed to achieve the goal: explanation of a certain set of phenomena and prediction of new ones. Let us express it by a simple formula:

$$R = N/P,$$

where R is the quality factor of a theory: the higher it is, the more elegant the theory as we see it; N is the number of explained phenomena, and P is the amount of information (in bits) used to formulate the first principles of the theory.

Now the question can be formulated in another form: why do theories with large values of R turn out to be correct? Accepting the multiple-universe theory, we can reduce the problem of elegance to the following question: why is the fraction of universes with a large R-factor large as compared with the fraction of other universes? In this case it would be not surprising that our civilization has just emerged in such a universe.

It is reasonable to suppose that the more complex the structure of a given universe, the more rarely such universes appear (the complexity of a universe is determined by the number of bits necessary to describe it). Yes, certainly intelligent life cannot appear in quite simple universes, which make a majority. There is some complexity threshold, such that intelligence cannot emerge below it. Their number is much smaller than that of simple universes, but only these ones are of interest. In other words, if the whole set of universes is classified by complexity levels, we obtain the sequence

$$K_1,\ K_2,\ K_3, \ldots, K_U,\ K_{U+1}, \ldots$$

The index denotes the complexity level: K_1 is the number of the simplest universes (described by a smallest number of bits), K_2 is the number of those slightly more involved, etc. The number of the simplest universes where intelligence is possible is denoted by K_U; K_{U+1} is the number of more complex universes which can be inhabited by intelligence. It has been supposed that the more complex are universes, the smaller is their fraction. This means that

$$K_1 \gg K_2 \gg K_3 \gg \cdots \gg K_U \gg K_{U+1} \gg \cdots$$

Thus, the probability of an intelligent civilization to find itself in a relatively simple universe is higher (and most probably much higher) than in a more complex one. And it is this circumstance that means that our Universe must be described by an elegant theory.

Bibliography

[1] B.P. Abbott *et al.* (LIGO Scientific Collaboration and Virgo Collaboration) Observation of gravitational waves from a binary black hole merger, *Phys. Rev. Lett.* **116**, 061102; arXiv:1602.03837.

[2] B.P. Abbott *et al.* (LIGO, Virgo and other collaborations), Multi-messenger observations of a binary neutron star merger, *Astroph. J.* **848**, (2), L12; arXiv:1710.05833.

[3] L.F. Abbott, E. Fahri and M. Wise, Particle production in the new inflationary cosmology, *Phys. Lett. B* **117**, 29 (1982).

[4] E. Abdalla, B. Cuadros-Melgar, A.B. Pavan and C. Molina, Stability and thermodynamics of brane black holes, *Nucl. Phys. B* **752**, 40–59 (2006); arXiv:gr-qc/0604033.

[5] A.A. Abdo *et al.*, Fermi large area telescope first source catalog, *Astrophys. J. Suppl.* **188**, 405–436 (2010).

[6] S.T. Abdyrakhmanov, K.A. Bronnikov and B.E. Meierovich, Uniqueness of RS2 type thick branes supported by a scalar field, *Grav. Cosmol.* **11**, 75 (2005).

[7] F.C. Adams *et al.*, Natural inflation: Particle physics models, power-law spectra for large-scale structure, and constraints from the Cosmic Background Explorer, *Phys. Rev. D* **47**, 426 (1993).

[8] S. Adler and R.B. Pearson, "No-hair" theorems for the Abelian Higgs and Goldstone models, *Phys. Rev. D* **18**, 2798 (1978).

[9] I. Affleck and M. Dine, A new mechanism for baryogenesis, *Nucl. Phys. B* **249**, 361 (1985).

[10] A. Aguirre, in: *Universe or Multiverse?*, ed. B.J. Carr (Cambridge University Press, 2005).

[11] P. Aguiar and P. Crawford, Dust-filled axially symmetric universes with cosmological constant, *Phys. Rev. D* **62**, 123511 (2000), arXiv:gr-qc/0009056.

[12] K. Akama, Pregeometry, *Lect. Notes Phys.* **176**, 267 (1982).
[13] K. Akama, Unification scenario of the gravitational and gauge forces via the higher dimensional soliton, *Progr. Theor. Phys.* **78**, 184 (1987).
[14] P.A.R. Ade *et al.* (Planck Collaboration), Planck-2013 results. XXII. Constraints on inflation, *Astron. Astrophys.* **571**, A22 (2014).
[15] S. Alexeyev, K. Rannu, and D. Gareeva, Possible observation manifestations of wormholes in the Brans–Dicke theory, *Sov. Phys. J. Exp. Theoret. Phys.* **113**, 628–636 (2011).
[16] Y. Ali-Haïmoud, Correlation function of high-threshold regions and application to the initial small-scale clustering of primordial black holes, *Phys. Rev. Lett.* **121**, 081304 (2018).
[17] I. Antoniadis, A possible new dimension at a few TeV, *Phys. Lett. B* **246**, 377 (1985).
[18] O. Arias, R. Cardenas and I. Quiros, Thick brane worlds arising from pure geometry, *Nucl. Phys. B* **643**, 187 (2002).
[19] C. Armendáriz-Picon, On a class of stable, traversable Lorentzian wormholes in classical general relativity, *Phys. Rev. D* **65**, 104010 (2002).
[20] R. Arnowitt and J. Dent, Gravitational forces on the branes, *Phys. Rev. D* **71**, 124024 (2005).
[21] I. Arraut, D. Batic, and M. Nowakowski, Maximal extension of the Schwarzschild spacetime inspired by noncommutative geometry, arXiv:1001.2226.
[22] A. Ashtekar, C. Beetle, O. Dreyer, S. Fairhurst *et al.*, Generic isolated horizons and their applications, *Phys. Rev. Lett.* **85**, 3564–3567 (2000); arXiv:gr-qc/0006006.
[23] A. Ashtekar, C. Beetle and J. Lewandowski, Geometry of generic isolated horizons, *Class. Quantum Grav.* **19**, 1195–1225 (2002); arXiv:gr-qc/0111067.
[24] E. Ayon-Beato, F. Canfora and J. Zanelli, Analytic selfgravitating skyrmions, cosmological bounces and AdS wormholes, *Phys. Lett. B* **752**, 201–205 (2016).
[25] E. Ayon-Beato and A. Garcia, Regular black hole in general relativity coupled to nonlinear electrodynamics, *Phys. Rev. Lett.* **80**, 5056 (1998).
[26] M. Banados, J. Silk, and S.M. West, Kerr black holes as particle accelerators to arbitrarily high energy, *Phys. Rev. Lett.* **103**, 111102 (2009).
[27] T. Banks and M. O'Loughlin, Classical and quantum production of cornucopions at energies below 10^{18} GeV, *Phys. Rev. D* **47**, 540 (1993).

[28] N. Barbosa-Cendejas and A. Herrera-Aguilar, 4D gravity localized in non-Z2-symmetric thick branes. *J. High Energy Phys.* **0510**, 101 (2005); *Phys. Rev. D* **73**, 084022 (2006).

[29] C. Barceló and M. Visser, Scalar fields, energy conditions, and traversable wormholes, *Phys. Lett.* **466B**, 127 (1999).

[30] C. Barceló and M. Visser, Scalar fields, energy conditions, and traversable wormholes, *Class. Quantum Grav.* **17**, 3843 (2000); arXiv:gr-qc/0003025.

[31] H. Bantilan *et al.*, Nonspherically symmetric collapse in asymptotically AdS spacetimes, *Phys. Rev. Lett.* **119**, 191103 (2017).

[32] J.M. Bardeen, Non-singular general-relativistic gravitational collapse. *Proc. Int. Conf. GR5*, Tbilisi, 1968, p. 174.

[33] J.M. Bardeen, B. Carter and S.W. Hawking, The four laws of black hole mechanics, *Commun. Math. Phys.* **31**, 161 (1973).

[34] J.D. Barrow and S. Cotsakis, Inflation and the conformal structure of higher-order gravity theories, *Phys. Lett. B* **214**, 515 (1988).

[35] J.D. Barrow and A.C. Ottewill, The stability of general relativistic cosmological theory, *J. Phys. A* **10**, 2757 (1983).

[36] D. Bazeia, F.A. Brito and J.R. Nascimento, Supergravity brane worlds and tachyon potentials, *Phys. Rev. D* **68**, 085007 (2003).

[37] R. Bean and J. Magueijo, Could supermassive black holes be quintessential primordial black holes? *Phys. Rev. D* **66**, 063505 (2002).

[38] J.D. Bekenstein, Extraction of energy and charge from a black hole, *Phys. Rev. D* **7**, 949 (1973).

[39] J.D. Bekenstein, Exact solutions of Einstein–conformal scalar equations, *Ann. Phys. (N.Y.)* **82**, 535 (1974).

[40] J.D. Bekenstein, Black holes: Classical properties, thermodynamics and heuristic quantization, in: *Cosmology and Gravitation*, ed. M. Novello (Atlantisciences, France, 2000), pp. 1–85.

[41] K.M. Belotsky, V.I. Dokuchaev, Y.N. Eroshenko, E.A. Esipova, M.Y. Khlopov, L.A. Khromykh, A.A. Kirillov, V.V. Nikulin, S.G. Rubin and I.V. Svadkovsky, Clusters of primordial black holes, *Eur. Phys. J. C* **79** (3), 246 (2019); arXiv:1807.06590.

[42] K.M. Belotsky, A.A. Kirillov, N.O. Nazarova and S.G. Rubin, Reionization effect enhancement due to primordial black holes, *Int. J. Mod. Phys. D* **26**, no. 09, 1750102 (2017); arXiv:1702.06338.

[43] C.L. Bennett *et al.*, Four-year COBE* DMR Cosmic Microwave Background observations: Maps and basic results, *Astrophys. J. Lett.* **464** L1 (1996).

[44] W. Berej, J. Matyjasek, D. Tryniecki and M. Woronowicz, Regular black holes in quadratic gravity, *Gen. Rel. Grav.* **38**, 885 (2006); arXiv:hep-th/0606185.

[45] J.C. Berengut and V.V. Flambaum, Astronomical and laboratory searches for space-time variation of fundamental constants, arXiv:1009.3693.

[46] V.A. Berezin. V.A. Kuzmin and I.I. Tkachev, Dynamics of bubbles in general relativity, *Phys. Rev. D* **36**, 2919–2944 (1987).

[47] O. Bergmann and R. Leipnik, Space-time structure of a static spherically symmetric scalar field, *Phys. Rev.* **107**, 1157 (1957).

[48] P.G. Bergmann, Comments on the scalar-tensor theory, *Int. J. Theor. Phys.* **1**, 25 (1968).

[49] A. Bernal and F.S. Guzmán, Scalar field dark matter: Nonspherical collapse and late-time behavior, *Phys. Rev. D* **74**, 063504 (2006).

[50] G.D. Birkhoff and R. Langer, *Relativity and Modern Physics* (Harvard University Press, Harvard, 1923).

[51] N.D. Birrell and P.C.W. Davies, *Quantum Fields in Curved Space* (Cambridge University Press, Cambridge, 1982).

[52] G.S. Bisnovatyi-Kogan and O.Yu. Tsupko, Strong gravitational lensing by Schwarzschild black holes, *Astrophysics* **51**, 99 (2008).

[53] E. Bittencourt, R. Klippert and G.B. Santos, Dynamical wormhole definitions confronted, *Class. Quantum Grav.* **35**, 155009 (2018); arXiv:1707.01078.

[54] M. Blagojevic, *Gravitation and Gauge Symmetries* (Institute of Physics, Bristol, 2002).

[55] U. Bleyer, M. Mohazzab and M. Rainer, Dynamics of dimensions in factor space cosmology, arXiv:gr-qc/9508035.

[56] N.M. Bocharova, K.A. Bronnikov and V.N. Melnikov, On an exact solution of the Einstein–scalar field equations, *Vestnik MGU Fiz., Astron.* No. 6, 706 (1970).

[57] C. G. Böhmer and N. Tamanini, Generalized hybrid metric-Palatini gravity, *Phys. Rev. D* **87**, 084031 (2013); arXiv:1302.2355.

[58] Christian G. Boehmer and Kevin Vandersloot, Stability of the Schwarzschild interior in loop, quantum gravity, *Phys. Rev. D* **78**, 067501 (2008); arXiv:0807.3042; Loop quantum dynamics of the Schwarzschild interior, *Phys. Rev. D* **76**, 104030 (2007); arXiv:0709.2129.

[59] C. Boehmer and T. Harko, Dynamical instability of fluid spheres in the presence of a cosmological constant, *Phys. Rev. D* **71**, 084026 (2005).

[60] A. Bogojevic and D. Stojkovic, A non-singular black hole, *Phys. Rev. D* **61**, 084011 (2000); arXiv:gr-qc/9804070.

[61] S.V. Bolokhov, K.A. Bronnikov, P.A. Korolyov, and M.V. Skvortsova, Visible, invisible and trapped ghosts as sources of wormholes and black universes, *J. Phys. Conf. Series* **675**, 012013 (2016); arXiv:1510.00546.

[62] S.V. Bolokhov, K.A. Bronnikov, and M.V. Skvortsova, Magnetic black universes and wormholes with a phantom scalar, *Class. Quantum Grav.* **29**, 245006 (2012); arXiv:1208.4619.

[63] S.V. Bolokhov, K.A. Bronnikov, and M.V. Skvortsova, The Schwarzschild singularity: a semiclassical bounce? *Grav. Cosmol.* **24**, 315–320 (2018); arXiv:1808.03717.

[64] S.V. Bolokhov, K.A. Bronnikov, and M.V. Skvortsova. Rotating cylinders with anisotropic fluids in general relativity, *Grav. Cosmol.* **25**, 122–130 (2019); arXiv:1904.06727.

[65] S.V. Bolokhov, K.A. Bronnikov, and M.V. Skvortsova, Cylindrical wormholes: A search for viable phantom-free models in GR, *Int. J. Mod. Phys. D* **28**, 1941008 (2019); arXiv:1903.09862.

[66] C. Bona, A new proof of the generalized Birkhoff theorem, *J. Math. Phys.* **29**, 1440 1988).

[67] N. Bostrom, *Anthropic Bias: Observation Selection Effects in Science and Philosophy* (Routledge, New York, 2002).

[68] R.H. Boyer and R.W. Lindquist, Maximal analytic extension of the Kerr metric, *J. Math. Phys.* **8**, 265 (1967).

[69] V. Bozza, Gravitational lensing in the strong field limit, *Phys. Rev. D* **66**, 103001 (2002).

[70] V.B. Braginsky and V.I. Panov, Test of the equivalence principle of the inertial and gravitational masses, *Zh. Eksp. Teor. Fiz.* **61**, 873 (1971).

[71] R.H. Brandenberger, H.A. Feldman and V.F. Mukhanov, Theory of cosmological perturbations. Part 1. Classical perturbations. Part 2. Quantum theory of perturbations. Part 3. Extensions, *Phys. Rep.* **215**, 203 (1992).

[72] G. Brando, J.C. Fabris, F.T. Falciano and O. Galkina, Stiff matter solution in Brans-Dicke theory and the general relativity limit, arXiv:1810.07860.

[73] C.H. Brans, Mach's principle and a relativistic theory of gravitation II, *Phys. Rev.* **125**, 2194 (1961).

[74] C. Brans and R.H. Dicke, Mach's principle and a relativistic theory of gravitation, *Phys. Rev.* **124**, 925 (1961).

[75] K.A. Bronnikov, Scalar–tensor theory and scalar charge, *Acta Phys. Polon* **B4**, 251–274 (1973).

[76] K.A. Bronnikov, Static fluid cylinders and plane layers in general relativity, *J. Phys. A: Math. Gen.* **12**, 201–207 (1979).

[77] K.A. Bronnikov, Gravitational and sound waves in stiff matter, *J. Phys. A: Math. Gen.* **13**, 3455–3463 (1980).

[78] K.A. Bronnikov, Extra dimensions, nonminimal couplings, horizons and wormholes, *Grav. Cosmol.* **2**, 221–226 (1996); arXiv:gr-qc/9703020.

[79] K.A. Bronnikov, Regular magnetic black holes and monopoles from nonlinear electrodynamics, *Phys. Rev. D* **63**, 044005 (2001); arXiv:gr-qc/0006014.

[80] K.A. Bronnikov, Scalar vacuum structure in general relativity and alternative theories. Conformal continuations, *Acta Phys. Polon. B* **32**, 3571–3592 (2001); arXiv:gr-qc/0110125.

[81] K.A. Bronnikov, Spherically symmetric false vacuum: no-go theorems and global structure, *Phys. Rev. D* **64**, 064013 (2001); arXiv:gr-qc/0104092.

[82] K.A. Bronnikov, Scalar-tensor gravity and conformal continuations, *J. Math. Phys.* **43**, 6096–6115 (2002); arXiv:gr-qc/0204001.

[83] K.A. Bronnikov, Rotating cylindrical wormholes: A no-go theorem. *J. Phys. Conf. Series* **675**, 012028 (2016).

[84] K.A. Bronnikov. Trapped ghosts as sources for wormholes and regular black holes. The stability problem, in: *Wormholes, Warp Drives and Energy Conditions*, ed. F.S.N. Lobo (Springer, 2017), p. 137–160 (Ref. [386] in this book).

[85] K.A. Bronnikov, Spherically symmetric black holes and wormholes in hybrid metric-Palatini gravity, *Grav. Cosmol.* **25**, 331 (2019); arXiv:1908.02012.

[86] K.A. Bronnikov and K.A. Baleevskikh, On gravitational lensing by symmetric and asymmetric wormholes, *Grav. Cosmol.* **25** (1) (2019); arXiv:1812.05704.

[87] K.A. Bronnikov, K.A. Baleevskikh and M.V. Skvortsova, Wormholes with fluid sources: a no-go theorem and new examples, *Phys. Rev. D* **96**, 124039 (2017); arXiv:1708.02324.

[88] K. A. Bronnikov, S. V. Bolokhov, M. V. Skvortsova, Hybrid metric-Palatini gravity: black holes, wormholes, singularities and instabilities, *Grav. Cosmol.* **26**, 212–227 (2020); arXiv:2006.00559.

[89] K.A. Bronnikov, S.V. Bolokhov and M.V. Skvortsova, A possible semiclassical bounce instead of a Schwarzschild singularity, *Int. J. Mod. Phys. A* **35**, 2040051 (2020); arXiv:1909.06888.

[90] K. A. Bronnikov, S. V. Bolokhov, M. V. Skvortsova Matter accretion versus semiclassical bounce in Schwarzschild interior, arXiv:2009.06330.

[91] K.A. Bronnikov and M.S. Chernakova, $f(R)$ theory of gravity and conformal continuations, *Izv. Vuzov, Fiz.* **46** (9) (2005); *Russ. Phys. J.* **48**, 940 (2005); arXiv:gr-qc/0503025.

[92] K.A. Bronnikov and M.S. Chernakova, Charged black holes and unusual wormholes in scalar-tensor gravity, *Grav. Cosmol.* **13**, 1 (49), 51 (2007).

[93] K.A. Bronnikov, M.S. Chernakova, J.C. Fabris, N. Pinto-Neto and M.E. Rodrigues, Cold black holes and conformal continuations, *Int. J. Mod. Phys. D* **17**, 25 (2008); arXiv:gr-qc/0609084.

[94] K.A. Bronnikov, S.V. Chervon and S.V. Sushkov, Wormholes supported by chiral fields, *Grav. Cosmol.* **15**, 241 (2009); arXiv: 0905.3804.

[95] K.A. Bronnikov, G. Clément, C.P. Constantindis and J.C. Fabris, Cold scalar-tensor black holes: causal structure, geodesics, stability, *Grav. Cosmol.* **4**, 128 (1998).

[96] K.A. Bronnikov, G. Clément, C.P. Constantindis and J.C. Fabris, Structure and stability of cold scalar-tensor black holes, *Phys. Lett. A* **243**, 121 (1998).

[97] K.A. Bronnikov, C.P. Constantinidis, R.L. Evangelista and J.C. Fabris, Electrically charged cold black holes in scalar-tensor theories, *Int. J. Mod. Phys. D* **8**, 481 (1999).

[98] K.A. Bronnikov, H. Dehnen and V.N. Melnikov, On a general class of brane-world black holes, *Phys. Rev. D* **68**, 024025 (2003).

[99] K.A. Bronnikov, H. Dehnen and V.N. Melnikov, Regular black holes and black universes, *Gen. Rel. Grav.* **39**, 973 (2007).

[100] K.A. Bronnikov, A. Dobosz and I.G. Dymnikova, Nonsingular vacuum cosmologies with a variable cosmological term, *Class. Quantum Grav.* **20**, 3797 (2003).

[101] K.A. Bronnikov and E.V. Donskoy, Possible black universes in a brane world, *Grav. Cosmol.* **16**, 42 (2010); arXiv:0910.4930.

[102] K.A. Bronnikov and E.V. Donskoy, Black universes with trapped ghosts, *Grav. Cosmol.* **17**, 31 (2011); arXiv:1110.6030.

[103] K.A. Bronnikov and E. Elizalde, Spherical systems in models of nonlocally corrected gravity, *Phys. Rev. D* **81**, 044032 (2010); arXiv:0910.3929.

[104] K.A. Bronnikov and J.C. Fabris, Weyl spacetimes and wormholes in D-dimensional Einstein and dilaton gravity, *Class. Quantum Grav.* **14**, 831–842 (1997).

[105] K.A. Bronnikov and J.C. Fabris, Regular phantom black holes, *Phys. Rev. Lett.* **96**, 251101 (2006).

[106] K.A. Bronnikov, J.C. Fabris, and A. Zhidenko, On the stability of scalar-vacuum space-times, *Eur. Phys. J. C* **71**, 1791 (2011).

[107] K.A. Bronnikov, S.B. Fadeev, and A.V. Michtchenko, Scalar fields in multidimensional gravity. No-hair and other no-go theorems, *Gen. Rel. Grav.* **35**, 505 (2003).

[108] K.A. Bronnikov and A.M. Galiakhmetov, Wormholes without exotic matter in Einstein–Cartan theory, *Grav. Cosmol.* **21**(4), 283–288 (2015); arXiv:1508.01114.

[109] K.A. Bronnikov and A.M. Galiakhmetov. Wormholes and black universes without phantom fields in Einstein–Cartan theory, *Phys. Rev. D* **94**, 124006 (2016); arXiv:1607.07791.

[110] K.A. Bronnikov and S.V Grinyok, Charged wormholes with non-minimally coupled scalar fields. Existence and stability, in: *Inquiring the Universe. Essays to celebrate Prof. Mario Novello's jubilee* (Frontier Group, Rio de Janeiro, 2003); arXiv:gr-qc/0205131.

[111] K.A. Bronnikov and S.V. Grinyok, Conformal continuations and wormhole instability in scalar-tensor gravity, *Grav. Cosmol.* **10**, 237 (2004); arXiv:gr-qc/0411063.

[112] K.A. Bronnikov and S.V. Grinyok, Electrically charged and neutral wormhole instability in scalar-tensor gravity, *Grav. Cosmol.* **11**, 75 (2005); arXiv:gr-qc/0509062.

[113] K.A. Bronnikov and A.V. Khodunov, Scalar field and gravitational instability, *Gen. Rel. Grav.* **11**, 13 (1979).

[114] K.A. Bronnikov, S.-W. Kim and M.V. Skvortsova. The Birkhoff theorem and string clouds, *Class. Quantum Gravity* **33**, 195006 (2016); arXiv:1604.04905.

[115] K.A. Bronnikov and Yu.N. Kireyev, Instability of black holes with scalar charge, *Phys. Lett. A* **67**, 95 (1978).

[116] K.A. Bronnikov and S.A. Kononogov, Possible variations of the fine structure constant α and their metrological significance, *Metrologia* **43**, R1 (2006).

[117] K.A. Bronnikov, S.A. Kononogov and V.N. Melnikov, Brane world corrections to Newton's law, *Gen. Rel. Grav.* **38**, 1215 (2006).

[118] K.A. Bronnikov, S.A. Kononogov, V.N. Melnikov and S.G. Rubin, Cosmologies from nonlinear multidimensional gravity with acceleration and slowly varying G, *Grav. Cosmol.* **14**, 230 (2008).

[119] K.A. Bronnikov, R.V. Konoplich and S.G. Rubin, Diversity of universes created by pure gravity, *Class. Quantum Grav.* **24**, 1261 (2007).

[120] K.A. Bronnikov and R.A. Konoplya, Echoes in brane worlds: Ringing at a black hole-wormhole transition, *Phys. Rev. D* **101**, 064004 (2020); arXiv:1912.05315.

[121] K.A. Bronnikov, R.A. Konoplya and A. Zhidenko, Instabilities of wormholes and regular black holes supported by a phantom scalar field, *Phys. Rev. D* **86**, 024028 (2012); arXiv:1205.2224.

[122] K.A. Bronnikov and M.A. Kovalchuk, On a generalization of Birkhoff's theorem, in: *Problems of Gravitation Theory and Particle Theory*, ed. K.P. Staniukovich, 10th issue (Atomizdat, M., 1979); *J. Phys. A: Math. Gen.* **13**, 187 (1980).

[123] K.A. Bronnikov and V.G. Krechet, Rotating cylindrical wormholes and energy conditions, *Int. J. Mod. Phys. A* **31** (2–3), 1641022 (2016); arXiv:1509.04665.

[124] K.A. Bronnikov and V.G. Krechet, Potentially observable cylindrical wormholes without exotic matter in GR, *Phys. Rev. D* **99**, 084051 (2019); arXiv:1807.03641.

[125] K.A. Bronnikov, V.G. Krechet and José P.S. Lemos, Rotating cylindrical wormholes, *Phys. Rev. D* **87**, 084060 (2013); arXiv:1303.2993.

[126] K.A. Bronnikov, V.G. Krechet and V.B. Oshurko, Rotating Melvin-like universes and wormholes in general relativity, *Symmetry* **2020**, 12, 1306 (2020); arXiv:2007.01145.

[127] K.A. Bronnikov and José P.S. Lemos, Cylindrical wormholes, *Phys. Rev. D* **79**, 104019 (2009).

[128] K. A. Bronnikov, L. N. Lipatova, I. D. Novikov, and A. A. Shatskiy, Example of a stable wormhole in general relativity, *Grav. Cosmol.* **19**, 269 (2013); arXiv:1312.6929.

[129] K.A. Bronnikov and B.E. Meierovich, A general thick brane supported by a scalar field, *Grav. Cosmol.* **9**, 313 (2003).

[130] K.A. Bronnikov and V.N. Melnikov, The Birkhoff theorem in multidimensional gravity, *Gen. Rel. Grav.* **27**, 465 (1995); arXiv:gr-qc/9406064.

[131] K.A. Bronnikov and V.N. Melnikov, P-brane black holes as stability islands, *Nucl. Phys. B* **584**, 434 (2000); arXiv:hep-th/0002200.

[132] K.A. Bronnikov and V.N. Melnikov. On observational predictions from multidimensional gravity, *Gen. Rel. Grav.* **33**, 1549–1578 (2001); arXiv:gr-qc/0103079.

[133] K.A. Bronnikov and V.N. Melnikov, Conformal frames and D-dimensional gravity, in: *Proceedings of the 18th Course of the School on Cosmology and Gravitation: The Gravitational Constant, Generalized Gravitational Theories and Experiments* (30 April–10 May 2003, Erice), eds. G.T. Gillies, V.N. Melnikov and V. de Sabbata (Kluwer, Dordrecht/Boston/London, 2004), pp. 39–64; arXiv:gr-qc/0310112.

[134] K.A. Bronnikov and S.G. Rubin, Self-stabilization of extra dimensions, *Phys. Rev. D* **73**, 124019 (2006).

[135] K.A. Bronnikov, Nilton Santos and Anzhong Wang, Cylindrical systems in general relativity (review), *Class. Quantum Grav.* **37**, 113002 (2020); arXiv:1901.06561.

[136] K.A. Bronnikov and G.N. Shikin, Spherically symmetric scalar vacuum: no-go theorems, black holes and solitons, *Grav. Cosmol.* **8**, 107 (2002), arXiv:gr-qc/0109027.

[137] K.A. Bronnikov, M.V. Skvortsova and A.A. Starobinsky, Notes on wormhole existence in scalar-tensor and F(R) gravity, *Grav. Cosmol.* **16**, 216 (2010); arXiv:1005.3262.

[138] K.A. Bronnikov and A.A. Starobinsky, No realistic wormholes from ghost-free scalar-tensor phantom dark energy, Pis'ma v ZhETF **85**(1), 3–8 (2007); *J. Exp. Theoret. Phys. Lett.* **85**(1), 1–5 (2007).

[139] K.A. Bronnikov and Sung-Won Kim, Possible wormholes in a brane world, *Phys. Rev. D* **67**, 064027 (2003).

[140] K.A. Bronnikov and S.V. Sushkov, Trapped ghosts: a new class of wormholes, *Class. Quantum Grav.* **27**, 095022 (2010); arXiv:1001.3511.

[141] K.A. Bronnikov, A.A. Popov and S.G. Rubin, Inhomogeneous compact extra dimensions and de Sitter cosmology, *Eur. Phys. J. C* **80**, 970 (2020); arXiv:2004.03277.

[142] T.A. Brun and J.B. Hartle, Entropy of classical histories, *Phys. Rev. E* **59**, 6370 (1999); arXiv:quant-ph/9808024.

[143] T. Bunch and P. Davies, Quantum field theory in de Sitter space: Renormalization by point splitting, *Proc. R. Soc. A* **360**, 117 (1978).

[144] R.V. Buniy and S.D.H. Hsu, Semi-classical wormholes are unstable, *Phys. Lett. B* **632**, 127 (2006); arXiv:hep-th/0504003.

[145] Alexander Burinskii, The Dirac–Kerr–Newman electron, *Grav. Cosmol.* **14**, 109 (2008); arXiv:hep-th/0507109.

[146] Alexander Burinskii, Gravity versus quantum theory: Is electron really pointlike? arXiv:1109.3872 (essay for GRF 2011 competition).

[147] A. Burinskii and S.R. Hildebrandt, New type of regular black holes and particlelike solutions from nonlinear electrodynamics, *Phys. Rev. D* **65**, 104017 (2002); *Grav. Cosmol.* **9**, 20 (2003); *Czech. J. Phys.* **53**, B283 (2003).

[148] Rong-Gen Cai and N. Ohta, Black holes in pure Lovelock gravities, *Phys. Rev. D* **74**, 064001 (2006); arXiv: hep-th/0604088.

[149] P. Callin and F. Ravndall, Higher order corrections to the Newtonian potential in the Randall-Sundrum model, *Phys. Rev. D* **70**, 104009 (2004).

[150] M. Campanelli and C.O. Lousto, Are black holes in Brans-Dicke theory precisely the same as a general relativity? *Int. J. Mod. Phys. D* **2**, 451 (1993).

[151] J. Campbell, *A Course of Differential Geometry* (Clarendon, Oxford, 1926).

[152] F. Canfora, N. Dimakis and A. Paliathanasis, Topologically nontrivial configurations in the 4d Einstein's nonlinear σ-model system, *Phys. Rev. D* **96**, 025021 (2017).

[153] S. Capozziello, T. Harko, T.S. Koivisto, F.S.N. Lobo, and G.J. Olmo, The virial theorem and the dark matter problem in hybrid metric-Palatini gravity, *J. Cosmol. Astro. Phys.* **07**, 024 (2013); arXiv:1212.5817.

[154] S. Capozziello, T. Harko, T.S. Koivisto, F.S.N. Lobo, and G.J. Olmo, Hybrid metric-Palatini gravity, *Universe* **1**, 199 (2015); arXiv:1508.04641.

[155] Salvatore Capozziello, Tiberiu Harko, Francisco S.N. Lobo, and Gonzalo J. Olmo, Hybrid modified gravity unifying local tests, galactic dynamics and late-time cosmic acceleration, *Int. J. Mod. Phys. D* **22**, 1342006 (2013); arXiv:1305.3756.

[156] S. Capozziello and M. de Laurentis, Extended theories of gravity, *Phys. Rep.* **509**, 167–321 (2011).

[157] S. Carlip, A note on real tunneling geometries, *Class. Quantum Grav.* **22**, 4381 (2005); arXiv:gr-qc/0508072.

[158] Bernard J. Carr and S.W. Hawking, Black holes in the early Universe, *Mon. Not. Roy. Astron. Soc.* **168**, 399–415 (1974).

[159] B.J. Carr, The primordial black hole mass spectrum, *Astrophys. J.* **201**, 1 (1975).

[160] Sean M. Carroll, James Geddes, Mark B. Hoffman, and Robert M. Wald, Classical stabilization of homogeneous extra dimensions, *Phys. Rev. D* **66**, 024036 (2002); arXiv: hep-th/0110149.

[161] B. Carr, K. Kohri, Y. Sendouda and J. Yokoyama, Constraints on primordial black holes, arXiv:2002.12778.

[162] S.M. Carroll, M. Hoffman and M. Trodden, Can the dark energy equation-of-state parameter w be less than −1? *Phys. Rev. D* **68**, 023509 (2003).

[163] E. Carugno *et al.*, Inflation in multidimensional quantum cosmology, *Phys. Rev. D* **53**, 6863 (1996).

[164] R. Casadio, A. Fabbri and L. Mazzacurati, New black holes in the brane world? *Phys. Rev. D* **65**, 084040 (2002).

[165] R. Casadio and L. Mazzacurati, Bulk shape of brane-world black holes, *Mod. Phys. Lett. A* **18**, 651 (2003); arXiv:gr-qc/0205129.

[166] C. Castro, A. Granik and M.S. El Naschie, Why we live in 3 dimensions, arXiv:hep-th/0004152.

[167] A. Chamblin, H.S. Reall, H. Shinkai and T. Shiromizu, Charged brane-world black holes, *Phys. Rev. D* **63**, 064015 (2001); arXiv:hep-th/0008177.

[168] K.C.K. Chan, J.H. Horne and R.B. Mann, Charged dilaton black holes with unusual asymptotics, *Nucl. Phys. B* **447**, 441 (1995).

[169] S. Chandrasekhar, *Mathematical Theory of Black Holes* (Clarendon Press, Oxford, 2006).

[170] G.F. Chapline, Cosmological effects of primordial black holes, *Nature* **253**, 251 (1975)

[171] V.M. Chechetkin, M.Yu. Khlopov, M.G. Sapozhnikov and Ya.B. Zeldovich, Astrophysical aspects of anti-proton interaction with He-4 (antimatter in the Universe), *Phys. Lett. B* **118**, 329 (1982).

[172] Hong Zhe Chen, Z. Fisher, J. Hernandez, R.C. Myers and Shan-Ming Ruan, Information flow in black hole evaporation, *J. High Energy Phys.* **03**, 152 (2020); arXiv:1911.03402.

[173] Gérard Clément, Júlio C. Fabris and Manuel E. Rodrigues, Phantom black holes in Einstein-Maxwell-dilaton theory, *Phys. Rev. D* **79**, 064021 (2009); arXiv:0901.4543.

[174] N.A. Chernikov and E.A. Tagirov, Quantum theory of scalar field in de Sitter space-time, *Ann. Inst. H. Poincare Phys. Theor. A* **9**, 109 (1968).

[175] T. Chiba and M. Yamaguchi, Runaway domain wall and space-time varying α, arXiv:1102.0105.

[176] T. Chiba, The constancy of the constants of Nature: updates, arXiv:1111.0092.

[177] Adrian Cho, Higgs boson makes its debut after decades-long search, *Science* **337** (6091), 141–143 (2012).

[178] S. Clesse and J. García-Bellido, Seven hints for primordial black hole dark matter, *Phys. Dark Univ.* **22**, 137–146 (2018).

[179] J.M. Cline, S. Geon, and G.D. Moore, The phantom menaced: Constraints on low-energy effective ghosts, *Phys. Rev. D* **70**, 043543 (2004).

[180] J.M. Cline, TASI Lectures on Early Universe Cosmology: Inflation, Baryogenesis and Dark Matter. *PoS* **2019**, *TASI2018*, 1.

[181] A.G. Cohen and D.B. Kaplan, Thermodynamic generation of the baryon asymmetry, *Phys. Lett. B* **199**, 251 (1987).

[182] A.G. Cohen and D.B. Kaplan, Spontaneous baryogenesis, *Nucl. Phys. B* **308**, 913 (1988).

[183] S. Coleman, Fate of the false vacuum: Semiclassical theory, *Phys. Rev. D* **15**, 2929 (1977).

[184] S. Coleman, The fate of the false vacuum. 2. First quantum corrections, *Phys. Rev. D* **16**, 1762 (1977).

[185] S. Coleman, Black holes as red herrings: Topological fluctuations and the loss of quantum coherence, *Nucl. Phys. B* **307**, 867 (1988).

[186] A.A. Coley, The dynamics of brane-world cosmological models. Einstein centennial review article, arXiv:astro-ph/0504226.

[187] E.J. Copeland, M. Sami and S. Tsujikawa, Dynamics of dark energy, *Int. J. Mod. Phys. D* **15**, 1753 (2006); arXiv:hep-th/0603057.

[188] J.G. Cramer, R.L. Forward, M.S. Morris, M. Visser, G. Benford, and G.A. Landis, Natural wormholes as gravitational lenses, *Phys. Rev. D* **51**, 3117 (1995).

[189] C. Csaki, J. Erlich, T.J. Hollowood and Y. Shirman, Universal aspects of gravity localized on thick branes, *Nucl. Phys. B* **581**, 309 (2000).

[190] N. Dadhich, S. Kar, S. Mukherjee and M. Visser, R = 0 spacetimes and self-dual Lorentzian wormholes, *Phys. Rev. D* **65**, 064004 (2002).

[191] Sergio Dain, Geometric inequalities for axially symmetric black holes, arXiv:1111.3615.

[192] P.C.W. Davies, Multiverse cosmological models, *Mod. Phys. Lett. A* **19**, 727 (2004).

[193] Bogdan Dănilă, Tiberiu Harko, Francisco S. N. Lobo, and Man Kwong Mak, Spherically symmetric static vacuum solutions in hybrid metric-Palatini gravity, *Phys. Rev. D* **99**, 064028 (2019); arXiv:1811.02742.

[194] F. Debbasch, L. Herrera, P.R.C.T. Pereira and N.O. Santos, Stationary cylindrical anisotropic fluid, *Gen. Rel. Grav.* **38**, 1825–1838 (2006).

[195] H. Deng, J. Garriga and A. Vilenkin, Primordial black hole and wormhole formation by domain walls, *J. Cosmol. Astropart. Phys.* **4**, 050 (2017).

[196] H. Dennhardt and O. Lechtenfeld, Scalar deformations of Schwarzschild holes and their stability, *Int. J. Mod. Phys. A* **13**, 741 (1998); arXiv:gr-qc/9612062.

[197] N. Deruelle and M. Sasaki, Newton's law on an Einstein "Gauss-Bonnet" brane, *Progr. Theor. Phys.* **110**, 441 (2003).

[198] B.S. DeWitt, *Dynamical Theory of Groups and Fields* (Gordon and Breach, NY, 1965).

[199] B.S. DeWitt and N. Graham, eds., *The Many-Worlds Interpretation of Quantum Mechanics* (Princeton UP, Princeton, 1973).

[200] Tushar Kanti Dey and Surajit Sen. Gravitational lensing by wormholes, *Mod. Phys. Lett. A* **23**, 953 (2008); arXiv:0806.4059.

[201] V.I. Dokuchaev, Birth and life of massive black holes, *Phys. Usp.* **34**, 447 (1991).

[202] V. Dokuchaev, Yu. Eroshenko and S. Rubin, Quasars formation around clusters of primordial black holes, *Grav. Cosmol.* **11**, 99 (2005).

[203] V. Dokuchaev, Yu. Eroshenko and S. Rubin, Origin of supermassive black holes, arXiv:0709.0070

[204] V.I. Dokuchaev and Yu.N. Eroshenko, Stochastic correlation model of glactic bulge velocity dispersions and central black holes masses, *Astron. Lett.* **27**, 759 (2001); arXiv:astro-ph/0202019;

[205] V.I. Dokuchaev, Yu.N. Eroshenko, and S.G. Rubin, Early formation of galaxies induced by clusters of black holes, *Astron. Rep.* **52**, 779–789 (2008); arXiv:0801.0885.

[206] V.I. Dokuchaev, Yu.N. Eroshenko, S.G. Rubin, and D.A. Samarchenko, Mechanism for the suppression of intermediate-mass black holes, *Astron. Lett.* **36**, 773–779 (2010); arXiv:1010.5325.

[207] A.D. Dolgov and A.D. Linde, Scalar field fluctuations in the expanding universe and the new inflationary universe scenario, *Phys. Lett. B* **116**, 329 (1982).

[208] A.D. Dolgov, Non-GUT baryogenesis, *Phys. Rep.* **222**, 309 (1992).

[209] A. Dolgov, K. Freese, R. Rangarajan and M. Srednicki, Baryogenesis during reheating in natural inflation and comments on spontaneous baryogenesis, *Phys. Rev. D* **56**, 6155 (1997).

[210] A. Dolgov and L. Silk, Baryon isocurvature fluctuations at small scales and baryonic dark matter, *Phys. Rev. D* **47**, 4244 (1993).

[211] A.D. Dolgov and S.H. Hansen, Equation of motion of a classical field with back reaction of produced particles, *Nucl. Phys. B* **54**, 408 (1999).

[212] A.D. Dolgov and F.R. Urban, Dynamical vacuum energy via adjustment mechanism, *Phys. Rev. D* **77**, 083503 (2008); arXiv:0801.3090.

[213] A.D. Dolgov, Primordial black holes around us now, long before, and far away, arXiv:2010.11824 [astro-ph.CO].

[214] G. Domènech and M. Sasaki, Conformal frame dependence of inflation, *J. Cosmol. Astro. Phys.* **04** (2015), 022 arXiv:1501.07699 [gr-qc].

[215] J.F. Donoghue, General relativity as an effective field theory: The leading quantum corrections, *Phys. Rev. D* **50**, 3874 (1994).

[216] A.G. Doroshkevich, N.S. Kardashev, D.I. Novikov, and I.D. Novikov, Passage of radiation through a wormhole, *Astron. Rep.* **52**, 616 (2008).

[217] A. Doroshkevich, J. Hansen, I. Novikov, and A. Shatskiy, Passage of radiation through wormholes, *Int. J. Mod. Phys. D* **18**, 1665 (2009).

[218] B.A. Dubrovin, S.P. Novikov and A.T. Fomenko, *Modern Geometry. Methods and Applications* (URSS, Moscow, 1988).

[219] M.J. Duff, L.B. Okun, and G. Veneziano, Trialogue on the number of fundamental constants, *J. High Energy Phys.* **03**, 023, 023 (2002), physics/0110060.

[220] J.P. Duruisseu and R. Kerner, *Gen. Rel. Grav.* **15**, 797 (1983).

[221] B.A. Dubrovin, A.T. Fomenko, and S.P. Novikov, *Modern Geometry: Methods and Applications* (Nauka, Moscow, 1979; Springer, New York, 1985).

[222] F. Duplessis, Y. Wang and R. Brandenberger, Multi-stream inflation in a landscape, *J. Cosmol. Astro. Phys.* 012 (2012).
[223] G.R. Dvali, G. Gabadadze and M. Porrati, Metastable gravitons and infinite volume extra dimensions, *Phys. Lett. B* **484**, 112 (2000).
[224] I. Dymnikova, Vacuum nonsingular black hole, *Gen. Rel. Grav.* **24**, 235 (1992).
[225] I.G. Dymnikova, A. Dobozh, M.L. Filchenkov and A. Gromov, Universes inside a Λ black hole, *Phys. Lett. B* **506**, 351 (2001).
[226] I. Dymnikova and M. Fil'chenkov, Quantum birth of a hot Universe II. Model parameter estimates from CMB temperature fluctuations, arXiv:gr-qc/0209065.
[227] D.A. Easson and R.H. Brandenberger, Universe generation from black hole interiors, *J. High Energy Phys.* **0106**, 024 (2001).
[228] E.F. Eiroa, E.R. de Celis and C. Simeone, Thin shells joining local cosmic string geometries, *Eur. Phys. J. C* **76**, 546 (2016).
[229] H. Ellis, Ether flow through a drainhole: a particle model in general relativity, *J. Math. Phys.* **14**, 104 (1973).
[230] G.F.R. Ellis, R. Maartens and M.A.H. MacCallum, *Relativistic Cosmology* (Cambridge University Press, Cambridge, UK, 2012)
[231] J. Ellis, N. Kaloper, K.A. Olive and J. Yokoyama, Topological R 4 inflation, *Phys. Rev. D* **59**, 103503 (1999).
[232] H. van Elst, J.E. Lidsey and R. Tavakol, Quantum cosmology and higher-order Lagrangian theories, *Class. Quantum Grav.* **11**, 2483 (1994).
[233] The Event Horizon Telescope Collaboration (April 10, 2019), First M87 Event Horizon Telescope results. I. The shadow of the supermassive black hole, *Astroph. J. Lett.* **875** (1), L1 (2019); arXiv:1906.11238.
[234] H. Everett, "Relative State" formulation of quantum mechanics, *Rev. Mod. Phys.* **29**, 454 (1957).
[235] C.W.F. Everitt *et al.*, Gravity Probe B: Final results of a space experiment to test general relativity, *Phys. Rev. Lett.* **106**, 221101 (2011); arXiv:1105.3456.
[236] V. Faraoni, Phantom cosmology with general potentials, *Class. Quantum Grav.* **22**, 3235 (2005).
[237] V. Faraoni, E. Gunzig and P. Nardone, Conformal transformations in classical gravitational theories and in cosmology, *Fund. Cosmic Phys.* 20, 121 (1999); arXiv:gr-qc/9811047.
[238] A. De Felice and S. Tsujikawa, f(R) theories, *Living Rev. Relativity,* **13**, 3 (2010); arXiv:1002.4928.
[239] Juan Fernandez-Gracia and Bartomeu Fiol, No-hair theorems for black holes in the Abelian Higgs model, *J. High Energy Phys.* **0911**, 054 (2009); arXiv:0906.2353.

[240] J.J. Ferrando and J.A. Sáez, Birkhoff theorem and conformal Killing-Yano tensors, *Gen. Rel. Grav.* **47**, 66 (2015).
[241] R.P. Feynman and A.R. Hibbs, *Quantum Mechanics and Path Integrals* (MacGraw-Hill, NY, 1965).
[242] L. Flamm, Beiträge zur Einsteinschen Gravitationstheorie, *Phys. Z.* **17**, 448 (1916).
[243] I.Z. Fisher, Scalar mesostatic field with regard for gravitational effects, *Zh. Eksp. Teor. Fiz.* **18**, 636 (1948); arXiv:gr-qc/9911008.
[244] S.D. Forghani, S.H. Mazharimousavi and M. Halilsoy, Cylindrical asymmetric thin-shell wormholes, *J. Cosmol. Astro. Phys.* **2019**, 067 (2919); arXiv:1807.05080.
[245] H. Firouzjahi, S. Sarangi and S.-H. Henry Tye, Spontaneous creation of inflationary universes and the cosmic landscape, arXiv:hep-th/0406107.
[246] J.L. Friedman, K. Schleich and D.M. Witt, Topological censorship, *Phys. Rev. Lett.* **71**, 1486-1489 (1993); Erratum: *ibid.* **75**, 1872 (1995).
[247] J. L. Friedman and A. Higuchi, Topological censorship and chronology protection, *Ann. Phys.* **15**, 109–128 (2006).
[248] B.N. Frolov, *Poincaré Gauge Theory of Gravity* (MGPU, Prometei, M., 2003).
[249] V.P. Frolov and I.D. Novikov, Physical effects in wormholes and time machines, *Phys. Rev. D* **42**, 1057 (1990).
[250] V.P. Frolov and I.D. Novikov, *Black Hole Physics: Basic Concepts and New Developments* (Kluwer Acad. Publ., 1997).
[251] M.J. Gander and G. Wanner, From Euler, Ritz, and Galerkin to modern computing, *SIAM Rev.* **54**, 627–666 (2012).
[252] V.A. Gani, A.E. Dmitriev and S.G. Rubin, Deformed compact extra space as dark matter candidate, *Int. J. Mod. Phys. D* **24**, 1545001 (2015).
[253] R. Gannouji, D. Polarski, A. Ranquet and A.A. Starobinsky, Scalar-tensor models of normal and phantom dark energy, *J. Cosmol. Astro. Phys.* **0609**, 016 (2006).
[254] J. Garriga and A. Vilenkin, Solution to the cosmological constant problems, *Phys. Rev. D* **64**, 023517 (2001); arXiv:hep-th/0011262.
[255] C. Germani and R. Maartens, Stars in the braneworld, *Phys. Rev. D* **64**, 124010 (2001).
[256] G.W. Gibbons and S.W. Hawking, Cosmological event horizons, thermodynamics, and particle creation, *Phys. Rev. D* **15**, 2738 (1977).

[257] K. Glampedakis, A.D. Johnson and D. Kennefick, Darboux transformation in black hole perturbation theory, *Phys. Rev. D* **96**, 024036 (2017); arXiv:1702.06459

[258] E.B. Gliner, Vacuum-like state of a medium and Friedmann cosmology, *Dokl. AN SSSR* **192**, 771 (1970).

[259] K. Gödel, An example of a new type of cosmological solution of Einstein's field equation of gravitation, *Rev. Mod. Phys.* **21**, 447 (1949).

[260] J.A. González, F.S. Guzmán and O. Sarbach, Instability of wormholes supported by a ghost scalar field. I. Linear stability analysis, *Class. Quantum Grav.* **26**, 015010 (2009); arXiv:0806.0608.

[261] M. Greene, J. Schwarz and E. Witten, *Superstring Theory.* (Cambridge UP, Cambridge, 1987).

[262] R. Gregory and R. Laflamme, Black strings and p-branes are unstable, *Phys. Rev. Lett.* **70**, 2837 (1993); arXiv:hep-th/9301052; *Nucl. Phys. B* **428**, 399 (1994); arXiv: hep-th/9494071.

[263] A.A. Grib, S.G. Mamayev and V.M. Mostepanenko, *Vacuum Quantum Effects in Strong Fields* (Friedmann Laboratory Publ., SPb, 1994).

[264] A.A. Grib and Yu.V. Pavlov, On the possible role of superheavy particles in the early Universe, in: *I.Ya. Pomeranchuk and Physics at the Turn of the Century, Proceedings of the International Conference*, eds. A.Berkov, N.Narozhny, and L.Okun (World Scientific, Singapore, 2003), pp. 406–412; arXiv:gr-qc/0009071.

[265] A.A. Grib and Yu.V. Pavlov, Quantum field theory in curved spacetime and the early Universe, *Grav. Cosmol. Suppl.* **8**, 148–153 (2002); arXiv:gr-qc/0206040.

[266] A.A. Grib and Yu.V. Pavlov, Cold dark matter and primordial superheavy particles, *Int. J. Mod. Phys. A* **17**, 4435 (2002); arXiv:gr-qc/0211015.

[267] A.A. Grib and Yu.V. Pavlov, Active galactic nuclei and transformation of dark matter into visible matter, *Grav. Cosmol.* **15**, 44 (2009); arXiv:0810.1724.

[268] A.A. Grib and Yu.V. Pavlov, On particles collisions in the vicinity of rotating black holes, *Pis'ma v ZhETF* **92**, 147 (2010) [*J. Exp. Theoret. Phys. Lett.* **92**, 125 (2010)]; *Astropart. Phys.* **34**, 581 (2011).

[269] A.A. Grib, Yu.V. Pavlov, and O.F. Piattella, High energy processes in the vicinity of Kerr's black hole horizon, arXiv:1105.1540.

[270] Ø. Grøn and S. Hervik, *Einstein's General Theory of Relativity* (Springer, NY, 2007).

[271] C. Gundlach, R.H. Price, and J. Pullin, Late time behavior of stellar collapse and explosions: 1. Linearized perturbations, *Phys. Rev. D* **49**, 883 (1994), arXiv:gr-qc/9307009.

[272] C. Gundlach, Critical phenomena in gravitational collapse, *Living Rev. Rel.* **2**, 4, 1999; arXiv:gr-qc/0001046.

[273] N. Günther, D. Nicole, and D. Wallace, Goldstone modes in vacuum decay and first-order phase transitions, *J. Phys. A* **13**, 1755 (1980).

[274] U. Günther and A. Zhuk, Remarks on dimensional reduction in multidimensional cosmological models, arXiv:gr-qc/0401003.

[275] U. Günther, P. Moniz and A. Zhuk, Multidimensional cosmology and asymptotical AdS, *Astrophys. Space Sci.* **283**, 679 (2003).

[276] A.H. Guth, The inflationary universe: A possible solution to the horizon and flatness problems, *Phys. Rev. D* **23**, 347 (1981).

[277] A.K. Guts, *Elements of a Theory of Time* (Nasledie, Omsk, 2004) (in Russian).

[278] J.C. Hafele and R.E. Keating, Around-the-world atomic clocks: Predicted relativistic time gains, *Science* **177**, 166 and 168 (1972).

[279] M. Hamermesh, *Group Theory and its Application to Physical Problems* (Addison–Wesley, Reading, Massachusetts, Palo Alto, London, 1964).

[280] T. Harko, T.S. Koivisto, F.S.N. Lobo and G.J. Olmo, Metric-Palatini gravity unifying local constraints and late-time cosmic acceleration, *Phys. Rev. D* **85**, 084016 (2012); arXiv:1110.1049.

[281] Tiberiu Harko, Zoltan Kovacs, Francisco S. N. Lobo, Thin accretion disks in stationary axisymmetric wormhole spacetimes, *Phys. Rev. D* **79**, 064001 (2009).

[282] T. Harko and F.S.N. Lobo, *Extensions of $f(R)$ Gravity:Curvature-Matter Couplings and Hybrid Metric-Palatini Theory* (Cambridge University Press, Cambridge, UK, 2018).

[283] J.B. Hartle and S.W. Hawking, Wave function of the Universe, *Phys. Rev. D* **28**, 2960 (1983).

[284] S.W. Hawking, Black hole explosions? *Nature* **248**, 30 (1974); *Commun. Math. Phys.* **43**, 199 (1975).

[285] S.W. Hawking and G.F.R. Ellis, *The Large Scale Structure of Space-Time* (Cambridge UP, Cambridge, 1973).

[286] S.W. Hawking and W. Israel, ed., *General Relativity, Einstein Centenary Survey* (Cambridge University Press, 2003).

[287] Sean A. Hayward, Formation and evaporation of non-singular black holes, *Phys. Rev. Lett.* **96**, 031103 (2006); arXiv:gr-qc/0506126.

[288] M. Heusler, *Black Hole Uniqueness Theorems* (Cambridge University Press, 1997).

[289] A. Hewish *et al.*, Observation of a rapidly pulsating radio source, *Nature* **217**, No. 5139, 709 (1968).

[290] D. Hochberg, A.A. Popov and S.V. Sushkov, Self-consistent wormhole solutions of semiclassical gravity, *Phys. Rev. Lett.* **78**, 2050 (1997).

[291] D. Hochberg and M. Visser, Geometric structure of the generic static traversable wormhole throat, *Phys. Rev. D* **56**, 4745 (1997).
[292] D. Hochberg and M. Visser, Dynamic wormholes, antitrapped surfaces, and energy conditions, *Phys. Rev. D* **58**, 044021 (1998).
[293] C.J. Hogan, in: *Universe or Multiverse*? ed., B.J. Carr (Cambridge University Press, 2005).
[294] R. Holman, E.W. Kolb, S.I. Vadas and Y. Wang, Extended inflation from higher-dimensional theories, *Phys. Rev. D* **43**, 995 (1991).
[295] P. Hořava and E. Witten, Heterotic and Type I string dynamics from eleven dimensions, *Nucl. Phys. B* **460**, 506 (1996); ibid. **475**, 94 (1996).
[296] G.W. Horndeski, Second-order scalar-tensor field equations in a four-dimensional space, *Int. J. Theor. Phys.* **10** (6), 363–384 (1974).
[297] Gary T. Horowitz, Creating naked singularities and negative energy, *Phys. Scripta* **T117**, 86 (2005); arXiv:hep-th/0312123.
[298] S.D.H. Hsu, *Phys. Lett. B* **644**, 67–71 (2007); Spacetime topology change and black hole information, arXiv:hep-th/0608175.
[299] X.M. Hu and Z.C. Wu, Quantum Kaluza-Klein cosmology, *Phys. Lett. B* **149**, 87 (1984).
[300] K. Inayoshi and Z. Haiman, Is there a maximum mass for black holes in galactic nuclei? *Astroph. J.* **828** (2), 110; arXiv:1601.02611.
[301] A. Ishibashi and H. Kodama, Stability of higher-dimensional Schwarzschild black holes, *Prog. Theor. Phys.* **110**, 901–919 (2003); arXiv:hep-th/0305185.
[302] A. Ishibashi and H. Kodama, Perturbations and stability of static black holes in higher dimensions, arXiv:1103.6148, to appear in *Prog. Theor. Phys. Suppl.*
[303] W. Israel, Singular hypersurfaces and thin shells in general relativity, *Nuovo Cim. B* **44**, 1–14 (1966); *ibid.* **48**, 463(E) (1967).
[304] C. Itzykson and J.B. Zuber, *Quantum Field Theory* (McGraw-Hill, NY, 1984).
[305] D.D. Ivanenko, P.I. Pronin and G.A. Sardanashvili, *Gauge Theory of Gravity* (MGU Press, M., 1985, in Russian).
[306] S. Iyer and C. M. Will, Black-hole normal modes: A WKB approach. I. Foundations and application of a higher-order WKB analysis of potential-barrier scattering, *Phys. Rev. D* **35**, 3621 (1987).
[307] Ted Jacobson and Thomas P. Sotiriou, Destroying black holes with test bodies, *J. Phys.: Conf. Ser.* **222**, 012041 (2010); arXiv:1006.1764.
[308] J.T. Jebsen, Über die allgemeinen kugelsymmetrische Lösungen der Einsteinschen Gravitationsgleichungen in Vakuum, *Ark. Mat. Ast. Fys.* (Stockholm) 15, nr.18 (1921).

[309] Nils Voje Johansen and Finn Ravndal, On the discovery of Birkhoff's theorem, *Gen. Rel. Grav.* **38**, 537 (2006); arXiv:physics/0508163.
[310] P. Jordan, *Schwertkraft und Weltall* (Vieweg, Braunschweig, 1955).
[311] S. Kachru, R. Kallosh, A. Linde and S. P. Trivedi, De Sitter vacua in string theory, *Phys. Rev. D* **68**, 046005 (2003).
[312] N. Kaloper *et al.*, Compact hyperbolic extra dimensions: Branes, Kaluza-Klein modes, and cosmology, *Phys. Rev. Lett.* **85**, 928 (2000).
[313] T. Kaluza, Zum Unitätsproblem der Physik, *Sitzungsber. d. Berl. Akad.*, S. 966 (1921).
[314] E. Kamke, *Differentialgleichungen: Lösungsmethoden und Lösungen.* I. Gewönliche Differentialgleichungen (6. verbesserte Auflage, Leipzig, 1959).
[315] A.B. Katok and B. Hasselblatt, *Introduction to the Modern Theory of Dynamical Systems* (Cambridge University Press, Cambridge, 1996; Faktorial, Moscow, 1999).
[316] G.L. Kane, M.J. Perry, and A. N. Zytkow, The beginning of the end of the anthropic principle, *New Astron.* **7**, 45 (2002).
[317] P. Kanti, Black poles in theories with large extra dimensions: A review, *Int. J. Mod. Phys. A* **19**, 4899 (2004).
[318] P. Kanti, B. Kleihaus and J. Kunz, Wormholes in dilatonic Einstein–Gauss–Bonnet theory, *Phys. Rev. Lett.* **107**, 271101 (2011); arXiv:1108.3003.
[319] N.S. Kardashev, I.D. Novikov and A.A. Shatskiy, Astrophysics of wormholes, *Int. J. Mod. Phys. D* **16**, 909 (2007); arXiv:astro-ph/0610441.
[320] R. Kerner, Cosmology without singularity and nonlinear gravitational Lagrangians, *Gen. Rel. Grav.* **14**, 453 (1982).
[321] R.P. Kerr, Gravitational field of a spinning mass as an example of algebraically special metrics, *Phys. Rev. Lett.* **11**, 237 (1963).
[322] S.V. Ketov, *Introduction to Quantum Theory of Strings and Superstrings* (Nauka, Novosibirsk, 1990, in Russian).
[323] S. Khalil and C. Muños, Lights, action: Optical tweezers, *Contemp. Phys.* **43**, 51 (2002).
[324] M.Yu. Khlopov, *Fundamentals of Cosmoparticle Physics* (URSS, M., 2004).
[325] M.Yu. Khlopov, R.V. Konoplich, R. Mignani, S.G. Rubin, and A.S. Sakharov, Physical origin, evolution and observational signature of diffused antiworld, *Astropart. Phys.* **12**, 367 (2000).
[326] M.Y. Khlopov, S.G. Rubin and A.S. Sakharov, Primordial structure of massive black hole clusters, *Astropart. Phys.* **23**, 265–277 (2005),
[327] M.Yu. Khlopov and S.G. Rubin, *Cosmological Pattern of Microphysics in the Inflationary Universe*, vol. 144 series: Fundamental

Theories of Physics (Kluwer Academic Publishers, Dordrecht-Boston-London, 2004).

[328] M. Khlopov, S. Rubin, and A. Sakharov, Possible origin of antimatter regions in the baryon dominated universe, *Phys. Rev. D* **62**, 083505 (2000).

[329] M.Y. Khlopov, S.G. Rubin and A.S. Sakharov, Strong primordial inhomogeneities and galaxy formation, *Grav. Cosmol.* **8**, 57–65 (2002); astro-ph/0202505.

[330] N. Khviengia, Z. Khviengia, H. Lü and C.N. Pope, Towards a field theory of F-theory, *Class. Quantum Grav.* **15**, 759 (1998).

[331] C. Kiefer, The semiclassical approximation to quantum gravity, in: *Canonical Gravity – From Classical to Quantum*, eds. J. Ehlers and H. Friedrich (Springer, Berlin, 1994).

[332] A. King, How big can a black hole grow, *Mon. Not. R. Astron. Soc.* **456**, L109–L112 (2016); arXiv:1511.08502.

[333] A.A. Kirillov, The nature of dark matter, *Phys. Lett. B* **632**, 453 (2006).

[334] A.A. Kirillov and E.P. Savelova, On the topological bias of discrete sources in the gas of wormholes, *Phys. Lett. B* **660**, 93 (2008).

[335] A.A. Kirillov and E.P. Savelova, Gas of wormholes in Euclidean quantum field theory, *Grav. Cosmol.* **14**, 256 (2008).

[336] A.A. Kirillov and E.P. Savelova, On the value of the cosmological constant in a gas of virtual wormholes, *Grav. Cosmol.* **19**, 92–100 (2013).

[337] O. Klein, Quantentheorie und fünfdimensionale Relativitätstheorie, *Z. Phys.* **37**, 895 (1926); **46**, 188 (1927).

[338] S. Kobayashi, K. Koyama and J. Soda, Thick brane worlds and their stability, *Phys. Rev. D* **65**, 064014 (2002).

[339] Tsutomu Kobayashi, Horndeski theory and beyond: A review, *Rep. Prog. Phys.* **82**, 086901 (2019); arXiv:1901.07183.

[340] H. Kodama, R.A. Konoplya, and A. Zhidenko, Gravitational stability of simply rotating Myers-Perry black holes: Tensorial perturbations, *Phys. Rev. D* **81**, 044007 (2010).

[341] L. Kofman, A. Linde, and A. Starobinsky, Reheating after Inflation, *Phys. Rev. Lett.* **73**, 3195 (1994).

[342] E.W. Kolb, First-order inflation, *Physica Scripta* **36**, 199 (1991).

[343] E. Komatsu *et al.*, Seven-year Wilkinson Microwave Anisotropy Probe (WMAP) observations: Cosmological interpretation, *Astrophys. J. Suppl.* **192**, 18 (2011).

[344] S.A. Kononogov, *Metrology and Fundamental Physical Constants* (Standartinform, M., 2008) (in Russian).

[345] S.A. Kononogov and V.N. Melnikov, Metrology and fundamental physical constants, *Izmer. Tekhnika* **6**, 1 (2005).

[346] N.P. Konopleva and V.N. Popov, *Gauge Fields* (Atomizdat, M., 1980) (in Russian).

[347] R.V. Konoplich, On probability of decay of metastable vacuum, *Sov. Nucl. Phys.* **32**, 1132 (1980).

[348] R.V. Konoplich, Calculation of quantum corrections to nontrivial classical solutions by means of the zeta function, *Teor. Mat. Fiz.* **73**, 1286 (1987).

[349] R.V. Konoplich, M.Yu. Khlopov, S.G. Rubin, and A.S. Sakharov, Formation of black holes in first order phase transitions, Preprint 1203, Rome1, Rome, 1998.

[350] R.V. Konoplich and S.G. Rubin, Decay probability for metastable vacuum in scalar theory (in Russian), *Yad. Fiz.* **37**, 1330 (1983); *Yad. Fiz.* **42**, 1282 (1985).

[351] R.V. Konoplich and S.G. Rubin, Decay of the metastable vacuum, *Yad. Fiz.* **44**, 558 (1986).

[352] R.A. Konoplya, Quasinormal behavior of the D-dimensional Schwarzschild black hole and the higher-order WKB approach, *Phys. Rev. D* **68**, 024018 (2003).

[353] R.A. Konoplya and C. Molina, The ringing wormholes, *Phys. Rev. D* **71**, 124009 (2005); arXiv:gr-qc/0504139.

[354] R.A. Konoplya and A. Zhidenko, Perturbations and quasi-normal modes of black holes in Einstein-Aether theory, *Phys. Lett. B* **644**, 186 (2007).

[355] R.A. Konoplya and A. Zhidenko, Passage of radiation through wormholes of arbitrary shape, *Phys. Rev. D* **81**, 124036 (2010); arXiv:1004.1284.

[356] R.A. Konoplya and A. Zhidenko, Quasinormal modes of black holes: From astrophysics to string theory, *Rev. Mod. Phys.* **83**, 793–836 (2011); arXiv:1102.4014.

[357] M.Yu. Konstantinov, Some problems of topology change description in the theory of space-time, *Int. J. Mod. Phys. D* **7**, 1 (1998).

[358] F. Kottler, Über die physikalischen Grundlagen der Einsteinschen Relativittstheorie, *Annalen der Physik* **4**, Bd.60, 401–461 (1918).

[359] S. Krasnikov, Traversable wormhole, *Phys. Rev. D* **62**, 084028 (2000).

[360] A.B. Krebs and S.G. Rubin, Instanton approach to the lattice polaron problem, *Phys. Rev. B* **49**, 11808 (1994).

[361] V.G. Krechet, Topological and physical effects of rotation and spin in relativistic gravity, *Izvestiya Vuzov. Fiz.* **10**, 57–62 (2007).

[362] V.G. Krechet and D.V. Sadovnikov, Spin–spin interaction in general relativity and induced geometries with nontrivial topology, *Grav. Cosmol.* **15**, 337–340 (2009).

[363] H. Kroger, G. Melkonian and S.G. Rubin, Singular points in scalar-tensor theory, *Gen. Rel. Grav.* **36**, 1649 (2004); arXiv:astro-ph/0310182.

[364] M.D. Kruskal, Maximal extension of Schwarzschild metric, *Phys. Rev.* **119**, 1743 (1960).

[365] L.D. Landau and E.M. Lifshitz, *Quantum Mechanics* (Izd. Fiz. Mat. Lit., M., 1963).

[366] L.D. Landau and E.M. Lifshitz, *Field Theory* (Nauka, M., 1973).

[367] G. Lazarides, Introduction to inflationary cosmology, arXiv:hep-ph/0204294.

[368] José P.S. Lemos, F.S.N. Lobo and S.Q. de Oliveira, Morris-Thorne wormholes with a cosmological constant, *Phys. Rev. D* **68**, 064004 (2003).

[369] José P.S. Lemos and Oleg B. Zaslavskii, Quasi-black holes: definition and general properties, *Phys. Rev. D* **76**, 084030 (2007); arXiv:0707.1094.

[370] José P.S. Lemos and Oleg B. Zaslavskii, The mass formula for quasi-black holes, *Phys. Rev. D* **78**, 124013 (2008); arXiv:0811.2778.

[371] José P. S. Lemos and Vilson T. Zanchin, Regular black holes: Electrically charged solutions, Reissner–Nordström outside a de Sitter core, *Phys. Rev. D* **83**, 124005 (2011); arXiv:1104.4790.

[372] W. Lerche, D. Lust and A. N. Schellekens, Chiral four-dimensional heterotic strings from selfdual lattices, *Nucl. Phys. B* **287**, 477 (1987).

[373] P. S. Letelier and E. Verdaguer. Anisotropic fluid with SU(2) type structure in general relativity: A model of localized matter, *J. Math. Phys.* **28**, 2431 (1987).

[374] T. Levi-Civita, ds2 einsteiniani in campi newtoniani. IX: L'analogo del potenziale logaritmico, *Rend. Accad. Lincei* **28**, 101–109 (1919).

[375] A. Lichnerovicz, *Theorie globale des connexions et des groupes d'holonomie* (Edizioni Cremonese, Roma, 1955).

[376] A.D. Linde, Quantum creation of the inflationary universe, *Lett. Nuovo Cim.* **39**, 401 (1984).

[377] A.D. Linde, *The Large-Scale Structure of the Universe* (Harwood Academic Publishers, London, 1990).

[378] A.D. Linde, *Particle Physics and Inflationary Cosmology* (Harvard Acad. Press, Geneva, 1990).

[379] A. Linde, Axions in inflationary cosmology, *Phys. Lett. B* **259**, 38 (1991).

[380] A.D. Linde, Inflation and quantum cosmology, *Physica Scripta* **36**, 35 (1991).

[381] A. Linde, From the big bang theory to the theory of a stationary universe, *Phys. Rev. D* **49**, 1783 (1994).

[382] B. Liu, L. McLerran and N. Turok, Bubble nucleation and growth at a baryon-number-producing electroweak phase transition, *Phys. Rev. D* **46**, 2668 (1992).

[383] M. Livio and M.J. Rees, Anthropic reasoning, *Science* **309**, 1022 (2005).

[384] F.S.N. Lobo, Phantom energy traversable wormholes, *Phys. Rev. D* **71**, 084011 (2005); arXiv:gr-qc/0502099.

[385] F.S.N. Lobo, Exotic solutions in general relativity: Traversable wormholes and "warp drive" spacetimes, in: *Classical and Quantum Gravity Research* (Nova Science Publishers, NY, 2008), p. 1–78.

[386] F.S.N. Lobo, ed., *Wormholes, Warp Drives and Energy Conditions* (Springer, Berlin, 2017).

[387] F.S.N. Lobo, M.E. Rodrigues, M.V. de S. Silva, A. Simpson, M. Visser Novel black-bounce geometries, arXiv:2009.12057.

[388] J.C. Long *et al.*, Upper limits to submillimetre-range forces from extra space-time dimensions, *Nature* **421**, 922 (2003).

[389] V. Lukash, Production of phonons in an isotropic universe, *Sov. Phys. J. Exp. Theoret. Phys.* 79 1601 (1980).

[390] Y. Lyakhova, A.A. Popov and S.G. Rubin, Classical evolution of subspaces,' *Eur. Phys. J. C* 78, no.9, 764 (2018); arXiv:1807.06235.

[391] D.H. Lyth and E.D. Stewart, More varieties of hybrid inflation, *Phys. Rev. D* **54**, 7186 (1996).

[392] D.H. Lyth and A. Riotto, Particle physics models of inflation and the cosmological density perturbation, *Phys. Rep.* 314, 1 (1999); arXiv: hep-ph/9807278.

[393] R. Maartens, Brane-world gravity, *Living Rev. Rel.* **7**, 7 (2004); arXiv:gr-qc/0312059.

[394] K. Maeda, Towards the Einstein-Hilbert action via conformal transformation, *Phys. Rev. D* **39**, 3159 (1989).

[395] K. Maeda, J.A. Stein-Schabes and T. Futamase, Inflation in a renormalizable cosmological model and the cosmic no-hair conjecture, *Phys. Rev. D* **39**, 2848 (1989).

[396] L. Magaard, Zur einbettung riemannscher Raume in Einstein-Raume und konformeuclidische Raume, PhD thesis, Kiel, 1963.

[397] G. Magnano and L.M. Sokolowski, Physical equivalence between nonlinear gravity theories and a general-relativistic self-gravitating scalar field, *Phys. Rev. D* **50**, 5039 (1994).

[398] M. Malquarti, E.J. Copeland, A.R. Liddle and M. Trodden, A new view of k-essence, *Phys. Rev. D* **67**, 123503 (2003).

[399] A.S. Mazumdar and S.K. Sethi, Extended inflation from Kaluza–Klein theories, *Phys. Rev. D* **46**, 5315–5320 (1992).

[400] Pawel O. Mazur, Black hole uniqueness theorems, arXiv:hep-th/0101012.

[401] Manasse R. Mbonye and Demosthenes Kazanas, A non-singular black hole model as a possible end-product of gravitational collapse, *Phys. Rev. D* **72**, 024016 (2005); arXiv:gr-qc/0506111.

[402] T. Matos and G. Miranda, Exact rotating magnetic traversable wormholes satisfying the energy conditions, arXiv:1507.02348.

[403] P.O. Mazur and E. Mottola, Gravitational condensate stars: an alternative to black holes, *Proc. Nat. Acad. Sci.* **101**, 9545 (2004); arXiv:gr-qc/0109035.

[404] Paul L. McFadden and Neil G. Turok, Effective theory approach to brane world black holes, *Phys. Rev. D* **71**, 086004 (2005).

[405] A. Melfo, N. Pantoja and A. Skirzewski, Thick domain wall spacetimes with and without reflection symmetry, *Phys. Rev. D* **67**, 105003 (2003).

[406] X.H. Meng and P. Wang, Modified Friedmann equations in R^{-1}-modified gravity, *Class. Quantum Grav.* **20**, 4949 (2003).

[407] Eckehard W. Mielke and Franz E. Schunck, Boson stars: Alternatives to primordial black holes? *Nucl. Phys. B* **564**, 185–203 (2000); arXiv:gr-qc/0001061.

[408] M.B. Mijić, M.S. Morris, and W.M. Suen, Initial conditions for $R+\epsilon R^2$ cosmology, *Phys. Rev. D* **39**, 1496 (1989).

[409] C.W. Misner, K.S. Thorne and J.A. Wheeler, *Gravitation* (Freeman, San Francisco, 1973).

[410] N.V. Mitskievich, *Physical Fields in General Relativity* (Nauka, M., 1969).

[411] N.V. Mitskievich, A.P. Yefremov and A.I. Nesterov, *Field Dynamics in General Relativity* (Energoatomizdat, M., 1985).

[412] Leonardo Modesto, Space-time structure of loop quantum black hole, arXiv:0811.2196.

[413] Leonardo Modesto, John W. Moffat and Piero Nicolini, Black holes in an ultraviolet complete quantum gravity, *Phys. Lett. B* **695**, 397 (2011); arXiv:1010.0680.

[414] Leonardo Modesto and Piero Nicolini, Charged rotating noncommutative black holes, *Phys. Rev. D* **82**, 104035 (2010); arXiv:1005.5605.

[415] Flavio Bombacigno, Fabio Moretti, and Giovanni Montani, Scalar modes in extended hybrid metric-Palatini gravity: weak field phenomenology, arXiv:1907.11949.

[416] M.S. Morris and K.S. Thorne, Wormholes in spacetime and their use for interstellar travel: A tool for teaching general relativity, *Am. J. Phys.* **56**, 395 (1988).

[417] M. Morris, K.S. Thorne and U. Yurtsever, Wormholes, time machines, and the Weak Energy Condition, *Phys. Rev. Lett.* **61**, 1446 (1988).

[418] Emil Mottola, New horizons in gravity: Dark energy and condensate stars, arXiv:1107.5086.

[419] H. Mouri and Y. Taniguchi, Runaway merging of black holes: Analytical constraint on the timescale, *Astroph. J.* **566**, L17 (2002); arXiv:astro-ph/0201102.

[420] D. Müller, H.V. Fagundes and R. Opher, Casimir energy in multiply connected static hyperbolic Universes, *Phys. Rev. D* **66**, 083507 (2002).

[421] S. Nasri, P.J. Silva, G.D. Starkman and M. Trodden, Radion stabilization in compact hyperbolic extra dimensions, *Phys. Rev. D* **66**, 045029 (2002).

[422] V. V. Nikulin and S. G. Rubin, Inflationary limits on the size of compact extra space, *Int. J. Mod. Phys. D* **28**, 1941004 (2019); arXiv:1903.05725 [gr-qc].

[423] H.P. Nilles, Supersymmetry, supergravity and particle physics, *Phys. Rep.* **110**, 1 (1984).

[424] S. Nojiri and S.D. Odintsov, Quantum de Sitter cosmology and phantom matter, *Phys. Lett. B* **562** 147 (2003).

[425] S. Nojiri and S.D. Odintsov, Modified gravity with negative and positive powers of curvature: Unification of inflation and cosmic acceleration, *Phys. Rev. D* **68**, 123512 (2003).

[426] S. Nojiri, S.D. Odintsov, Introduction to modified gravity and gravitational alternative for dark energy, *Int. J. Geom. Meth. Mod. Phys.* 4, 115 (2007); arXiv:hep-th/0601213.

[427] S. Nojiri, S.D. Odintsov and S. Ogushi, Friedmann-Robertson-Walker brane cosmological equations from the five-dimensional bulk (A)dS black hole, *Int. J. Mod. Phys. A* **17**, 4809 (2002).

[428] K. Nordtvedt, Post-Newtonian metric for a general class of scalar-tensor gravitational theories and observational consequences, *Astroph. J.* **161**, 1059 (1970).

[429] I. Novikov, A.G. Polnarev, A. Starobinsky and Ya.B. Zel'dovich, Primordial black holes, *Astron. Astrophys.* **80**, 104 (1979).

[430] I.D. Novikov, The time machine and self-consistent evolutions in problems with self-interaction. Preprint Nordita-90/38A. Nordisk Inst. f. Teoretisk Fysik, Kobenhavn (1990).

[431] I.D. Novikov, N.S. Kardashev, and A.A. Shatskii, The multicomponent Universe and the astrophysics of wormholes, *Physics Uspekhi* **50**, 965 (2007).

[432] U. Nucamendi and M. Salgado, Scalar hairy black holes and solitons in asymptotically flat spacetimes, *Phys. Rev. D* **68**, 044026 (2003).

[433] H. Ochiai and K. Sato, Dynamical properties of Euclidean solutions in a multidimensional cosmological model, *Prog. Theor. Phys.* **103**, 893 (2000).

[434] S.D. Odintsov and V.K. Oikonomou, The early-time cosmology with stiff era from modified gravity, *Phys. Rev. D* **96**, 104059 (2017); arXiv:1711.04571.

[435] I. Olasagasti and A. Vilenkin, Gravity of higher-dimensional global defects, *Phys. Rev. D* **62**, 044014 (2000).

[436] K.A. Olive, M. Peloso, and J-P. Uzan, The wall of fundamental constants. arXiv:1011.1504.

[437] J.R. Oppenheimer and G. Volkoff, On massive neutron cores, *Phys. Rev.* **55**, 374 (1939).

[438] J.P. Ostriker and P.J.E. Peebles, A numerical study of the stability of flattened galaxies: or, can cold galaxies survive? *Astroph. J.* **186**, 467 (1973).

[439] Attilio Palatini, (1919), Deduzione invariantiva delle equazioni gravitazionali dal principio di Hamilton, *Rendiconti del Circolo Matematico di Palermo* **1**, 43, 203–212 (1919).

[440] B. Paul, Removing the Ostrogradski ghost from degenerate gravity theories, *Phys. Rev. D* **96**, 044035 (2017).

[441] M. Pavšič, Einstein's gravity from a first order lagrangian in an embedding space, *Phys. Lett. A* **116**, 1 (1986).

[442] A. Peltola and G. Kunstatter, Complete, single-horizon quantum corrected black hole spacetime, *Phys. Rev. D* **79**, 061501 (2009); arXiv:0811.3240.

[443] R. Penrose, Conformal treatment of the infinity, in: *Relativity, Groups and Topology*, eds. C. DeWitt and B. DeWitt (Gordon and Breach, London, 1964), p. 565.

[444] R. Penney, Generalization of the Reissner-Nordström solution to the Einstein field equations, *Phys. Rev.* **182**, 1383 (1969).

[445] R. Penrose, Gravitational collapse: The role of general relativity, *Rev. del Nuovo Cim.* **1**, 252 (1969).

[446] A.M. Perelomov, Schrödinger equation spectrum and Korteweg-De Vries type invariants, *Ann. Inst. Henrí Poincaré A* **24**, 161 (1976).

[447] P.M. Petriakova, Sub-Planckian evolution of the Universe, arXiv:2010.09918.

[448] O.P.F. Piedra and J. de Oliveira, Fermion perturbations in string theory black holes, *Class. Quantum Grav.* 28, 085023 (2011).

[449] A. Polnarev and M. Khlopov, Cosmology, primordial black holes, and supermassive particles, *Phys. Usp.* **145**, 369 (1985).

[450] V.N. Ponomarev, A.O. Barvinsky and Yu.N. Obukhov, *Geometrodynamic Methods and Gauge Approach to the Theory of Gravitational Interactions* (Energoatomizdat, M., 1985).

[451] J. Polchinski, *String Theory* (Cambridge UP, Cambridge, 1998).
[452] R.V. Pound and G.A. Rebka, Jr., Apparent weight of photons, *Phys. Rev. Lett.* **4**, 337 (1960).
[453] R.V. Pound and J.L. Snider, Effect of gravity on gamma radiation, *Phys. Rev.* **140**, B788 (1960).
[454] Alexei Pozanenko and Alexander Shatskiy. On a possible connection between wormholes and soft gamma repeaters, *Grav. Cosmol.* **16**, 259 (2010); arXiv:1007.3620.
[455] F. Quevedo, Lectures on string/brane cosmology, *Class. Quantum Grav.* **19**, 5721 (2002).
[456] F. Rahaman, M. Kalam, and S. Chakraborty. Gravitational lensing by stable C-field wormhole, *Chin. J. Phys.* **45**, 518 (2007); arXiv:0705.0740.
[457] M. Rainer and A.I. Zhuk, Einstein and Brans–Dicke frames in multidimensional cosmology, *Gen. Rel. Grav.* **32**, 79 (2000).
[458] R. Rajaraman, *Solitons and Instantons. An Introduction to Solitons And Instantons in Quantum Field Theory* (North-Holland, Amsterdam, New-York, Oxford, 1982).
[459] L. Randall and R. Sundrum, A large mass hierarchy from a small extra dimension, *Phys. Rev. Lett.* **83**, 3370 (1999).
[460] L. Randall and R. Sundrum, An alternative to compactification, *Phys. Rev. Lett.* **83**, 3690 (1999).
[461] P.K. Rashevsky, *Riemannian Geometry and Tensor Analysis* (Nauka, M., 1967).
[462] M.J. Rees. *Our Cosmic Habitat* (Princeton Univ. Press, Princeton, 2002).
[463] M. Rees, R. Ruffini and J.A. Wheeler. *Black Holes, Gravitational Waves and Cosmology: An Introduction to Current Problems* (Gordon and Breach, NY, London, Paris, 1974).
[464] M. Regis, M. Serone and P. Ullio, A dark matter candidate from an extra (non-universal) dimension, *Phys.Lett. B* 663, 250 (2008); arXiv:hep-ph/0612286.
[465] S.-J. Rey, Dynamics of inflationary phase transition, *Nucl. Phys. B* **284**, 706 (1987).
[466] A.G. Riess, Observational evidence from supernovae for an accelerating universe and a cosmological constant, *Astron. J.* **116**, 1009 (1998).
[467] W. Rindler, *Essential Relativity* (Springer, NY, 1979).
[468] T.G. Rizzo, Collider production of TeV scale black holes and higher-curvature gravity, *J. High Energy Phys.* **0506**, 079 (2005); arXiv:hep-ph/0503163; H. Davoudiasl and T.G. Rizzo, Signatures of spherical compactification at the LHC, *Phys. Rev. D* **76**, 055009 (2007); arXiv:hep-ph/0702078.

[469] V.I. Rodichev, *Gravitation Theory in an Orthogonal Hedron* (Nauka, M., 1974).
[470] E. Roessl and M. Shaposhnikov, Localizing gravity on a 't Hooft-Polyakov monopole in seven dimensions, *Phys. Rev. D* **66**, 084008 (2002); arXiv:hep-th/0205320.
[471] Marek Rogatko, Uniqueness theorem for static black hole solutions of sigma-models in higher dimensions, *Class. Quantum Grav.* **19**, L151 (2002); arXiv:hep-th/0207187.
[472] Marek Rogatko, Uniqueness theorem for stationary axisymmetric black holes in Einstein–Maxwell–axion-dilaton gravity, *Phys. Rev. D* **82**, 044017 (2010); arXiv:1007.4374.
[473] João Luís Rosa, José P. S. Lemos, and Francisco S. N. Lobo, Wormholes in generalized hybrid metric-Palatini gravity obeying the matter null energy condition everywhere. *Phys. Rev. D* **98**, 064054 (2018); arXiv:1808.08975.
[474] V.A. Rubakov, Large and infinite extra dimensions, *Physics-Uspekhi* **171**, 9, 913 (2001).
[475] V.A. Rubakov and M.E. Shaposhnikov, Do we live inside a domain wall? *Phys. Lett. B* **125**, 136 (1983).
[476] S. Rubin, Effect of massive fields on inflation, *J. Exp. Theoret. Phys. Lett.* **74**, 275 (2001).
[477] S.G. Rubin, Fine tuning of parameters of the universe, *Chaos Solitons Fractals* **14**, 891, (2002); arXiv:astro-ph/0207013.
[478] S.G. Rubin, Massive primordial black holes in hybrid inflation, *I. Ya. Pomeranchuk and Physics at the Turn of the Century* (Moscow, 2003), p. 413; arXiv:astro-ph/0511181.
[479] S.G. Rubin, Origin of universes with different properties, *Grav. Cosmol.* **9**, 243 (2003).
[480] S. Rubin, M. Khlopov and A. Sakharov, Primordial black holes from non-equilibrium second order phase transition, *Grav. Cosmol.* **6**, (Suppl), 51–58 (2000).
[481] S. Rubin, A. Sakharov, and M.Yu. Khlopov, The formation of primary galactic nuclei during phase transitions in the early universe, *Sov. Phys. J. Exp. Theoret. Phys.* **92**, 921 (2001).
[482] S.G. Rubin. Cascade birth of universes in multidimensional spaces, *Sov. Phys. J. Exp. Theoret. Phys.* **106**, 714 (2008).
[483] S.G. Rubin. On the origin of gauge symmetries and fundamental constants, *Sov. Phys. J. Exp. Theoret. Phys.* **109**, 961 (2009).
[484] S.G. Rubin, Scalar field localization on deformed extra space, *Eur. Phys. J. C* **75**, 333 (2015); arXiv:1503.05011.
[485] S.G. Rubin, Interpenetrating subspaces as a funnel to extra space, *Phys. Lett. B* **759**, 622–625 (2016).

[486] J. Saavedra and Y. Vasquez, Effective gravitational equations on brane world with induced gravity described by f(R) term, *J. Cosmol. Astro. Phys.* **0904**, 013 (2009); arXiv:0803.1823.
[487] M. Safonova, D.F. Torres and G.E. Romero. Microlensing by natural wormholes: theory and simulations, *Phys. Rev. D* **65**, 023001 (2002).
[488] V. Sahni and A. Starobinsky, The case for a positive cosmological Lambda-term, *Int. J. Mod. Phys. D* **9**, 373 (2000).
[489] V. Sahni and A. Starobinsky, Reconstructing dark energy, *Int. J. Mod. Phys. D* **15**, 2105 (2006); arXiv:astro-ph/0610026.
[490] Franz E. Schunck and Eckehard W. Mielke, Topical review: General relativistic boson stars, *Class. Quantum Grav.* **20**, R301–R356 (2003); arXiv:0801.0307.
[491] B.F. Schutz and C.M. Will, Black hole normal modes — a semianalytic approach, *Astrophys. J. Lett.* 291, L33 (1985).
[492] K. Schwarzschild, *Über das Gravitationsfeld eines Massenpunktes nach der Einsteinschen Theorie, Sitzungsberichte der Königlich Preussischen Akademie der Wissenschaften* **1**, 189–196 (1916).
[493] K. Schwarzschild, On the gravitational field of a sphere of incompressible fluid according to Einstein's theory, *Sitzber. Deut. Akad. Wiss. Berlin, Kl. Math.-Phys. Tech.*, 424 (1916); arXiv: physics/9912033.
[494] S.S. Seahra and P.S. Wesson, Application of the Campbell-Magaard theorem to higher-dimensional physics, *Class. Quantum Grav.* **20**, 1321 (2003); arXiv:gr-qc/0302015.
[495] A. Sen, Rolling tachyon, *J. High Energy Phys.* **0204**, 048 (2002); **0207**, 065 (2002).
[496] M.R. Setare and A. Sepehri, Stability of cylindrical thin shell wormhole during evolution of universe from inflation to late-time acceleration, *J. High Energy Phys.* **2015**, 79 (2015); arXiv:1412.8666.
[497] S. Shankaranarayanan and N. Dadhich, Non-singular black-holes on the brane, *Int. J. Mod. Phys. D* **13**, 1095 (2004); arXiv:gr-qc/0306111.
[498] I.I. Shapiro *et al.*, Viking relativity experiment — Verification of signal retardation by solar gravity, *Astroph. J.* **234**, L219 (1979).
[499] H.-a. Shinkai and S.A. Hayward, Fate of the first traversible wormhole: Black-hole collapse or inflationary expansion, *Phys. Rev. D* **66**, 044005 (2002); arXiv:gr-qc/0205041.
[500] T. Shiromizu, K. Maeda and M. Sasaki, The Einstein equations on the 3-brane world, *Phys. Rev. D* **62**, 024012 (2000).
[501] Y. Shtanov, J. Traschen and R. Brandenberger, Universe reheating after inflation, *Phys. Rev. D* **51**, 5438 (1995); arXiv:hep-ph/9407247.

[502] P. Sikivie, Axions, domain walls, and the early Universe, *Phys. Rev. Lett.* **48**, 1156 (1982).

[503] Alex Simpson and Matt Visser, Black-bounce to traversable wormhole, *J. Cosmol. Astro. Phys.* **1902**, 042 (2019); arXiv:1812.07114.

[504] W. de Sitter, On the curvature of space, *Proc. Kon. Ned. Akad Wet.* **20**, 229 (1917).

[505] Anais Smailagic and Euro Spallucci, "Kerr" black hole: the Lord of the String, arXiv:1003.3918.

[506] M.N. Smolyakov and I.P. Volobuev, Linearized gravity, Newtonian limit and light deflection in RS1 model; arXiv:hep-th/0208025.

[507] L.M. Sokolowski, Stability of a metric f(R) gravity theory implies the Newtonian limit, *Acta Phys. Polon.* B 39, 2879 (2008).

[508] T. P. Sotiriou and V. Faraoni, f(R) theories of gravity, *Rev. Mod. Phys.* **82**, 451 (2010); arXiv:0805.1726.

[509] Euro Spallucci and Stefano Ansoldi, Regular black holes in UV self-complete quantum gravity, *Phys. Lett. B* **701**, 471 (2011); arXiv:1101.2760.

[510] D.N. Spergel *et al.*, Wilkinson Microwave Anisotropy Probe (WMAP) three year results: Implications for cosmology, *Astrophys. J Supp.* **170**, 377 (2007); arXiv:astro-ph/0603449.

[511] K.P. Staniukovich and V.N. Melnikov, *Hydrodynamics, Fields and Constants in Gravitation Theory* (Energoatomizdat, M., 1983).

[512] G.D. Starkman, D. Stojkovic and M. Trodden, Large extra dimensions and cosmological problems, *Phys. Rev. D* 63, 103511 (2001); arXiv:hep-th/0012226.

[513] A. Starobinsky, Relic gravitation radiation spectrum and initial state of the universe, *J. Exp. Theoret. Phys. Lett.* **30**, 682 (1979).

[514] A.A. Starobinsky, A new type of isotropic cosmological models without singularity, *Phys. Lett.* **1980**, *B91*, 99–102,

[515] A.A. Starobinsky, Can the effective gravitational constant become negative? *Pis'ma v Astron. Zh.* **7**, 67 (1981); *Sov. Astron. Lett.* **7**, 361 (1981).

[516] A.A. Starobinsky, Dynamics of phase transition in the new inflationary universe scenario and generation of perturbations, *Phys. Lett. B* 117,175 (1982).

[517] A.A. Starobinsky, Disappearing cosmological constant in f(R) gravity, *J. Exp. Theoret. Phys. Lett.* 86, 157 (2007).

[518] A. Strominger, Inverse-dimensional expansion in quantum gravity, *Phys. Rev. D* **24**, 3082 (1981).

[519] W.-M. Suen and K. Young, Wave function of the Universe as a leaking system, *Phys. Rev. D* **39**, 2201 (1989).

[520] M. Sullivan *et al.*, SNLS3: Constraints on Dark Energy combining the Supernova Legacy Survey three year data with other probes, *Astrophys. J.* 737, 102 (2011); arXiv:1104.1444.
[521] S. Sushkov, Wormholes supported by a phantom energy, *Phys. Rev. D* **71**, 043520 (2005); arXiv:gr-qc/0502084.
[522] Sergey V. Sushkov and Oleg B. Zaslavskii, Horizon closeness bounds for static black hole mimickers, *Phys. Rev. D* **79**, 067502 (2009); arXiv:0903.1510.
[523] L. Susskind, The anthropic landscape of string theory, arXiv: hep-th/0302219.
[524] J.L. Synge, The gravitational field of a particle, *Proc. Roy. Irish Acad. A* **53**, 83 (1950).
[525] J.L. Synge, *Relativity: The General Theory* (North Holland, Amsterdam, 1960).
[526] P. Szekeres, On the singularities of a Riemannian manifold, *Publ. Mat. Debrecen* **7**, 285 (1960).
[527] F. Tamburini and I. Licata, General Relativistic Wormhole Connections from Planck-Scales and the ER=EPR Conjecture, *Entropy* **2020**, 22 (1), 3–18 (2020); arXiv:1912.12424.
[528] F.R. Tangherlini, Schwarzschild field in n dimensions and the dimensionality of space problem, *Nuovo Cim.* **27**, 636 (1963).
[529] M. Tegmark *et al.*, Dimensionless constants, cosmology and other dark matters, *Phys. Rev. D* **73**, 023505 (2006).
[530] J.M. Tejeiro and A. Larranaga, Gravitational lensing by asymptotically flat wormholes, arXiv:gr-qc/0505054.
[531] R.C. Tolman, Effect of inhomogeneity on cosmological models, *Proc. Nat. Acad. Sci. USA* **20**, 410 (1934).
[532] A.A. Tseytlin and C. Vafa, Elements of string cosmology, *Nucl. Phys. B* **372**, 443 (1992).
[533] Naoki Tsukamoto, Strong deflection limit analysis and gravitational lensing of an Ellis wormhole, *Phys. Rev. D* **94**, 124001 (2016); arXiv:1607.07022.
[534] J.-Ph. Uzan, The fundamental constants and their variation: observational and theoretical status, *Rev. Mod. Phys.* **75**, 403 (2003).
[535] J.-Ph. Uzan, Varying constants, gravitation and cosmology, *Living Rev. Rel.*; arXiv:1009.5514.
[536] A. Vilenkin, Creation of universes from nothing, *Phys. Lett. B* **117**, 25 (1982).
[537] A. Vilenkin, Quantum cosmology and the initial state of the Universe, *Phys. Rev. D* **37**, 888 (1988).
[538] A. Vilenkin and L. Ford, Gravitational effects upon cosmological phase transitions, *Phys. Rev. D* **26**, 1231 (1982).

[539] M. Visser, *Lorentzian Wormholes: From Einstein to Hawking* (AIP, Woodbury, 1995).
[540] M. Visser and D.L. Wiltshire, On-brane data for braneworld stars, *Phys. Rev. D* **67**, 104004 (2003).
[541] Salvatore Vitale The first five years of gravitational-wave astrophysics, arXiv:2011.03563.
[542] Yu.S. Vladimirov, *Reference Frames in Gravitation Theory* (Energoizdat, M., 1982).
[543] Yu.S. Vladimirov, *Geometrophysics* (BINOM, M., 2005).
[544] D. Vollick, Negative energies on the brane, *Gen. Rel. Grav.* **34**, 1 (2002).
[545] D. Vollick, Spherically symmetric solutions and dark matter on the brane, *Gen. Rel. Grav.* **34**, 471 (2002).
[546] G.E. Volovik, *Proceedings of From Quantum to Emergent Gravity: Theory and Phenomenology*, Trieste, Italy, 2007 (Trieste, 2007), PoS(QG-Ph).
[547] G.E. Volovik, $\hbar$ as parameter of Minkowski metric in effective theory, *JETP Letters* **90**, 697 (2010); arXiv:0904.1965 [gr-qc]
[548] R. Wagoner, Scalar-tensor theory and gravitational waves, *Phys. Rev. D* **1**, 3209 (1970).
[549] R. Wald, *General Relativity* (University of Chicago Press, Chicago, 1984).
[550] Robert M. Wald, Gravitational collapse and cosmic censorship, arXiv:gr-qc/9710068.
[551] A. Wang and P.S. Letelier, Dynamical wormholes and energy conditions, *Prog. Theor. Phys.* **94**, L137–142 (1995).
[552] J.K. Webb *et al.*, A search for time variation of the fine structure constant, *Phys. Rev. Lett.* **82**, 884 (1999).
[553] J.K. Webb *et al.*, Evidence for spatial variation of the fine structure constant, arXiv:1008.3907.
[554] S. Weinberg, *Gravitation and Cosmology: Principles and Applications of GR*, (Wiley, NY, 1972).
[555] S. Weinberg, Anthropic bound on the cosmological constant, *Phys. Rev. Lett.* **59**, 2607 (1987).
[556] S. Weinberg, The cosmological constant problem, arXiv:astro-ph/0005265.
[557] S. Weinstein, Anthropic reasoning in multiverse cosmology and string theory, *Class. Quantum Grav.* **23**, 4231 (2006).
[558] U. Weiss and W. Haeffner, Complex-time path integrals beyond the stationary-phase approximation: Decay of metastable states and quantum statistical metastability, *Phys. Rev. D* **27**, 2916 (1983).
[559] J.A. Wheeler, Geons, *Phys. Rev.* **97**, 511 (1955).

[560] J.A. Wheeler, in: *Relativity, Groups and Topology*, eds. B.S. and C.M. DeWitt (Gordon and Breach, NY, 1963).
[561] J.A. Wheeler, Our Universe: The known and the unknown, *Am. Sci.* **59**, 1 (1968).
[562] B. Whitt, Fourth-order gravity as general relativity plus matter, *Phys. Lett. B* **145**, 176 (1984).
[563] C. Will, *Theory and Experiment in Gravitational Physics* (Cambridge University Press, Cambridge, 1981).
[564] C. Will, Finally, results from Gravity Probe-B, *Physics* **4**, 43 (2011); arXiv:1106.1198.
[565] Clifford M. Will, The confrontation between general relativity and experiment, *Living Rev. Relativity* **17**, 4 (2014); arXiv:1403.7377.
[566] D.L. Wiltshire, in: *Cosmology: The Physics of the Universe*, eds B. Robson, N. Visvanathan and W.S. Woolcock (World Scientific, Singapore, 1996), pp. 473–531.
[567] E. Witten, Strong coupling expansion of Calabi-Yau compactification, *Nucl. Phys. B* **471**, 135 (1996).
[568] R.P. Woodard, The theorem of Ostrogradsky, arXiv:1506.02210.
[569] P. Worden, Equivalence principle tests in earth orbit, *Acta Astronautica* **5**, 27 (1978).
[570] D. Yamauchi and M. Sasaki, Brane world in arbitrary dimensions without Z2 symmetry, *Prog. Theor. Phys.* **118**, 245 (2007); arXiv:0705.2443.
[571] J. Yokoyama, Chaotic new inflation and formation of primordial black holes, *Phys. Rev. D* **58**, 083510 (1998).
[572] Artyom Yurov and Valerian Yurov, A look at the generalized Darboux transformations for the quasinormal spectra in Schwarzschild black hole perturbation theory: just how general should it be? *Phys. Lett. A*, 383, 2571–2578 (2019); arXiv:1809.10279.
[573] N.A. Zaitsev and S.M. Kolesnikov, Self-consistent interaction of scalar and tensor gravitational fileds, in: *Problems of Gravitation Theory and Particle Theory*, eds. by K. P. Staniukovich and G. A. Sokolik (issue 4, Atomizdat, Moscow, 1970, in Russian), pp. 24–50.
[574] V. Zakharov, V. Manakov, V. Novikov, and D. Pitaevsky, *Theory of Solitons: The Inverse Problem Method* (Nauka, M., 1980, in Russian).
[575] O.B. Zaslavskii, Acceleration of particles by black holes – general explanation, *Class. Quantum Grav.* **28**, 105010 (2011).
[576] Ya.B. Zel'dovich and I.D. Novikov, The hypothesis of cores retarded during expansion and the hot cosmological model, *Sov. Astron.* **10**, 602 (1967).

[577] Ya.B. Zel'dovich and M.Yu. Khlopov, On the concentration of relic magnetic monopoles in the Universe, *Phys. Lett. B* **79**, 239 (1978).
[578] A.L. Zel'manov, *Chronomtric Invariants* (American Research Press, Rehoboth, New Mexico, 2006).
[579] A.L. Zel'manov and V.G. Agakov, *Elements of General Relativity* (Nauka, M., 1989).
[580] D. Zipoy, Topology of some spheroidal metrics, *J. Math. Phys.* **7**, 1137–1143 (1966).

Index

Lightning Source UK Ltd.
Milton Keynes UK
UKHW022206250621
386173UK00002B/118

9 789811 233449